Student Solutions Manual

Trigonometry

SECOND EDITION

James Stewart
McMaster University and University of Toronto

Lothar Redlin
The Pennsylvania State University

Saleem Watson
California State University, Long Beach

Australia • Brazil • Japan • Korea • Mexico • Singapore • Spain • United Kingdom • United States

ISBN-13: 978-1-133-10352-3
ISBN-10: 1-133-10352-9

Brooks/Cole
20 Davis Drive
Belmont, CA 94002-3098
USA

Cengage Learning is a leading provider of customized learning solutions with office locations around the globe, including Singapore, the United Kingdom, Australia, Mexico, Brazil, and Japan. Locate your local office at: **www.cengage.com/global**

Cengage Learning products are represented in Canada by Nelson Education, Ltd.

To learn more about Brooks/Cole, visit **www.cengage.com/brookscole**

Purchase any of our products at your local college store or at our preferred online store **www.cengagebrain.com**

Printed in the United States of America
1 2 3 4 5 6 7 16 15 14 13 12

Chapter 1 Fundamentals

1.1 Coordinate Geometry

1. The point that is 3 units to the right of the y -axis and 5 units below the x -axis has coordinates $\left(3,-5\right)$.

3. The point midway between $\left(a,b\right)$ and $\left(c,d\right)$ is $\left(\frac{a+c}{2},\frac{b+d}{2}\right)$. So the point midway between $\left(1,2\right)$ and $\left(7,10\right)$ is $\left(\frac{1+7}{2},\frac{2+10}{2}\right)=\left(\frac{8}{2},\frac{12}{2}\right)=\left(4,6\right)$.

5. (a) To find the x -intercept(s) of the graph of an equation we set y equal to 0 in the equation and solve for x : $2\left(0\right)=x+1 \quad \Leftrightarrow \quad x=-1$, so the x -intercept of $2y=x+1$ is -1.

(b) To find the y -intercept(s) of the graph of an equation we set x equal to 0 in the equation and solve for y : $2y=0+1 \quad \Leftrightarrow \quad y=\frac{1}{2}$, so the y -intercept of $2y=x+1$ is $\frac{1}{2}$.

7.

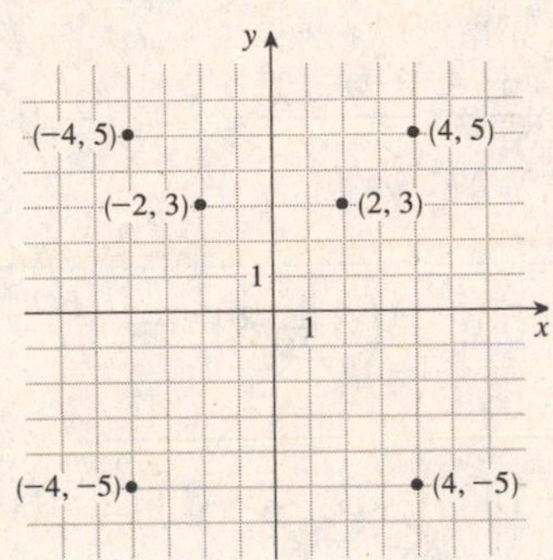

9. The two points are $\left(0,2\right)$ and $\left(3,0\right)$.

(a) $d=\sqrt{\left(3-0\right)^2+\left(0-\left(-2\right)\right)^2}=\sqrt{3^2+2^2}=\sqrt{9+4}=\sqrt{13}$

(b) midpoint: $\left(\frac{3+0}{2},\frac{0+2}{2}\right)=\left(\frac{3}{2},1\right)$

11. The two points are $\left(-3,3\right)$ and $\left(5,-3\right)$.

(a) $d=\sqrt{\left(-3-5\right)^2+\left(3-\left(-3\right)\right)^2}=\sqrt{\left(-8\right)^2+6^2}=\sqrt{64+36}=\sqrt{100}=10$

(b) midpoint: $\left(\frac{-3+5}{2},\frac{3+\left(-3\right)}{2}\right)=\left(1,0\right)$

13. (a)

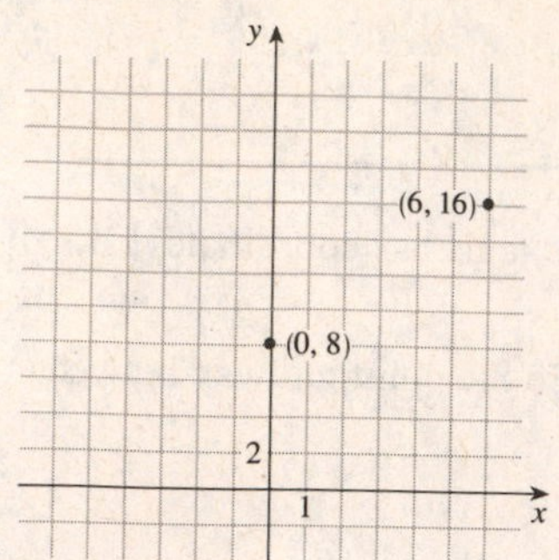

(b)

$$d = \sqrt{(0-6)^2 + (8-16)^2}$$
$$= \sqrt{(-6)^2 + (-8)^2} = \sqrt{100} = 10$$

(c) midpoint: $\left(\frac{0+6}{2}, \frac{8+16}{2}\right) = (3,12)$

15. (a)

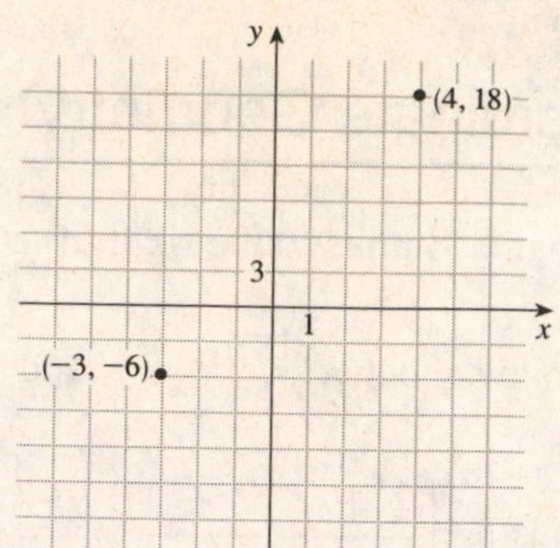

(b)

$$d = \sqrt{(-3-4)^2 + (-6-18)^2}$$
$$= \sqrt{(-7)^2 + (-24)^2} = \sqrt{49+576}$$
$$= \sqrt{625} = 25$$

(c) midpoint: $\left(\frac{-3+4}{2}, \frac{-6+18}{2}\right) = \left(\frac{1}{2}, 6\right)$

17. (a)

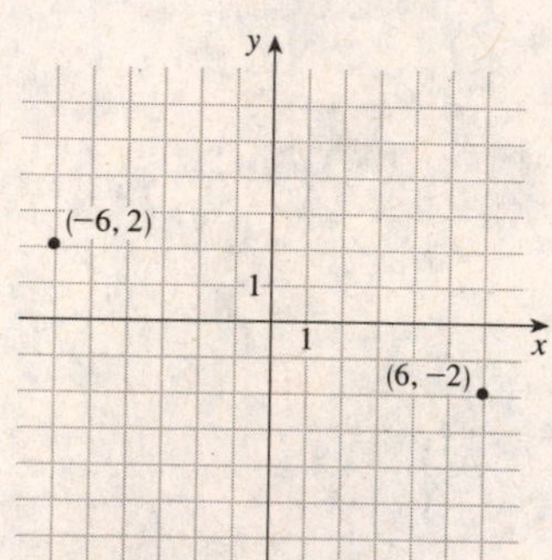

(b)

$$d = \sqrt{(6-(-6))^2 + (-2-2)^2}$$
$$= \sqrt{12^2 + (-4)^2}$$
$$= \sqrt{144+16} = \sqrt{160} = 4\sqrt{10}$$

(c) midpoint: $\left(\frac{6-6}{2}, \frac{-2+2}{2}\right) = (0,0)$

19.

$$d(A,B) = \sqrt{(1-5)^2 + (3-3)^2} = \sqrt{(-4)^2} = 4$$

.

$$d(A,C) = \sqrt{(1-1)^2 + (3-(-3))^2} = \sqrt{(6)^2} = 6$$

. So the area is $4 \cdot 6 = 24$.

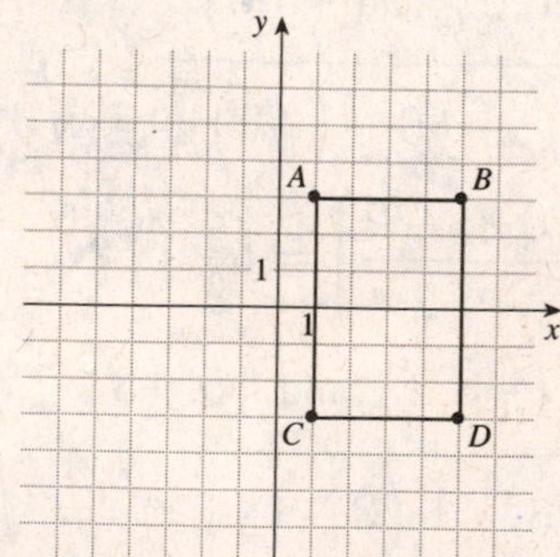

21. From the graph, the quadrilateral $ABCD$ has a pair of parallel sides, so $ABCD$ is a trapezoid. The area is $\left(\frac{b_1 + b_2}{2}\right)h$. From the graph we see that $b_1 = d(A,B) = \sqrt{(1-5)^2 + (0-0)^2} = \sqrt{4^2} = 4$; $b_2 = d(C,D) = \sqrt{(4-2)^2 + (3-3)^2} = \sqrt{2^2} = 2$; and h is the difference in y-coordinates is $|3-0| = 3$. Thus the area of the trapezoid is $\left(\frac{4+2}{2}\right)3 = 9$.

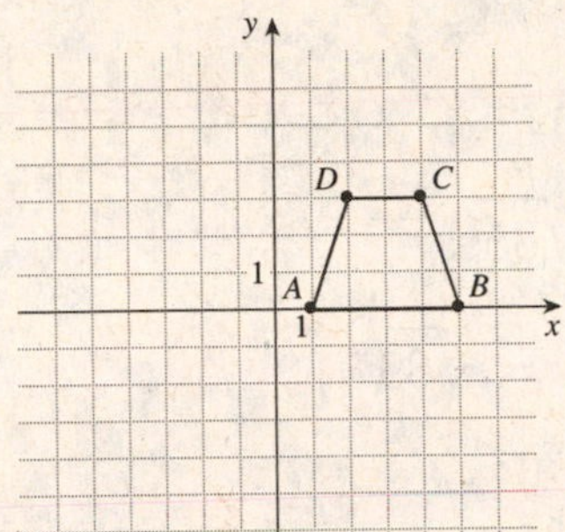

23.

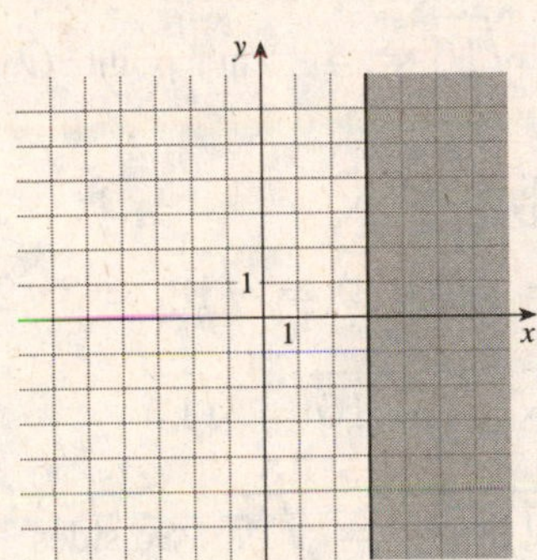

25.

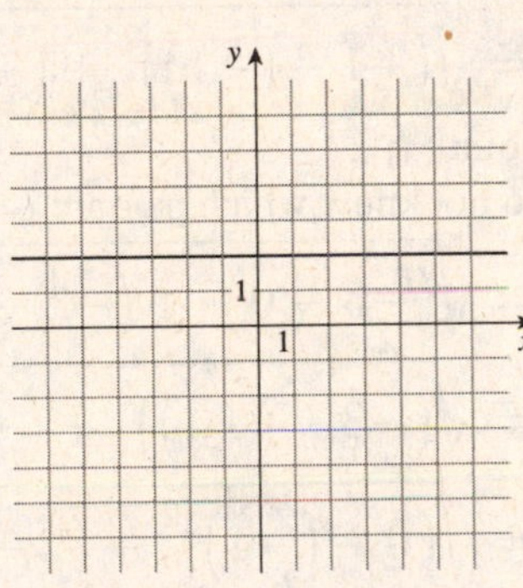

27.

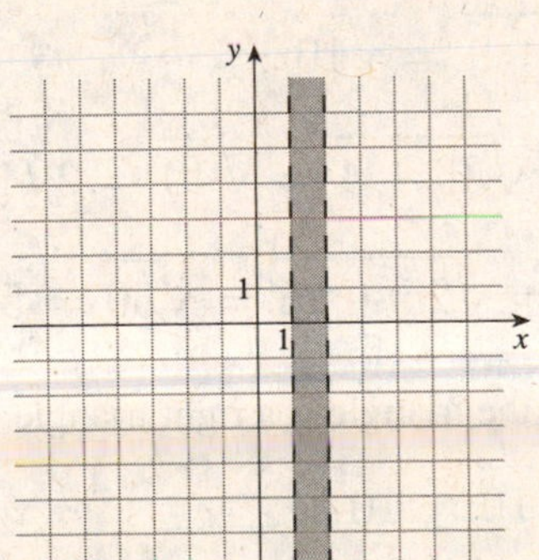

29.

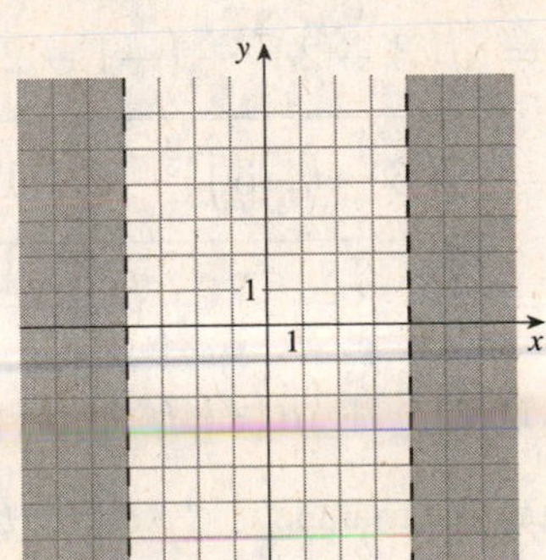

31.

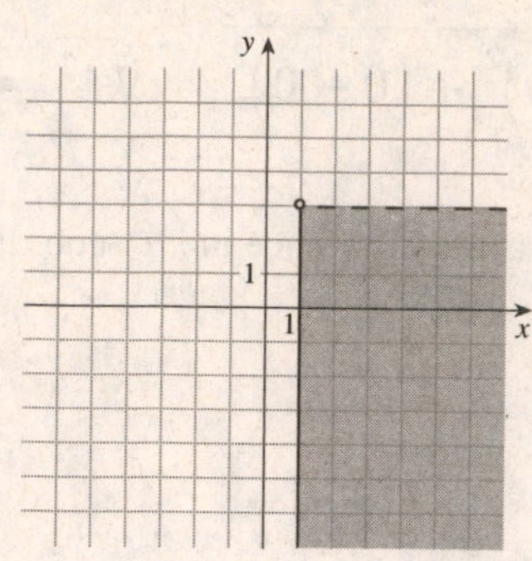

33. $d(0,A)=\sqrt{(6-0)^2+(7-0)^2}=\sqrt{6^2+7^2}=\sqrt{36+49}=\sqrt{85}$.

$d(0,B)=\sqrt{(-5-0)^2+(8-0)^2}=\sqrt{(-5)^2+8^2}=\sqrt{25+64}=\sqrt{89}$.

Thus point $A(6,7)$ is closer to the origin.

35. $d(P,R)=\sqrt{(-1-3)^2+(-1-1)^2}=\sqrt{(-4)^2+(-2)^2}=\sqrt{16+4}=\sqrt{20}=2\sqrt{5}$.

$d(Q,R)=\sqrt{(-1-(-1))^2+(-1-3)^2}=\sqrt{0+(-4)^2}=\sqrt{16}=4$. Thus point $Q(-1,3)$ is closer to point R.

37. Since we do not know which pair are isosceles, we find the length of all three sides.

$d(A,B)=\sqrt{(-3-0)^2+(-1-2)^2}=\sqrt{(-3)^2+(-3)^2}=\sqrt{9+9}=\sqrt{18}=3\sqrt{2}$.

$d(C,B)=\sqrt{(-3-(-4))^2+(-1-3)^2}=\sqrt{1^2+(-4)^2}=\sqrt{1+16}=\sqrt{17}$.

$d(A,C)=\sqrt{(0-(-4))^2+(2-3)^2}=\sqrt{4^2+(-1)^2}=\sqrt{16+1}=\sqrt{17}$. So sides AC and CB have the same length.

39. (a) Here we have $A=(2,2)$, $B=(3,-1)$, and $C=(-3,-3)$. So

$d(A,B)=\sqrt{(3-2)^2+(-1-2)^2}=\sqrt{1^2+(-3)^2}=\sqrt{1+9}=\sqrt{10}$;

$d(C,B)=\sqrt{(3-(-3))^2+(-1-(-3))^2}=\sqrt{6^2+2^2}=\sqrt{36+4}=\sqrt{40}=2\sqrt{10}$;

$d(A,C)=\sqrt{(-3-2)^2+(-3-2)^2}=\sqrt{(-5)^2+(-5)^2}=\sqrt{25+25}=\sqrt{50}=5\sqrt{2}$.

Since $[d(A,B)]^2+[d(C,B)]^2=[d(A,C)]^2$, we conclude that the triangle is a right triangle.

(b) The area of the triangle is $\frac{1}{2}\cdot d(C,B)\cdot d(A,B)=\frac{1}{2}\cdot\sqrt{10}\cdot 2\sqrt{10}=10$.

41. We show that all sides are the same length (its a rhombus) and then show that the diagonals are equal. Here we have $A = (-2,9)$, $B = (4,6)$, $C = (1,0)$, and $D = (-5,3)$. So

$d(A,B) = \sqrt{(4-(-2))^2 + (6-9)^2} = \sqrt{6^2 + (-3)^2} = \sqrt{36+9} = \sqrt{45}$;

$d(B,C) = \sqrt{(1-4)^2 + (0-6)^2} = \sqrt{(-3)^2 + (-6)^2} = \sqrt{9+36} = \sqrt{45}$;

$d(C,D) = \sqrt{(-5-1)^2 + (3-0)^2} = \sqrt{(-6)^2 + (-3)^2} = \sqrt{36+9} = \sqrt{45}$;

$d(D,A) = \sqrt{(-2-(-5))^2 + (9-3)^2} = \sqrt{3^2 + 6^2} = \sqrt{9+36} = \sqrt{45}$. So the points form a rhombus. Also

$d(A,C) = \sqrt{(1-(-2))^2 + (0-9)^2} = \sqrt{3^2 + (-9)^2} = \sqrt{9+81} = \sqrt{90} = 3\sqrt{10}$,

and $d(B,D) = \sqrt{(-5-4)^2 + (3-6)^2} = \sqrt{(-9)^2 + (-3)^2} = \sqrt{81+9} = \sqrt{90} = 3\sqrt{10}$.

Since the diagonals are equal, the rhombus is a square.

43. Let $P = (0,y)$ be such a point. Setting the distances equal we get

$\sqrt{(0-5)^2 + (y-(-5))^2} = \sqrt{(0-1)^2 + (y-1)^2} \Leftrightarrow$

$\sqrt{25 + y^2 + 10y + 25} = \sqrt{1 + y^2 - 2y + 1} \Rightarrow y^2 + 10y + 50 = y^2 - 2y + 2 \Leftrightarrow$

$12y = -48 \Leftrightarrow y = -4$. Thus, the point is $P = (0,-4)$. Check:

$\sqrt{(0-5)^2 + (-4-(-5))^2} = \sqrt{(-5)^2 + 1^2} = \sqrt{25+1} = \sqrt{26}$;

$\sqrt{(0-1)^2 + (-4-1)^2} = \sqrt{(-1)^2 + (-5)^2} = \sqrt{25+1} = \sqrt{26}$.

45. As indicated by Example 3, we must find a point $S(x_1, y_1)$ such that the midpoints of PR and of QS are the same. Thus $\left(\frac{4+(-1)}{2}, \frac{2+(-4)}{2}\right) = \left(\frac{x_1+1}{2}, \frac{y_1+1}{2}\right)$. Setting the x-coordinates equal, we get $\frac{4+(-1)}{2} = \frac{x_1+1}{2} \Leftrightarrow 4-1 = x_1+1 \Leftrightarrow x_1 = 2$. Setting the y-coordinates equal, we get $\frac{2+(-4)}{2} = \frac{y_1+1}{2} \Leftrightarrow 2-4 = y_1+1 \Leftrightarrow y_1 = -3$. Thus $S = (2,-3)$.

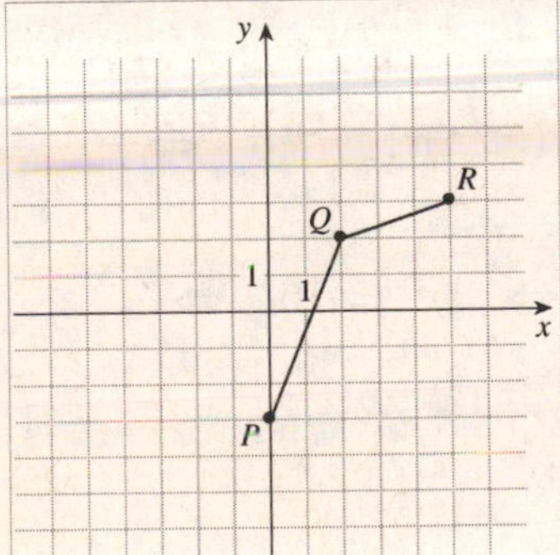

47. (a)

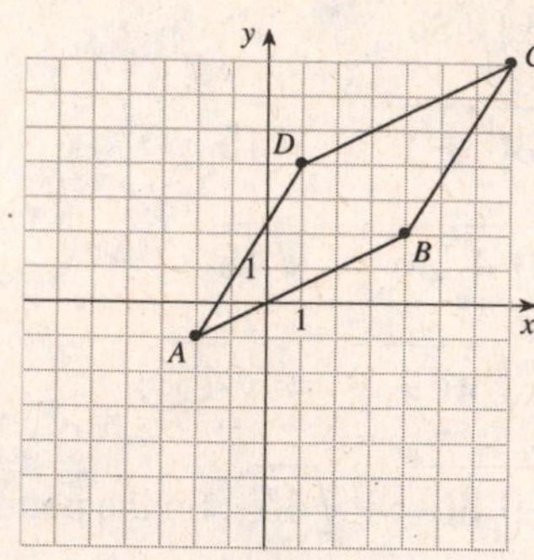

(b) The midpoint of AC is $\left(\frac{-2+7}{2},\frac{-1+7}{2}\right)=\left(\frac{5}{2},3\right)$, the midpoint of BD is $\left(\frac{4+1}{2},\frac{2+4}{2}\right)=\left(\frac{5}{2},3\right)$.

(c) Since the they have the same midpoint, we conclude that the diagonals bisect each other.

49. $(0,0):\ 0-2(0)-1\overset{?}{=}0 \ \Leftrightarrow\ -1\overset{?}{=}0$. No.

$(1,0):\ 1-2(0)-1\overset{?}{=}0 \ \Leftrightarrow\ -1+1\overset{?}{=}0$. Yes.

$(-1,-1):\ (-1)-2(-1)-1\overset{?}{=}0 \ \Leftrightarrow\ -1+2-1\overset{?}{=}0$. Yes.

So $(1,0)$ and $(-1,-1)$ are points on the graph of this equation.

51. $(0,-2):\ (0)^2+(0)(-2)+(-2)^2\overset{?}{=}4 \ \Leftrightarrow\ 0+0+4\overset{?}{=}4$. Yes.

$(1,-2):\ (1)^2+(1)(-2)+(-2)^2\overset{?}{=}4 \ \Leftrightarrow\ 1-2+4\overset{?}{=}4$. No.

$(2,-2):\ (2)^2+(2)(-2)+(-2)^2\overset{?}{=}4 \ \Leftrightarrow\ 4-4+4\overset{?}{=}4$. Yes.

So $(0,-2)$ and $(2,-2)$ are points on the graph of this equation.

53. To find x -intercepts, set $y=0$. This gives $0=4x-x^2 \ \Leftrightarrow\ 0=x(4-x) \ \Leftrightarrow\ 0=x$ or $x=4$, so the x -intercept are 0 and 4. To find y -intercepts, set $x=0$. This gives $y=4(0)-0^2 \ \Leftrightarrow\ y=0$, so the y -intercept is 0.

55. To find x -intercepts, set $y=0$. This gives $x^4+0^2-x(0)=16 \ \Leftrightarrow\ x^4=16 \ \Leftrightarrow$ $x=\pm 2$. So the x -intercept are -2 and 2.

To find y -intercepts, set $x=0$. This gives $0^4+y^2-(0)y=16 \ \Leftrightarrow\ y^2=16 \ \Leftrightarrow$ $y=\pm 4$. So the y -intercept are -4 and 4.

57. $y = -x + 4$

x	y
-4	8
-2	6
0	4
1	3
2	2
3	1
4	0

When $y = 0$ we get $x = 4$. So the x -intercept is 4 , and $x = 0 \Rightarrow y = 4$, so the y -intercept is 4 .

x -axis symmetry: $\left(-y\right) = -x + 4 \Leftrightarrow$ $y = x - 4$, which is not the same as $y = -x + 4$, so the graph is not symmetric with respect to the x -axis.

y -axis symmetry: $y = -\left(-x\right) + 4 \Leftrightarrow$ $y = x + 4$, which is not the same as $y = -x + 4$, so the graph is not symmetric with respect to the y -axis.

Origin symmetry: $\left(-y\right) = -\left(-x\right) + 4 \Leftrightarrow$ $y = -x - 4$, which is not the same as $y = -x + 4$, so the graph is not symmetric with respect to the origin.

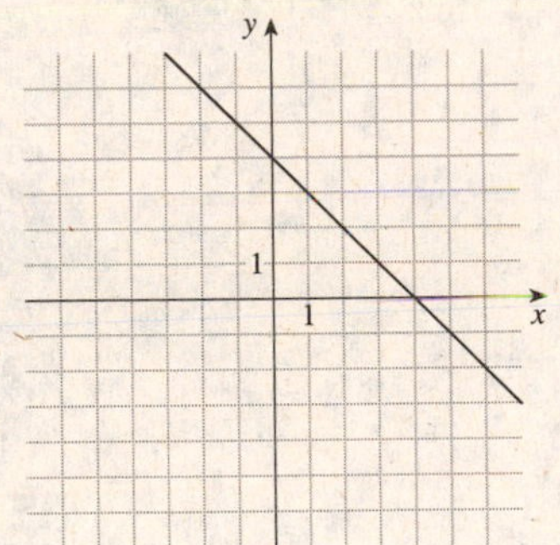

59. $2x - y = 6$

x	y
-1	-8
0	-6
1	-4
2	-2
3	0
4	2
5	4

When $y = 0$ we get $2x = 6$ So the x -intercept is 3 . When $x = 0$ we get $-y = 6$ so the y -intercept is -6 .

x -axis symmetry: $2x - \left(-y\right) = 6 \Leftrightarrow$ $2x + y = 6$, which is not the same, so the graph is not symmetric with respect to the x -axis.

y -axis symmetry: $2\left(-x\right) - \quad y = 6 \Leftrightarrow$ $2x + y = -6$, so the graph is not symmetric with respect to the y -axis.

Origin symmetry: $2\left(-x\right) - \left(-y\right) = 6 \Leftrightarrow$ $-2x + y = 6$, which not he same, so the graph is not symmetric with respect to the origin.

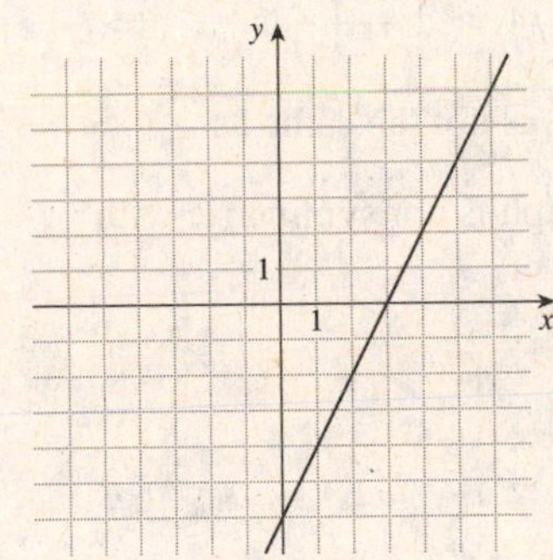

61. $y = 1 - x^2$

x	y
-3	-8
-2	-3
-1	0
0	1
1	0
2	-3
3	-8

$y = 0 \Rightarrow 0 = 1 - x^2 \Leftrightarrow x^2 = 1 \Rightarrow$ $x = \pm 1$, so the x-intercepts are 1 and -1, and $x = 0 \Rightarrow y = 1 - (0)^2 = 1$, so the y-intercept is 1.

x-axis symmetry: $(-y) = 1 - x^2 \Leftrightarrow$ $-y = 1 - x^2$, which is not the same as $y = 1 - x^2$, so the graph is not symmetric with respect to the x-axis.

y-axis symmetry: $y = 1 - (-x)^2 \Leftrightarrow$ $y = 1 - x^2$, so the graph is symmetric with respect to the y-axis.

Origin symmetry: $(-y) = 1 - (-x)^2 \Leftrightarrow$ $-y = 1 - x^2$ which is not the same as $y = 1 - x^2$. The graph is not symmetric with respect to the origin.

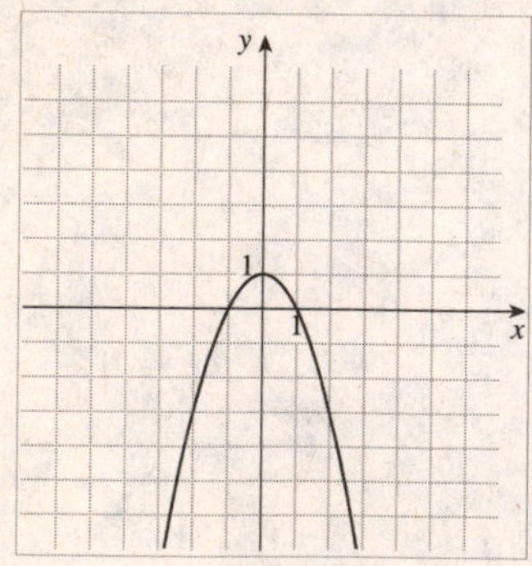

63. $4y = x^2 \Leftrightarrow y = \frac{1}{4}x^2$

x	y
-6	9
-4	4
-2	1
0	0
2	1
4	4
3	2

$y = 0 \Rightarrow 0 = \frac{1}{4}x^2 \Leftrightarrow x^2 = 0 \Rightarrow$ $x = 0$, so the x-intercept is 0, and $x = 0$ $\Rightarrow y = \frac{1}{4}(0)^2 = 0$, so the y-intercept is 0.

x-axis symmetry: $(-y) = \frac{1}{4}x^2$, which is not the same as $y = \frac{1}{4}x^2$, so the graph is not symmetric with respect to the x-axis.

y-axis symmetry: $y = \frac{1}{4}(-x)^2 \Leftrightarrow$ $y = \frac{1}{4}x^2$, so the graph is symmetric with respect to the y-axis.

Origin symmetry: $(-y) = \frac{1}{4}(-x)^2 \Leftrightarrow$ $-y = \frac{1}{4}x^2$, which is not the same as $y = \frac{1}{4}x^2$, so the graph is not symmetric with respect to the origin.

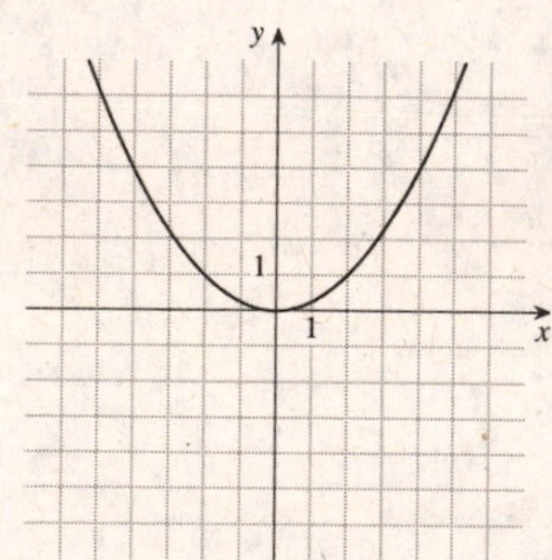

69. $y = \sqrt{4 - x^2}$. Since the radicand (the inside of the square root) cannot be negative, we must have $4 - x^2 \geq 0 \Leftrightarrow x^2 \leq 4 \Leftrightarrow |x| \leq 2$.

x	y
-2	0
-1	$\sqrt{3}$
0	4
1	$\sqrt{3}$
2	0

$y = 0 \Rightarrow 0 = \sqrt{4 - x^2} \Leftrightarrow$ $4 - x^2 = 0 \Leftrightarrow x^2 = 4 \Rightarrow x = \pm 2$, so the x -intercepts are -2 and 2 , and $x = 0$ $\Rightarrow y = \sqrt{4 - (0)^2} = \sqrt{4} = 2$, so the y -intercept is 2 . Since $y \geq 0$, the graph is not symmetric with respect to the x -axis.

y -axis symmetry:

$y = \sqrt{4 - (-x)^2} = \sqrt{4 - x^2}$, so the graph is symmetric with respect to the y -axis. Also, since $y \geq 0$ the graph is not symmetric with respect to the origin.

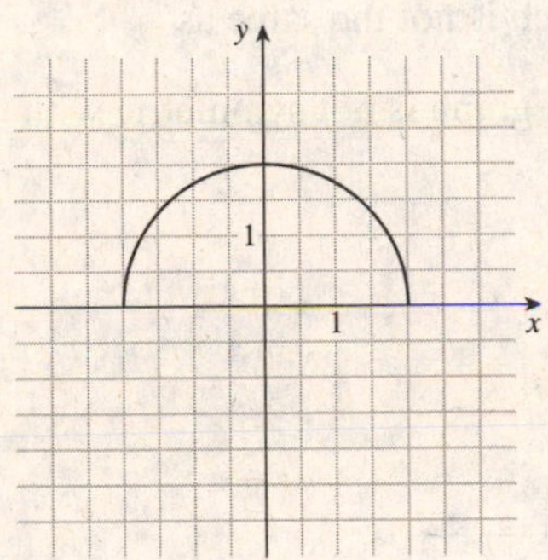

71. Solve for x in terms of y : $x + y^2 = 4$ $\Leftrightarrow x = 4 - y^2$

x	y
-12	-4
-5	-3
0	-2
3	-1
4	0
3	1
0	2
-5	3
-12	4

$y = 0 \Rightarrow x + 0^2 = 4 \Leftrightarrow x = 4$, so the x -intercept is 4 , and $x = 0 \Rightarrow$ $0 + y^2 = 4 \Rightarrow y = \pm 2$, so the y -intercepts are -2 and 2 .

x -axis symmetry: $x + (-y)^2 = 4 \Leftrightarrow$ $x + y^2 = 4$, so the graph is symmetric with respect to the x -axis.

y -axis symmetry: $(-x) + y^2 = 4 \Leftrightarrow$ $-x + y^2 = 4$, which is not the same, so the graph is not symmetric with respect to the y -axis.

Origin symmetry: $(-x) + (-y)^2 = 4 \Leftrightarrow$ $-x + y^2 = 4$, which is not the same as $x + y^2 = 4$, so the graph is not symmetric with respect to the origin.

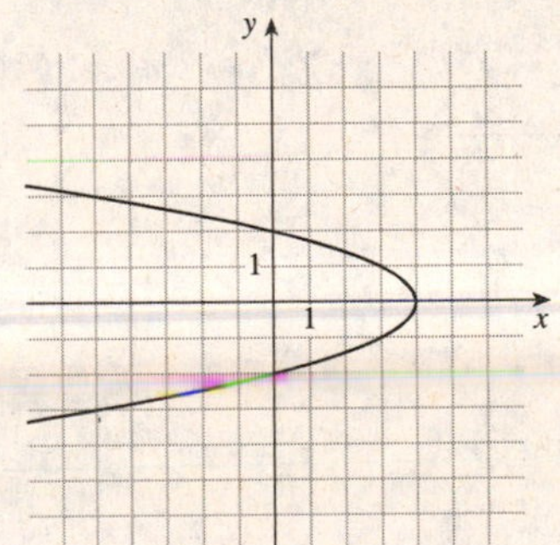

73. $y = 16 - x^4$

x	y
-3	-65
-2	0
-1	15
0	16
1	15
2	0
3	-65

$y = 0 \Rightarrow 0 = 16 - x^4 \Rightarrow x^4 = 16$ $\Rightarrow x^2 = 4 \Rightarrow x = \pm 2$, so the x-intercepts are ± 2, and so $x = 0 \Rightarrow y = 16 - 0^4 = 16$, so the y-intercept is 16.

x-axis symmetry: $(-y) = 16 - x^4 \Leftrightarrow y = -16 + x^4$, which is not the same as $y = 16 - x^4$, so the graph is not symmetric with respect to the x-axis. y-axis symmetry: $y = 16 - (-x)^4 = 16 - x^4$, so the graph is symmetric with respect to the y-axis.

Origin symmetry: $(-y) = 16 - (-x)^4 \Leftrightarrow -y = 16 - x^4$, which is not the same as $y = 16 - x^4$, so the graph is not symmetric with respect to the origin.

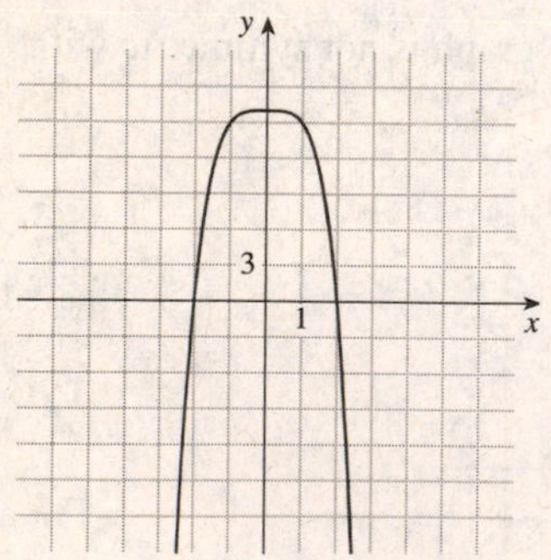

75. $y = 4 - |x|$

x	y
-6	-2
-4	0
-2	2
0	4
2	2
4	0
6	-2

$y = 0 \Rightarrow 0 = 4 - |x| \Leftrightarrow |x| = 4$ $\Rightarrow x = \pm 4$, so the x-intercepts are -4 and 4, and $x = 0 \Rightarrow y = 4 - |0| = 4$, so the y-intercept is 4.

x-axis symmetry: $(-y) = 4 - |x| \Leftrightarrow y = -4 + |x|$, which is not the same as $y = 4 - |x|$, so the graph is not symmetric with respect to the x-axis.

y-axis symmetry: $y = 4 - |-x| = 4 - |x|$, so the graph is symmetric with respect to the y-axis.

Origin symmetry: $(-y) = 4 - |-x| \Leftrightarrow y = -4 + |x|$, which is not the same as $y = 4 - |x|$, so the graph is not symmetric with respect to the origin.

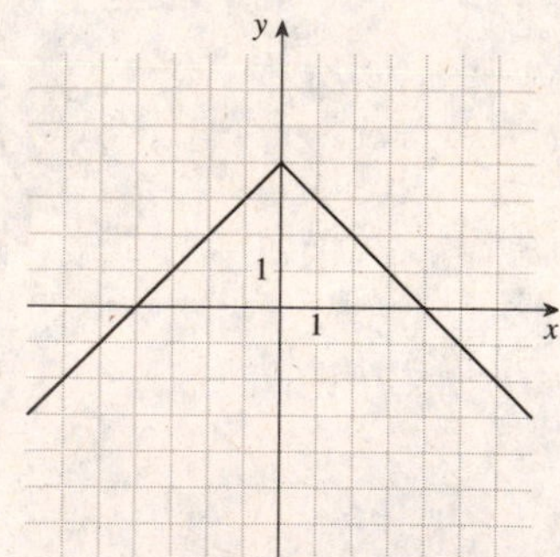

77. x -axis symmetry: $(-y) = x^4 + x^2 \iff y = -x^4 - x^2$, which is not the same as $y = x^4 + x^2$, so the graph is not symmetric with respect to the x -axis.

y -axis symmetry: $y = (-x)^4 + (-x)^2 = x^4 + x^2$, so the graph is symmetric with respect to the y -axis.

Origin symmetry: $(-y) = (-x)^4 + (-x)^2 \iff -y = x^4 + x^2$, which is not the same as $y = x^4 + x^2$, so the graph is not symmetric with respect to the origin.

79. x -axis symmetry: $x^2(-y)^2 + x(-y) = 1 \iff x^2y^2 - xy = 1$, which is not the same as $x^2y^2 + xy = 1$, so the graph is not symmetric with respect to the x -axis.

y -axis symmetry: $(-x)^2 y^2 + (-x)y = 1 \iff x^2y^2 - xy = 1$, which is not the same as $x^2y^2 + xy = 1$, so the graph is not symmetric with respect to the y -axis.

Origin symmetry: $(-x)^2(-y)^2 + (-x)(-y) = 1 \iff x^2y^2 + xy = 1$, so the graph is symmetric with respect to the origin.

81. x -axis symmetry: $(-y) = x^3 + 10x \iff y = -x^3 - 10x$, which is not the same as $y = x^3 + 10x$, so the graph is not symmetric with respect to the x -axis.

y -axis symmetry: $y = (-x)^3 + 10(-x) \iff y = -x^3 - 10x$, which is not the same as $y = x^3 + 10x$, so the graph is not symmetric with respect to the y -axis.

Origin symmetry: $(-y) = (-x)^3 + 10(-x) \iff -y = -x^3 - 10x \iff y = x^3 + 10x$, so the graph is symmetric with respect to the origin.

83. Symmetric with respect to the y -axis.

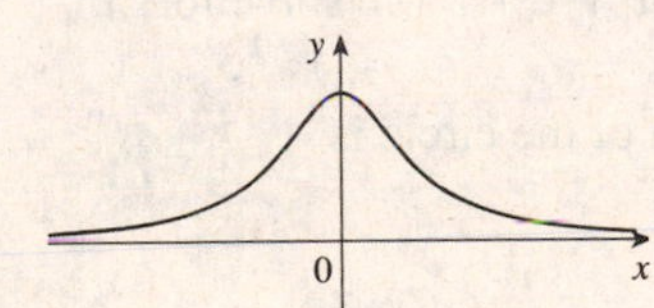

85. Symmetric with respect to the origin.

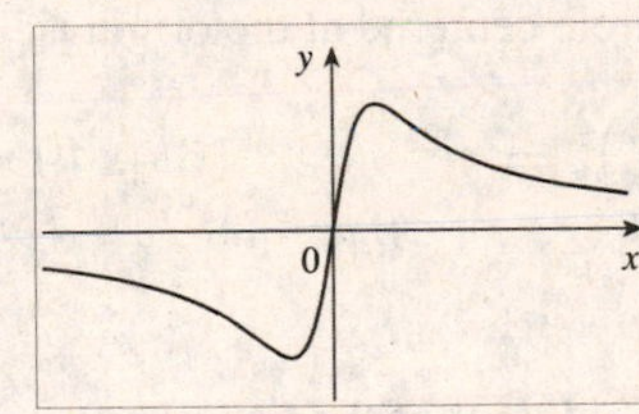

87. $x^2 + y^2 = 9$ has center $(0,0)$ and radius 3.

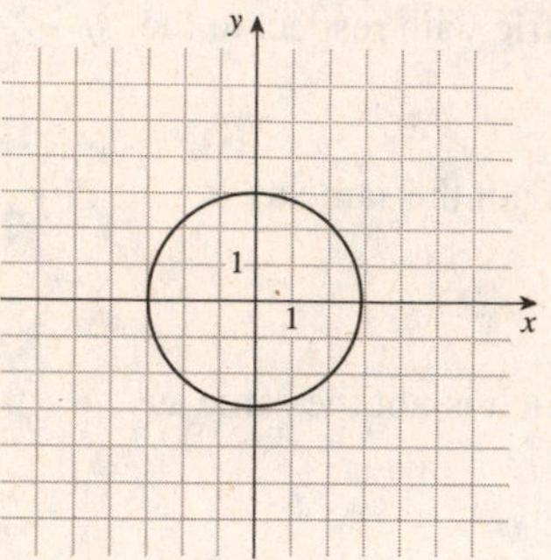

89. $(x-3)^2 + y^2 = 16$ has center $(3,0)$ and radius 4.

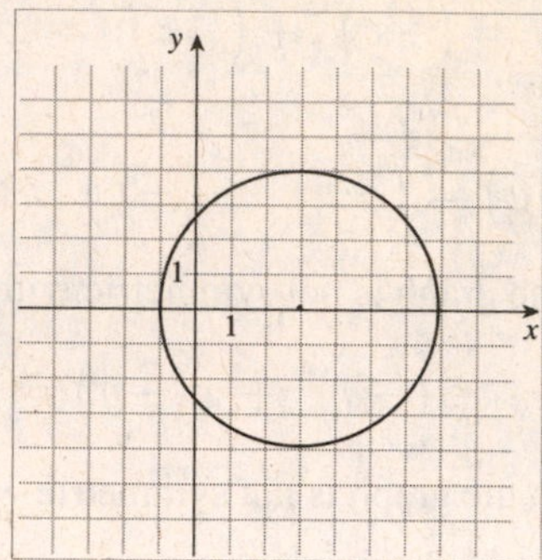

91. $(x+3)^2 + (y-4)^2 = 25$ has center $(-3,4)$ and radius 5.

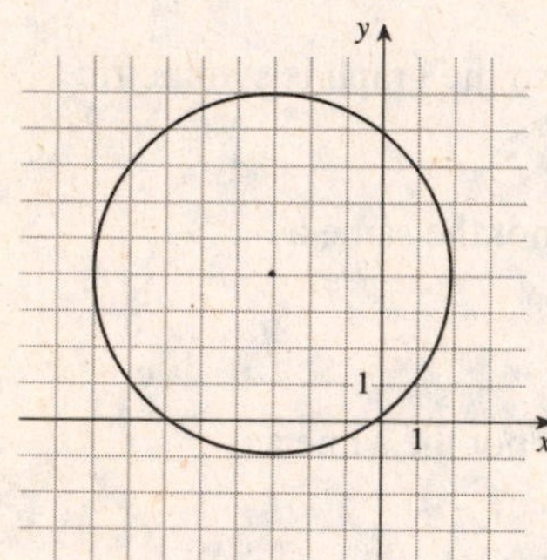

93. Using $h = 2$, $k = -1$, and $r = 3$, we get $(x-2)^2 + (y-(-1))^2 = 3^2 \Leftrightarrow (x-2)^2 + (y+1)^2 = 9$.

95. The equation of a circle centered at the origin is $x^2 + y^2 = r^2$. Using the point $(4,7)$ we solve for r^2. This gives $(4)^2 + (7)^2 = r^2 \Leftrightarrow 16 + 49 = 65 = r^2$. Thus, the equation of the circle is $x^2 + y^2 = 65$.

97. The center is at the midpoint of the line segment, which is $\left(\dfrac{-1+5}{2}, \dfrac{1+9}{2}\right) = (2,5)$. The radius is one half the diameter, so $r = \frac{1}{2}\sqrt{(-1-5)^2 + (1-9)^2} = \frac{1}{2}\sqrt{36+64} = \frac{1}{2}\sqrt{100} = 5$. Thus, the equation of the circle is $(x-2)^2 + (y-5)^2 = 5^2$ or $(x-2)^2 + (y-5)^2 = 25$.

99. Since the circle is tangent to the x-axis, it must contain the point $(7,0)$, so the radius is the change in the y-coordinates. That is, $r = |-3-0| = 3$. So the equation of the circle is $(x-7)^2 + (y-(-3))^2 = 3^2$, which is $(x-7)^2 + (y+3)^2 = 9$.

101. From the figure, the center of the circle is at $(-2,2)$. The radius is the change in the y-coordinates, so $r=|2-0|=2$. Thus the equation of the circle is $(x-(-2))^2+(y-2)^2=2^2$, which is $(x+2)^2+(y-2)^2=4$.

103. Completing the square gives $x^2+y^2-4x+10y+13=0 \Leftrightarrow$
$x^2-4x+\left(\frac{-4}{2}\right)^2+y^2+10y+\left(\frac{10}{2}\right)^2=-13+\left(\frac{4}{2}\right)^2+\left(\frac{10}{2}\right)^2 \Leftrightarrow$
$x^2-4x+4+y^2+10y+25=-13+4+25 \Leftrightarrow (x-2)^2+(y+5)^2=16$.
Thus, the center is $(2,-5)$, and the radius is 4.

105. Completing the square gives $x^2+y^2-\frac{1}{2}x+\frac{1}{2}y=\frac{1}{8} \Leftrightarrow$
$x^2-\frac{1}{2}x+\left(\frac{-1/2}{2}\right)^2+y^2+\frac{1}{2}y+\left(\frac{1/2}{2}\right)^2=\frac{1}{8}+\left(\frac{-1/2}{2}\right)^2+\left(\frac{1/2}{2}\right)^2 \Leftrightarrow$
$x^2-\frac{1}{2}x+\frac{1}{16}+y^2+\frac{1}{2}y+\frac{1}{16}=\frac{1}{8}+\frac{1}{16}+\frac{1}{16}=\frac{2}{8}=\frac{1}{4} \Leftrightarrow \left(x-\frac{1}{4}\right)^2+\left(y+\frac{1}{4}\right)^2=\frac{1}{4}$. Thus, the circle has center $\left(\frac{1}{4},-\frac{1}{4}\right)$ and radius $\frac{1}{2}$.

107. Completing the square gives $2x^2+2y^2-3x=0 \Leftrightarrow x^2+y^2-\frac{3}{2}x=0 \Leftrightarrow$
$\left(x-\frac{3}{4}\right)^2+y^2=\left(\frac{3}{4}\right)^2=\frac{9}{16}$.
Thus, the circle has center $\left(\frac{3}{4},0\right)$ and radius $\frac{3}{4}$.

109. $\{(x,y) \mid x^2+y^2\le 1\}$. This is the set of points inside (and on) the circle $x^2+y^2=1$.

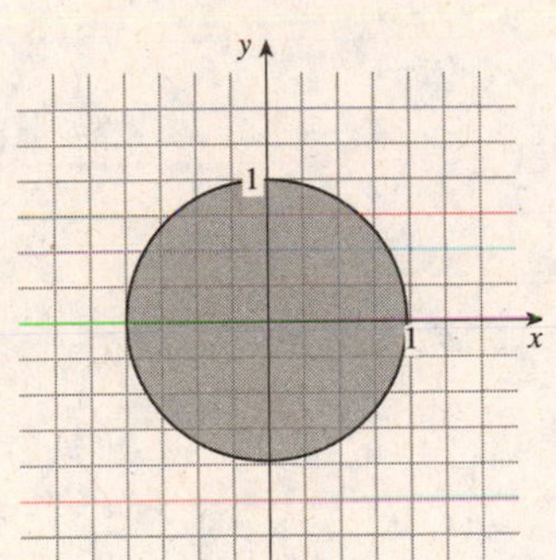

111. Completing the square gives $x^2+y^2-4y-12=0$
$\Leftrightarrow x^2+y^2-4y+\left(\frac{-4}{2}\right)^2=12+\left(\frac{-4}{2}\right)^2 \Leftrightarrow x^2+(y-2)^2=16$. Thus, the center is $(0,2)$, and the radius is 4. So the circle $x^2+y^2=4$, with center $(0,0)$ and radius 2, sits completely inside the larger circle. Thus, the area is $\pi 4^2-\pi 2^2=16\pi-4\pi=12\pi$.

113. (a) $d(A,B)=\sqrt{3^2+4^2}=\sqrt{25}=5$.

(b) We want the distances from $C=(4,2)$ to $D=(11,26)$. The walking distance is $|4-11|+|2-26|=7+24=31$ blocks. Straight-line distance is $\sqrt{(4-11)^2+(2-26)^2}=\sqrt{7^2+24^2}=\sqrt{625}=25$ blocks.

(c) The two points are on the same avenue or the same street.

115. (a) Closest: 2 Mm. Farthest: 8 Mm.

(b) When $y=2$ we have $\dfrac{(x-3)^2}{25}+\dfrac{2^2}{16}=1 \Leftrightarrow \dfrac{(x-3)^2}{25}+\frac{1}{4}=1 \Leftrightarrow \dfrac{(x-3)^2}{25}=\frac{3}{4} \Leftrightarrow$ $(x-3)^2=\frac{75}{4}$. Taking the square root of both sides we get $x-3=\pm\sqrt{\frac{75}{4}}=\pm\frac{5\sqrt{3}}{2} \Leftrightarrow$ $x=3\pm\frac{5\sqrt{3}}{2}$. So $x=3-\frac{5\sqrt{3}}{2}\approx-1.33$ or $x=3+\frac{5\sqrt{3}}{2}\approx 7.33$. The distance from $(-1.33,2)$ to the center $(0,0)$ is $d=\sqrt{(-1.33-0)^2+(2-0)^2}=\sqrt{5.7689}\approx 2.40$ Mm. The distance from $(7.33,2)$ to the center $(0,0)$ is $d=\sqrt{(7.33-0)^2+(2-0)^2}=\sqrt{57.7307}\approx 7.60$ Mm.

117. (a) The point $(3,7)$ is reflected to the point $(-3,7)$.

(b) The point (a,b) is reflected to the point $(-a,b)$.

(c) Since the point $(-a,b)$ is the reflection of (a,b), the point $(-4,-1)$ is the reflection of $(4,-1)$.

(d) $A=(3,3)$, so $A'=(-3,3)$; $B=(6,1)$, so $B'=(-6,1)$; and $C=(1,-4)$, so $C'=(-1,-4)$.

119. We need to find a point $S(x_1, y_1)$ such that $PQRS$ is a parallelogram. As indicated by Example 3, this will be the case if the diagonals PR and QS bisect each other. So the midpoints of PR and QS are the same. Thus $\left(\frac{0+5}{2}, \frac{-3+3}{2}\right) = \left(\frac{x_1+2}{2}, \frac{y_1+2}{2}\right)$. Setting the x -coordinates equal, we get

$$\frac{0+5}{2} = \frac{x_1+2}{2} \Leftrightarrow 0+5 = x_1+2 \Leftrightarrow x_1 = 3 .$$

Setting the y -coordinates equal, we get $\frac{-3+3}{2} = \frac{y_1+2}{2} \Leftrightarrow -3+3 = y_1+2 \Leftrightarrow y_1 = -2$.

Thus $S = (3, -2)$.

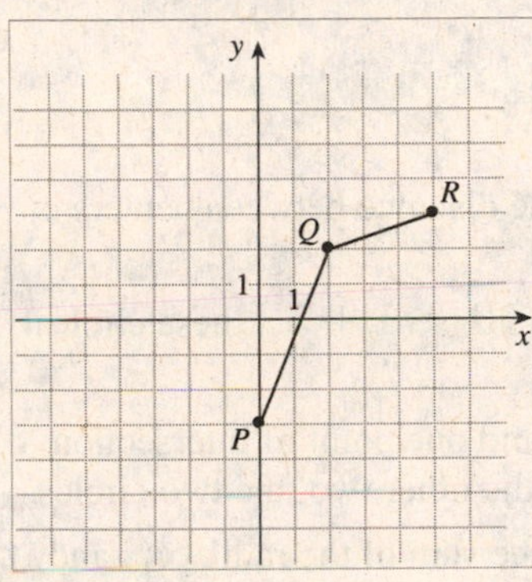

121. (a) (i) $(x-2)^2+(y-1)^2=9$, the center is at $(2,1)$, and the radius is 3 .

$(x-6)^2+(y-4)^2=16$, the center is at $(6,4)$, and the radius is 4 . The distance between centers is $\sqrt{(2-6)^2+(1-4)^2}=\sqrt{(-4)^2+(-3)^2}=\sqrt{16+9}=\sqrt{25}=5$. Since $5<3+4$, these circles intersect.

(ii) $x^2+(y-2)^2=4$, the center is at $(0,2)$, and the radius is 2 . $(x-5)^2+(y-14)^2=9$, the center is at $(5,14)$, and the radius is 3 . The distance between centers is $\sqrt{(0-5)^2+(2-14)^2}=\sqrt{(-5)^2+(-12)^2}=\sqrt{25+144}=\sqrt{169}=13$. Since $13>2+3$, these circles do not intersect.

(iii) $(x-3)^2+(y+1)^2=1$, the center is at $(3,-1)$, and the radius is 1 .

$(x-2)^2+(y-2)^2=25$, the center is at $(2,2)$, and the radius is 5 . The distance between centers is $\sqrt{(3-2)^2+(-1-2)^2}=\sqrt{1^2+(-3)^2}=\sqrt{1+9}=\sqrt{10}$. Since $\sqrt{10}<1+5$, these circles intersect.

(b) As shown in the diagram, if two circles intersect, then the centers of the circles and one point of intersection form a triangle. So because in any triangle each side has length less than the sum of the other two, the two circles will intersect only if the distance between their centers, d , is less than or equal to the sum of the radii, r_1 and r_2 . That is, the circles will intersect if $d \le r_1+r_2$.

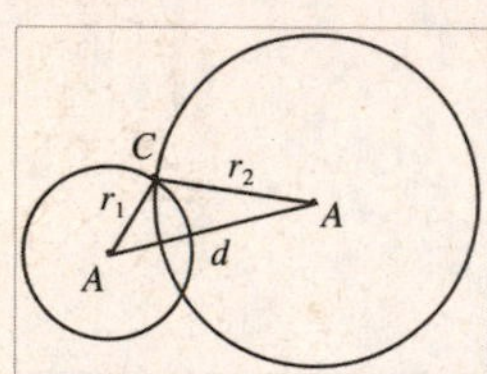

122. (a) Symmetric about the x -axis.

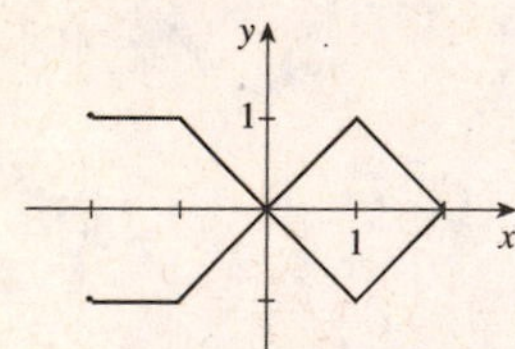

(b) Symmetric about the y -axis.

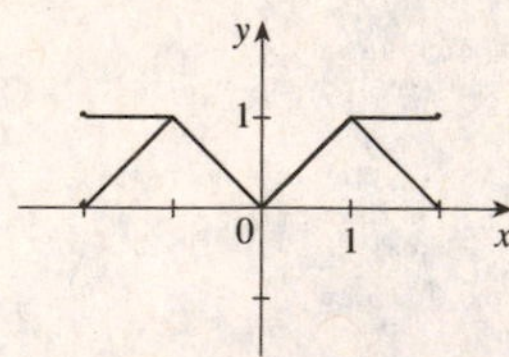

(c) Symmetric about the origin.

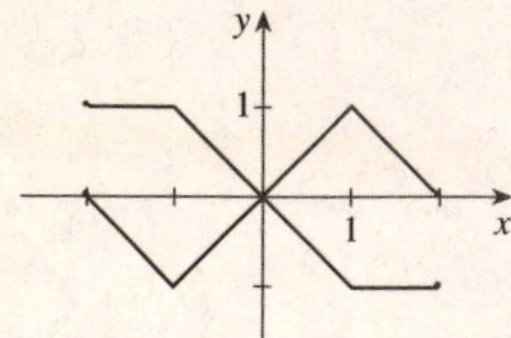

1.2 Lines

1. We find the steepness or slope of a line passing through two points by dividing the difference in the y -coordinates of these points by the difference in the x -coordinates. So the line passing through the points $(0,1)$ and $(2,5)$ has slope $\frac{5-1}{2-0}=2$.

3. The point-slope form of the equation of the line with slope 3 passing through the point $(1,2)$ is $y-2=3(x-1)$.

5. $m=\frac{y_2-y_1}{x_2-x_1}=\frac{2-0}{4-0}=\frac{2}{4}=\frac{1}{2}$

7. $m=\frac{y_2-y_1}{x_2-x_1}=\frac{0-2}{-10-2}=\frac{-2}{-12}=\frac{1}{6}$

9. $m=\frac{y_2-y_1}{x_2-x_1}=\frac{4-3}{2-4}=\frac{1}{-2}=-\frac{1}{2}$

11. $m=\frac{y_2-y_1}{x_2-x_1}=\frac{6-(-3)}{-1-1}=\frac{9}{-2}=-\frac{9}{2}$

13. For ℓ_1, we find two points, $(-1,2)$ and $(0,0)$ that lie on the line. Thus the slope of ℓ_1 is $m=\frac{y_2-y_1}{x_2-x_1}=\frac{2-0}{-1-0}=-2$. For ℓ_2, we find two points $(0,2)$ and $(2,3)$. Thus, the slope of ℓ_2 is $m=\frac{y_2-y_1}{x_2-x_1}=\frac{3-2}{2-0}=\frac{1}{2}$. For ℓ_3 we find the points $(2,-2)$ and $(3,1)$. Thus, the slope of ℓ_3 is $m=\frac{y_2-y_1}{x_2-x_1}=\frac{1-(-2)}{3-2}=3$. For ℓ_4, we find the points $(-2,-1)$ and $(2,-2)$. Thus, the slope of ℓ_4 is $m=\frac{y_2-y_1}{x_2-x_1}=\frac{-2-(-1)}{2-(-2)}=\frac{-1}{4}=-\frac{1}{4}$.

15. First we find two points $(0,4)$ and $(4,0)$ that lie on the line. So the slope is $m=\frac{0-4}{4-0}=-1$. Since the y -intercept is 4, the equation of the line is $y=mx+b=-1x+4$. So $y=-x+4$, or $x+y-4=0$.

17. We choose the two intercepts as points, $(0,-3)$ and $(2,0)$. So the slope is $m=\frac{0-(-3)}{2-0}=\frac{3}{2}$. Since the y -intercept is -3, the equation of the line is $y=mx+b=\frac{3}{2}x-3$, or $3x-2y-6=0$.

19. Using the equation $y-y_1=m(x-x_1)$, we get $y-3=5(x-2)$ $\Leftrightarrow$ $-5x+y=-7$ $\Leftrightarrow$ $5x-y-7=0$.

21. Using the equation $y-y_1=m(x-x_1)$, we get $y-7=\frac{2}{3}(x-1)$ $\Leftrightarrow$ $3y-21=2x-2$ $\Leftrightarrow$ $-2x+3y=19$ $\Leftrightarrow$ $2x-3y+19=0$.

23. First we find the slope, which is $m = \dfrac{y_2 - y_1}{x_2 - x_1} = \dfrac{6-1}{1-2} = \dfrac{5}{-1} = -5$. Substituting into $y - y_1 = m(x - x_1)$, we get $y - 6 = -5(x-1) \Leftrightarrow y - 6 = -5x + 5 \Leftrightarrow$ $5x + y - 11 = 0$.

25. Using $y = mx + b$, we have $y = 3x + (-2)$ or $3x - y - 2 = 0$.

27. We are given two points, $(1,0)$ and $(0,-3)$. Thus, the slope is

$m = \dfrac{y_2 - y_1}{x_2 - x_1} = \dfrac{-3-0}{0-1} = \dfrac{-3}{-1} = 3$. Using the y -intercept, we have $y = 3x + (-3)$ or $y = 3x - 3$ or $3x - y - 3 = 0$.

29. Since the equation of a horizontal line passing through (a,b) is $y = b$, the equation of the horizontal line passing through $(4,5)$ is $y = 5$.

31. Since $x + 2y = 6 \Leftrightarrow 2y = -x + 6 \Leftrightarrow y = -\frac{1}{2}x + 3$, the slope of this line is $-\frac{1}{2}$. Thus, the line we seek is given by $y - (-6) = -\frac{1}{2}(x-1) \Leftrightarrow 2y + 12 = -x + 1 \Leftrightarrow$ $x + 2y + 11 = 0$.

33. Any line parallel to $x = 5$ will have undefined slope and be of the form $x = a$. Thus the equation of the line is $x = -1$.

35. First find the slope of $2x + 5y + 8 = 0$. This gives $2x + 5y + 8 = 0 \Leftrightarrow 5y = -2x - 8 \Leftrightarrow$ $y = -\frac{2}{5}x - \frac{8}{5}$. So the slope of the line that is perpendicular to $2x + 5y + 8 = 0$ is

$m = -\dfrac{1}{-2/5} = \frac{5}{2}$. The equation of the line we seek is $y - (-2) = \frac{5}{2}(x - (-1)) \Leftrightarrow$ $2y + 4 = 5x + 5 \Leftrightarrow 5x - 2y + 1 = 0$.

37. First find the slope of the line passing through $(2,5)$ and $(-2,1)$. This gives

$m = \dfrac{1-5}{-2-2} = \dfrac{-4}{-4} = 1$, and so the equation of the line we seek is $y - 7 = 1(x-1) \Leftrightarrow$ $x - y + 6 = 0$.

39. (a)

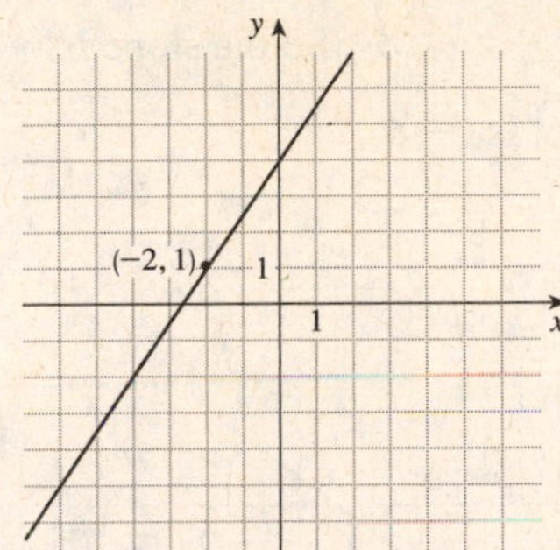

(b) $y-1=\frac{3}{2}\left(x-(-2)\right) \Leftrightarrow$

$2y-2=3\left(x+2\right) \Leftrightarrow$

$2y-2=3x+6 \Leftrightarrow 3x-2y+8=0$

.

41.

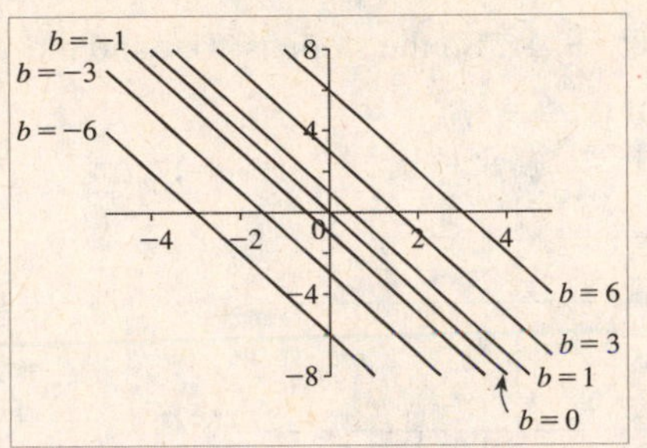

$y=-2x+b$, $b=0$, ± 1 , ± 3 , ± 6 . They have the same slope, so they are parallel.

43.

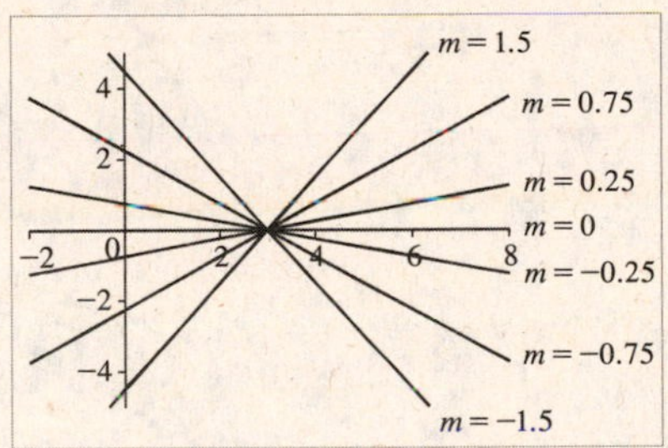

$y=m\left(x-3\right)$, $m=0$, ± 0.25 , ± 0.75 , ± 1.5 . Each of the lines contains the point $\left(3,0\right)$ because the point $\left(3,0\right)$ satisfies each equation $y=m\left(x-3\right)$. Since $\left(3,0\right)$ is on the x -axis, we could also say that they all have the same x -intercept.

45. $x+y=3 \Leftrightarrow y=-x+3$. So the slope is -1 , and the y -intercept is 3 .

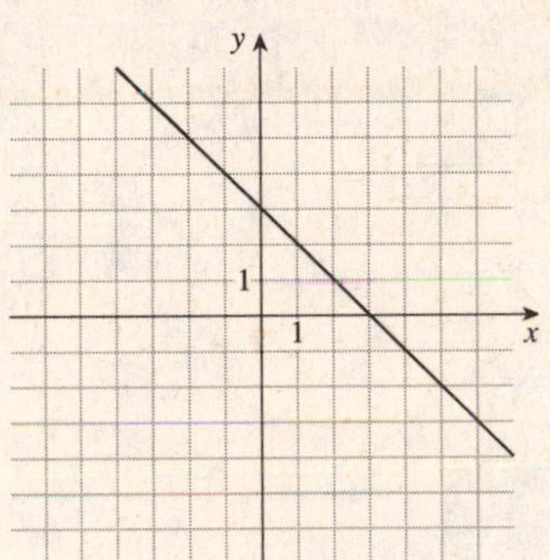

47. $x+3y=0 \Leftrightarrow 3y=-x \Leftrightarrow$

$y=-\frac{1}{3}x$. So the slope is $-\frac{1}{3}$, and the y -intercept is 0 .

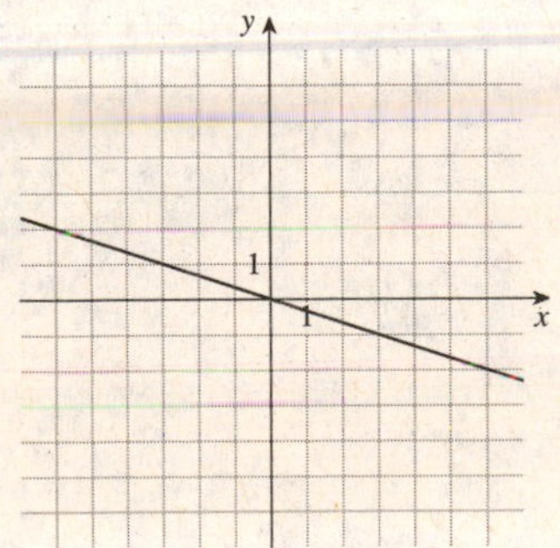

49. $\frac{1}{2}x-\frac{1}{3}y+1=0 \Leftrightarrow$

$-\frac{1}{3}y=-\frac{1}{2}x-1 \Leftrightarrow y=\frac{3}{2}x+3$. So the slope is $\frac{3}{2}$, and the y -intercept is 3 .

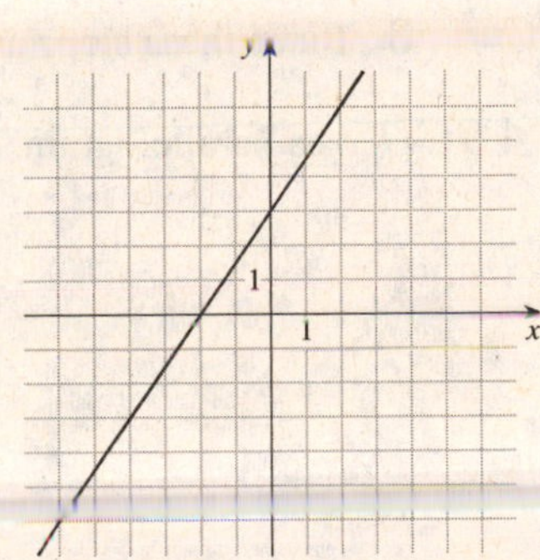

51. $y = 4$ can also be expressed as $y = 0x + 4$. So the slope is 0, and the y-intercept is 4.

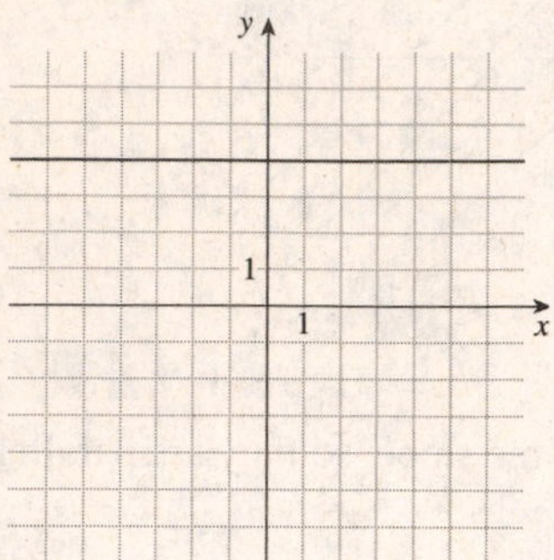

53. $3x - 4y = 12 \Leftrightarrow -4y = -3x + 12$

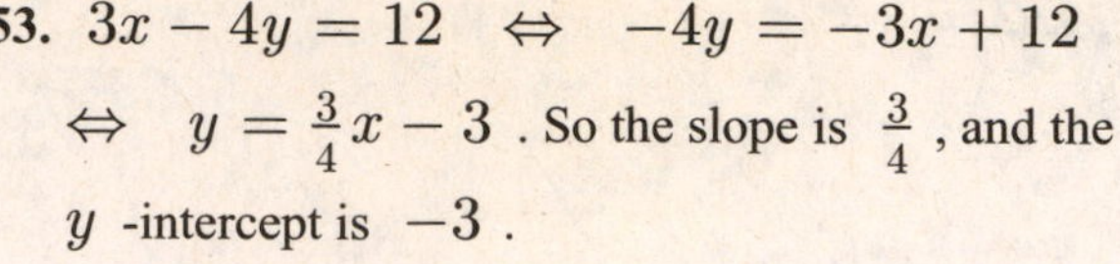

$\Leftrightarrow y = \frac{3}{4}x - 3$. So the slope is $\frac{3}{4}$, and the y-intercept is -3.

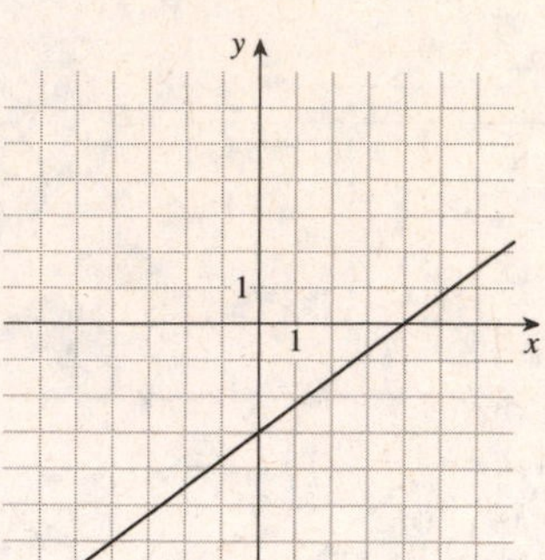

55. $3x + 4y - 1 = 0 \Leftrightarrow 4y = -3x + 1$ $\Leftrightarrow y = -\frac{3}{4}x + \frac{1}{4}$. So the slope is $-\frac{3}{4}$, and the y-intercept is $\frac{1}{4}$.

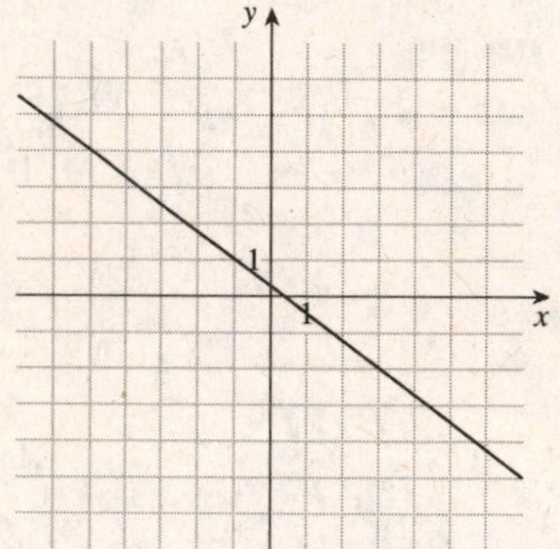

57. We first plot the points to find the pairs of points that determine each side. Next we find the slopes of opposite sides. The slope of AB is $\frac{4-1}{7-1} = \frac{3}{6} = \frac{1}{2}$, and the slope of DC is $\frac{10-7}{5-(-1)} = \frac{3}{6} = \frac{1}{2}$. Since these slope are equal, these two sides are parallel. The slope of AD is $\frac{7-1}{-1-1} = \frac{6}{-2} = -3$, and the slope of BC is $\frac{10-4}{5-7} = \frac{6}{-2} = -3$. Since these slope are equal, these two sides are parallel. Hence $ABCD$ is a parallelogram.

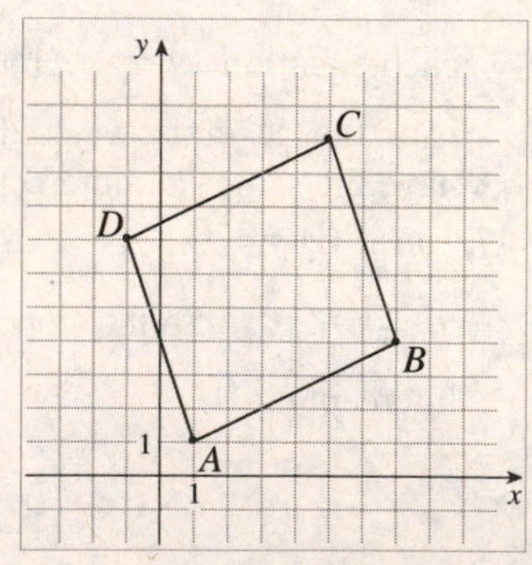

59. We first plot the points to find the pairs of points that determine each side. Next we find the slopes of opposite sides. The slope of AB is $\frac{3-1}{11-1}=\frac{2}{10}=\frac{1}{5}$ and the slope of DC is $\frac{6-8}{0-10}=\frac{-2}{-10}=\frac{1}{5}$. Since these slope are equal, these two sides are parallel. Slope of AD is $\frac{6-1}{0-1}=\frac{5}{-1}=-5$, and the slope of BC is $\frac{3-8}{11-10}=\frac{-5}{1}=-5$. Since these slope are equal, these two sides are parallel. Since $(\text{slope of } AB)\times(\text{slope of } AD)=\frac{1}{5}\times(-5)=-1$, the first two sides are each perpendicular to the second two sides. So the sides form a rectangle.

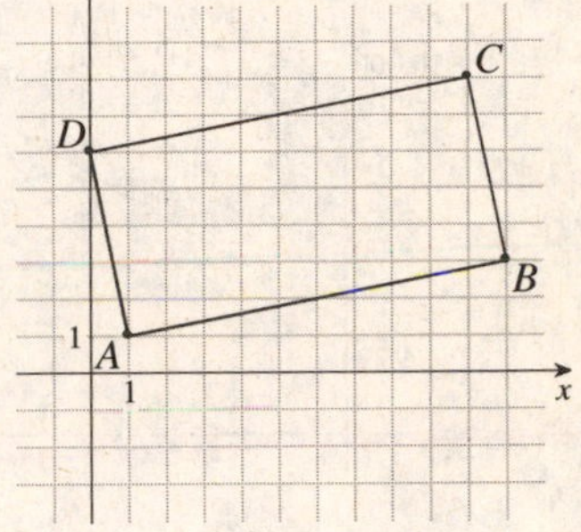

61. We need the slope and the midpoint of the line AB. The midpoint of AB is $\left(\frac{1+7}{2},\frac{4-2}{2}\right)=(4,1)$, and the slope of AB is $m=\frac{-2-4}{7-1}=\frac{-6}{6}=-1$. The slope of the perpendicular bisector will have slope $\frac{-1}{m}=\frac{-1}{-1}=1$. Using the point-slope form, the equation of the perpendicular bisector is $y-1=1(x-4)$ or $x-y-3=0$.

63. (a) We start with the two points $(a,0)$ and $(0,b)$. The slope of the line that contains them is $\frac{b-0}{0-a}=-\frac{b}{a}$. So the equation of the line containing them is $y=-\frac{b}{a}x+b$ (using the slope-intercept form). Dividing by b (since $b\neq 0$) gives $\frac{y}{b}=-\frac{x}{a}+1 \Leftrightarrow \frac{x}{a}+\frac{y}{b}=1$.

(b) Setting $a=6$ and $b=-8$, we get $\frac{x}{6}+\frac{y}{-8}=1 \Leftrightarrow 4x-3y=24 \Leftrightarrow 4x-3y-24=0$.

65. Let h be the change in your horizontal distance, in feet. Then $-\frac{6}{100}=\frac{-1000}{h} \Leftrightarrow h=\frac{100,000}{6}\approx 16,667$. So the change in your horizontal distance is about $16,667$ feet.

67. (a) The slope is $0.0417D=0.0417(200)=8.34$. It represents the increase in dosage for each one-year increase in the child's age.

(b) When $a=0$, $c=8.34(0+1)=8.34$ mg.

69. (a)

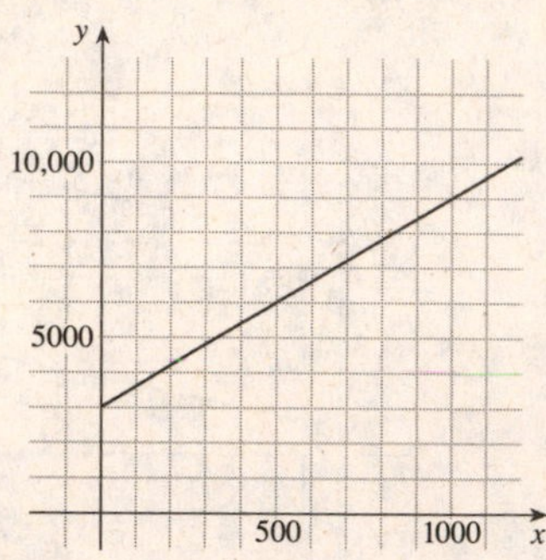

(b) The slope is the cost per toaster oven, $\$6$. The y -intercept, $\$3000$, is the monthly fixed cost --- the cost that is incurred no matter how many toaster ovens are produced.

71. (a) Using n in place of x and t in place of y , we find that the slope is $\frac{t_2 - t_1}{n_2 - n_1} = \frac{80 - 70}{168 - 120} = \frac{10}{48} = \frac{5}{24}$. So the linear equation is $t - 80 = \frac{5}{24}(n - 168) \Leftrightarrow$ $t - 80 = \frac{5}{24}n - 35 \Leftrightarrow t = \frac{5}{24}n + 45$.

(b) When $n = 150$, the temperature is approximately given by $t = \frac{5}{24}(150) + 45 = 76.25°$ F $\approx 76°$ F.

73. (a) We are given $\frac{\text{change in pressure}}{\text{10 feet change in depth}} = \frac{4.34}{10} = 0.434$. Using P for pressure and d for depth, and using the point $P = 15$ when $d = 0$, we have $P - 15 = 0.434(d - 0) \Leftrightarrow$ $P = 0.434d + 15$.**(c)** The slope represents the increase in pressure per foot of descent. The y -intercept represents the pressure at the surface.**(d)** When $P = 100$, then $100 = 0.434d + 15 \Leftrightarrow$ $0.434d = 85 \Leftrightarrow d = 195.9$ ft. Thus the pressure is 100 lb/in ^{3} at a depth of approximately 196 ft.

(b)

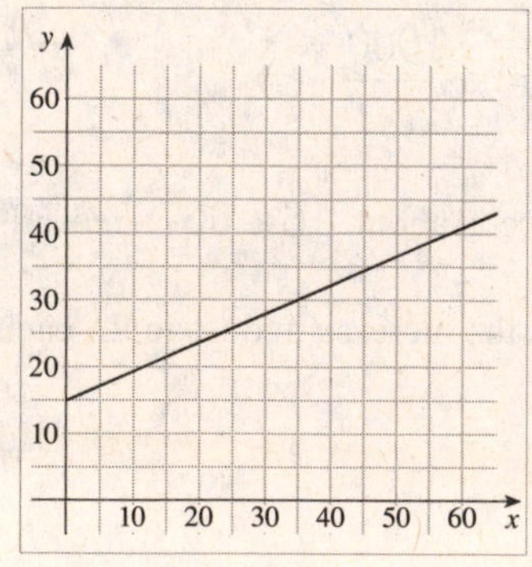

75. (a) Using d in place of x and C in place of y, we find the slope to be

$$\frac{C_2 - C_1}{d_2 - d_1} = \frac{460 - 380}{800 - 480} = \frac{80}{320} = \frac{1}{4}$$. So the linear equation is $C - 460 = \frac{1}{4}(d - 800) \Leftrightarrow$

$C - 460 = \frac{1}{4}d - 200 \Leftrightarrow C = \frac{1}{4}d + 260$.

(b) Substituting $d = 1500$ we get $C = \frac{1}{4}(1500) + 260 = 635$. Thus, the cost of driving 1500 miles is $\$635$.

(d) The y -intercept represents the fixed cost, $\$260$.

(e) It is a suitable model because you have fixed monthly costs such as insurance and car payments, as well as costs that occur as you drive, such as gasoline, oil, tires, etc., and the cost of these for each additional mile driven is a constant.

(c)

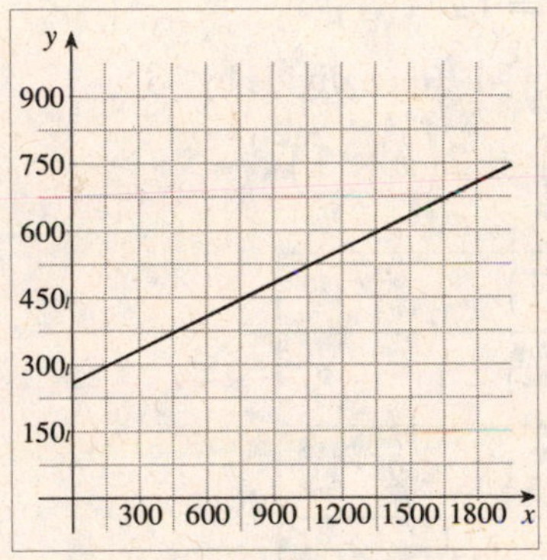

The slope of the line represents the cost per mile, $\$0.25$.

77. Slope is the rate of change of one variable per unit change in another variable. So if the slope is positive, then the temperature is rising. Likewise, if the slope is negative then the temperature is decreasing. If the slope is 0 , then the temperature is not changing.

1.3 What Is a Function?

1. If a function f is given by the formula $y = f(x)$, then $f(a)$ is the *value* of f at $x = a$.

3. (a) $f(x) = x^2 - 3x$ and $g(x) = \dfrac{x-5}{x}$ have 5 in their domain because they are defined when $x = 5$. However, $h(x) = \sqrt{x-10}$ is undefined when $x = 5$ because $\sqrt{5-10} = \sqrt{-5}$, so 5 is not in the domain of h.

(b) $f(5) = 5^2 - 3(5) = 25 - 15 = 10$ and $g(5) = \dfrac{5-5}{5} = \dfrac{0}{5} = 0$.

5. $f(x) = 2(x+3)$

7. $f(x) = (x-5)^2$

9. Square, then add 2.

11. Subtract 4, then divide by 3.

13. Machine diagram for $f(x) = \sqrt{x-1}$.

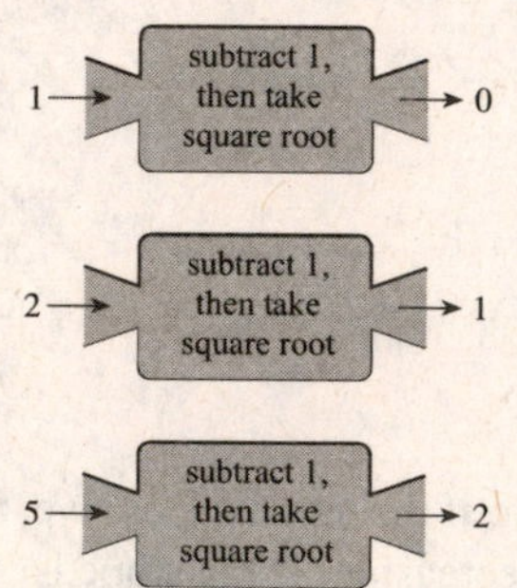

15. $f(x) = 2(x-1)^2$

x	$f(x)$
-1	$2(-1-1)^2 = 8$
0	$2(-1)^2 = 2$
1	$2(1-1)^2 = 0$
2	$2(2-1)^2 = 2$
3	$2(3-1)^2 = 8$

17. $f(x) = x^2 - 6$; $f(-3) = (-3)^2 - 6 = 9 - 6 = 3$;
$f(3) = 3^2 - 6 = 9 - 6 = 3$; $f(0) = 0^2 - 6 = -6$;
$f\left(\frac{1}{2}\right) = \left(\frac{1}{2}\right)^2 - 6 = \frac{1}{4} - 6 = -\frac{23}{4}$; $f(10) = 10^2 - 6 = 100 - 6 = 94$.

19. $f(x) = 2x + 1$; $f(1) = 2(1) + 1 = 3$; $f(-2) = 2(-2) + 1 = -3$;
$f\left(\frac{1}{2}\right) = 2\left(\frac{1}{2}\right) + 1 = 2$; $f(a) = 2(a) + 1 = 2a + 1$;
$f(-a) = 2(-a) + 1 = -2a + 1$; $f(a+b) = 2(a+b) + 1 = 2a + 2b + 1$.

21. $g(x)=\dfrac{1-x}{1+x}$; $g(2)=\dfrac{1-(2)}{1+(2)}=\dfrac{-1}{3}=-\dfrac{1}{3}$; $g(-2)=\dfrac{1-(-2)}{1+(-2)}=\dfrac{3}{-1}=-3$;

$g\left(\dfrac{1}{2}\right)=\dfrac{1-\left(\frac{1}{2}\right)}{1+\left(\frac{1}{2}\right)}=\dfrac{\frac{1}{2}}{\frac{3}{2}}=\dfrac{1}{3}$; $g(a)=\dfrac{1-(a)}{1+(a)}=\dfrac{1-a}{1+a}$;

$g(a-1)=\dfrac{1-(a-1)}{1+(a-1)}=\dfrac{1-a+1}{1+a-1}=\dfrac{2-a}{a}$; $g(-1)=\dfrac{1-(-1)}{1+(-1)}=\dfrac{2}{0}$, so

$g(-1)$ is not defined.

23. $f(x)=2x^2+3x-4$; $f(0)=2(0)^2+3(0)-4=-4$;

$f(2)=2(2)^2+3(2)-4=8+6-4=10$;

$f(-2)=2(-2)^2+3(-2)-4=8-6-4=-2$;

$f(\sqrt{2})=2(\sqrt{2})^2+3(\sqrt{2})-4=4+3\sqrt{2}-4=3\sqrt{2}$;

$f(x+1)=2(x+1)^2+3(x+1)-4=2x^2+4x+2+3x+3-4=2x^2+7x+1$

; $f(-x)=2(-x)^2+3(-x)-4=2x^2-3x-4$.

25. $f(x)=2|x-1|$; $f(-2)=2|-2-1|=2(3)=6$;

$f(0)=2|0-1|=2(1)=2$; $f\left(\frac{1}{2}\right)=2\left|\frac{1}{2}-1\right|=2\left(\frac{1}{2}\right)=1$;

$f(2)=2|2-1|=2(1)=2$; $f(x+1)=2|(x+1)-1|=2|x|$;

$f(x^2+2)=2|(x^2+2)-1|=2|x^2+1|=2x^2+2$ (since $x^2+1>0$).

27. Since $-2<0$, we have $f(-2)=(-2)^2=4$. Since $-1<0$, we have

$f(-1)=(-1)^2=1$. Since $0\geq 0$, we have $f(0)=0+1=1$. Since $1\geq 0$, we have

$f(1)=1+1=2$. Since $2\geq 0$, we have $f(2)=2+1=3$.

29. Since $-4\leq -1$, we have $f(-4)=(-4)^2+2(-4)=16-8=8$. Since $-\frac{3}{2}\leq -1$,

we have $f\left(-\frac{3}{2}\right)=\left(-\frac{3}{2}\right)^2+2\left(-\frac{3}{2}\right)=\frac{9}{4}-3=-\frac{3}{4}$. Since $-1\leq -1$, we have

$f(-1)=(-1)^2+2(-1)=1-2=-1$. Since $-1<0\leq 1$, we have $f(0)=0$.

Since $25>1$, we have $f(25)=-1$.

31. $f(x+2)=(x+2)^2+1=x^2+4x+4+1=x^2+4x+5$;

$f(x)+f(2)=x^2+1+(2)^2+1=x^2+1+4+1=x^2+6$.

33. $f(x^2)=x^2+4$; $[f(x)]^2=[x+4]^2=x^2+8x+16$.

35. $f(a) = 3(a) + 2 = 3a + 2$; $f(a+h) = 3(a+h) + 2 = 3a + 3h + 2$;

$$\frac{f(a+h) - f(a)}{h} = \frac{(3a + 3h + 2) - (3a + 2)}{h} = \frac{3a + 3h + 2 - 3a - 2}{h} = \frac{3h}{h} = 3$$

37. $f(a) = 5$; $f(a+h) = 5$; $\dfrac{f(a+h) - f(a)}{h} = \dfrac{5-5}{h} = 0$.

39. $f(a) = \dfrac{a}{a+1}$; $f(a+h) = \dfrac{a+h}{a+h+1}$;

$$\frac{f(a+h) - f(a)}{h} = \frac{\dfrac{a+h}{a+h+1} - \dfrac{a}{a+1}}{h} = \frac{\dfrac{(a+h)(a+1)}{(a+h+1)(a+1)} - \dfrac{a(a+h+1)}{(a+h+1)(a+1)}}{h}$$

$$= \frac{\dfrac{(a+h)(a+1) - a(a+h+1)}{(a+h+1)(a+1)}}{h} = \frac{a^2 + a + ah + h - (a^2 + ah + a)}{h(a+h+1)(a+1)}$$

$$= \frac{1}{(a+h+1)(a+1)}$$

41. $f(a) = 3 - 5a + 4a^2$;

;

$$f(a+h) = 3 - 5(a+h) + 4(a+h)^2 = 3 - 5a - 5h + 4(a^2 + 2ah + h^2)$$
$$= 3 - 5a - 5h + 4a^2 + 8ah + 4h^2$$

.

$$\frac{f(a+h) - f(a)}{h} = \frac{(3 - 5a - 5h + 4a^2 + 8ah + 4h^2) - (3 - 5a + 4a^2)}{h}$$

$$= \frac{3 - 5a - 5h + 4a^2 + 8ah + 4h^2 - 3 + 5a - 4a^2}{h} = \frac{-5h + 8ah + 4h^2}{h}$$

$$= \frac{h(-5 + 8a + 4h)}{h} = -5 + 8a + 4h$$

43. $f(x) = 2x$. Since there is no restrictions, the domain is the set of real numbers, $(-\infty, \infty)$.

45. $f(x) = 2x$. The domain is restricted by the exercise to $[-1, 5]$.

47. $f(x) = \dfrac{1}{x-3}$. Since the denominator cannot equal 0 we have $x - 3 \neq 0 \quad \Leftrightarrow \quad x \neq 3$.

Thus the domain is $\{x \mid x \neq 3\}$. In interval notation, the domain is $(-\infty, 3) \cup (3, \infty)$.

49. $f(x)=\dfrac{x+2}{x^2-1}$. Since the denominator cannot equal 0 we have $x^2-1\neq 0 \Leftrightarrow x^2\neq 1$ $\Rightarrow x\neq \pm 1$. Thus the domain is $\{x \mid x\neq \pm 1\}$. In interval notation, the domain is $(-\infty,-1)\cup(-1,1)\cup(1,\infty)$.

51. $f(x)=\sqrt{x-5}$. We require $x-5\geq 0 \Leftrightarrow x\geq 5$. Thus the domain is $\{x \mid x\geq 5\}$. The domain can also be expressed in interval notation as $[5,\infty)$.

53. $f(t)=\sqrt[3]{t-1}$. Since the odd root is defined for all real numbers, the domain is the set of real numbers, $(-\infty,\infty)$.

55. $h(x)=\sqrt{2x-5}$. Since the square root is defined as a real number only for nonnegative numbers, we require that $2x-5\geq 0 \Leftrightarrow 2x\geq 5 \Leftrightarrow x\geq \frac{5}{2}$. So the domain is $\{x \mid x\geq \frac{5}{2}\}$. In interval notation, the domain is $\left[\frac{5}{2},\infty\right)$.

57. $g(x)=\dfrac{\sqrt{2+x}}{3-x}$. We require $2+x\geq 0$, and the denominator cannot equal 0. Now $2+x\geq 0 \Leftrightarrow x\geq -2$, and $3-x\neq 0 \Leftrightarrow x\neq 3$. Thus the domain is $\{x \mid x\geq -2 \text{ and } x\neq 3\}$, which can be expressed in interval notation as $[-2,3)\cup(3,\infty)$.

59. $g(x)=\sqrt[4]{x^2-6x}$. Since the input to an even root must be nonnegative, we have $x^2-6x\geq 0 \Leftrightarrow x(x-6)\geq 0$. We make a table:

	$(-\infty,0)$	$(0,6)$	$(6,\infty)$
Sign of x	$-$	$+$	$+$
Sign of $x-6$	$-$	$-$	$+$
Sign of $x(x-6)$	$+$	$-$	$+$

Thus the domain is $(-\infty,0]\cup[6,\infty)$.

61. $f(x)=\dfrac{3}{\sqrt{x-4}}$. Since the input to an even root must be nonnegative and the denominator cannot equal 0, we have $x-4>0 \Leftrightarrow x>4$. Thus the domain is $(4,\infty)$.

63. $f(x)=\dfrac{(x+1)^2}{\sqrt{2x-1}}$. Since the input to an even root must be nonnegative and the denominator cannot equal 0, we have $2x-1>0 \Leftrightarrow x>\frac{1}{2}$. Thus the domain is $\left(\frac{1}{2},\infty\right)$.

65. To evaluate $f(x)$, divide the input by 3 and add $\frac{2}{3}$ to the result.

(a) $f(x) = \dfrac{x}{3} + \dfrac{2}{3}$

(b)

x	$f(x)$
2	$\frac{4}{3}$
4	2
6	$\frac{8}{3}$
8	$\frac{10}{3}$

(c)

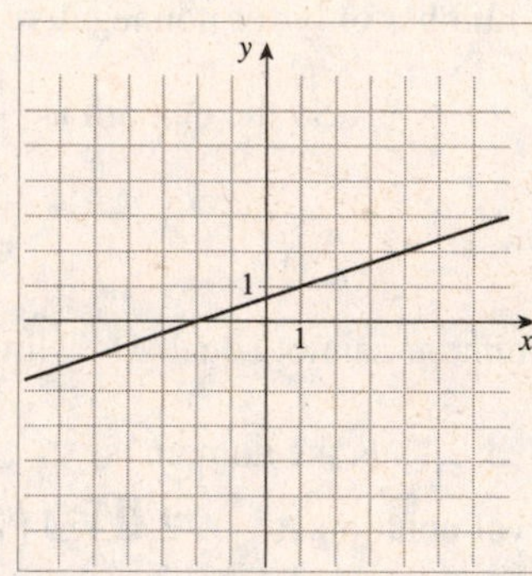

67. Let $T(x)$ be the amount of sales tax charged in Lemon County on a purchase of x dollars. To find the tax, take 8% of the purchase price.

(a) $T(x) = 0.08x$

(b)

x	$T(x)$
2	0.16
4	0.32
6	0.48
8	0.64

(c)

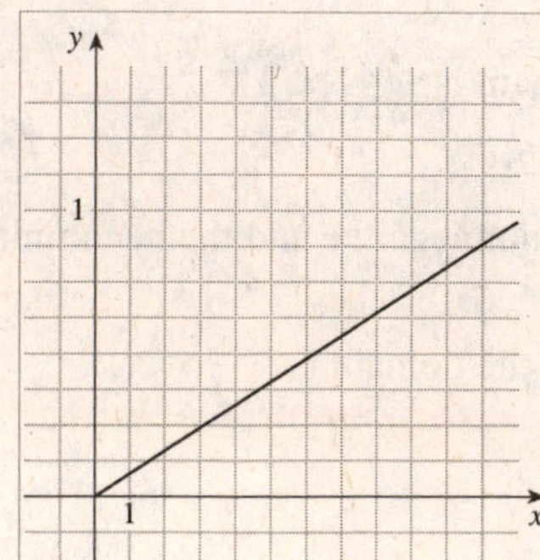

69. (a)

$$C(10) = 1500 + 3(10) + 0.02(10)^2 + 0.0001(10)^3 = 1500 + 30 + 2 + 0.1 = 1532.1$$

$$C(100) = 1500 + 3(100) + 0.02(100)^2 + 0.0001(100)^3 = 1500 + 300 + 200 + 100 = 2100$$

(b) $C(10)$ represents the cost of producing 10 yards of fabric and $C(100)$ represents the cost of producing 100 yards of fabric.

(c) $C(0) = 1500 + 3(0) + 0.02(0)^2 + 0.0001(0)^3 = 1500$

71. (a) $V(0) = 50\left(1 - \frac{0}{20}\right)^2 = 50$ and $V(20) = 50\left(1 - \frac{20}{20}\right)^2 = 0$.

(b) $V(0) = 50$ represents the volume of the full tank at time $t = 0$, and $V(20) = 0$ represents the volume of the empty tank twenty minutes later.

(c)

x	$V(x)$
0	50
5	28.125
10	12.5
15	3.125
20	0

73. (a) $v(0.1) = 18500(0.25 - 0.1^2) = 4440$, $v(0.4) = 18500(0.25 - 0.4^2) = 1665$.

(b) They tell us that the blood flows much faster (about 2.75 times faster) 0.1 cm from the center than 0.1 cm from the edge.

(c)

r	$v(r)$
0	4625
0.1	4440
0.2	3885
0.3	2960
0.4	1665
0.5	0

75. (a) $L(0.5c) = 10\sqrt{1 - \dfrac{(0.5c)^2}{c^2}} \approx 8.66$ m, $L(0.75c) = 10\sqrt{1 - \dfrac{(0.75c)^2}{c^2}} \approx 6.61$ m, and $L(0.9c) = 10\sqrt{1 - \dfrac{(0.9c)^2}{c^2}} \approx 4.36$ m.

(b) It will appear to get shorter.

77. (a) $C(75) = 75 + 15 = \$90$; $C(90) = 90 + 15 = \$105$; $C(100) = \$100$; and $C(105) = \$105$.

(b) The total price of the books purchased, including shipping.

79. (a) $F(x) = \begin{cases} 15(40-x) & \text{if } 0 < x < 40 \\ 0 & \text{if } 40 \le x \le 65 \\ 15(x-65) & \text{if } x > 65 \end{cases}$

(b) $F(30) = 15(40-10) = 15 \cdot 10 = \$150$; $F(50) = \$0$; and $F(75) = 15(75-65)15 \cdot 10 = \$150$.

(c) The fines for violating the speed limits on the freeway.

81.

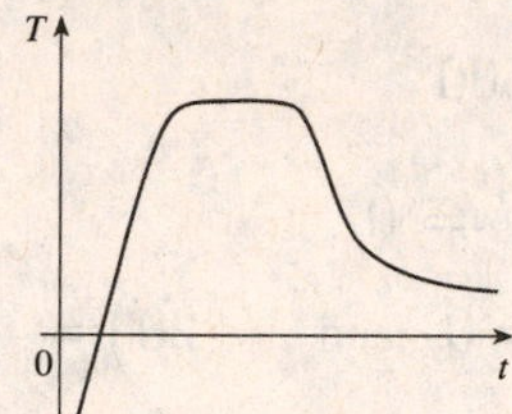

83.

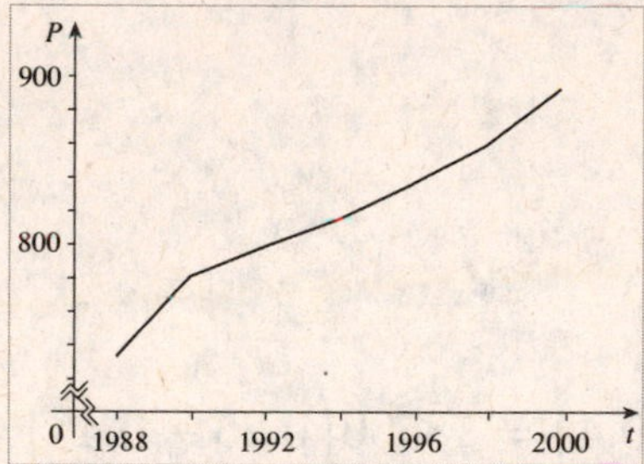

85. Answers will vary.

1.4 Graphs of Functions

1. To graph the function f we plot the points $(x, f(x))$ in a coordinate plane. To graph $f(x) = x^3 + 2$ we plot the points $(x, x^3 + 2)$. So, the point $(2, 2^3 + 2) = (2, 10)$ is on the graph of f. The height of the graph of f above the x-axis when $x = 2$ is 10.

3. If the point $(2,3)$ is on the graph of f, then $f(2) = 3$.

5.

x	$f(x) = 2$
-9	2
-6	2
-3	2
0	2
3	2
6	2

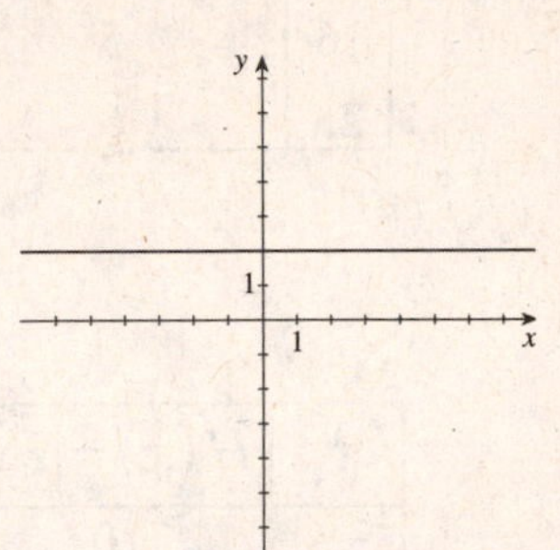

7.

x	$f(x) = 2x - 4$
-1	-6
0	-4
1	-2
2	0
3	2
4	4
5	6

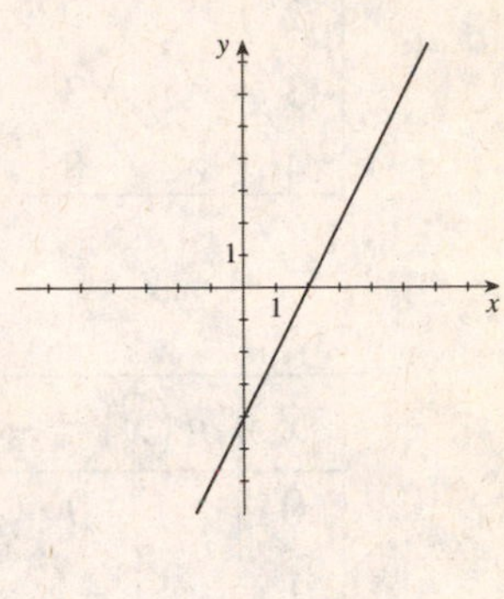

9.

x	$f(x) = -x + 3$, $-3 \le x \le 3$
-3	6
-2	5
0	3
1	2
2	1
3	0

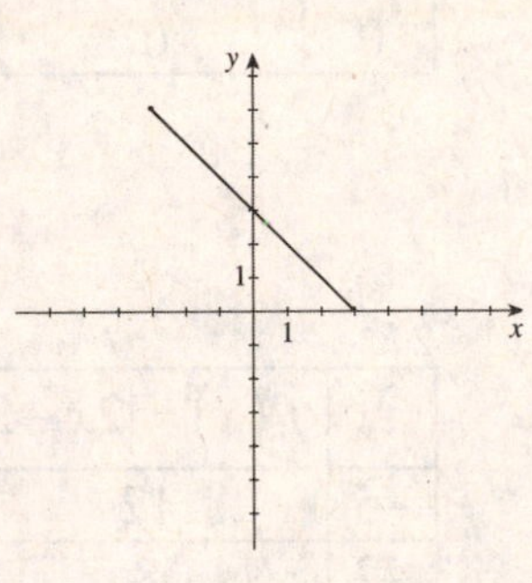

11.

x	$f(x) = -x^2$
± 4	-16
± 3	-9
± 2	-4
± 1	-1
0	0

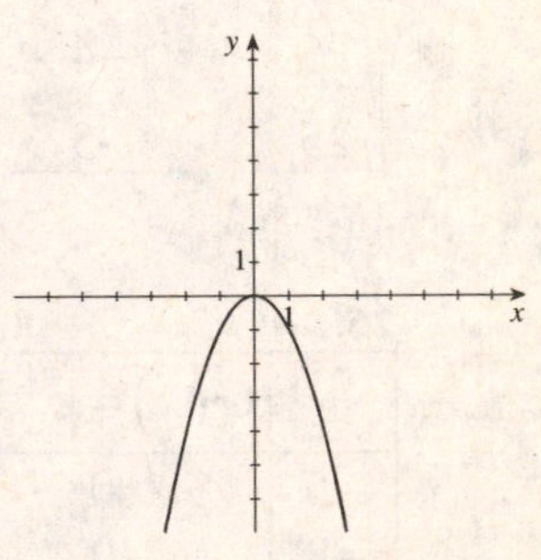

13.

x	$h(x) = 16 - x^2$
± 6	-20
± 4	0
± 2	12
0	16

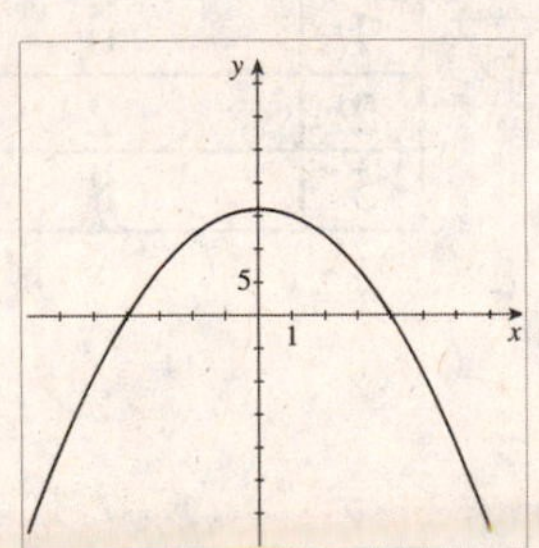

15.

x	$g(x) = x^3 - 8$
-3	-35
-2	-16
-1	-9
0	-8
1	-7
2	0
3	19

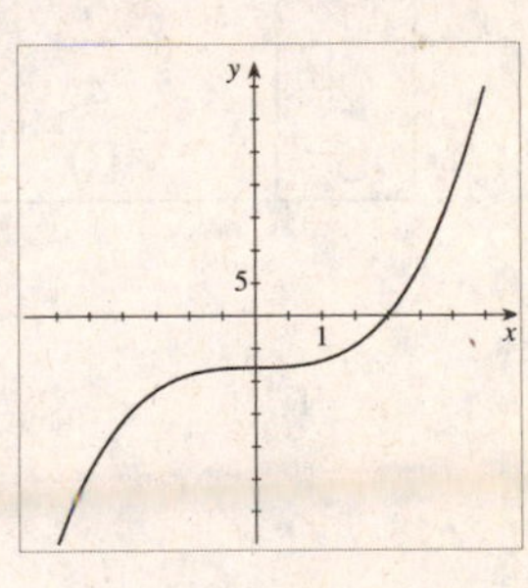

17.

x	$g(x)=x^2-2x$
-2	8
-1	3
0	0
1	-1
2	0
3	3
4	8

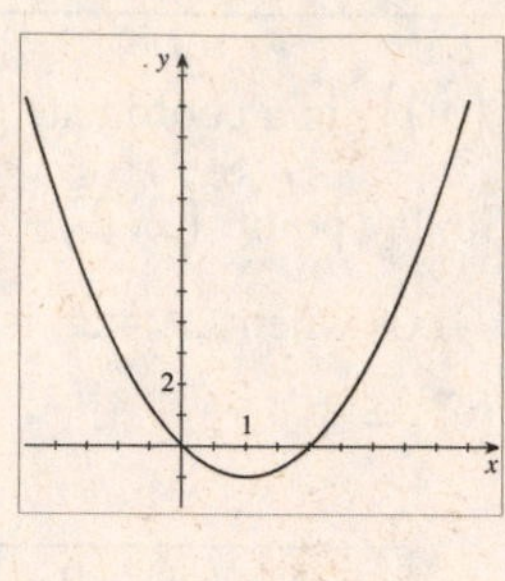

19.

x	$f(x)=1+\sqrt{x}$
0	1
1	2
4	3
9	4
16	5
25	6

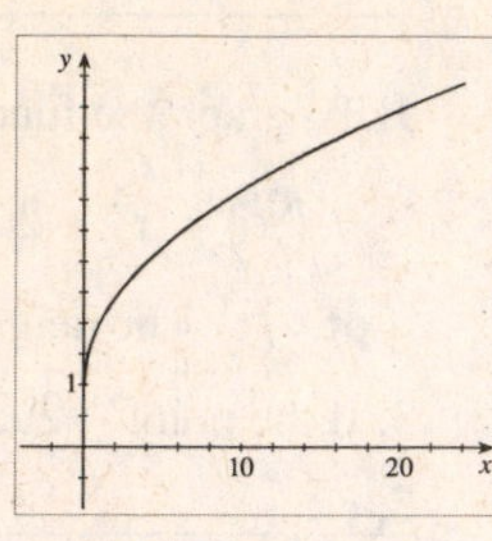

21.

x	$g(x)=-\sqrt{x}$
0	0
1	-1
4	-2
9	-3
16	-4
25	-5

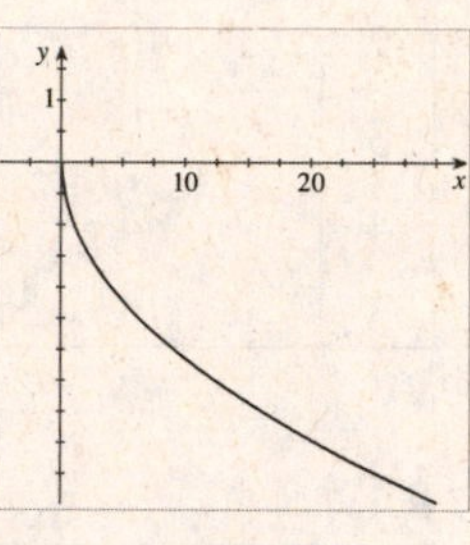

23.

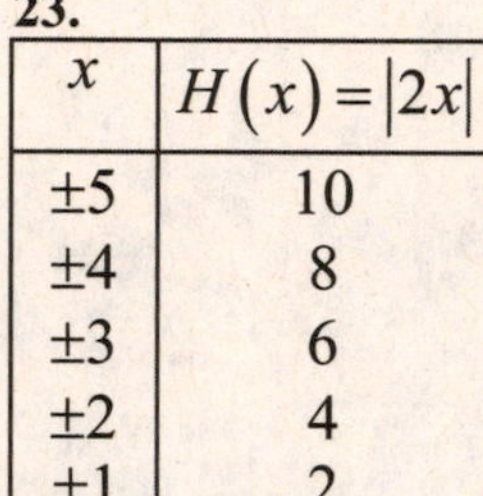

x	$H(x)=\lvert 2x\rvert$
±5	10
±4	8
±3	6
±2	4
±1	2
0	0

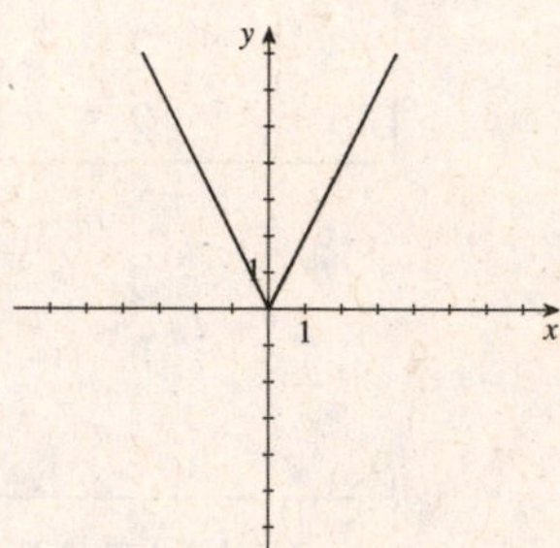

25.

x	$G(x)=\lvert x\rvert+x$
-5	0
-2	0
0	0
1	2
2	4
5	10

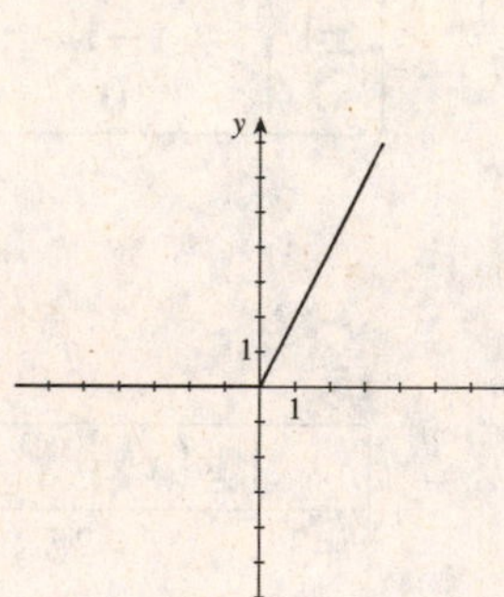

27.

x	$f(x)=\lvert 2x-2\rvert$
-5	12
-2	8
0	2
1	0
2	2
5	8

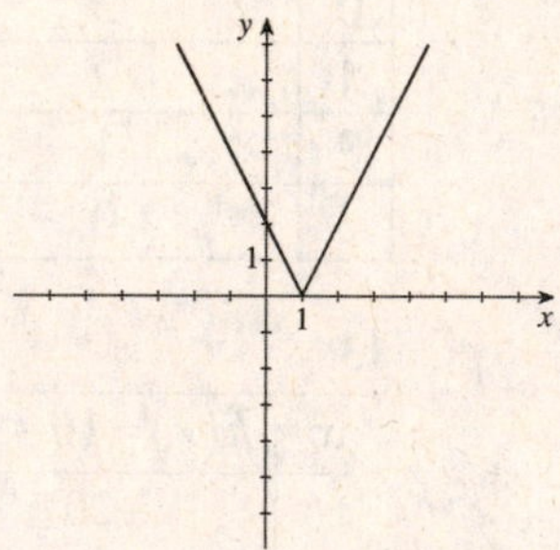

29. $f(x)=8x-x^2$

(a) $[-5,5]$ by $[-5,5]$

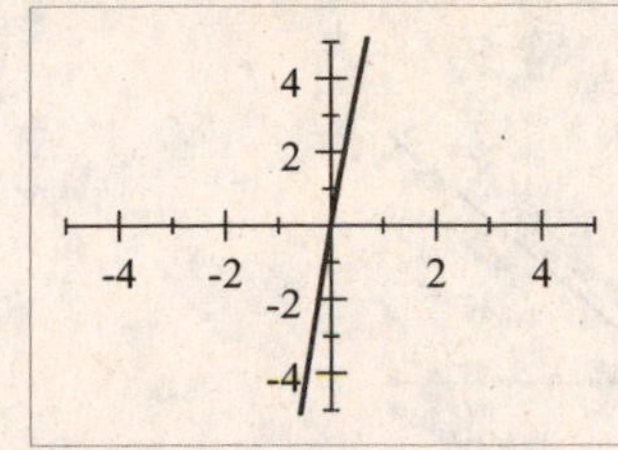

(b) $[-10,10]$ by $[-10,10]$

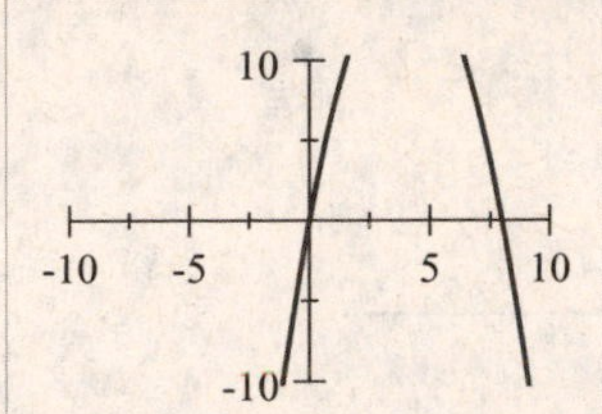

(c) $[-2,10]$ by $[-5,20]$

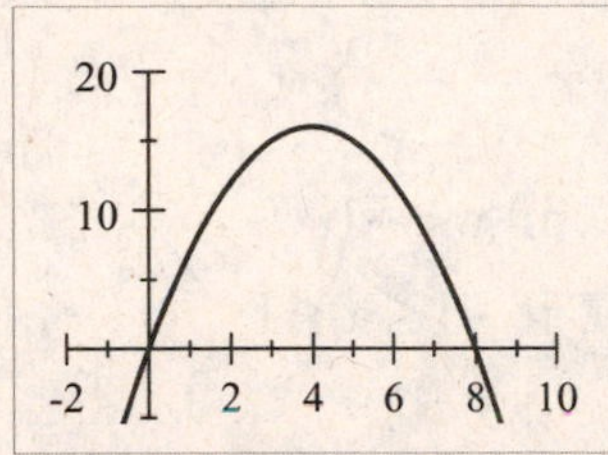

(d) $[-10,10]$ by $[-100,100]$

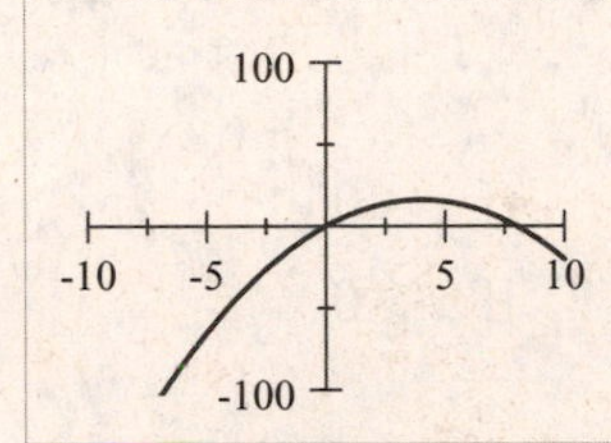

The viewing rectangle in part (c) produces the most appropriate graph of the equation.

31. $h(x)=x^3-5x-4$

(a) $[-2,2]$ by $[-2,2]$

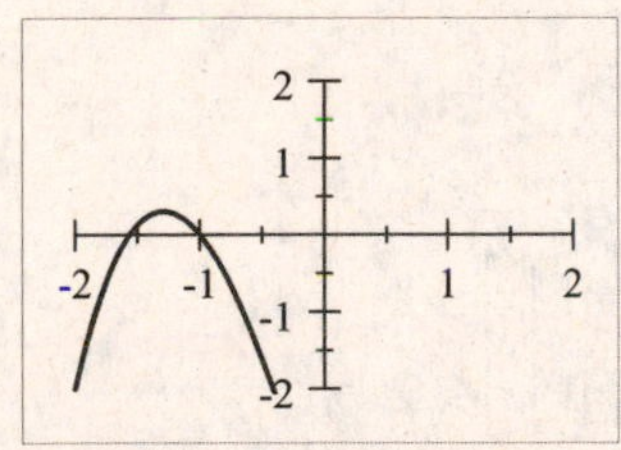

(b) $[-3,3]$ by $[-10,10]$

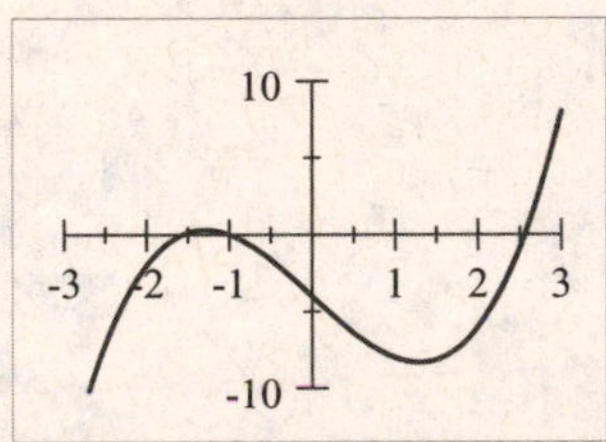

(c) $[-3,3]$ by $[-10,5]$

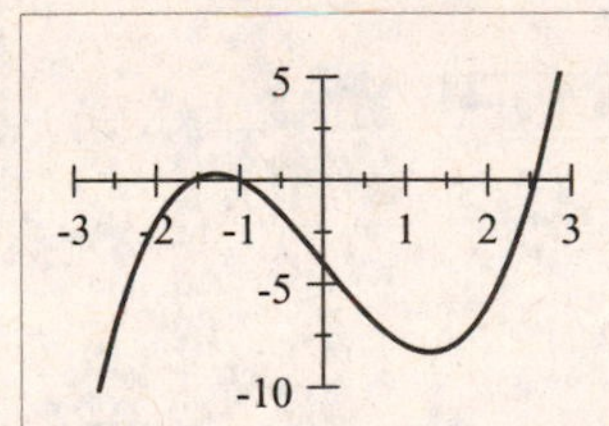

(d) $[-10,10]$ by $[-10,10]$

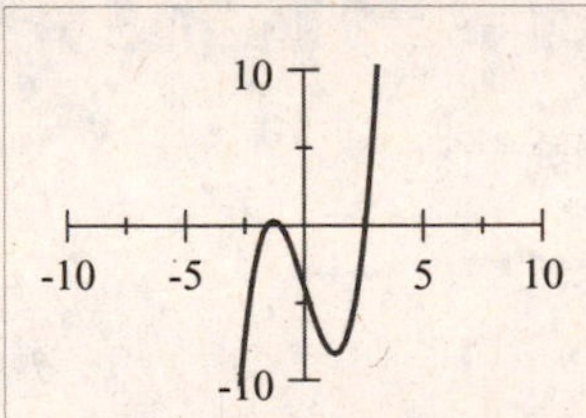

The viewing rectangle in part (c) produces the most appropriate graph of the equation.

33. $f(x)=\begin{cases}0 & \text{if } x<2\\ 1 & \text{if } x\ge 2\end{cases}$

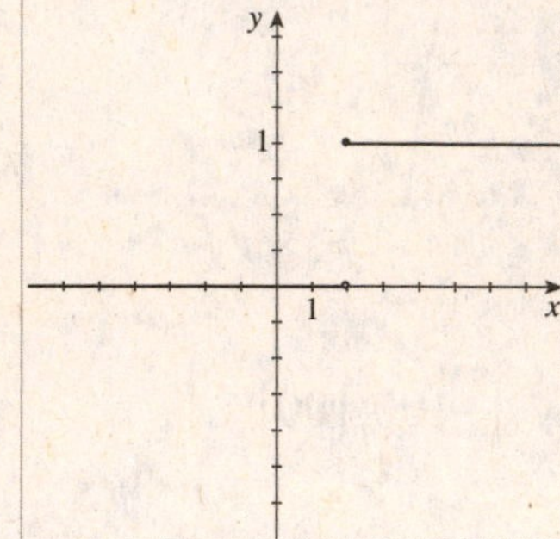

35. $f(x)=\begin{cases}3 & \text{if } x<2\\ x-1 & \text{if } x\ge 2\end{cases}$

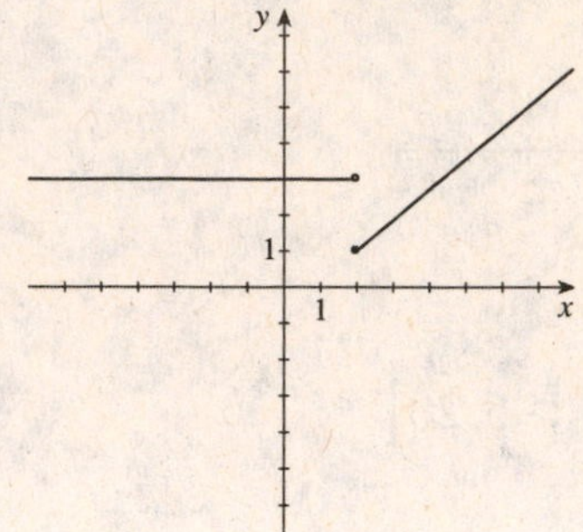

37. $f(x)=\begin{cases}x & \text{if } x\le 0\\ x+1 & \text{if } x>0\end{cases}$

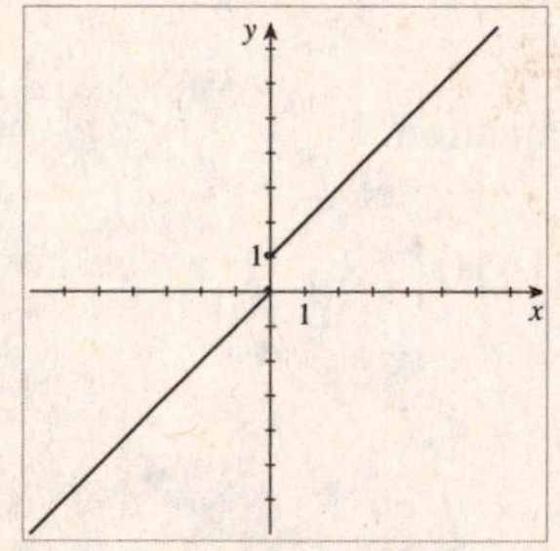

39. $f(x)=\begin{cases}-1 & \text{if } x<-1\\ 1 & \text{if } -1\le x\le 1\\ -1 & \text{if } x>1\end{cases}$

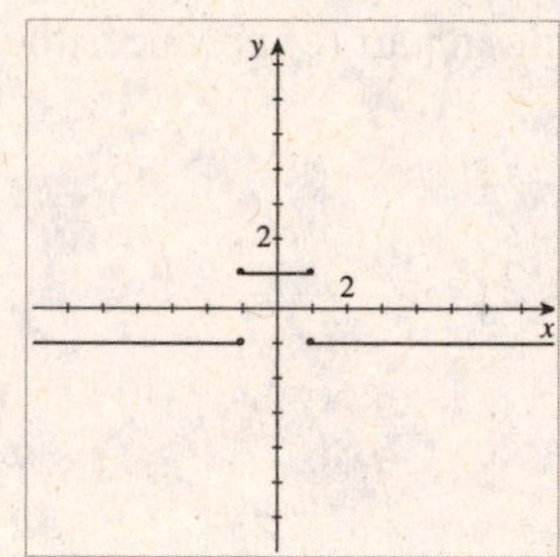

41. $f(x)=\begin{cases}2 & \text{if } x\le -1\\ x^2 & \text{if } x>-1\end{cases}$

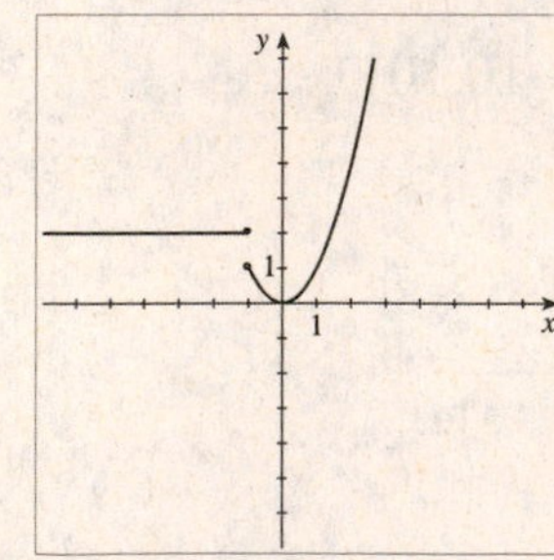

43. $f(x)=\begin{cases}0 & \text{if } |x|\le 2\\ 3 & \text{if } |x|>2\end{cases}$

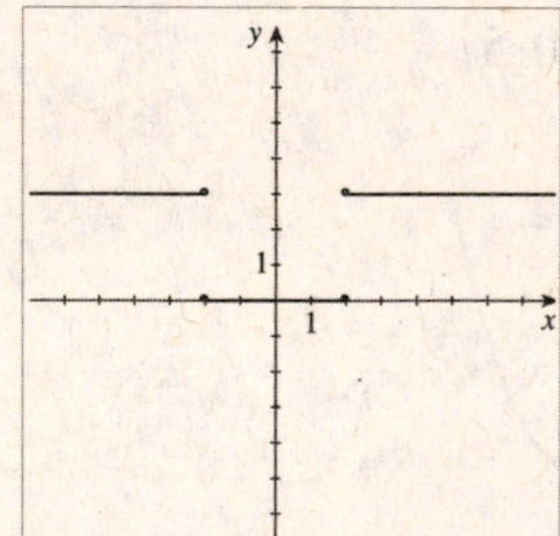

Contents

45. $f(x)=\begin{cases}4 & \text{if } x<-2\\ x^2 & \text{if } -2\le x\le 2\\ -x+6 & \text{if } x>2\end{cases}$

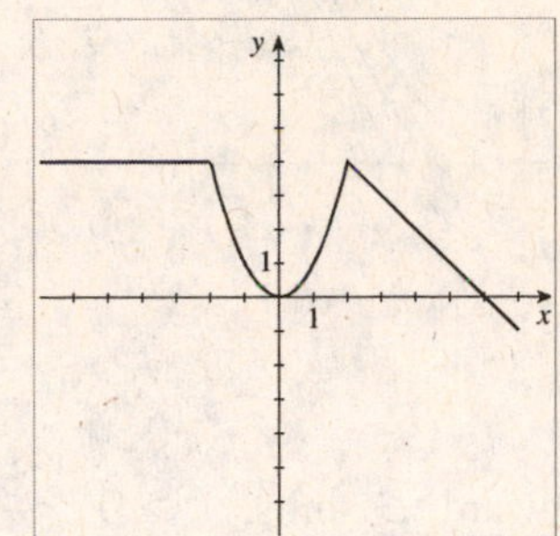

47. $f(x)=\begin{cases}x+2 & \text{if } x\le -1\\ x^2 & \text{if } x>-1\end{cases}$

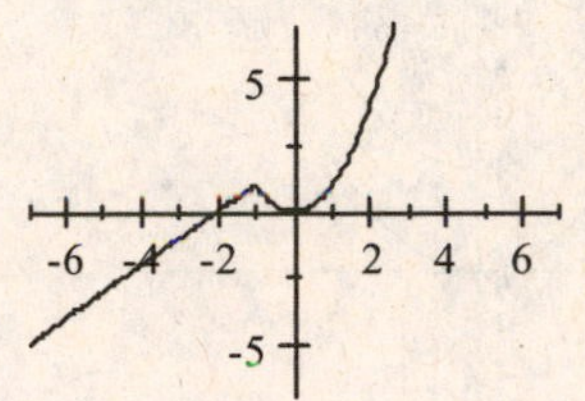

49. $f(x)=\begin{cases}-2 & \text{if } x<-2\\ x & \text{if } -2\le x\le 2\\ 2 & \text{if } x>2\end{cases}$

51. The curves in parts (a) and (c) are graphs of a function of x, by the Vertical Line Test.

53. The given curve is the graph of a function of x, by the Vertical Line Test. Domain: $[-3,2]$. Range: $[-2,2]$.

55. No, the given curve is not the graph of a function of x, by the Vertical Line Test.

57. Solving for y in terms of x gives $x^2+2y=4 \Leftrightarrow 2y=4-x^2 \Leftrightarrow y=2-\frac{1}{2}x^2$. This defines y as a function of x.

59. Solving for y in terms of x gives $x=y^2 \Leftrightarrow y=\pm\sqrt{x}$. The last equation gives two values of y for a given value of x. Thus, this equation does not define y as a function of x.

61. Solving for y in terms of x gives $x+y^2=9 \Leftrightarrow y^2=9-x \Leftrightarrow y=\pm\sqrt{9-x}$. The last equation gives two values of y for a given value of x. Thus, this equation does not define y as a function of x.

63. Solving for y in terms of x gives $x^2y+y=1 \Leftrightarrow y(x^2+1)=1 \Leftrightarrow y=\dfrac{1}{x^2+1}$. This defines y as a function of x.

65. Solving for y in terms of x gives $2|x|+y=0 \Leftrightarrow y=-2|x|$. This defines y as a function of x.

67. Solving for y in terms of x gives $x=y^3 \Leftrightarrow y=\sqrt[3]{x}$. This defines y as a function of x.

68. Solving for y in terms of x gives $x=y^4 \Leftrightarrow y=\pm\sqrt[4]{x}$. The last equation gives two values of y for any positive value of x. Thus, this equation does not define y as a function of x.

69. (a) $f(x)=x^2+c$, for $c=0$, 2, 4, and 6.

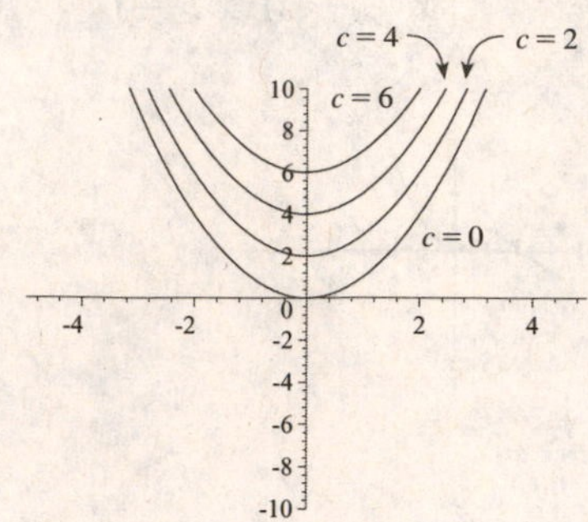

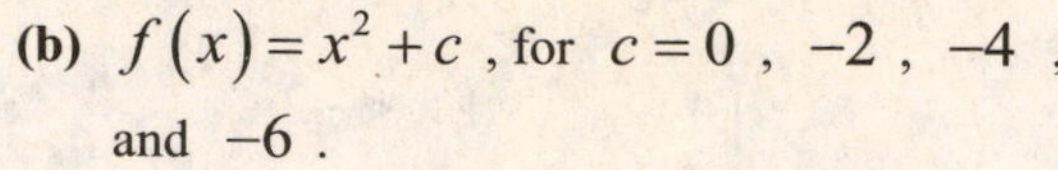

(b) $f(x)=x^2+c$, for $c=0$, -2, -4, and -6.

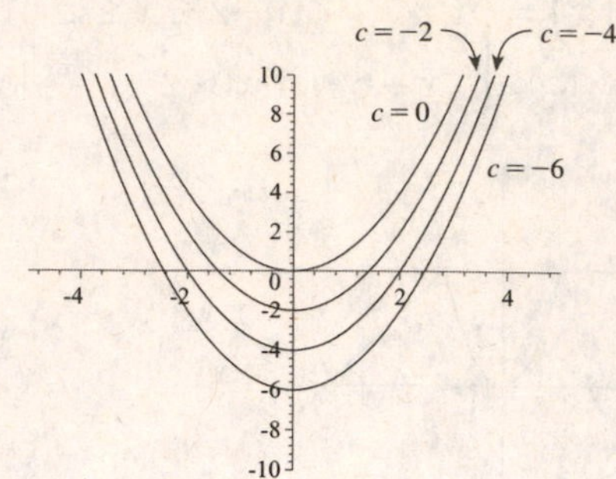

(c) The graphs in part (a) are obtained by shifting the graph of $f(x)=x^2$ upward c units, $c>0$.

The graphs in part (b) are obtained by shifting the graph of $f(x)=x^2$ downward c units.

71. (a) $f(x)=(x-c)^3$, for $c=0, 2, 4,$ and 6.

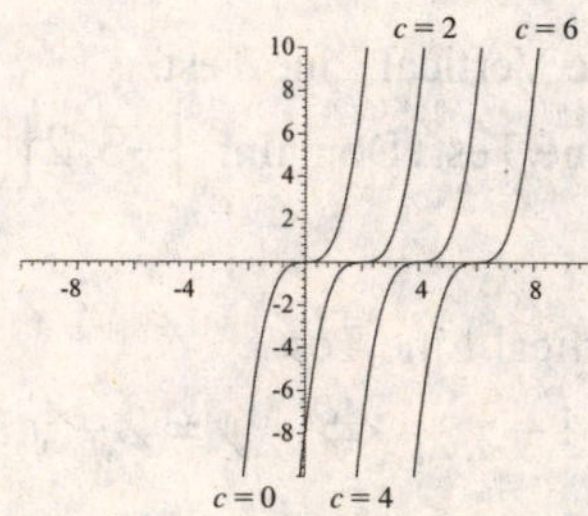

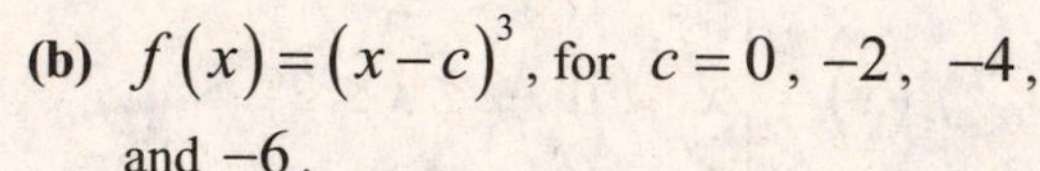

(b) $f(x)=(x-c)^3$, for $c=0, -2, -4,$ and -6.

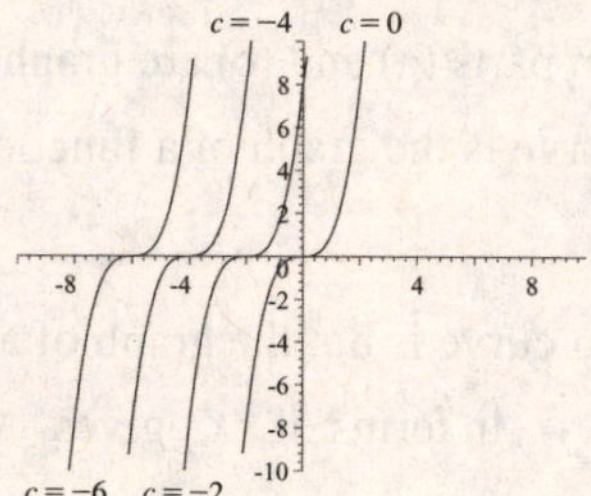

(c) The graphs in part (a) are obtained by shifting the graph of $f(x)=x^3$ to the right c units, $c>0$.

The graphs in part (b) are obtained by shifting the graph of $f(x)=x^3$ to the left $|c|$ units, $c<0$.

73. (a) $f(x)=x^c$, for $c=\frac{1}{2}$, $\frac{1}{4}$, and $\frac{1}{6}$.

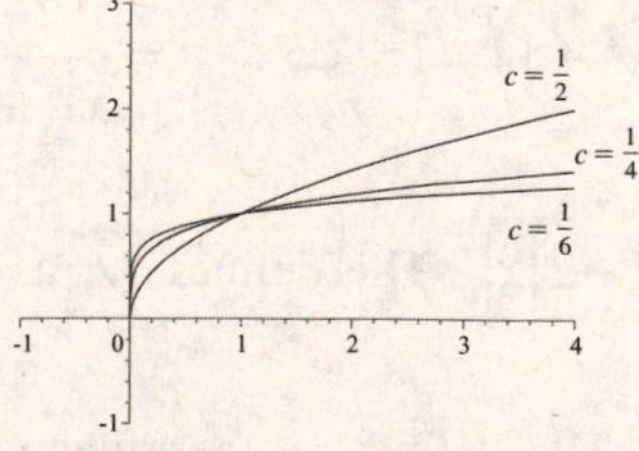

(b) $f(x)=x^c$, for $c=1$, $\frac{1}{3}$, and $\frac{1}{5}$.

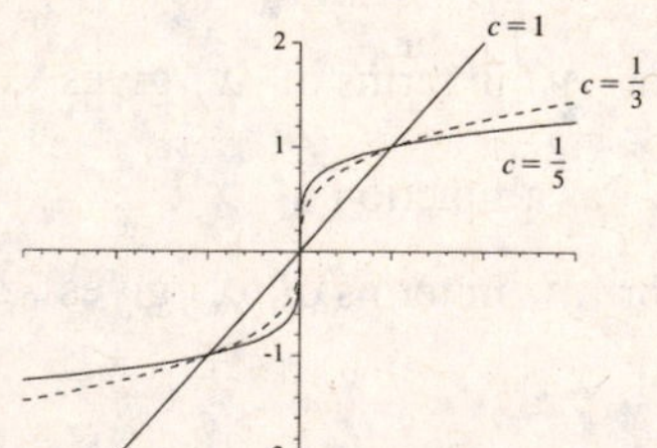

(c) Graphs of even roots are similar to $y=\sqrt{x}$, graphs of odd roots are similar to $y=\sqrt[3]{x}$. As c increases, the graph of $y=\sqrt[c]{x}$ becomes steeper near $x=0$ and flatter when $x>1$.

75. The slope of the line segment joining the points $(-2,1)$ and $(4,-6)$ is $m=\dfrac{-6-1}{4-(-2)}=-\frac{7}{6}$.

Using the point-slope form, we have $y-1=-\frac{7}{6}(x+2) \Leftrightarrow y=-\frac{7}{6}x-\frac{7}{3}+1 \Leftrightarrow$ $y=-\frac{7}{6}x-\frac{4}{3}$. Thus the function is $f(x)=-\frac{7}{6}x-\frac{4}{3}$ for $-2\le x\le 4$.

77. First solve the circle for y : $x^2+y^2=9 \Leftrightarrow y^2=9-x^2 \Rightarrow y=\pm\sqrt{9-x^2}$. Since we seek the top half of the circle, we choose $y=\sqrt{9-x^2}$. So the function is $f(x)=\sqrt{9-x^2}$, $-3\le x\le 3$.

79. We graph $T(r)=\dfrac{0.5}{r^2}$ for $10\le r\le 100$.

As the balloon is inflated, the skin gets thinner, as we would expect.

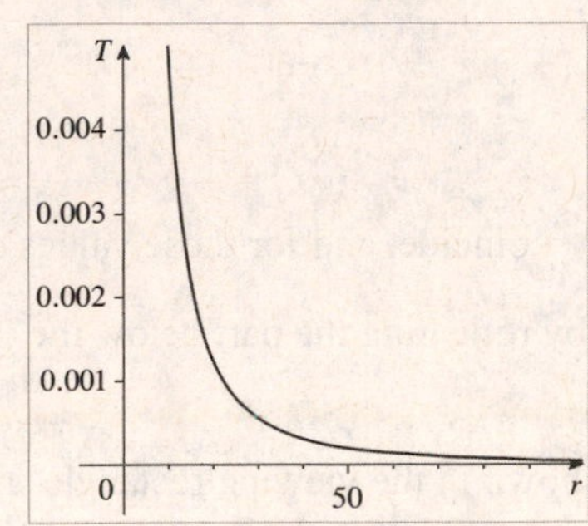

81. (a)

$$E(x)=\begin{cases}6.00+0.10x & \text{if } 0\le x\le 300\\ 36.00+0.06(x-300) & \text{if } 300<x\end{cases}$$

(b)

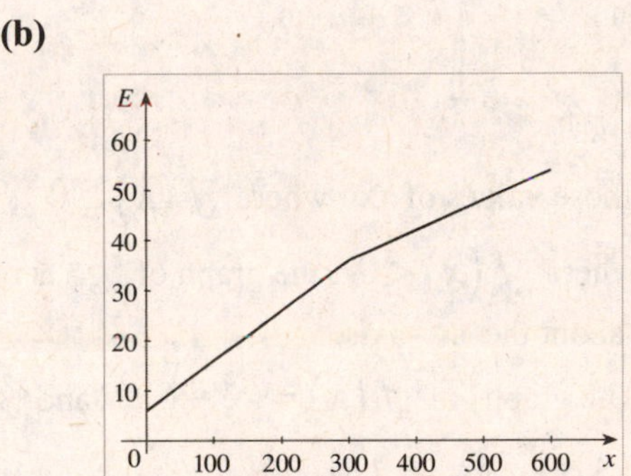

83. $$P(x)=\begin{cases}0.44 & \text{if } 0<x\le 1\\ 0.61 & \text{if } 1<x\le 2\\ 0.78 & \text{if } 2<x\le 3\\ 0.95 & \text{if } 3<x\le 3.5\end{cases}$$

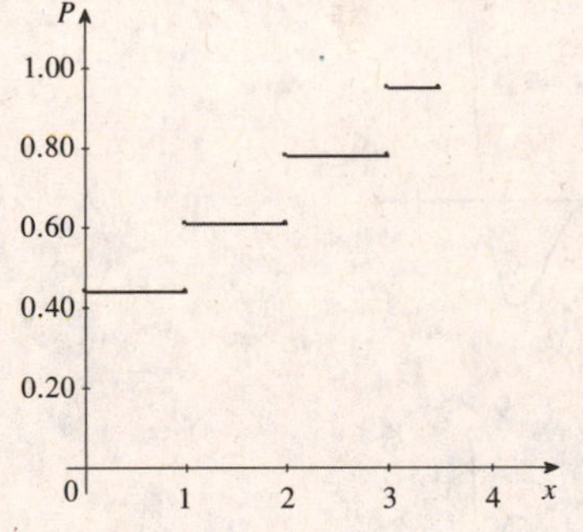

85. Answers will vary. Some examples are almost anything we purchase based on weight, volume, length, or time, for example gasoline. Although the amount delivered by the pump is continuous, the amount we pay is rounded to the penny. An example involving time would be the cost of a telephone call.

87. (a) The graphs of $f(x)=x^2+x-6$ and $g(x)=|x^2+x-6|$ are shown in the viewing rectangle $[-10,10]$ by $[-10,10]$.

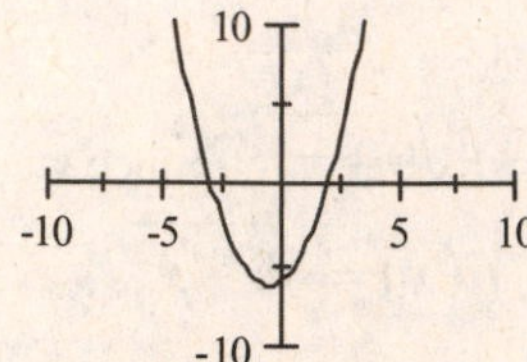

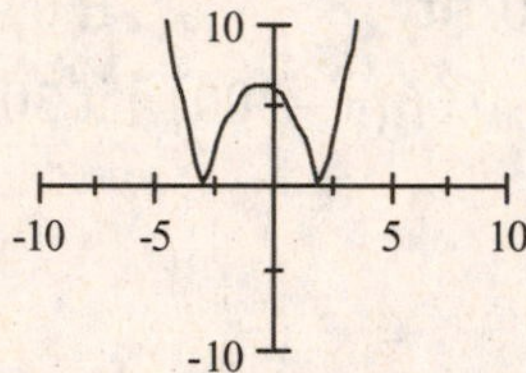

For those values of x where $f(x)\geq 0$, the graphs of f and g coincide, and for those values of x where $f(x)<0$, the graph of g is obtained from that of f by reflecting the part below the x-axis about the x-axis.

(b) The graphs of $f(x)=x^4-6x^2$ and $g(x)=|x^4-6x^2|$ are shown in the viewing rectangle $[-5,5]$ by $[-10,15]$.

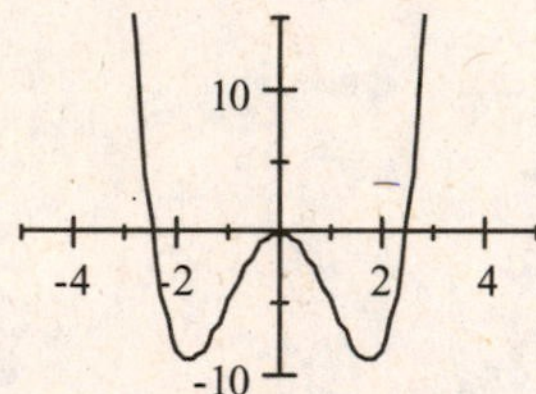

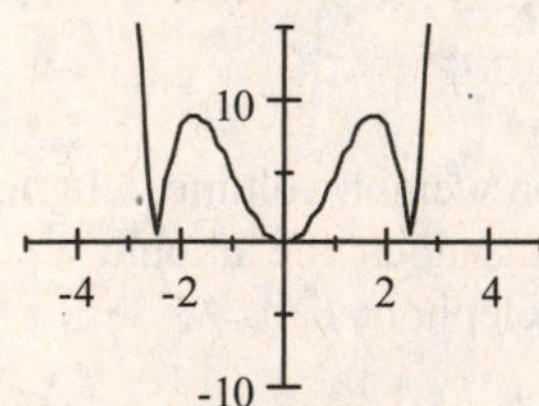

For those values of x where $f(x)\geq 0$, the graphs of f and g coincide, and for those values of x where $f(x)<0$, the graph of g is obtained from that of f by reflecting the part below the x-axis above the x-axis.

(c) In general, if $g(x)=|f(x)|$, then for those values of x where $f(x)\geq 0$, the graphs of f and g coincide, and for those values of x where $f(x)<0$, the graph of g is obtained from that of f by reflecting the part below the x-axis above the x-axis.

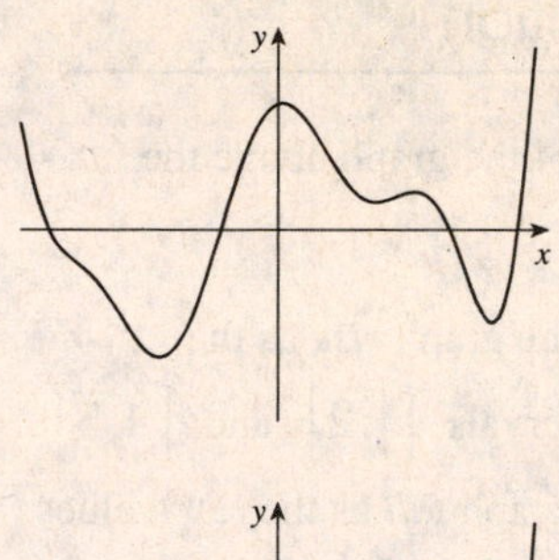
y
x

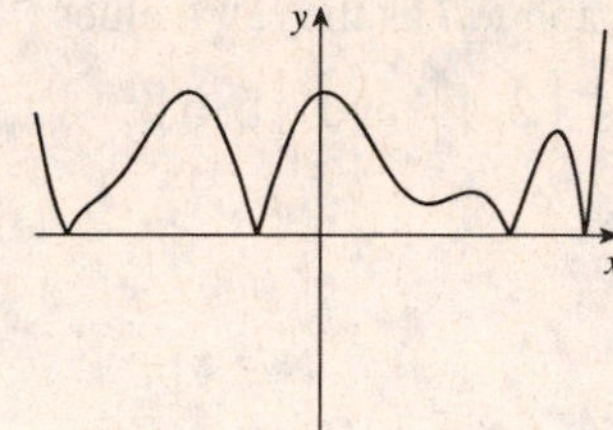
y
x

1.5 Getting Information from the Graph of a Function

1. To find a function value $f(a)$ from the graph of f we find the height of the graph above the x-axis at $x = a$. From the graph of f we see that $f(3) = 4$.

3. (a) If f is increasing on an interval, then the y-values of the points on the graph *rise* as the x-values increase. From the graph of f we see that f is increasing on the intervals $[1,2]$ and $[4,5]$.

(b) If f is decreasing on an interval, then y-values of the points on the graph *fall* as the x-values increase. From the graph of f we see that f is decreasing on the intervals $[2,4]$ and $[5,6]$.

5. (a) $h(-2) = 1$, $h(0) = -1$, $h(2) = 3$, and $h(3) = 4$.

(b) Domain: $[-3,4]$. Range: $[-1,4]$.

(c) $h(-3) = 3$, $h(2) = 3$, and $h(4) = 3$, so $h(x) = 3$ when $x = -3$, $x = 2$, or $x = 4$.

(d) The graph of h lies below or on the horizontal line $y = 3$ when $-3 \leq x \leq 2$ or $x = 4$, so $h(x) \leq 3$ for those values of x.

7. (a) $g(-4) = 3$, $g(-2) = 2$, $g(0) = -2$, $g(2) = 1$, and $g(4) = 0$.

(b) Domain: $[-4,4]$. Range: $[-2,3]$.

9. (a)

(b) Domain: $(-\infty,\infty)$; Range: $(-\infty,\infty)$

11. (a)

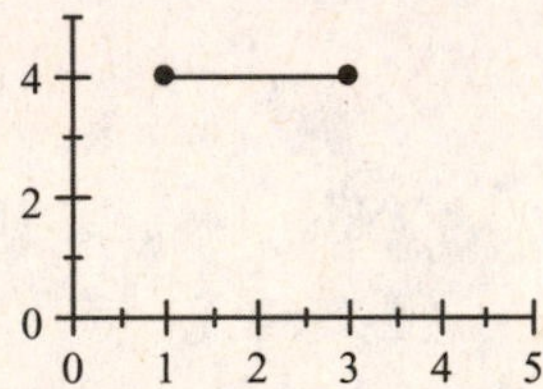

(b) Domain: $[1,3]$; Range: $\{4\}$

13. (a)

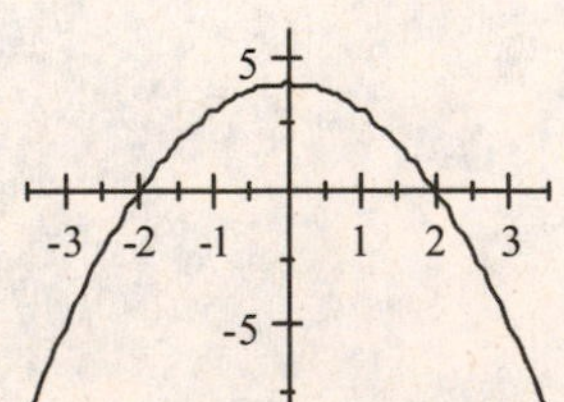

(b) Domain: $(-\infty,\infty)$; Range: $(-\infty,4]$

15. (a)

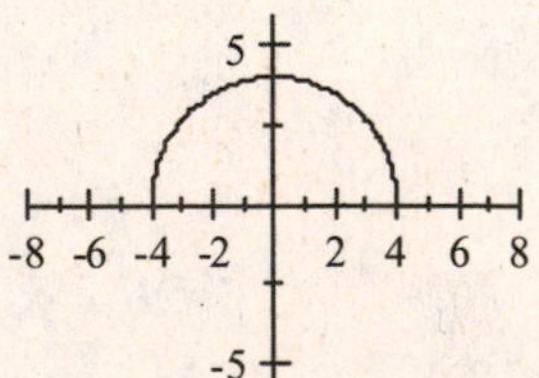

(b) Domain: $[-4,4]$; Range: $[0,4]$

17. (a)

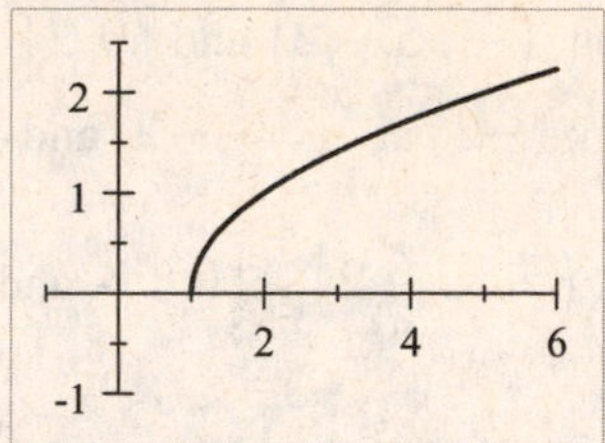

(b) Domain: $[1,\infty)$; Range: $[0,\infty)$

19. (a) The function is increasing on $[-1,1]$ and $[2,4]$.

(b) The function is decreasing on $[1,2]$.

21. (a) The function is increasing on $[-2,-1]$ and $[1,2]$.

(b) The function is decreasing on $[-3,-2]$, $[-1,1]$, and $[2,3]$.

23. (a) $f(x) = x^2 - 5x$ is graphed in the viewing rectangle $[-2,7]$ by $[-10,10]$

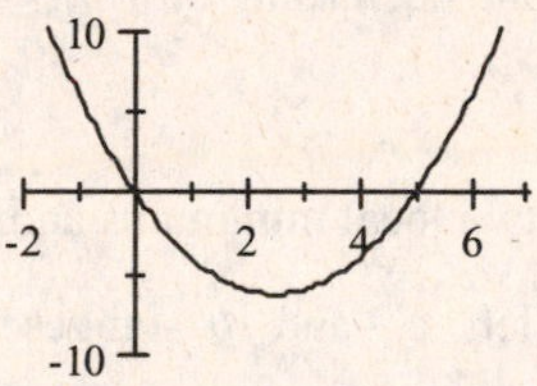

.

(b) The function is increasing on $[2.5,\infty)$. It is decreasing on $(-\infty,2.5]$.

25. (a) $f(x) = 2x^3 - 3x^2 - 12x$ is graphed in the viewing rectangle $[-3,5]$ by $[-25,20]$.

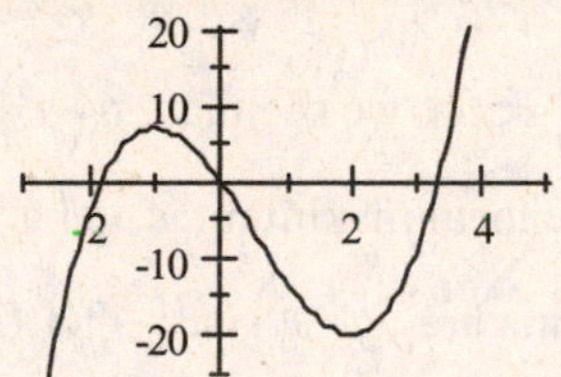

(b) The function is increasing on $(-\infty,-1]$ and $[2,\infty)$. It is decreasing on $[-1,2]$.

27. (a) $f(x) = x^3 + 2x^2 - x - 2$ is graphed in the viewing rectangle $[-5,5]$ by $[-3,3]$.

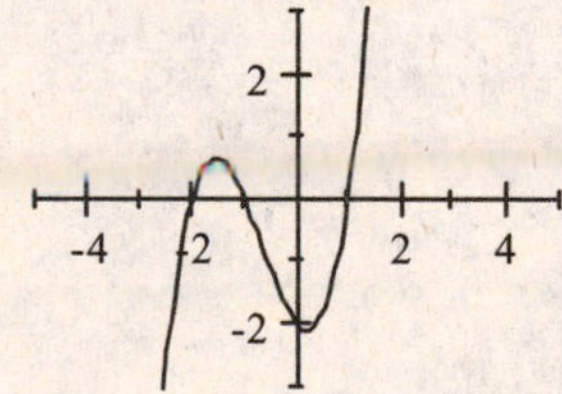

(b) The function is increasing on $(-\infty,-1.55]$ and $[0.22,\infty)$. It is decreasing on $[-1.55,0.22]$.

29. (a) $f(x) = x^{2/5}$ is graphed in the viewing rectangle $[-10,10]$ by $[-5,5]$

.

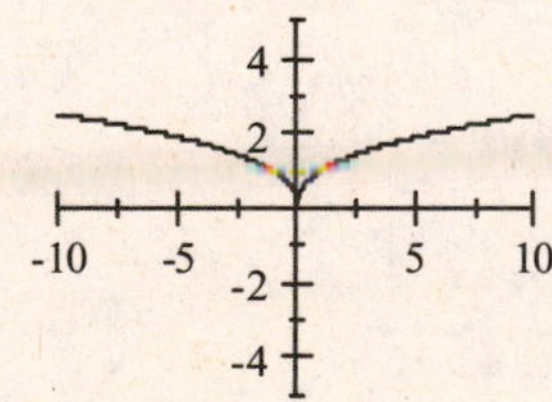

(b) The function is increasing on $[0,\infty)$. It is decreasing on $(-\infty,0]$.

31. (a) Local maximum: 2 at $x = 0$. Local minimum: -1 at $x = -2$ and 0 at $x = 2$.

(b) The function is increasing on $[-2,0]$ and $[2,\infty)$ and decreasing on $(-\infty,-2]$ and $[0,2]$.

33. (a) Local maximum: 0 at $x = 0$ and 1 at $x = 3$. Local minimum: -2 at $x = -2$ and -1 at $x = 1$.

(b) The function is increasing on $[-2,0]$ and $[1,3]$ and decreasing on $(-\infty,-2]$, $[0,1]$, and $[3,\infty)$.

35. (a) In the first graph, we see that $f(x) = x^3 - x$ has a local minimum and a local maximum. Smaller x - and y -ranges show that $f(x)$ has a local maximum of about 0.38 when $x \approx -0.58$ and a local minimum of about -0.38 when $x \approx 0.58$.

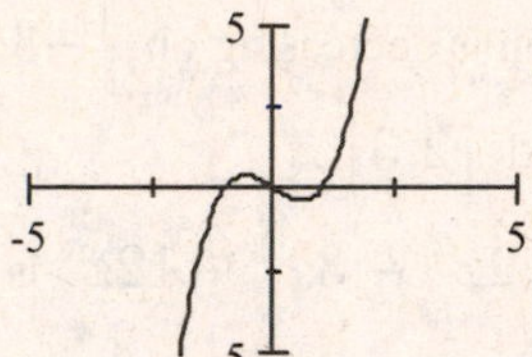

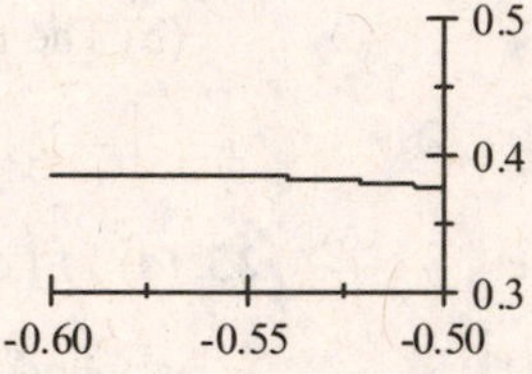

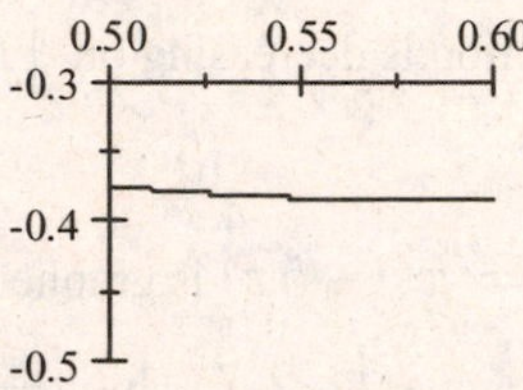

(b) The function is increasing on $(-\infty,-0.58]$ and $[0.58,\infty)$ and decreasing on $[-0.58,0.58]$.

37. (a) In the first graph, we see that $g(x) = x^4 - 2x^3 - 11x^2$ has two local minimums and a local maximum. The local maximum is $g(x) = 0$ when $x = 0$. Smaller x - and y -ranges show that local minima are $g(x) \approx -13.61$ when $x \approx -1.71$ and $g(x) \approx -73.32$ when $x \approx 3.21$.

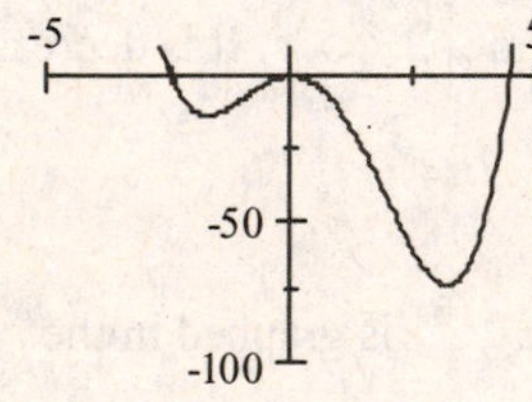

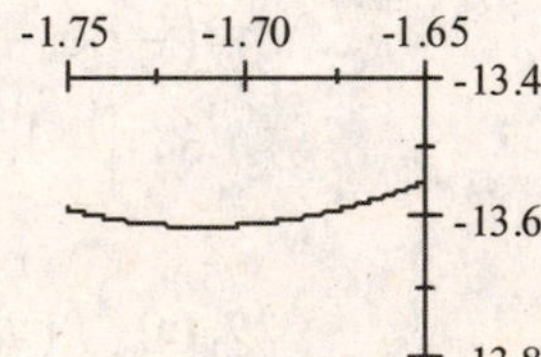

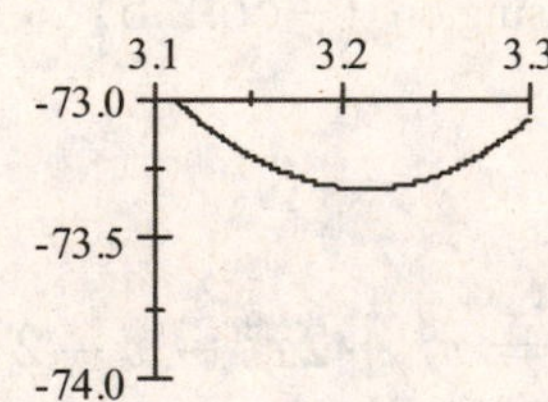

(b) The function is increasing on $[-1.71,0]$ and $[3.21,\infty)$ and decreasing on $(-\infty,-1.71]$ and $[0,3.21]$.

39. **(a)** In the first graph, we see that $U(x) = x\sqrt{6-x}$ has only a local maximum. Smaller x - and y -ranges show that $U(x)$ has a local maximum of about 5.66 when $x \approx 4.00$.

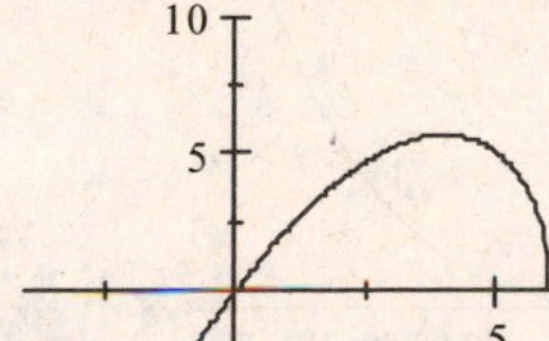

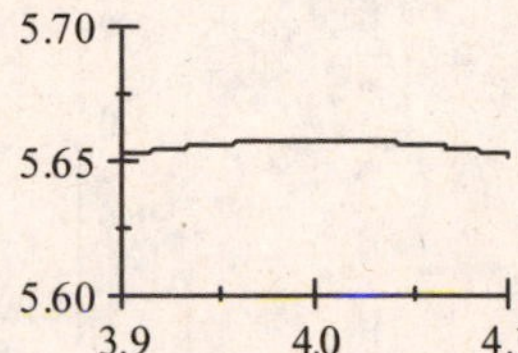

(b) The function is increasing on $(-\infty, 4.00]$ and decreasing on $[4.00, 6]$.

41. **(a)** In the first graph, we see that $V(x) = \dfrac{1-x^2}{x^3}$ has a local minimum and a local maximum. Smaller x - and y -ranges show that $V(x)$ has a local maximum of about 0.38 when $x \approx -1.73$ and a local minimum of about -0.38 when $x \approx 1.73$.

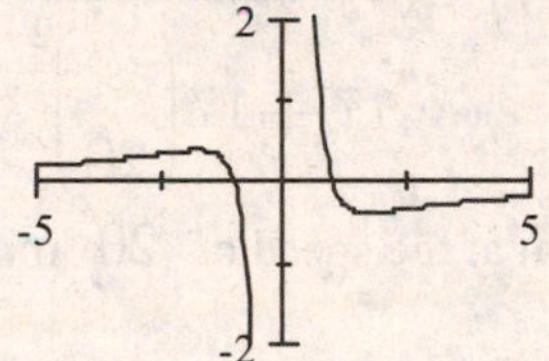

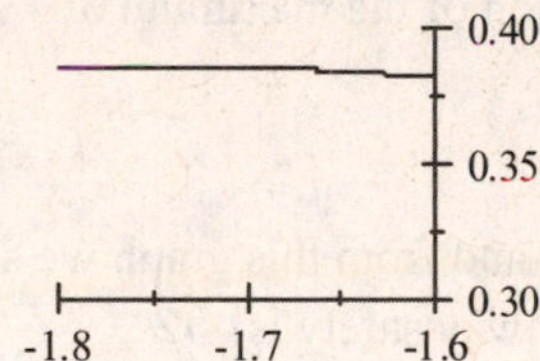

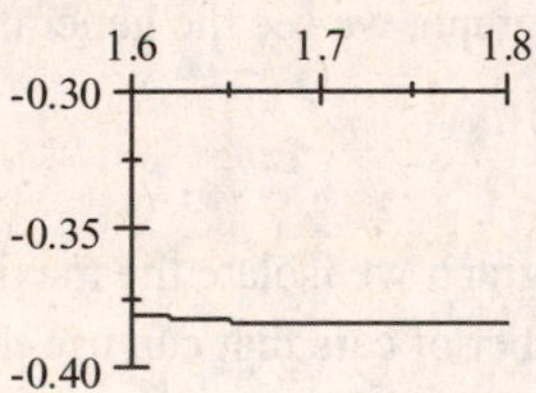

(b) The function is increasing on $(-\infty, -1.73]$ and $[1.73, \infty)$ and decreasing on $[-1.73, 0)$ and $(0, 1.73]$.

43. **(a)** At 6 a.m. the graph shows that the power consumption is about 500 megawatts. Since $t = 18$ represents 6 p.m., the graph shows that the power consumption at 6 p.m. is about 725 megawatts.
(b) The power consumption is lowest between 3 a.m. and 4 a.m..
(c) The power consumption is highest just before 12 noon.

45. **(a)** This person appears to be gaining weight steadily until the age of 21 when this person's weight gain slows down. The person continues to gain weight until the age of 30, at which point this person experiences a sudden weight loss. Weight gain resumes around the age of 32, and the person dies at about age 68. Thus, the person's weight W is increasing on $[0, 30]$ and $[32, 68]$ and decreasing on $[30, 32]$
(b) The sudden weight loss could be due to a number of reasons, among them major illness, a weight loss program, etc.

47. **(a)** The function W is increasing on $[0, 150]$ and $[300, \infty)$ and decreasing on $[150, 300]$.
(b) W has a local maximum at $x = 150$ and a local minimum at $x = 300$.

49. Runner A won the race. All runners finished the race. Runner B fell, but got up and finished the race.

(a)

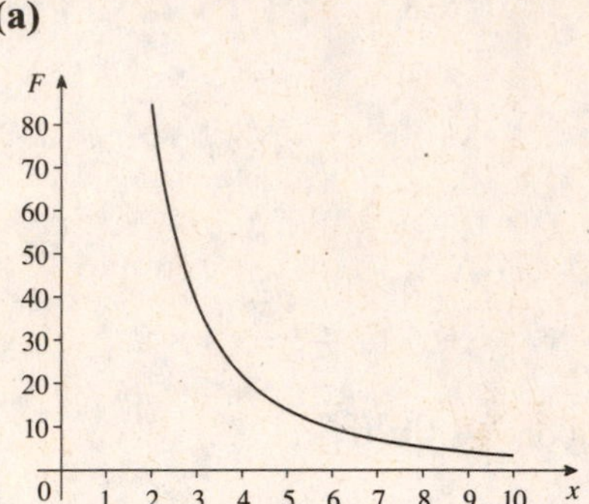

(b) As the distance x increases, the gravitational attraction F decreases. The rate of decrease is rapid at first, and slows as the distance increases.

51. (a)

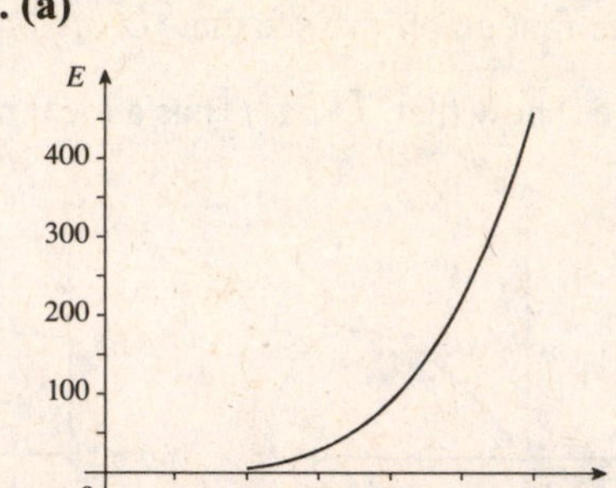

(b) As the temperature T increases, the energy E increases. The rate of increase gets larger as the temperature increases.

53. In the first graph, we see the general location of the maximum of $N(s) = \dfrac{88s}{17 + 17\left(\dfrac{s}{20}\right)^2}$. In the second graph we isolate the maximum, and from this graph we see that at the speed of 20 mi/h the largest number of cars that can use the highway safely is 52.

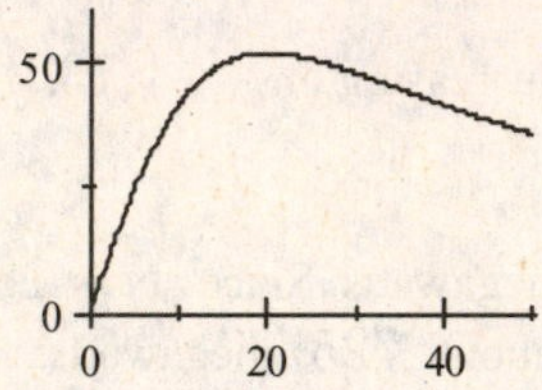

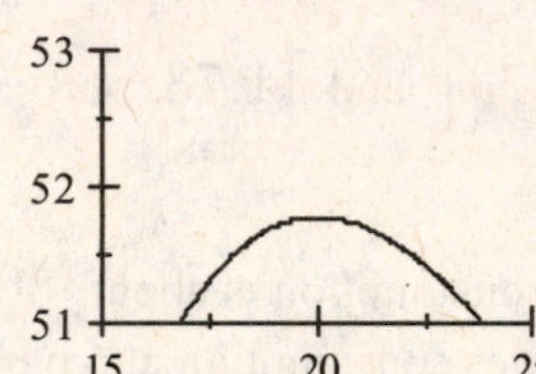

55. In the first graph, we see the general location of the maximum of $v(r) = 3.2(1 - r)r^2$ is around $r = 0.7$ cm. In the second graph, we isolate the maximum, and from this graph we see that at the maximum velocity is approximately 0.47 when $r \approx 0.67$ cm.

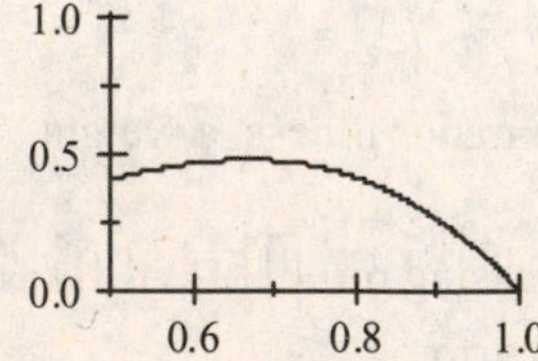

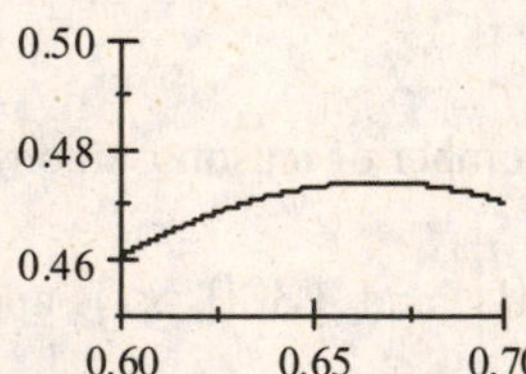

57. Numerous answers are possible.

1.6 Transformations of Functions

1. (a) The graph of $y = f(x)+3$ is obtained from the graph of $y = f(x)$ by shifting *upward* 3 units.

(b) The graph of $y = f(x+3)$ is obtained from the graph of $y = f(x)$ by shifting *left* 3 units.

3. (a) The graph of $y = -f(x)$ is obtained from the graph of $y = f(x)$ by reflecting in the x *-axis*.

(b) The graph of $y = f(-x)$ is obtained from the graph of $y = f(x)$ by reflecting in the y *-axis*.

5. (a) The graph of $y = f(x)-5$ can be obtained by shifting the graph of $y = f(x)$ downward 5 units.

(b) The graph of $y = f(x-5)$ can be obtained by shifting the graph of $y = f(x)$ to the right 5 units.

7. (a) The graph of $y = -f(x)$ can be obtained by reflecting the graph of $y = f(x)$ in the x -axis.

(b) The graph of $y = f(-x)$ can be obtained by reflecting the graph of $y = f(x)$ in the y -axis.

9. (a) The graph of $y = -f(x)+5$ can be obtained by reflecting the graph of $y = f(x)$ in the x -axis, then shifting the resulting graph upward 5 units.

(b) The graph of $y = 3f(x)-5$ can be obtained by stretching the graph of $y = f(x)$ vertically by a factor of 3 , then shifting the resulting graph downward 5 units.

11. (a) The graph of $y = 2f(x+1)-3$ can be obtained by shifting the graph of $y = f(x)$ to the left 1 unit, stretching it vertically by a factor of 2 , and shifting it downward 3 units.

(b) The graph of $y = 2f(x-1)+3$ can be obtained by shifting the graph of $y = f(x)$ to the right 1 unit, stretching it vertically by a factor of 2 , and shifting it upward 3 units.

13. (a) The graph of $y = f(4x)$ can be obtained by shrinking the graph of $y = f(x)$ horizontally by a factor of $\frac{1}{4}$.

(b) The graph of $y = f(\frac{1}{4}x)$ can be obtained by stretching the graph of $y = f(x)$ horizontally by a factor of 4 .

15. (a) The graph of $g(x) = (x+2)^2$ is obtained by shifting the graph of $f(x)$ to the left 2 units.

(b) The graph of $g(x) = x^2 + 2$ is obtained by shifting the graph of $f(x)$ upward 2 units.

17. (a) The graph of $g(x) = |x+2| - 2$ is obtained by shifting the graph of $f(x)$ to the left 2 units and downward 2 units.

(b) The graph of $g(x) = g(x) = |x-2| + 2$ is obtained from by shifting the graph of $f(x)$ to the right 2 units and upward 2 units.

19. (a)

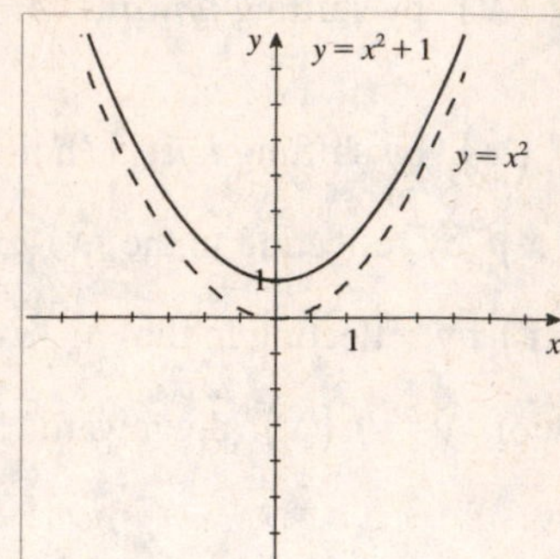

(b)

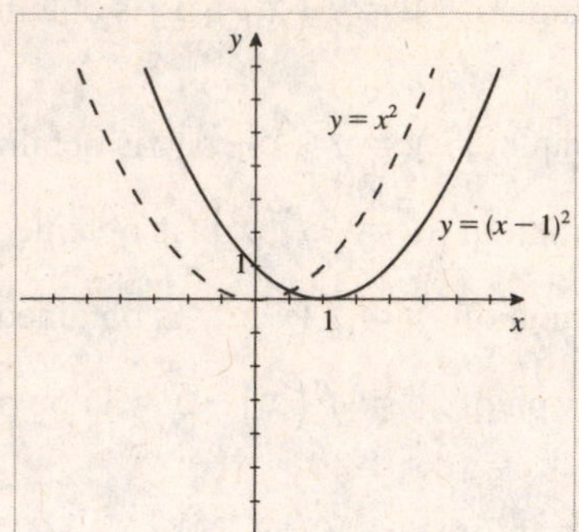

(c)

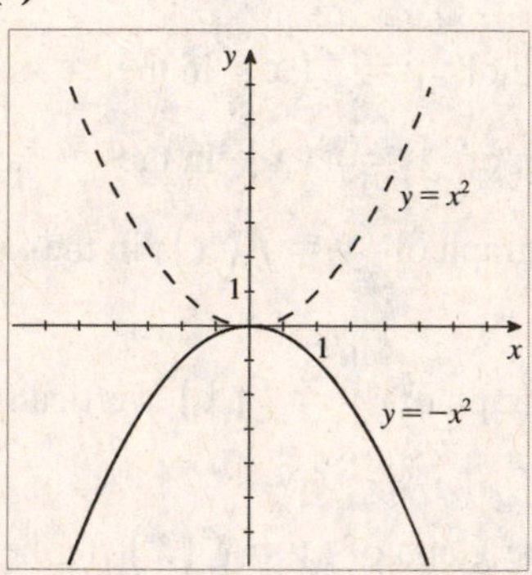

(d)

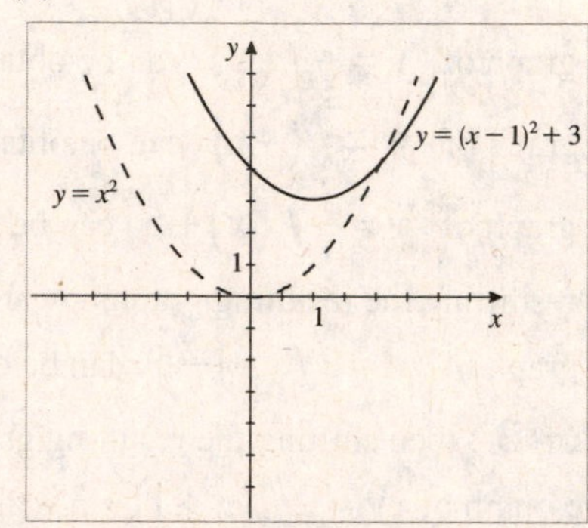

21. $f(x) = x^2 - 1$. Shift the graph of $y = x^2$ downward 1 unit.

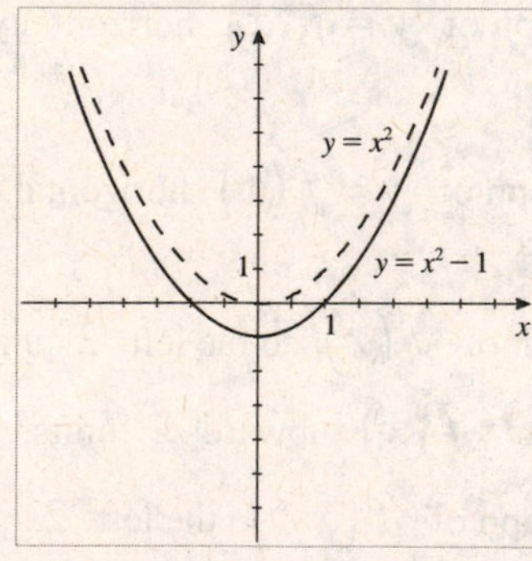

23. $f(x) = \sqrt{x} + 1$. Shift the graph of $y = \sqrt{x}$ upward 1 unit.

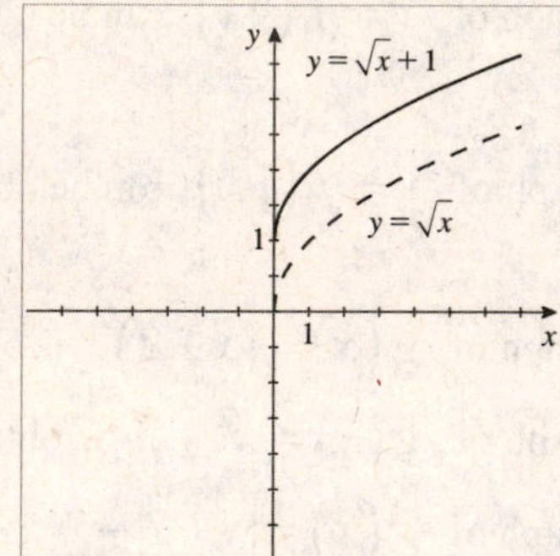

25. $f(x) = (x-5)^2$. Shift the graph of $y = x^2$ to the right 5 units.

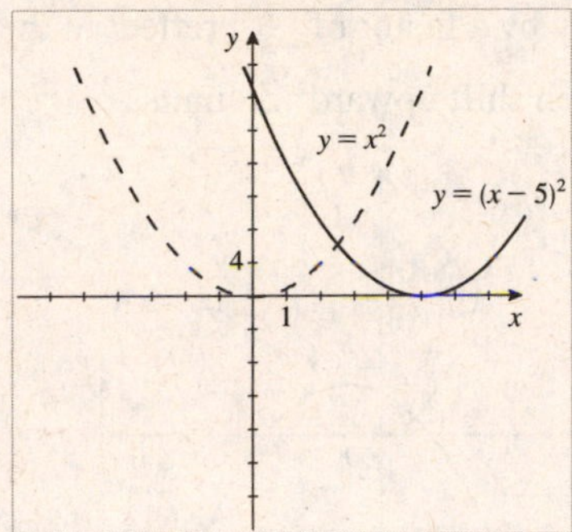

27. $f(x) = \sqrt{x+4}$. Shift the graph of $y = \sqrt{x}$ to the left 4 units.

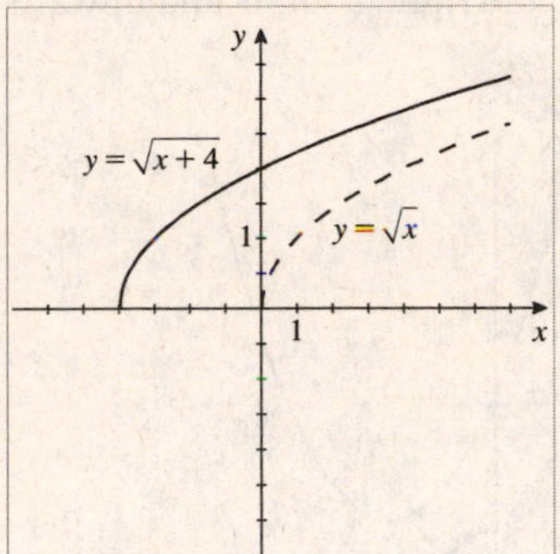

29. $f(x) = -x^3$. Reflect the graph of $y = x^3$ in the x-axis.

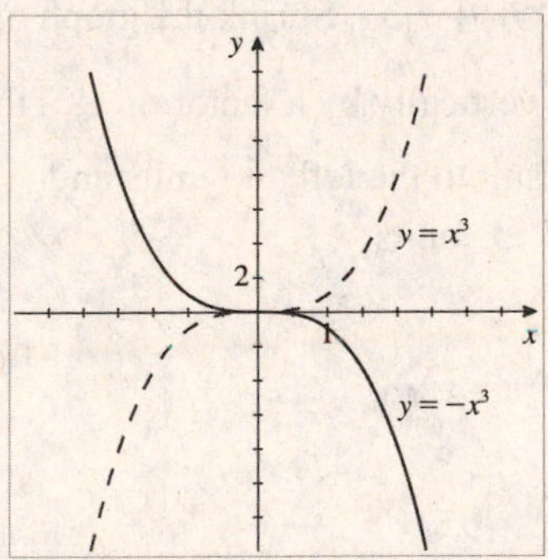

31. $y = \sqrt[4]{-x}$. Reflect the graph of $y = \sqrt[4]{x}$ in the y-axis.

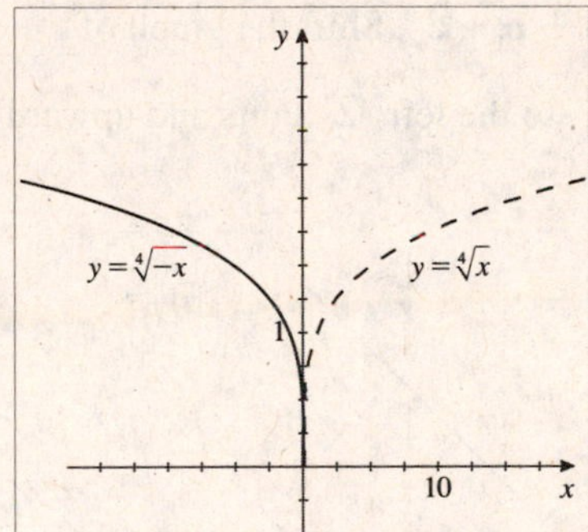

33. $y = \frac{1}{4}x^2$. Shrink the graph of $y = x^2$ vertically by a factor of $\frac{1}{4}$.

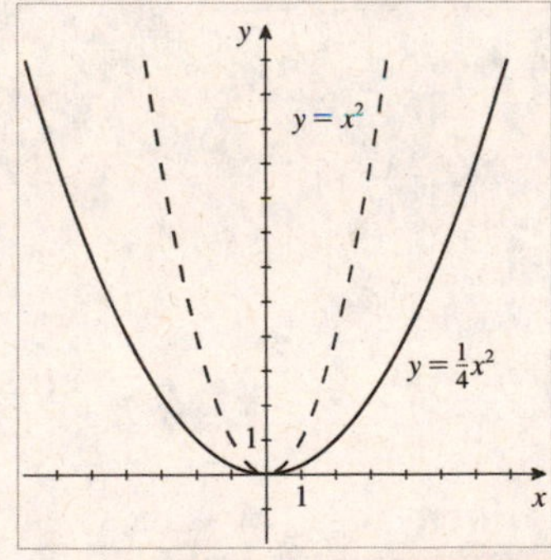

35. $y = 3|x|$. Stretch the graph of $y = |x|$ vertically by a factor of 3.

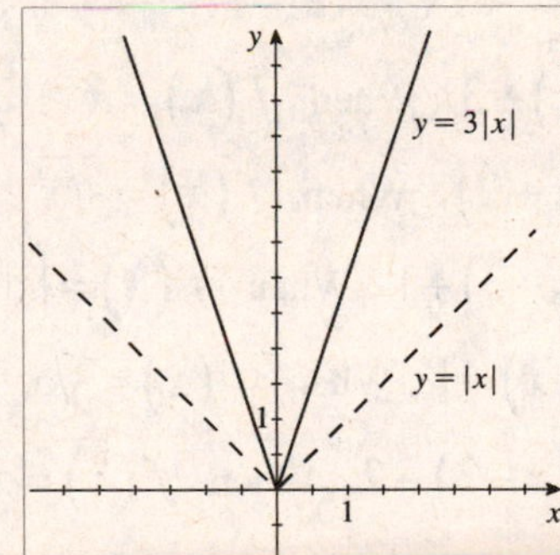

37. $y=(x-3)^2+5$. Shift the graph of $y=x^2$ to the right 3 units and upward 5 units.

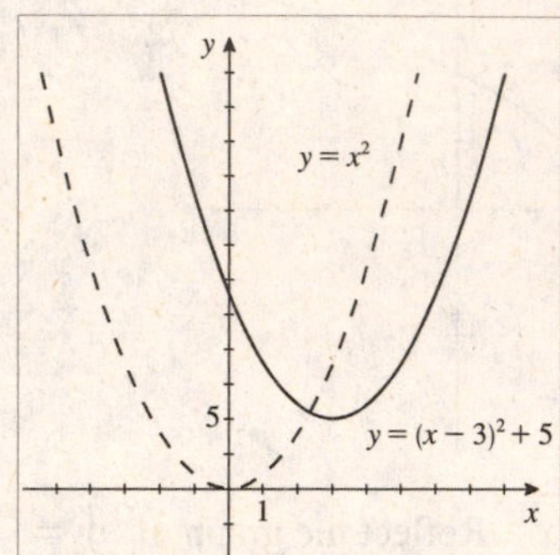

39. $y=3-\frac{1}{2}(x-1)^2$. Shift the graph of $y=x^2$ to the right one unit, shrink vertically by a factor of $\frac{1}{2}$, reflect in the x -axis, then shift upward 3 units.

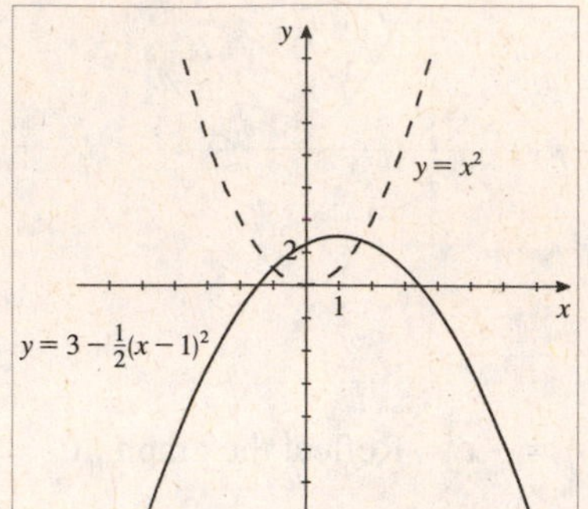

41. $y=|x+2|+2$. Shift the graph of $y=|x|$ to the left 2 units and upward 2 units.

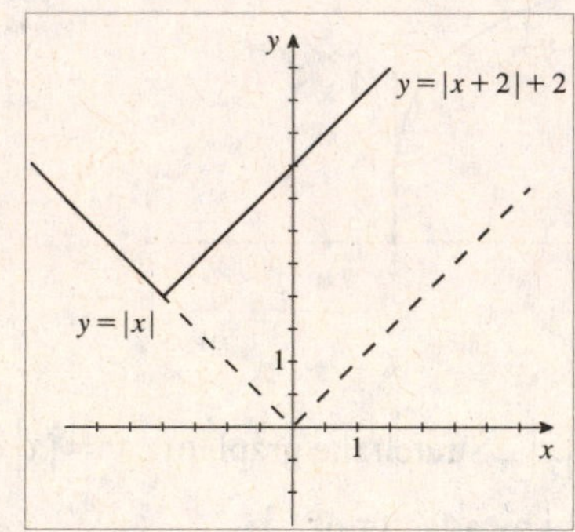

43. $y=\frac{1}{2}\sqrt{x+4}-3$. Shrink the graph of $y=\sqrt{x}$ vertically by a factor of $\frac{1}{2}$, then shift the result to the left 4 units and downward 3 units.

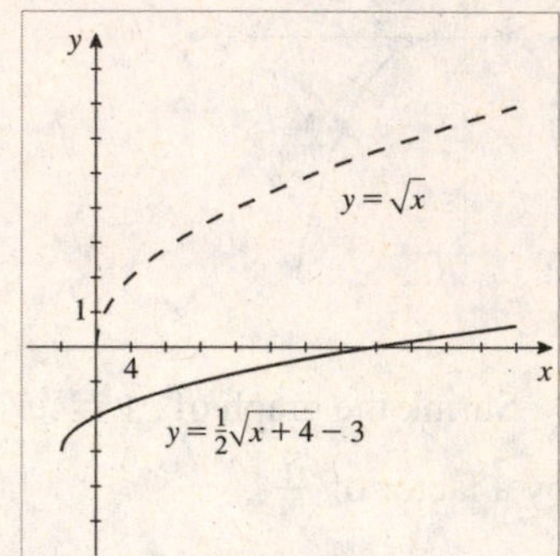

45. $y=f(x)+3$. When $f(x)=x^2$, $y=x^2+3$.

47. $y=f(x+2)$. When $f(x)=\sqrt{x}$, $y=\sqrt{x+2}$.

49. $y=f(x-3)+1$. When $f(x)=|x|$, $y=|x-3|+1$.

51. $y=f(-x)+1$. When $f(x)=\sqrt[4]{x}$, $y=\sqrt[4]{-x}+1$.

53. $y=2f(x-3)-2$. When $f(x)=x^2$, $y=2(x-3)^2-2$.

55. $g(x)=f(x-2)=(x-2)^2=x^2-4x+4$

57. $g(x)=f(x+1)+2=|x+1|+2$

59. $g(x)=-f(x+2)=-\sqrt{x+2}$

61. (a) $y=f(x-4)$ is graph #3.

(b) $y=f(x)+3$ is graph #1.

(c) $y=2f(x+6)$ is graph #2.

(d) $y=-f(2x)$ is graph #4.

63. (a) $y = f(x-2)$

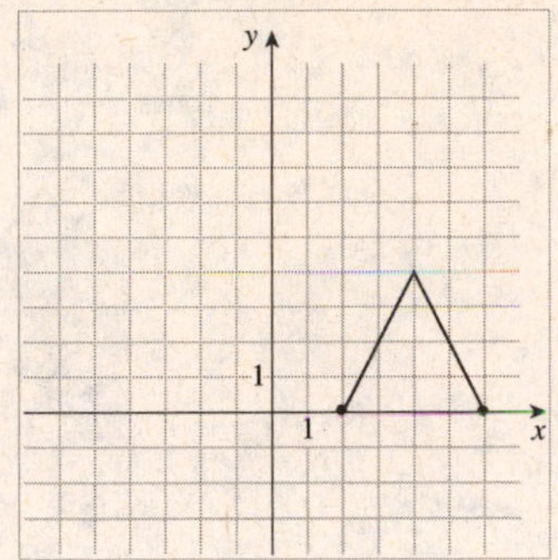

(b) $y = f(x) - 2$

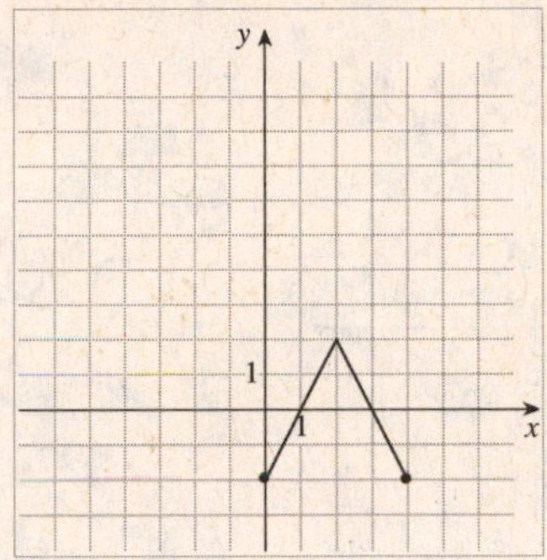

(c) $y = 2f(x)$

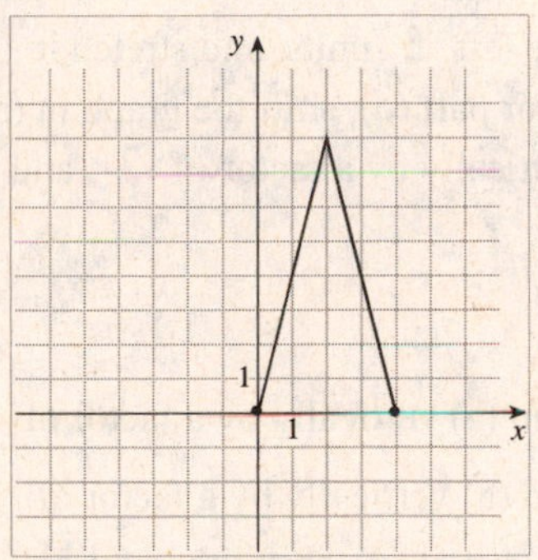

(d) $y = -f(x) + 3$

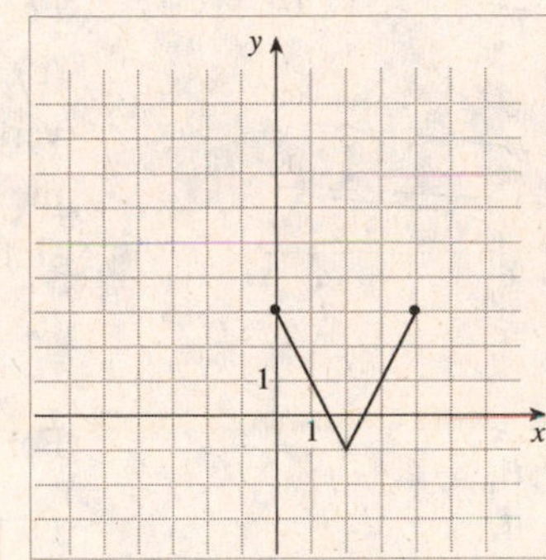

(e) $y = f(-x)$

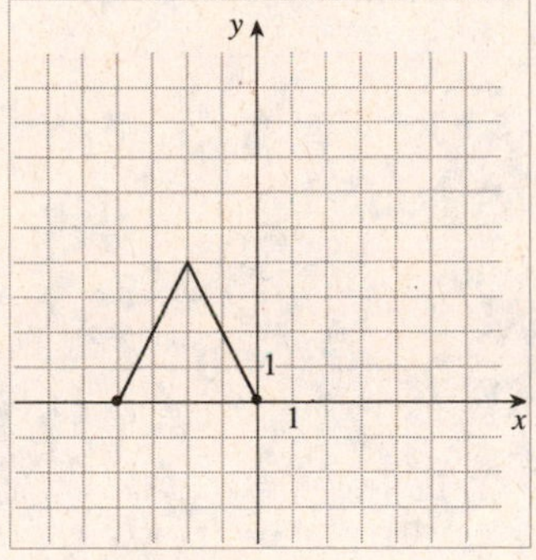

(f) $y = \frac{1}{2} f(x-1)$

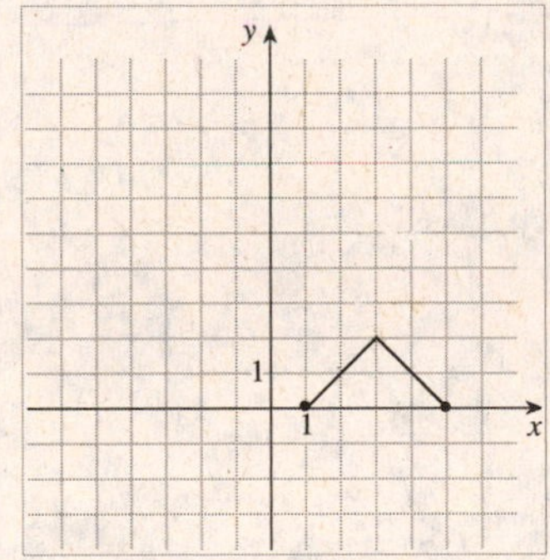

65. (a) $y = g(2x)$

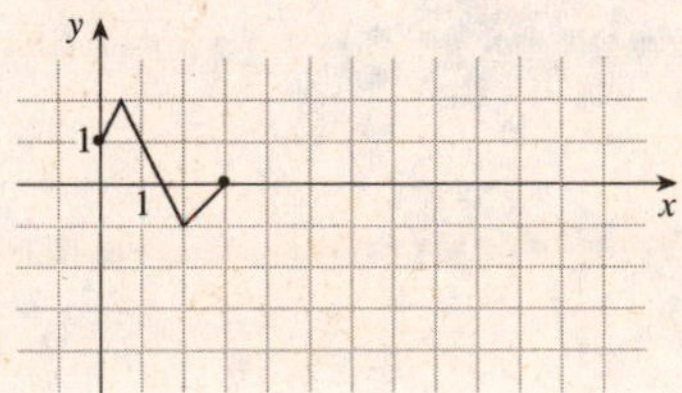

(b) $y = g\left(\frac{1}{2}x\right)$

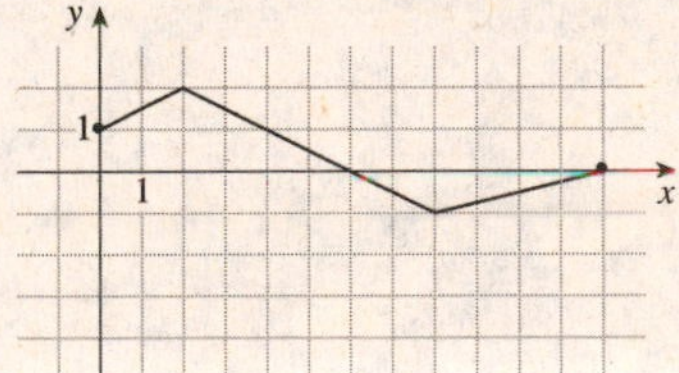

67. $y = [[2x]]$

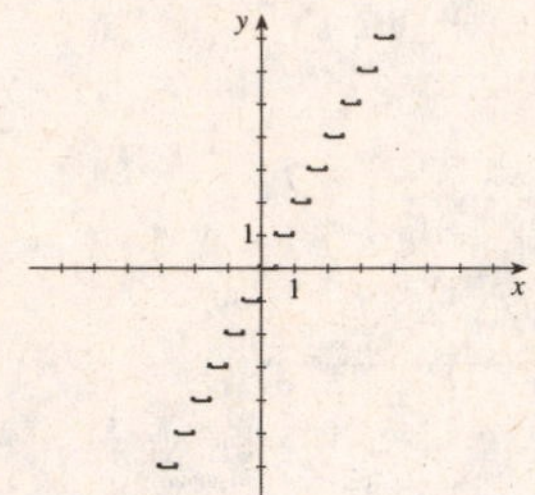

69.

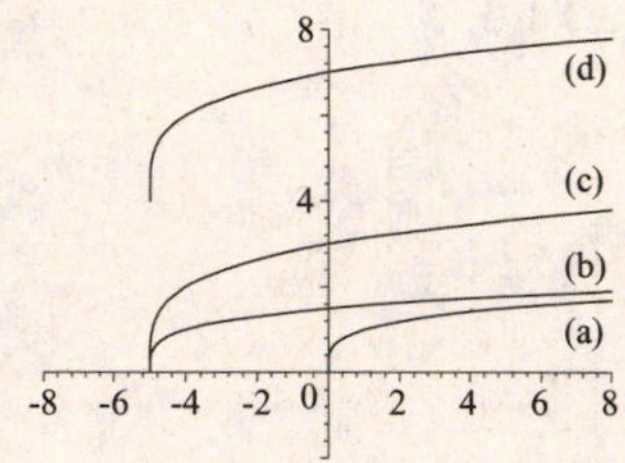

For part (b), shift the graph in (a) to the left 5 units; for part (c), shift the graph in (a) to the left 5 units, and stretch it vertically by a factor of 2 ; for part (d), shift the graph in (a) to the left 5 units, stretch it vertically by a factor of 2 , and then shift it upward 4 units.

71.

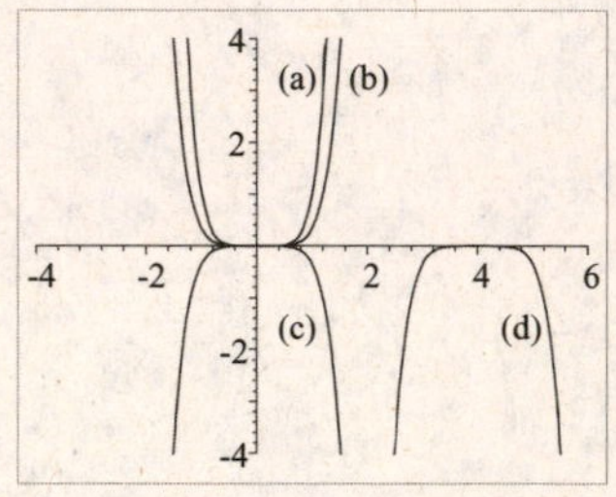

For part (b), shrink the graph in (a) vertically by a factor of $\frac{1}{3}$; for part (c), shrink the graph in (a) vertically by a factor of $\frac{1}{3}$, and reflect it in the x -axis; for part (d), shift the graph in (a) to the right 4 units, shrink vertically by a factor of $\frac{1}{3}$, and then reflect it in the x -axis.

73. (a) $y = f(x) = \sqrt{2x - x^2}$

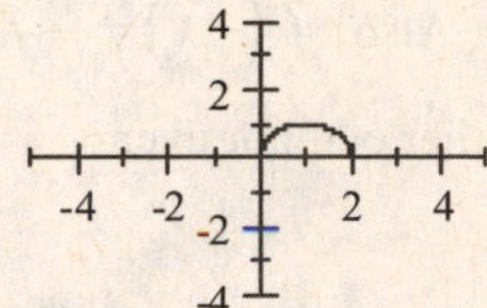

(b)

$y = f(2x) = \sqrt{2(2x) - (2x)^2} = \sqrt{4x - 4x^2}$

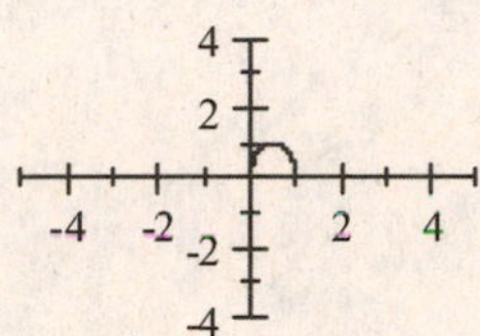

(c)

$y = f\left(\tfrac{1}{2}x\right) = \sqrt{2\left(\tfrac{1}{2}x\right) - \left(\tfrac{1}{2}x\right)^2} = \sqrt{x - \tfrac{1}{4}x^2}$

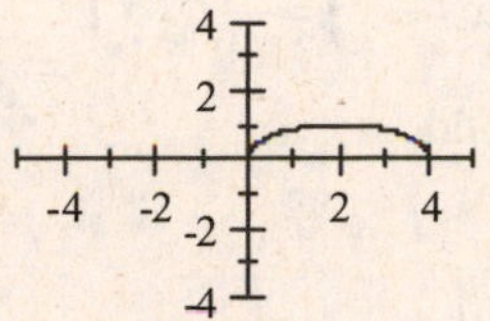

The graph in part (b) is obtained by horizontally shrinking the graph in part (a) by a factor of $\frac{1}{2}$ (so the graph is half as wide). The graph in part (c) is obtained by horizontally stretching the graph in part (a) by a factor of 2 (so the graph is twice as wide).

75. $f(x) = x^4$.

$f(-x) = (-x)^4 = x^4 = f(x)$. Thus $f(x)$ is even.

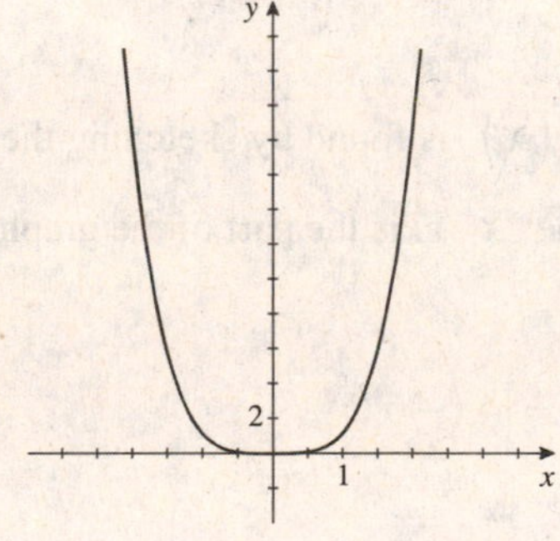

77. $f(x) = x^2 + x$.

$f(-x) = (-x)^2 + (-x) = x^2 - x$. Thus $f(-x) \neq f(x)$. Also, $f(-x) \neq -f(x)$, so $f(x)$ is neither odd nor even.

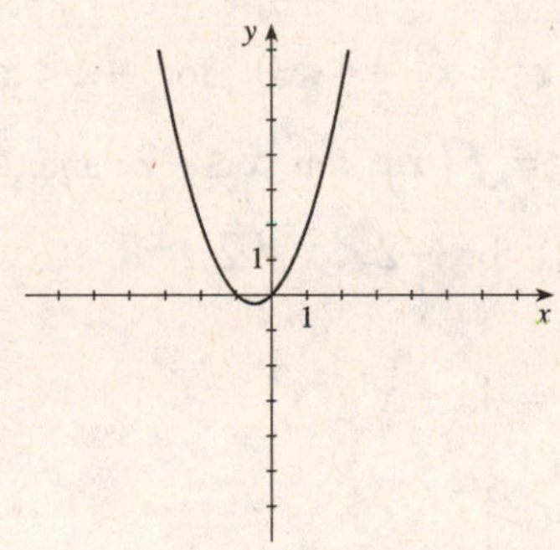

79. $f(x) = x^3 - x$.

$$f(-x) = (-x)^3 - (-x) = -x^3 + x$$
$$= -(x^3 - x) = -f(x)$$

Thus $f(x)$ is odd.

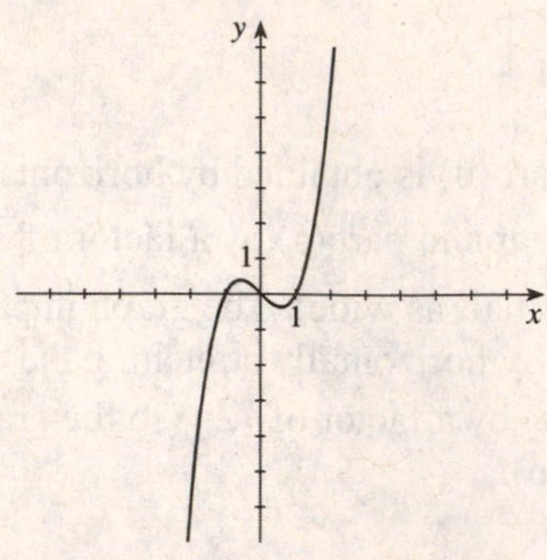

81. $f(x) = 1 - \sqrt[3]{x}$.

$f(-x) = 1 - \sqrt[3]{(-x)} = 1 + \sqrt[3]{x}$. Thus $f(-x) \neq f(x)$. Also $f(-x) \neq -f(x)$, so $f(x)$ is neither odd nor even.

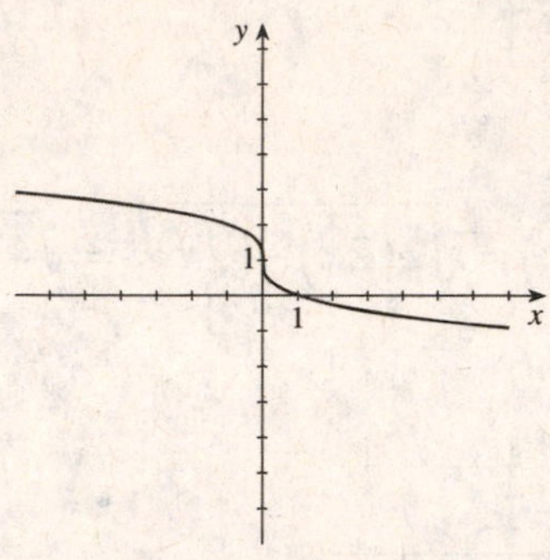

83. (a) Even

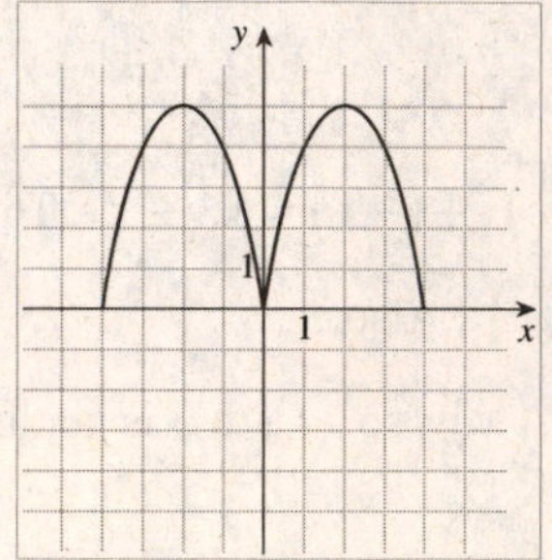

(b) Odd

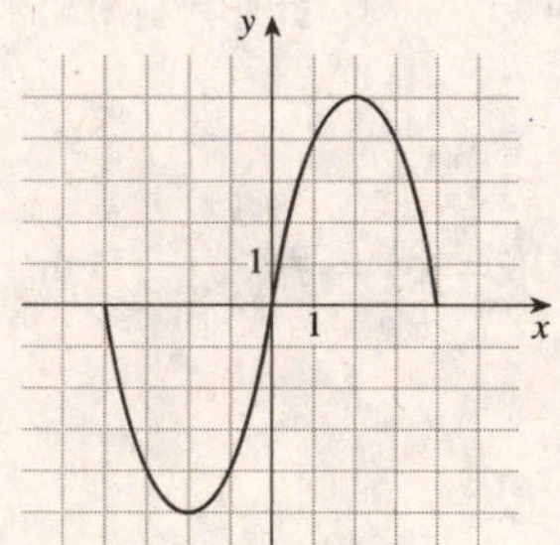

85. Since $f(x) = x^2 - 4 < 0$, for $-2 < x < 2$, the graph of $y = g(x)$ is found by sketching the graph of $y = f(x)$ for $x \leq -2$ and $x \geq 2$, then reflecting in the x-axis the part of the graph of $y = f(x)$ for $-2 < x < 2$.

87. (a) $f(x) = 4x - x^2$

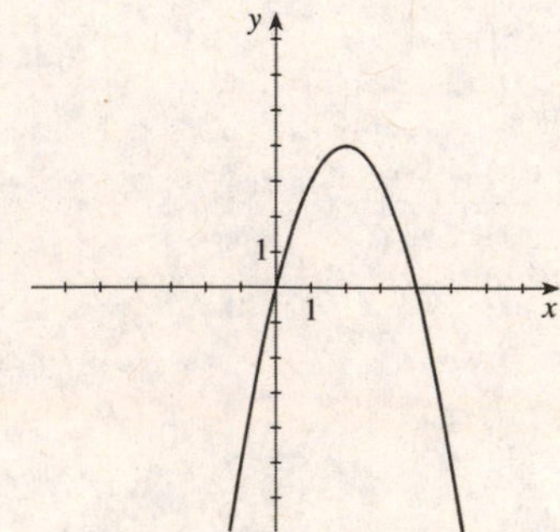

(b) $f(x) = |4x - x^2|$

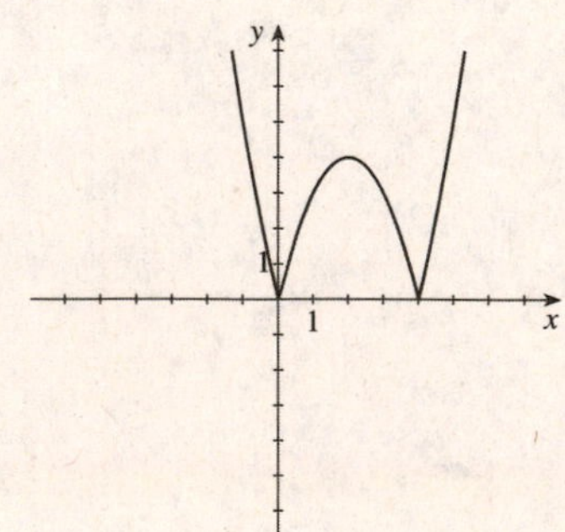

89. (a) The graph of $y = t^2$ must be shrunk vertically by a factor of 0.01 and shifted vertically 4 units upward to obtain the graph of $y = f(t)$.

(b) The graph of $y = f(t)$ must be shifted horizontally 10 units to the left to obtain the graph of $y = g(t)$. So $g(t) = f(t+10) = 4 + 0.01(t+10)^2 = 5 + 0.2t + 0.01t^2$.

91. f even implies $f(-x) = f(x)$; g even implies $g(-x) = g(x)$; f odd implies $f(-x) = -f(x)$; and g odd implies $g(-x) = -g(x)$ If f and g are both even, then $(f+g)(-x) = f(-x) + g(-x) = f(x) + g(x) = (f+g)(x)$ and $f+g$ is even. If f and g are both odd, then $(f+g)(-x) = f(-x) + g(-x) = -f(x) - g(x) = -(f+g)(x)$ and $f+g$ is odd. If f odd and g even, then $(f+g)(-x) = f(-x) + g(-x) = -f(x) + g(x)$, which is neither odd nor even.

93. $f(x) = x^n$ is even when n is an even integer and $f(x) = x^n$ is odd when n is an odd integer. These names were chosen because polynomials with only terms with odd powers are odd functions, and polynomials with only terms with even powers are even functions.

1.7 Combining Functions

1. From the graphs of f and g in the figure, we find $(f+g)(2)=f(2)+g(2)=3+5=8$, $(f-g)(2)=f(2)-g(2)=3-5=-2$, $(fg)(2)=f(2)g(2)=3\cdot5=15$, and $\left(\frac{f}{g}\right)(2)=\frac{f(2)}{g(2)}=\frac{3}{5}$.

3. If the rule of the function f is add one and the rule of the function g is multiply by 2 then the rule of $f\circ g$ is *multiply by* 2 *, then add one* and the rule of $g\circ f$ is *add one, then multiply by* 2.

5. $f(x)=x-3$ has domain $(-\infty,\infty)$. $g(x)=x^2$ has domain $(-\infty,\infty)$. The intersection of the domains of f and g is $(-\infty,\infty)$.

$(f+g)(x)=(x-3)+(x^2)=x^2+x-3$, and the domain is $(-\infty,\infty)$.

$(f-g)(x)=(x-3)-(x^2)=-x^2+x-3$, and the domain is $(-\infty,\infty)$.

$(fg)(x)=(x-3)(x^2)=x^3-3x^2$, and the domain is $(-\infty,\infty)$.

$\left(\frac{f}{g}\right)(x)=\frac{x-3}{x^2}$, and the domain is $\{x\mid x\neq0\}$.

7. $f(x)=\sqrt{4-x^2}$, has domain $[-2,2]$. $g(x)=\sqrt{1+x}$, has domain $[-1,\infty)$. The intersection of the domains of f and g is $[-1,2]$.

$(f+g)(x)=\sqrt{4-x^2}+\sqrt{1+x}$, and the domain is $[-1,2]$.

$(f-g)(x)=\sqrt{4-x^2}-\sqrt{1+x}$, and the domain is $[-1,2]$.

$(fg)(x)=\sqrt{4-x^2}\sqrt{1+x}=\sqrt{-x^3-x^2+4x+4}$, and the domain is $[-1,2]$.

$\left(\frac{f}{g}\right)(x)=\frac{\sqrt{4-x^2}}{\sqrt{1+x}}=\sqrt{\frac{4-x^2}{1+x}}$, and the domain is $(-1,2]$.

9. $f(x)=\dfrac{2}{x}$ has domain $x\neq 0$. $g(x)=\dfrac{4}{x+4}$, has domain $x\neq -4$. The intersection of the domains of f and g is $\{x \mid x\neq 0,-4\}$; in interval notation, this is $(-\infty,-4)\cup(-4,0)\cup(0,\infty)$.

$(f+g)(x)=\dfrac{2}{x}+\dfrac{4}{x+4}=\dfrac{2}{x}+\dfrac{4}{x+4}=\dfrac{2(3x+4)}{x(x+4)}$, and the domain is $(-\infty,-4)\cup(-4,0)\cup(0,\infty)$.

$(f-g)(x)=\dfrac{2}{x}-\dfrac{4}{x+4}=-\dfrac{2(x-4)}{x(x+4)}$, and the domain is $(-\infty,-4)\cup(-4,0)\cup(0,\infty)$.

$(fg)(x)=\dfrac{2}{x}\cdot\dfrac{4}{x+4}=\dfrac{8}{x(x+4)}$, and the domain is $(-\infty,-4)\cup(-4,0)\cup(0,\infty)$.

$\left(\dfrac{f}{g}\right)(x)=\dfrac{\frac{2}{x}}{\frac{4}{x+4}}=\dfrac{x+4}{2x}$, and the domain is $(-\infty,-4)\cup(-4,0)\cup(0,\infty)$.

11. $f(x)=\sqrt{x}+\sqrt{1-x}$. The domain of $\sqrt{x}$ is $[0,\infty)$, and the domain of $\sqrt{1-x}$ is $(-\infty,1]$. Thus the domain is $(-\infty,1]\cap[0,\infty)=[0,1]$.

13. $h(x)=(x-3)^{-1/4}=\dfrac{1}{(x-3)^{1/4}}$. Since $1/4$ is an even root and the denominator can not equal 0, $x-3>0 \Leftrightarrow x>3$. So the domain is $(3,\infty)$.

15.

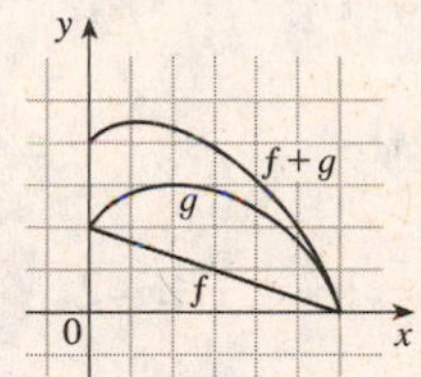

17.

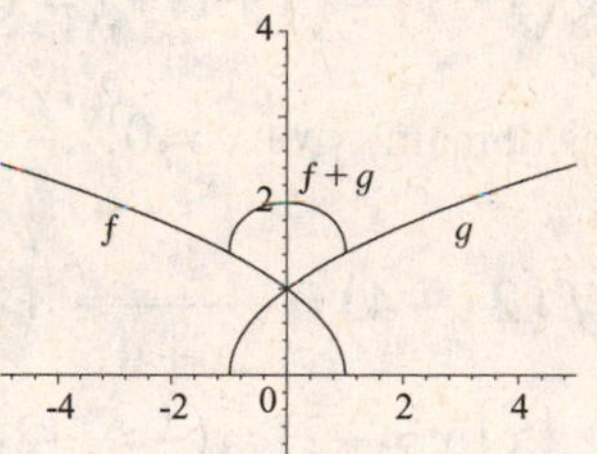

19.

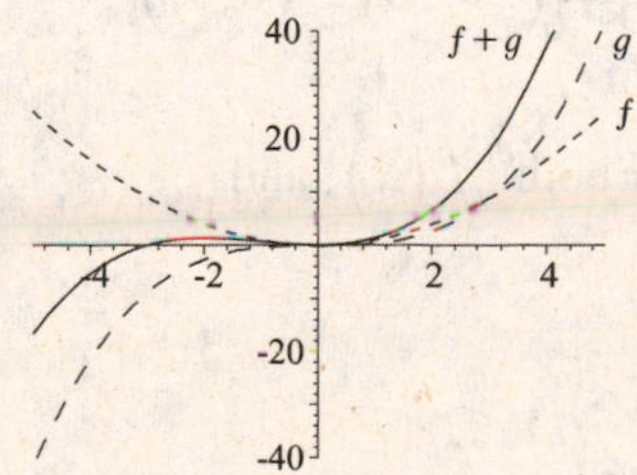

21. (a) $f(g(0))=f\left(2-(0)^2\right)=f(2)=3(2)-5=1$

(b) $g(f(0))=g(3(0)-5)=g(-5)=2-(-5)^2=-23$

23. (a) $(f \circ g)(-2) = f(g(-2)) = f\left(2-(-2)^2\right) = f(-2) = 3(-2)-5 = -11$

(b) $(g \circ f)(-2) = g(f(-2)) = g(3(-2)-5) = g(-11) = 2-(-11)^2 = -119$

25. (a) $(f \circ g)(x) = f(g(x)) = f\left(2-x^2\right) = 3\left(2-x^2\right)-5 = 6-3x^2-5 = 1-3x^2$

(b)

$(g \circ f)(x) = g(f(x)) = g(3x-5) = 2-(3x-5)^2 = 2-\left(9x^2-30x+25\right) = -9x^2+30x-23$

27. $f(g(2)) = f(5) = 4$

29. $(g \circ f)(4) = g(f(4)) = g(2) = 5$

31. $(g \circ g)(-2) = g(g(-2)) = g(1) = 4$

33. $f(x) = 2x+3$, has domain $(-\infty,\infty)$; $g(x) = 4x-1$, has domain $(-\infty,\infty)$.

$(f \circ g)(x) = f(4x-1) = 2(4x-1)+3 = 8x+1$, and the domain is $(-\infty,\infty)$.

$(g \circ f)(x) = g(2x+3) = 4(2x+3)-1 = 8x+11$, and the domain is $(-\infty,\infty)$.

$(f \circ f)(x) = f(2x+3) = 2(2x+3)+3 = 4x+9$, and the domain is $(-\infty,\infty)$.

$(g \circ g)(x) = g(4x-1) = 4(4x-1)-1 = 16x-5$, and the domain is $(-\infty,\infty)$.

35. $f(x) = x^2$, has domain $(-\infty,\infty)$; $g(x) = x+1$, has domain $(-\infty,\infty)$.

$(f \circ g)(x) = f(x+1) = (x+1)^2 = x^2+2x+1$, and the domain is $(-\infty,\infty)$.

$(g \circ f)(x) = g\left(x^2\right) = \left(x^2\right)+1 = x^2+1$, and the domain is $(-\infty,\infty)$.

$(f \circ f)(x) = f\left(x^2\right) = \left(x^2\right)^2 = x^4$, and the domain is $(-\infty,\infty)$.

$(g \circ g)(x) = g(x+1) = (x+1)+1 = x+2$, and the domain is $(-\infty,\infty)$.

37. $f(x) = \dfrac{1}{x}$, has domain $\{x \mid x \neq 0\}$; $g(x) = 2x+4$, has domain $(-\infty,\infty)$.

$(f \circ g)(x) = f(2x+4) = \dfrac{1}{2x+4}$. $(f \circ g)(x)$ is defined for $2x+4 \neq 0 \iff x \neq -2$.

So the domain is $\{x \mid x \neq -2\} = (-\infty,-2) \cup (-2,\infty)$.

$(g \circ f)(x) = g\left(\dfrac{1}{x}\right) = 2\left(\dfrac{1}{x}\right)+4 = \dfrac{2}{x}+4$, the domain is $\{x \mid x \neq 0\} = (-\infty,0) \cup (0,\infty)$.

$(f \circ f)(x) = f\left(\dfrac{1}{x}\right) = \dfrac{1}{\left(\dfrac{1}{x}\right)} = x$. $(f \circ f)(x)$ is defined whenever both $f(x)$ and

$f(f(x))$ are defined; that is, whenever $\{x \mid x \neq 0\} = (-\infty,0) \cup (0,\infty)$.

$(g \circ g)(x) = g(2x+4) = 2(2x+4)+4 = 4x+8+4 = 4x+12$, and the domain is $(-\infty,\infty)$

39. $f(x)=|x|$, has domain $(-\infty,\infty)$; $g(x)=2x+3$, has domain $(-\infty,\infty)$

$(f\circ g)(x)=f(2x+4)=|2x+3|$, and the domain is $(-\infty,\infty)$.

$(g\circ f)(x)=g(|x|)=2|x|+3$, and the domain is $(-\infty,\infty)$.

$(f\circ f)(x)=f(|x|)=||x||=|x|$, and the domain is $(-\infty,\infty)$.

$(g\circ g)(x)=g(2x+3)=2(2x+3)+3=4x+6+3=4x+9$. Domain is $(-\infty,\infty)$.

41. $f(x)=\dfrac{x}{x+1}$, has domain $\{x\mid x\neq -1\}$; $g(x)=2x-1$, has domain $(-\infty,\infty)$

$(f\circ g)(x)=f(2x-1)=\dfrac{2x-1}{(2x-1)+1}=\dfrac{2x-1}{2x}$, and the domain is $\{x\mid x\neq 0\}=(-\infty,0)\cup(0,\infty)$.

$(g\circ f)(x)=g\left(\dfrac{x}{x+1}\right)=2\left(\dfrac{x}{x+1}\right)-1=\dfrac{2x}{x+1}-1$, and the domain is $\{x\mid x\neq -1\}=(-\infty,-1)\cup(-1,\infty)$

$(f\circ f)(x)=f\left(\dfrac{x}{x+1}\right)=\dfrac{\frac{x}{x+1}}{\frac{x}{x+1}+1}\cdot\dfrac{x+1}{x+1}=\dfrac{x}{x+x+1}=\dfrac{x}{2x+1}$. $(f\circ f)(x)$ is defined whenever both $f(x)$ and $f(f(x))$ are defined; that is, whenever $x\neq -1$ and $2x+1\neq 0$ $\Rightarrow$ $x\neq -\frac{1}{2}$, which is $(-\infty,-1)\cup(-1,-\frac{1}{2})\cup(-\frac{1}{2},\infty)$.

$(g\circ g)(x)=g(2x-1)=2(2x-1)-1=4x-2-1=4x-3$, and the domain is $(-\infty,\infty)$.

43. $f(x)=\dfrac{x}{x+1}$, has domain $\{x\mid x\neq -1\}$; $g(x)=\dfrac{1}{x}$ has domain $\{x\mid x\neq 0\}$.

$(f\circ g)(x)=f\left(\dfrac{1}{x}\right)=\dfrac{\frac{1}{x}}{\frac{1}{x}+1}=\dfrac{1}{x\left(\frac{1}{x}+1\right)}=\dfrac{1}{x+1}$. $(f\circ g)(x)$ is defined whenever both $g(x)$ and $f(g(x))$ are defined, so the domain is $\{x\mid x\neq -1,0\}$.

$(g\circ f)(x)=g\left(\dfrac{x}{x+1}\right)=\dfrac{1}{\frac{x}{x+1}}=\dfrac{x+1}{x}$. $(g\circ f)(x)$ is defined whenever both $f(x)$ and $g(f(x))$ are defined, so the domain is $\{x\mid x\neq -1,0\}$.

$(f\circ f)(x)=f\left(\dfrac{x}{x+1}\right)=\dfrac{\frac{x}{x+1}}{\frac{x}{x+1}+1}=\dfrac{x}{(x+1)\left(\frac{x}{x+1}+1\right)}=\dfrac{x}{2x+1}$. $(f\circ f)(x)$ is defined whenever both $f(x)$ and $f(f(x))$ are defined, so the domain is $\{x\mid x\neq -1,-\frac{1}{2}\}$.

$(g\circ g)(x)=g\left(\dfrac{1}{x}\right)=\dfrac{1}{\frac{1}{x}}=x$. $(g\circ g)(x)$ is defined whenever both $g(x)$ and $g(g(x))$ are defined, so the domain is $\{x\mid x\neq 0\}$.

45. $(f\circ g\circ h)(x)=f(g(h(x)))=f(g(x-1))=f\left(\sqrt{x-1}\right)=\sqrt{x-1}-1$

47. $(f\circ g\circ h)(x)=f(g(h(x)))=f(g(\sqrt{x}))=f(\sqrt{x}-5)=(\sqrt{x}-5)^4+1$

For Exercises 49--58, many answers are possible.

49. $F(x)=(x-9)^5$. Let $f(x)=x^5$ and $g(x)=x-9$, then $F(x)=(f\circ g)(x)$.

51. $G(x)=\dfrac{x^2}{x^2+4}$. Let $f(x)=\dfrac{x}{x+4}$ and $g(x)=x^2$, then $G(x)=(f\circ g)(x)$.

53. $H(x)=|1-x^3|$. Let $f(x)=|x|$ and $g(x)=1-x^3$, then $H(x)=(f\circ g)(x)$.

55. $F(x)=\dfrac{1}{x^2+1}$. Let $f(x)=\dfrac{1}{x}$, $g(x)=x+1$, and $h(x)=x^2$, then $F(x)=(f\circ g\circ h)(x)$.

57. $G(x)=(4+\sqrt[3]{x})^9$. Let $f(x)=x^9$, $g(x)=4+x$, and $h(x)=\sqrt[3]{x}$, then $G(x)=(f\circ g\circ h)(x)$.

59. The price per sticker is $0.15-0.000002x$ and the number sold is x, so the revenue is $R(x)=(0.15-0.000002x)x=0.15x-0.000002x^2$.

61. (a) Because the ripple travels at a speed of 60 cm/s, the distance traveled in t seconds is the radius, so $g(t)=60t$.

(b) The area of a circle is πr^2, so $f(r)=\pi r^2$.

(c) $f\circ g=\pi(g(t))^2=\pi(60t)^2=3600\pi t^2$ cm^2. This function represents the area of the ripple as a function of time.

63. Let r be the radius of the spherical balloon in centimeters. Since the radius is increasing at a rate of 2 cm/s, the radius is $r=2t$ after t seconds. Therefore, the surface area of the balloon can be written as $S=4\pi r^2=4\pi(2t)^2=4\pi(4t^2)=16\pi t^2$.

65. (a) $f(x)=0.90x$

(b) $g(x)=x-100$

(c) $f\circ g=f(x-100)=0.90(x-100)=0.90x-90$. $f\circ g$ represents applying the $\$100$ coupon, then the 10% discount. $g\circ f=g(0.90x)=0.90x-100$. $g\circ f$ represents applying the 10% discount, then the $\$100$ coupon. So applying the 10% discount, then the $\$100$ coupon gives the lower price.

67. $A(x)=1.05x.$ $(A\circ A)(x)=A(A(x))=A(1.05x)=1.05(1.05x)=(1.05)^2x$.
$(A\circ A\circ A)(x)=A(A\circ A(x))=A((1.05)^2x)=1.05[(1.05)^2x]=(1.05)^3x$.
$(A\circ A\circ A\circ A)(x)=A(A\circ A\circ A(x))=A((1.05)^3x)=1.05[(1.05)^3x]=(1.05)^4x$. A represents the amount in the account after 1 year; $A\circ A$ represents the amount in the account after 2 years; $A\circ A\circ A$ represents the amount in the account after 3 years; and $A\circ A\circ A\circ A$ represents the amount in the account after 4 years. We can see that if we compose n copies of A, we get $(1.05)^n x$.

69. $g(x)=2x+1$ and $h(x)=4x^2+4x+7$.

Method 1: Notice that $(2x+1)^2=4x^2+4x+1$. We see that adding 6 to this quantity gives $(2x+1)^2+6=4x^2+4x+1+6=4x^2+4x+7$, which is $h(x)$. So let $f(x)=x^2+6$, and we have $(f\circ g)(x)=(2x+1)^2+6=h(x)$.

Method 2: Since $g(x)$ is linear and $h(x)$ is a second degree polynomial, $f(x)$ must be a second degree polynomial, that is, $f(x)=ax^2+bx+c$ for some a , b , and c . Thus $f(g(x))=f(2x+1)=a(2x+1)^2+b(2x+1)+c \Leftrightarrow$ $4ax^2+4ax+a+2bx+b+c=4ax^2+(4a+2b)x+(a+b+c)=4x^2+4x+7$. Comparing this with $f(g(x))$, we have $4a=4$ (the x^2 coefficients), $4a+2b=4$ (the x coefficients), and $a+b+c=7$ (the constant terms) $\Leftrightarrow$ $a=1$ and $2a+b=2$ and $a+b+c=7$ $\Leftrightarrow$ $a=1$, $b=0,c=6$. Thus $f(x)=x^2+6$.

$f(x)=3x+5$ and $h(x)=3x^2+3x+2$.

Note since $f(x)$ is linear and $h(x)$ is quadratic, $g(x)$ must also be quadratic. We can then use trial and error to find $g(x)$. Another method is the following: We wish to find g so that $(f\circ g)(x)=h(x)$. Thus $f(g(x))=3x^2+3x+2 \Leftrightarrow 3(g(x))+5=3x^2+3x+2$ $\Leftrightarrow 3(g(x))=3x^2+3x-3 \Leftrightarrow g(x)=x^2+x-1$.

70. If $g(x)$ is even, then $h(-x)=f(g(-x))=f(g(x))=h(x)$. So yes, h is always an even function.

If $g(x)$ is odd, then h is not necessarily an odd function. For example, if we let $f(x)=x-1$ and $g(x)=x^3$, g is an odd function, but $h(x)=(f\circ g)(x)=f(x^3)=x^3-1$ is not an odd function.

If $g(x)$ is odd and f is also odd, then

$h(-x)=(f\circ g)(-x)=f(g(-x))=f(-g(x))=-f(g(x))=-(f\circ g)(x)=-h(x)$.

So in this case, h is also an odd function.

If $g(x)$ is odd and f is even, then

$h(-x)=(f\circ g)(-x)=f(g(-x))=f(-g(x))=f(g(x))=(f\circ g)(x)=h(x)$, so in this case, h is an even function.

1.8 One-to-One Functions and Their Inverses

1. A function f is one-to-one if different inputs produce *different* outputs. You can tell from the graph that a function is one-to-one by using the *Horizontal Line* Test.

3. (a) Proceeding backward through the description of f, we can describe f^{-1} as follows: Take the third root, subtract 5, then divide by 3.

(b) $f(x)=(3x+5)^3$ and $f^{-1}(x)=\dfrac{\sqrt[3]{x}-5}{3}$.

5. By the Horizontal Line Test, f is not one-to-one.

7. By the Horizontal Line Test, f is one-to-one.

9. By the Horizontal Line Test, f is not one-to-one.

11. $f(x)=-2x+4$. If $x_1\neq x_2$, then $-2x_1\neq -2x_2$ and $-2x_1+4\neq -2x_2+4$. So f is a one-to-one function.

13. $g(x)=\sqrt{x}$. If $x_1\neq x_2$, then $\sqrt{x_1}\neq\sqrt{x_2}$ because two different numbers cannot have the same square root. Therefore, g is a one-to-one function.

15. $h(x)=x^2-2x$. Since $h(0)=0$ and $h(2)=(2)-2(2)=0$ we have $h(0)=h(2)$. So f is not a one-to-one function.

17. $f(x)=x^4+5$. Every nonzero number and its negative have the same fourth power. For example, $(-1)^4=1=(1)^4$, so $f(-1)=f(1)$. Thus f is not a one-to-one function.

19. $f(x)=\dfrac{1}{x^2}$. Every nonzero number and its negative have the same square. For example, $\dfrac{1}{(-1)^2}=1=\dfrac{1}{(1)^2}$, so $f(-1)=f(1)$. Thus f is not a one-to-one function.

21. (a) $f(2)=7$. Since f is one-to-one, $f^{-1}(7)=2$.

(b) $f^{-1}(3)=-1$. Since f is one-to-one, $f(-1)=3$.

23. $f(x)=5-2x$. Since f is one-to-one and $f(1)=5-2(1)=3$, then $f^{-1}(3)=1$. (Find 1 by solving the equation $5-2x=3$.)

25. $f(g(x))=f(x+6)=(x+6)-6=x$ for all x.

$g(f(x))=g(x-6)=(x-6)+6=x$ for all x. Thus f and g are inverses of each other.

27. $f(g(x))=f\left(\dfrac{x+5}{2}\right)=2\left(\dfrac{x+5}{2}\right)-5=x+5-5=x$ for all x.

$g(f(x))=g(2x-5)=\dfrac{(2x-5)+5}{2}=x$ for all x. Thus f and g are inverses of each other.

29. $f(g(x))=f\left(\dfrac{1}{x}\right)=\dfrac{1}{1/x}=x$ for all $x\neq 0$. Since $f(x)=g(x)$, we also have $g(f(x))=x$ for all $x\neq 0$. Thus f and g are inverses of each other.

31. $f(g(x))=f\left(\sqrt{x+4}\right)=\left(\sqrt{x+4}\right)^2-4=x+4-4=x$ for all $x\geq -4$.

$g(f(x))=g\left(x^2-4\right)=\sqrt{\left(x^2-4\right)+4}=\sqrt{x^2}=x$ for all $x\geq 0$. Thus f and g are inverses of each other.

33. $f(g(x))=f\left(\dfrac{1}{x}+1\right)=\dfrac{1}{\left(\dfrac{1}{x}+1\right)-1}=x$ for all $x\neq 0$.

$g(f(x))=g\left(\dfrac{1}{x-1}\right)=\dfrac{1}{\left(\dfrac{1}{x-1}\right)}+1=(x-1)+1=x$ for all $x\neq 1$. Thus f and g are inverses of each other.

35. $f(g(x))=f\left(\dfrac{2x+2}{x-1}\right)=\dfrac{\frac{2x+2}{x-1}+2}{\frac{2x+2}{x-1}-2}=\dfrac{2x+2+2(x-1)}{2x+2-2(x-1)}=\dfrac{4x}{4}=x$ for all $x\neq 1$.

$g(f(x))=g\left(\dfrac{x+2}{x-2}\right)=\dfrac{2\left(\frac{x+2}{x-2}\right)+2}{\frac{x+2}{x-2}-1}=\dfrac{2(x+2)+2(x-2)}{x+2-1(x-2)}=\dfrac{4x}{4}=x$ for all $x\neq 2$. Thus f and g are inverses of each other.

37. $f(x)=2x+1$. $y=2x+1 \Leftrightarrow 2x=y-1 \Leftrightarrow x=\frac{1}{2}(y-1)$. So $f^{-1}(x)=\frac{1}{2}(x-1)$.

39. $f(x)=4x+7$. $y=4x+7 \Leftrightarrow 4x=y-7 \Leftrightarrow x=\frac{1}{4}(y-7)$. So $f^{-1}(x)=\frac{1}{4}(x-7)$.

41. $f(x)=5-4x^3$. $y=5-4x^3 \Leftrightarrow 4x^3=5-y \Leftrightarrow x^3=\frac{1}{4}(5-y) \Leftrightarrow x=\sqrt[3]{\frac{1}{4}(5-y)}$. So $f^{-1}(x)=\sqrt[3]{\frac{1}{4}(5-x)}$.

43. $f(x)=\dfrac{1}{x+2}$. $y=\dfrac{1}{x+2} \Leftrightarrow x+2=\dfrac{1}{y} \Leftrightarrow x=\dfrac{1}{y}-2$. So $f^{-1}(x)=\dfrac{1}{x}-2$.

45. $f(x)=\dfrac{x}{x+4}$. $y=\dfrac{x}{x+4} \Leftrightarrow y(x+4)=x \Leftrightarrow xy+4y=x \Leftrightarrow x-xy=4y \Leftrightarrow x(1-y)=4y \Leftrightarrow x=\dfrac{4y}{1-y}$. So $f^{-1}(x)=\dfrac{4x}{1-x}$.

47. $f(x)=\dfrac{2x+5}{x-7}$. $y=\dfrac{2x+5}{x-7} \Leftrightarrow y(x-7)=2x+5 \Leftrightarrow xy-7y=2x+5 \Leftrightarrow xy-2x=7y+5 \Leftrightarrow x(y-2)=7y+5 \Leftrightarrow x=\dfrac{7y+5}{y-2}$. So $f^{-1}(x)=\dfrac{7x+5}{x-2}$.

49. $f(x)=\dfrac{1+3x}{5-2x}$. $y=\dfrac{1+3x}{5-2x} \Leftrightarrow y(5-2x)=1+3x \Leftrightarrow 5y-2xy=1+3x \Leftrightarrow 3x+2xy=5y-1 \Leftrightarrow x(3+2y)=5y-1 \Leftrightarrow x=\dfrac{5y-1}{2y+3}$. So $f^{-1}(x)=\dfrac{5x-1}{2x+3}$.

51. $f(x)=\sqrt{2+5x}$, $x\geq -\frac{2}{5}$. $y=\sqrt{2+5x}, y\geq 0$ $\Leftrightarrow$ $y^2=2+5x$ $\Leftrightarrow$ $5x=y^2-2$ $\Leftrightarrow$ $x=\frac{1}{5}(y^2-2)$ and $y\geq 0$. So $f^{-1}(x)=\frac{1}{5}(x^2-2)$, $x\geq 0$.

53. $f(x)=4-x^2$, $x\geq 0$. $y=4-x^2$ $\Leftrightarrow$ $x^2=4-y$ $\Leftrightarrow$ $x=\sqrt{4-y}$. So $f^{-1}(x)=\sqrt{4-x}$, $x\leq 4$. (Note that $x\geq 0$ $\Rightarrow$ $f(x)\leq 4$.)

55. $f(x)=4+\sqrt[3]{x}$. $y=4+\sqrt[3]{x}$ $\Leftrightarrow$ $\sqrt[3]{x}=y-4$ $\Leftrightarrow$ $x=(y-4)^3$. So $f^{-1}(x)=(x-4)^3$.

57. $f(x)=1+\sqrt{1+x}$. $y=1+\sqrt{1+x}$, $y\geq 1$ $\Leftrightarrow$ $\sqrt{1+x}=y-1$ $\Leftrightarrow$ $1+x=(y-1)^2$ $\Leftrightarrow$ $x=(y-1)^2-1=y^2-2y$. So $f^{-1}(x)=x^2-2x$, $x\geq 1$.

59. $f(x)=x^4$, $x\geq 0$. $y=x^4$, $y\geq 0$ $\Leftrightarrow$ $x=\sqrt[4]{y}$. So $f^{-1}(x)=\sqrt[4]{x}$, $x\geq 0$.

60. $f(x)=1-x^3$. $y=1-x^3$ $\Leftrightarrow$ $x^3=1-y$ $\Leftrightarrow$ $x=\sqrt[3]{1-y}$. So $f^{-1}(x)=\sqrt[3]{1-x}$.

61. (a), (b) $f(x)=3x-6$

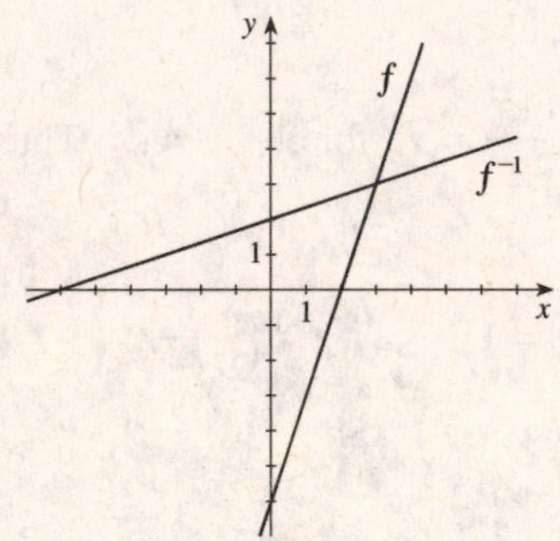

(c) $f(x)=3x-6$. $y=3x-6$ $\Leftrightarrow$ $3x=y+6$ $\Leftrightarrow$ $x=\frac{1}{3}(y+6)$. So $f^{-1}(x)=\frac{1}{3}(x+6)$.

63. (a), (b) $f(x)=\sqrt{x+1}$

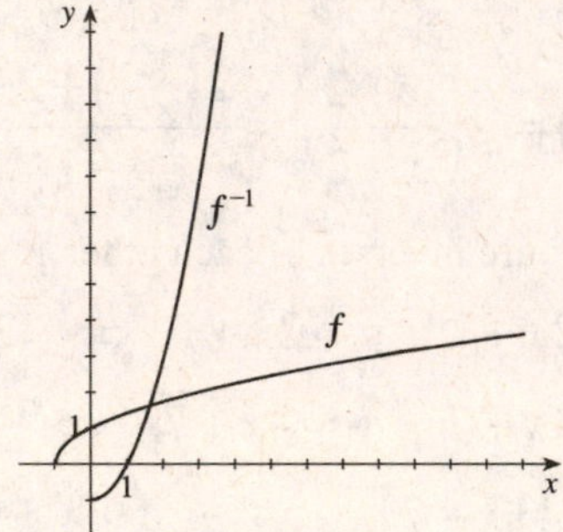

(c) $f(x)=\sqrt{x+1}$, $x\geq -1$. $y=\sqrt{x+1}$, $y\geq 0$ $\Leftrightarrow$ $y^2=x+1$ $\Leftrightarrow$ $x=y^2-1$ and $y\geq 0$. So $f^{-1}(x)=x^2-1$, $x\geq 0$.

65. $f(x)=x^3-x$. Using a graphing device and the Horizontal Line Test, we see that f is not a one-to-one function. For example, $f(0)=0=f(-1)$.

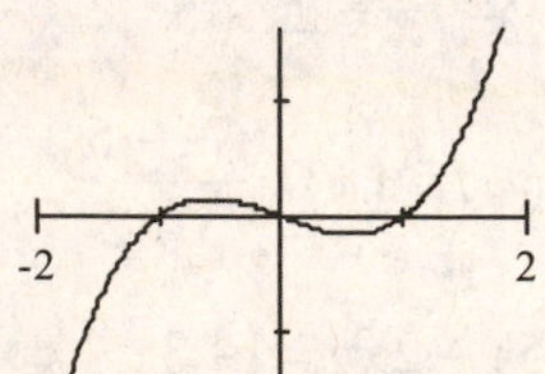

67. $f(x)=\dfrac{x+12}{x-6}$. Using a graphing device and the Horizontal Line Test, we see that f is a one-to-one function.

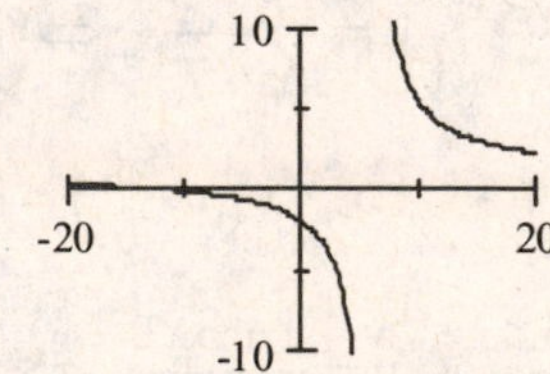

69. $f(x) = |x| - |x-6|$. Using a graphing device and the Horizontal Line Test, we see that f is not a one-to-one function. For example $f(0) = -6 = f(-2)$.

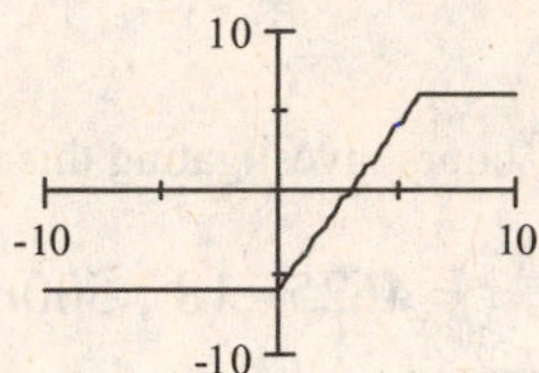

71. (a) $y = f(x) = 2 + x \Leftrightarrow x = y - 2$. So $f^{-1}(x) = x - 2$.

(b)

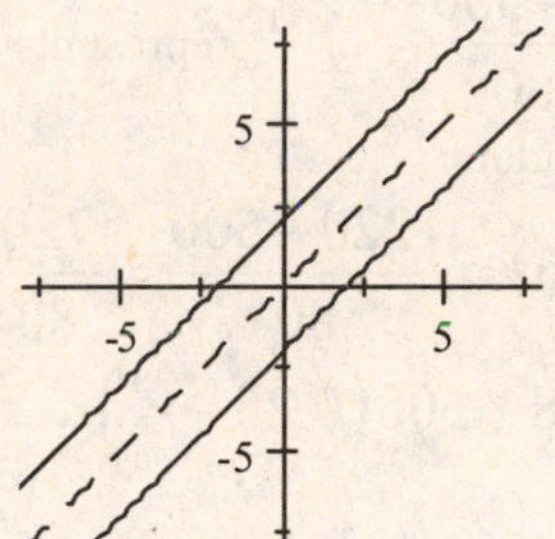

73. (a) $y = g(x) = \sqrt{x+3}$, $y \geq 0 \Leftrightarrow x + 3 = y^2$, $y \geq 0 \Leftrightarrow x = y^2 - 3$, $y \geq 0$. So $g^{-1}(x) = x^2 - 3$, $x \geq 0$.

(b)

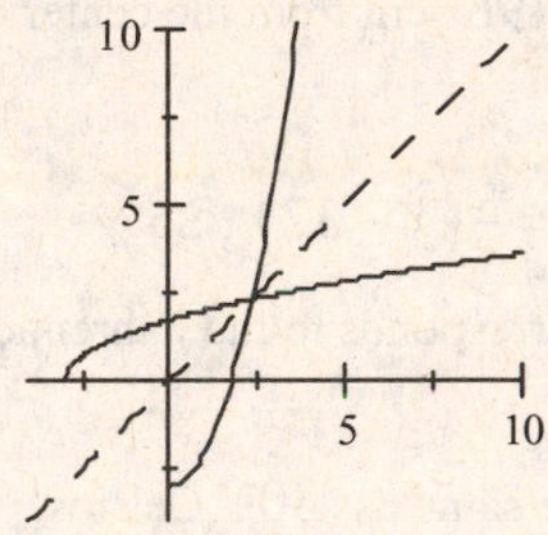

75. If we restrict the domain of $f(x)$ to $[0, \infty)$, then $y = 4 - x^2 \Leftrightarrow x^2 = 4 - y \Rightarrow x = \sqrt{4-y}$ (since $x \geq 0$, we take the positive square root). So $f^{-1}(x) = \sqrt{4-x}$.

If we restrict the domain of $f(x)$ to $(-\infty, 0]$, then $y = 4 - x^2 \Leftrightarrow x^2 = 4 - y \Rightarrow x = -\sqrt{4-y}$ (since $x \leq 0$, we take the negative square root). So $f^{-1}(x) = -\sqrt{4-x}$.

77. If we restrict the domain of $h(x)$ to $[-2, \infty)$, then $y = (x+2)^2 \Rightarrow x + 2 = \sqrt{y}$ (since $x \geq -2$, we take the positive square root) $\Leftrightarrow x = -2 + \sqrt{y}$. So $h^{-1}(x) = -2 + \sqrt{x}$.

If we restrict the domain of $h(x)$ to $(-\infty, -2]$, then $y = (x+2)^2 \Rightarrow x + 2 = -\sqrt{y}$ (since $x \leq -2$, we take the negative square root) $\Leftrightarrow x = -2 - \sqrt{y}$. So $h^{-1}(x) = -2 - \sqrt{x}$.

79.

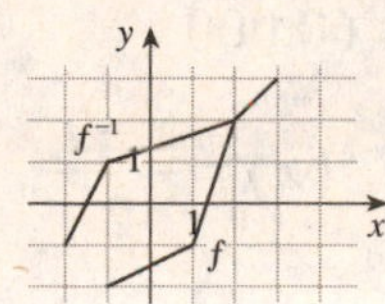

81. (a) $f(x)=500+80x$.

(b) $f(x)=500+80x$. $y=500+80x \Leftrightarrow 80x=y-500 \Leftrightarrow x=\dfrac{y-500}{80}$. So $f^{-1}(x)=\dfrac{x-500}{80}$. f^{-1} represents the number of hours of investigation the investigate spends on a case for x dollars.

(c) $f^{-1}(1220)=\dfrac{1220-500}{80}=\dfrac{720}{80}=9$. The investigator spent 9 hours investigating this case.

83. (a) $v(r)=18\text{ , }500(0.25-r^2)$. $t=18\text{ , }500(0.25-r^2) \Leftrightarrow t=4625-18\text{ , }500r^2$ $\Leftrightarrow 18500r^2=4625-t \Leftrightarrow r^2=\dfrac{4625-t}{18{,}500} \Rightarrow r=\pm\sqrt{\dfrac{4625-t}{18{,}500}}$. Since r represents a distance, $r\geq 0$, so $v^{-1}(t)=\sqrt{\dfrac{4625-t}{18{,}500}}$. v^{-1} represents the radial distance from the center of the vein at which the blood has velocity v .

(b) $v^{-1}(30)=\sqrt{\dfrac{4625-30}{18{,}500}}\approx 0.498$ cm. The velocity is 30 at 0.498 cm from the center of the artery or vein.

85. (a) $F(x)=\frac{9}{5}x+32$. $y=\frac{9}{5}x+32 \Leftrightarrow \frac{9}{5}x=y-32 \Leftrightarrow x=\frac{5}{9}(y-32)$. So $F^{-1}(x)=\frac{5}{9}(x-32)$. F^{-1} represents the Celsius temperature that corresponds to the Fahrenheit temperature of F .

(b) $F^{-1}(86)=\frac{5}{9}(86-32)=\frac{5}{9}(54)=30$. So 86° Fahrenheit is the same as 30° Celsius.

87. (a) $f(x)=\begin{cases} \text{if } x>20{,}000 \\ 0.1x & , \\ \text{if } 0\leq x\leq 20{,}000 & 2000+0.2(x-20{,}000) \end{cases}$

(b) We will find the inverse of each piece of the function f .

$f_1(x)=0.1x$. $y=0.1x \Leftrightarrow x=10y$. So $f_1^{-1}(x)=10x$.

$f_2(x)=2000+0.2(x-20{,}000)=0.2x-2000$. $y=0.2x-2000 \Leftrightarrow$

$0.2x=y+2000 \Leftrightarrow x=5y+10{,}000$. So $f_2^{-1}(x)=5x+10{,}000$.

Since $f(0)=0$ and $f(20{,}000)=2000$ we have

$f^{-1}(x)=\begin{cases} \text{if } x>2000 \\ 10x & , \\ \text{if } 0\leq x\leq 2000 & 5x+10{,}000 \end{cases}$

It represents the taxpayer's income.

(c) $f^{-1}(10{,}000)=5(10{,}000)+10{,}000=60{,}000$. The required income is € $60{,}000$.

89. $f(x)=7+2x$. $y=7+2x \Leftrightarrow 2x=y-7 \Leftrightarrow x=\dfrac{y-7}{2}$. So $f^{-1}(x)=\dfrac{x-7}{2}$.

f^{-1} is the number of toppings on a pizza that costs x dollars.

91. (a) $f(x)=\dfrac{2x+1}{5}$ is multiply by 2, add 1, and then divide by 5. So the reverse is multiply by 5, subtract 1, and then divide by 2 or $f^{-1}(x)=\dfrac{5x-1}{2}$. Check:

$$f\circ f^{-1}(x)=f\left(\frac{5x-1}{2}\right)=\frac{2\left(\frac{5x-1}{2}\right)+1}{5}=\frac{5x-1+1}{5}=\frac{5x}{5}=x \text{ and}$$

$$f^{-1}\circ f(x)=f^{-1}\left(\frac{2x+1}{5}\right)=\frac{5\left(\frac{2x+1}{5}\right)-1}{2}=\frac{2x+1-1}{2}=\frac{2x}{2}=x.$$

(b) $f(x)=3-\dfrac{1}{x}=\dfrac{-1}{x}+3$ is take the negative reciprocal and add 3. Since the reverse of take the negative reciprocal is take the negative reciprocal, $f^{-1}(x)$ is subtract 3 and take the negative reciprocal, that is, $f^{-1}(x)=\dfrac{-1}{x-3}$. Check:

$$f\circ f^{-1}(x)=f\left(\frac{-1}{x-3}\right)=3-\frac{1}{\frac{-1}{x-3}}=3-\left(1\cdot\frac{x-3}{-1}\right)=3+x-3=x \text{ and}$$

$$f^{-1}\circ f(x)=f^{-1}\left(3-\frac{1}{x}\right)=\frac{-1}{\left(3-\frac{1}{x}\right)-3}=\frac{-1}{-\frac{1}{x}}=-1\cdot\frac{x}{-1}=x.$$

(c) $f(x)=\sqrt{x^3+2}$ is cube, add 2, and then take the square root. So the reverse is square, subtract 2, then take the cube root or $f^{-1}(x)=\sqrt[3]{x^2-2}$. Domain for $f(x)$ is $\left[-\sqrt[3]{2},\infty\right)$; domain for $f^{-1}(x)$ is $[0,\infty)$. Check:

$f\circ f^{-1}(x)=f\left(\sqrt[3]{x^2-2}\right)=\sqrt{\left(\sqrt[3]{x^2-2}\right)^3+2}=\sqrt{x^2-2+2}=\sqrt{x^2}=x$ (on the appropriate domain) and $f^{-1}\circ f(x)=f^{-1}\left(\sqrt{x^3+2}\right)=\sqrt[3]{\left(\sqrt{x^3+2}\right)^2-2}=\sqrt[3]{x^3+2-2}=\sqrt[3]{x^3}=x$ (on the appropriate domain).

(d) $f(x)=(2x-5)^3$ is double, subtract 5, and then cube. So the reverse is take the cube root, add 5, and divide by 2 or $f^{-1}(x)=\frac{\sqrt[3]{x}+5}{2}$ Domain for both $f(x)$ and $f^{-1}(x)$ is $(-\infty,\infty)$.

Check: $f\circ f^{-1}(x)=f\left(\frac{\sqrt[3]{x}+5}{2}\right)=\left[2\left(\frac{\sqrt[3]{x}+5}{2}\right)-5\right]^3=\left(\sqrt[3]{x}+5-5\right)^3=\left(\sqrt[3]{x}\right)^3=\sqrt[3]{x^3}=x$

and $f^{-1}\circ f(x)=f^{-1}\left((2x-5)^3\right)=\frac{\sqrt[x]{(2x-5)^3}+5}{2}=\frac{(2x-5)+5}{2}=\frac{2x}{2}=x$.

In a function like $f(x)=3x-2$, the variable occurs only once and it easy to see how to reverse the operations step by step. But in $f(x)=x^3+2x+6$, you apply two different operations to the variable x (cubing and multiplying by 2) and then add 6, so it is not possible to reverse the operations step by step.

93. (a) We find $g^{-1}(x)$: $y=2x+1 \Leftrightarrow 2x=y-1 \Leftrightarrow x=\frac{1}{2}(y-1)$. So $g^{-1}(x)=\frac{1}{2}(x-1)$. Thus

$f(x)=h\circ g^{-1}(x)=h\left(\frac{1}{2}(x-1)\right)=4\left[\frac{1}{2}(x-1)\right]^2+4\left[\frac{1}{2}(x-1)\right]+7=x^2-2x+1+2x-2+7=x^2+6$

.

(b) $f\circ g=h \Leftrightarrow f^{-1}\circ f\circ g=f^{-1}\circ h \Leftrightarrow I\circ g=f^{-1}\circ h \Leftrightarrow g=f^{-1}\circ h$. Note that we compose with f^{-1} on the left on each side of the equation. We find f^{-1}: $y=3x+5$ $\Leftrightarrow 3x=y-5 \Leftrightarrow x=\frac{1}{3}(y-5)$. So $f^{-1}(x)=\frac{1}{3}(x-5)$. Thus

$g(x)=f^{-1}\circ h(x)=f^{-1}\left(3x^2+3x+2\right)=\frac{1}{3}\left[\left(3x^2+3x+2\right)-5\right]=\frac{1}{3}\left[3x^2+3x-3\right]=x^2+x-1$

Chapter 1 Review

1. **(a)**

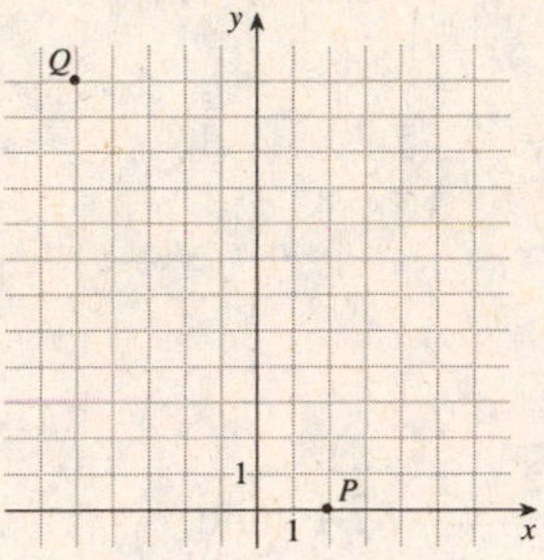

(b) The distance from P to Q is

$$d(P,Q) = \sqrt{(-5-2)^2 + (12-0)^2} = \sqrt{49+144}$$

(c) The midpoint is
$\left(\dfrac{-5+2}{2}, \dfrac{12+0}{2}\right) = \left(-\dfrac{3}{2}, 6\right)$.

(d) The line has slope $m = \dfrac{12-0}{-5-2} = -\frac{12}{7}$,
and has equation $y - 0 = -\frac{12}{7}(x-2) \Leftrightarrow$
$y = -\frac{12}{7}x + \frac{24}{7} \Leftrightarrow$
$12x + 7y - 24 = 0$.

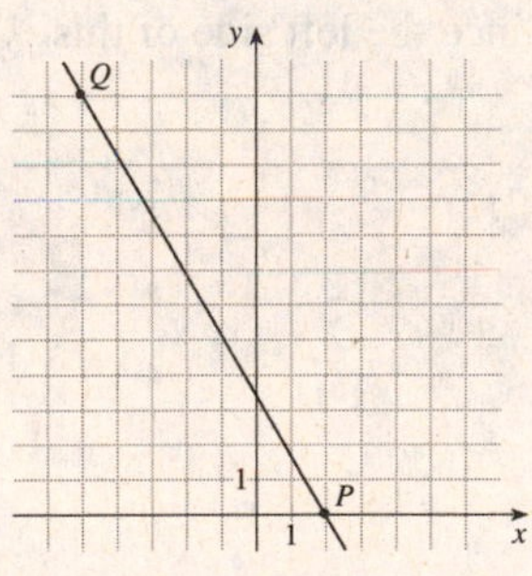

(e) The radius of this circle was found in part (b). It is $r = d(P,Q) = \sqrt{193}$. So an equation is $(x-2)^2 + (y-0)^2 = \left(\sqrt{193}\right)^2 \Leftrightarrow$
$(x-2)^2 + y^2 = 193$.

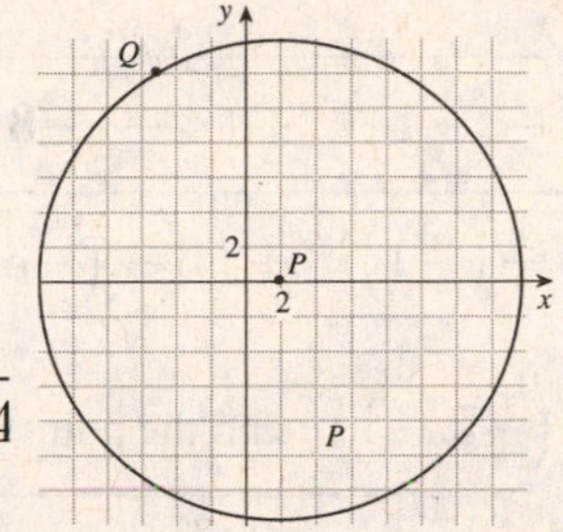

3.

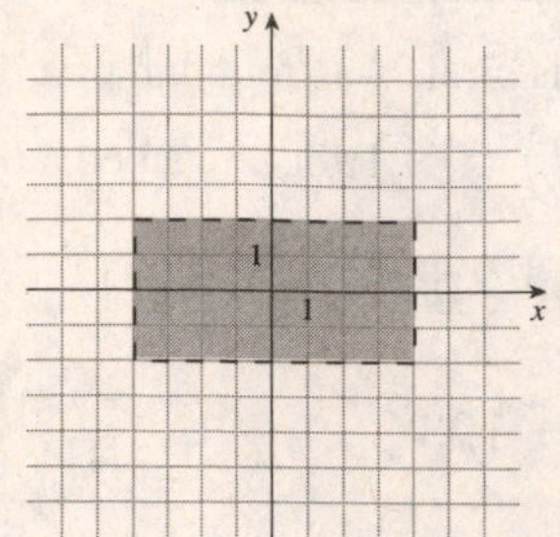

5. $d(A,C)=\sqrt{(4-(-1))^2+(4-(-3))^2}=\sqrt{(4+1)^2+(4+3)^2}=\sqrt{74}$ and

$d(B,C)=\sqrt{(5-(-1))^2+(3-(-3))^2}=\sqrt{(5+1)^2+(3+3)^2}=\sqrt{72}$. Therefore, B is closer to C .

7. The center is $C=(-5,-1)$, and the point $P=(0,0)$ is on the circle. The radius of the circle is

$r=d(P,C)=\sqrt{(0-(-5))^2+(0-(-1))^2}=\sqrt{(0+5)^2+(0+1)^2}=\sqrt{26}$. Thus, the equation of the circle is $(x+5)^2+(y+1)^2=26$.

9. $x^2+y^2+2x-6y+9=0 \Leftrightarrow (x^2+2x)+(y^2-6y)=-9 \Leftrightarrow$

$(x^2+2x+1)+(y^2-6y+9)=-9+1+9 \Leftrightarrow (x+1)^2+(y-3)^2=1$. This equation represents a circle with center at $(-1,\ 3)$ and radius 1 .

11. $x^2+y^2+72=12x \Leftrightarrow (x^2-12x)+y^2=-72 \Leftrightarrow$

$(x^2-12x+36)+y^2=-72+36 \Leftrightarrow (x-6)^2+y^2=-36$. Since the left side of this equation must be greater than or equal to zero, this equation has no graph.

13. $y = 2 - 3x$

x	y
-2	8
0	2
$\frac{2}{3}$	0

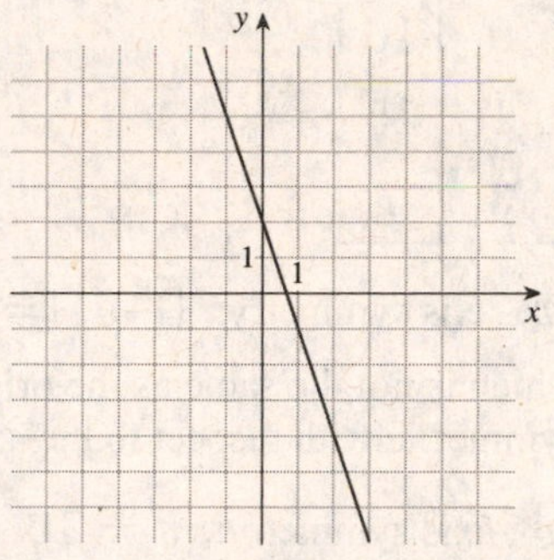

x -axis symmetry: $\left(-y\right) = 2 - 3x \iff y = -2 + 3x$, which is not the same as the original equation, so the graph is not symmetric with respect to the x -axis.

y -axis symmetry: $y = 2 - 3\left(-x\right) \iff y = 2 + 3x$, which is not the same as the original equation, so the graph is not symmetric with respect to the y -axis.

Origin symmetry: $\left(-y\right) = 2 - 3\left(-x\right) \iff -y = 2 + 3x$
$\iff y = -2 - 3x$, which is not the same as the original equation, so the graph is not symmetric with respect to the origin.
Hence the graph has no symmetry.

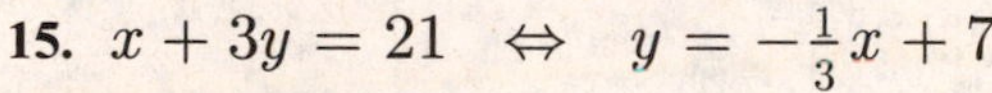

15. $x + 3y = 21 \iff y = -\frac{1}{3}x + 7$

x	y
-3	8
0	7
21	0

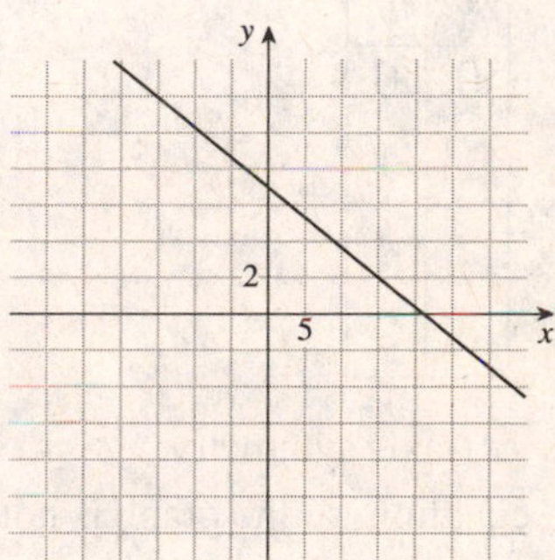

x -axis symmetry: $x + 3\left(-y\right) = 21 \iff x - 3y = 21$, which is not the same as the original equation, so the graph is not symmetric with respect to the x -axis.

y -axis symmetry: $\left(-x\right) + 3y = 21 \iff x - 3y = -21$, which is not the same as the original equation, so the graph is not symmetric with respect to the y -axis.

Origin symmetry: $\left(-x\right) + 3\left(-y\right) = 21 \iff x + 3y = -21$, which is not the same as the original equation, so the graph is not symmetric with respect to the origin. Hence the graph has no symmetry.

17. $y = 16 - x^2$

x	y
-3	7
-1	15
0	16
1	15
3	7

x-axis symmetry: $(-y) = 16 - x^2 \Leftrightarrow y = -16 + x^2$, which is not the same as the original equation, so the graph is not symmetric with respect to the x-axis.

y-axis symmetry: $y = 16 - (-x)^2 \Leftrightarrow y = 16 - x^2$, which is the same as the original equation, so its is symmetric with respect to the y-axis.

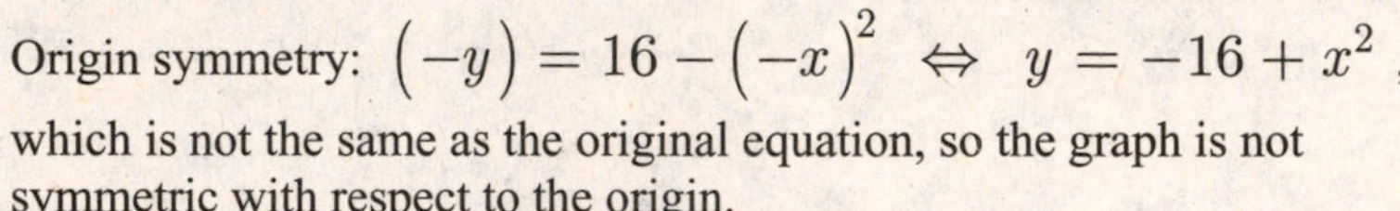

Origin symmetry: $(-y) = 16 - (-x)^2 \Leftrightarrow y = -16 + x^2$, which is not the same as the original equation, so the graph is not symmetric with respect to the origin.

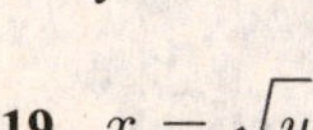

19. $x = \sqrt{y}$

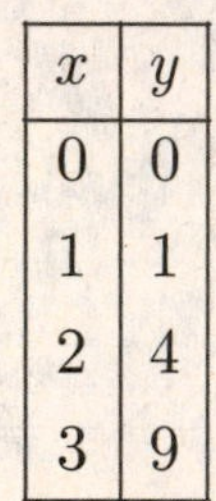

x	y
0	0
1	1
2	4
3	9

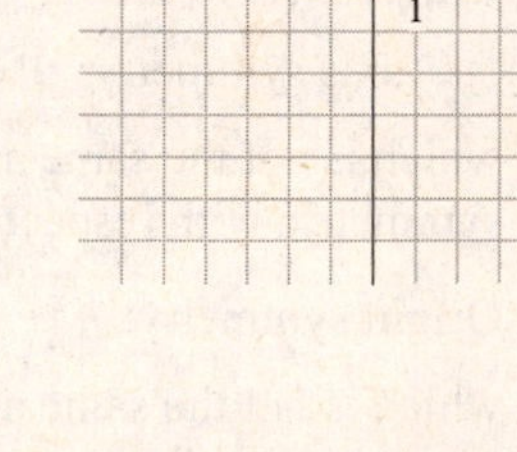

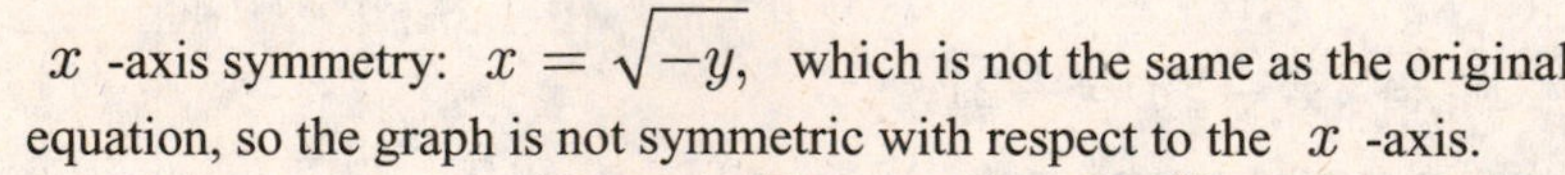

x-axis symmetry: $x = \sqrt{-y}$, which is not the same as the original equation, so the graph is not symmetric with respect to the x-axis.

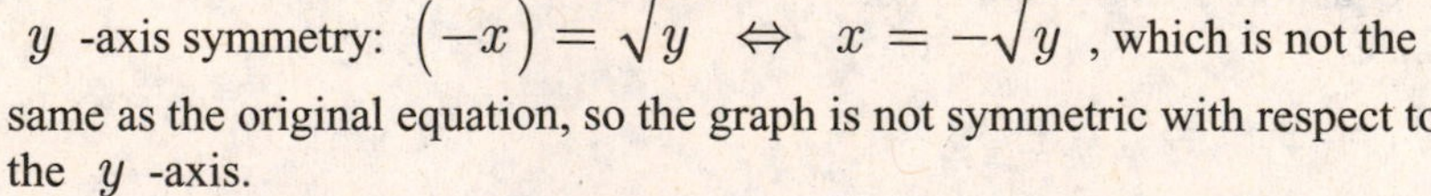

y-axis symmetry: $(-x) = \sqrt{y} \Leftrightarrow x = -\sqrt{y}$, which is not the same as the original equation, so the graph is not symmetric with respect to the y-axis.

Origin symmetry: $(-x) = \sqrt{-y}$, which is not the same as the original equation, so the graph is not symmetric with respect to the origin. Hence, the graph has no symmetry.

21. The line has slope $m = \dfrac{-4+6}{2+1} = \dfrac{2}{3}$, and so, by the point-slope formula, the equation is $y + 4 = \frac{2}{3}(x-2) \Leftrightarrow y = \frac{2}{3}x - \frac{16}{3} \Leftrightarrow 2x - 3y - 16 = 0$.

23. The x-intercept is 4, and the y-intercept is 12, so the slope is $m = \dfrac{12-0}{0-4} = -3$. Therefore, by the slope-intercept formula, the equation of the line is $y = -3x + 12 \Leftrightarrow 3x + y - 12 = 0$.

25. We first find the slope of the line $3x + 15y = 22$. This gives $3x + 15y = 22 \Leftrightarrow$ $15y = -3x + 22 \Leftrightarrow y = -\frac{1}{5}x + \frac{22}{15}$. So this line has slope $m = -\frac{1}{5}$, as does any line parallel to it. Then the parallel line passing through the origin has equation $y - 0 = -\frac{1}{5}(x - 0) \Leftrightarrow$ $x + 5y = 0$.

27. Here the center is at $(0,0)$, and the circle passes through the point $(-5,12)$, so the radius is $r = \sqrt{(-5-0)^2 + (12-0)^2} = \sqrt{25+144} = \sqrt{169} = 13$. The equation of the circle is $x^2 + y^2 = 13^2 \Leftrightarrow x^2 + y^2 = 169$. The line shown is the tangent that passes through the point $(-5,12)$, so it is perpendicular to the line through the points $(0,0)$ and $(-5,12)$. This line has slope $m_1 = \dfrac{12-0}{-5-0} = -\dfrac{12}{5}$. The slope of the line we seek is $m_2 = -\dfrac{1}{m_1} = -\dfrac{1}{-12/5} = \dfrac{5}{12}$. Thus, an equation of the tangent line is $y - 12 = \frac{5}{12}(x+5) \Leftrightarrow y - 12 = \frac{5}{12}x + \frac{25}{12} \Leftrightarrow$ $y = \frac{5}{12}x + \frac{169}{12} \Leftrightarrow 5x - 12y + 169 = 0$.

29. (a) The slope, 0.3, represents the increase in length of the spring for each unit increase in weight w. The S -intercept is the resting or natural length of the spring.

(b) When $w = 5$, $S = 0.3(5) + 2.5 = 1.5 + 2.5 = 4.0$ inches.

31. Square, then subtract 5 can be represented by the function $f(x) = x^2 - 5$.

33. $f(x) = 3(x+10)$: Add 10, then multiply by 3.

35. $g(x) = x^2 - 4x$

x	$g(x)$
-1	5
0	0
1	-3
2	-4
3	-3

37. $C(x) = 5000 + 30x - 0.001x^2$

(a) $C(1000) = 5000 + 30(1000) - 0.001(1000)^2 = \$34,000$ and

$C(10{,}000) = 5000 + 30(10{,}000) - 0.001(10{,}000)^2 = \$205,000$.

(b) From part (a), we see that the total cost of printing 1000 copies of the book is $\$34,000$ and the total cost of printing $10,000$ copies is $\$205,000$.

(c) $C(0) = 5000 + 30(0) - 0.001(0)^2 = \$5000$. This represents the fixed costs associated with getting the print run ready.

39. $f(x)=x^2-4x+6$; $f(0)=(0)^2-4(0)+6=6$; $f(2)=(2)^2-4(2)+6=2$;
$f(-2)=(-2)^2-4(-2)+6=18$; $f(a)=(a)^2-4(a)+6=a^2-4a+6$;
$f(-a)=(-a)^2-4(-a)+6=a^2+4a+6$;
$f(x+1)=(x+1)^2-4(x+1)+6=x^2+2x+1-4x-4+6=x^2-2x+3$;
$f(2x)=(2x)^2-4(2x)+6=4x^2-8x+6$;
$2f(x)-2=2(x^2-4x+6)-2=2x^2-8x+12-2=2x^2-8x+10$.

41. By the Vertical Line Test, figures (b) and (c) are graphs of functions. By the Horizontal Line Test, figure (c) is the graph of a one-to-one function.

43. Domain: We must have $x+3\geq 0 \Leftrightarrow x\geq -3$. In interval notation, the domain is $[-3,\infty)$.

Range: For x in the domain of f , we have $x\geq -3 \Leftrightarrow x+3\geq 0 \Leftrightarrow \sqrt{x+3}\geq 0 \Leftrightarrow f(x)\geq 0$. So the range is $[0,\infty)$.

45. $f(x)=7x+15$. The domain is all real numbers, $(-\infty,\infty)$.

47. $f(x)=\sqrt{x+4}$. We require $x+4\geq 0 \Leftrightarrow x\geq -4$. Thus the domain is $[-4,\infty)$.

49. $f(x)=\frac{1}{x}+\frac{1}{x+1}+\frac{1}{x+2}$. The denominators cannot equal 0 , therefore the domain is $\{x\,|\,x\neq 0,-1,-2\}$.

51. $h(x)=\sqrt{4-x}+\sqrt{x^2-1}$. We require the expression inside the radicals be nonnegative. So $4-x\geq 0 \Leftrightarrow 4\geq x$; also $x^2-1\geq 0 \Leftrightarrow (x-1)(x+1)\geq 0$. We make a table:

Interval	$(-\infty,-1)$	$(-1,1)$	$(1,\infty)$
Sign of $x-1$	$-$	$-$	$+$
Sign of $x+1$	$-$	$+$	$+$
Sign of $(x-1)(x+1)$	$+$	$-$	$+$

Thus the domain is $(-\infty,4]\cap\{(-\infty,-1]\cup[1,\infty)\}=(-\infty,-1]\cup[1,4]$.

53. $f(x)=1-2x$

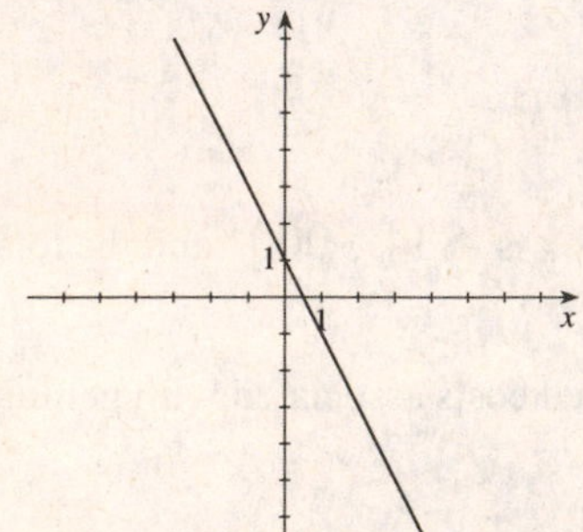

55. $f(t)=1-\frac{1}{2}t^2$

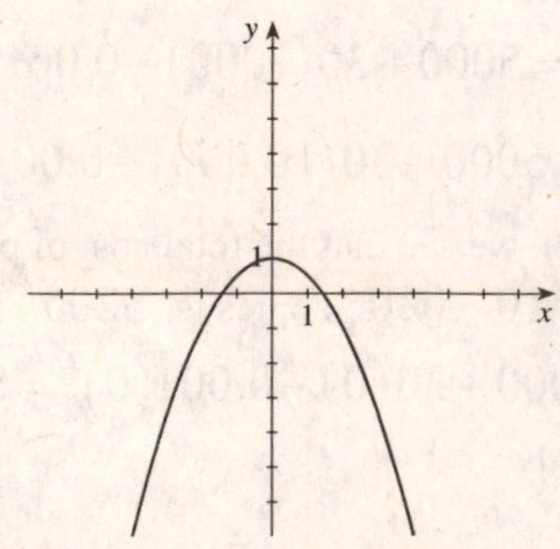

57. $f(x) = x^2 - 6x + 6$

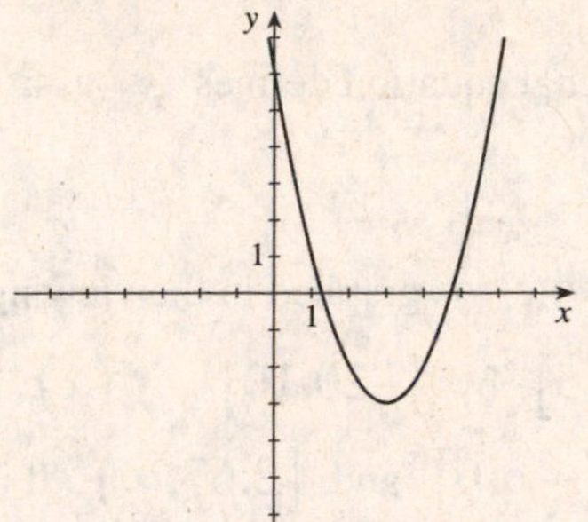

59. $g(x) = 1 - \sqrt{x}$

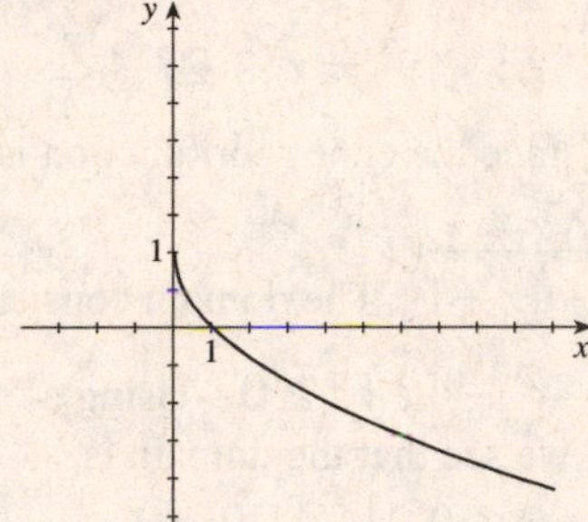

61. $h(x) = \frac{1}{2}x^3$

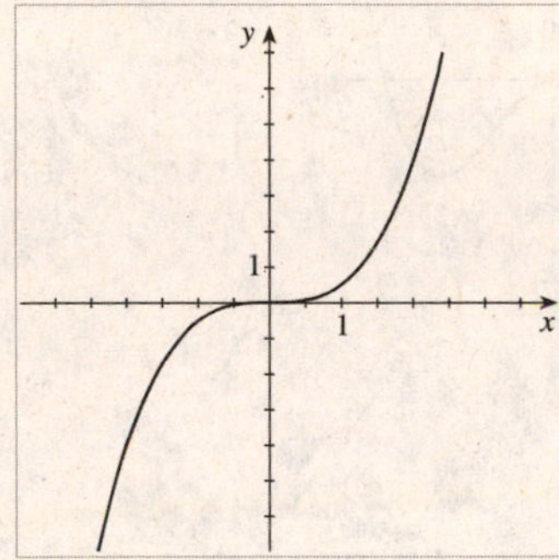

63. $h(x) = \sqrt[3]{x}$

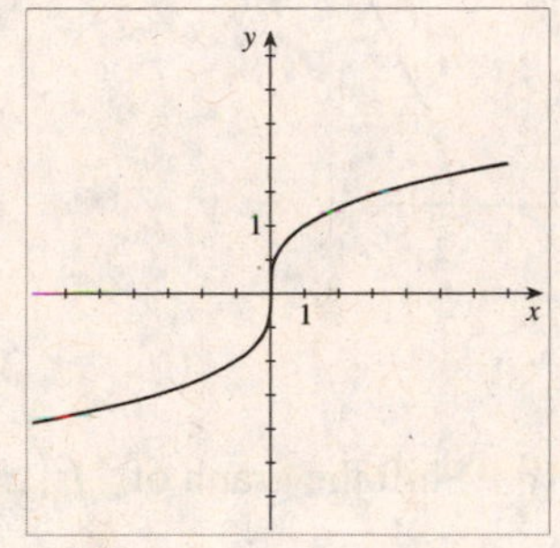

65. $g(x) = \dfrac{1}{x^2}$

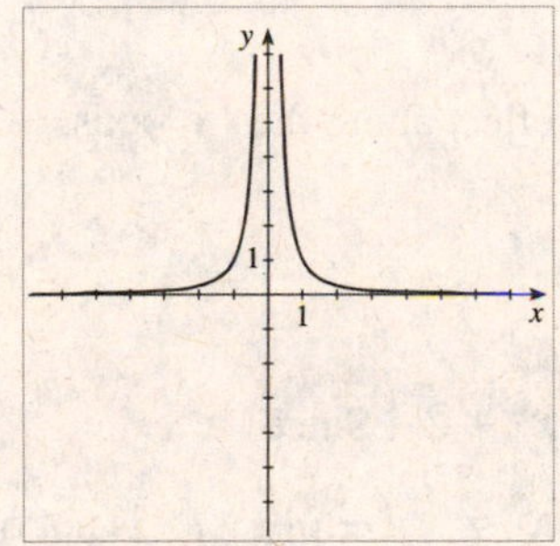

67. $f(x) = \begin{cases} 1 - x & \text{if } x < 0 \\ 1 & \text{if } x \geq 0 \end{cases}$

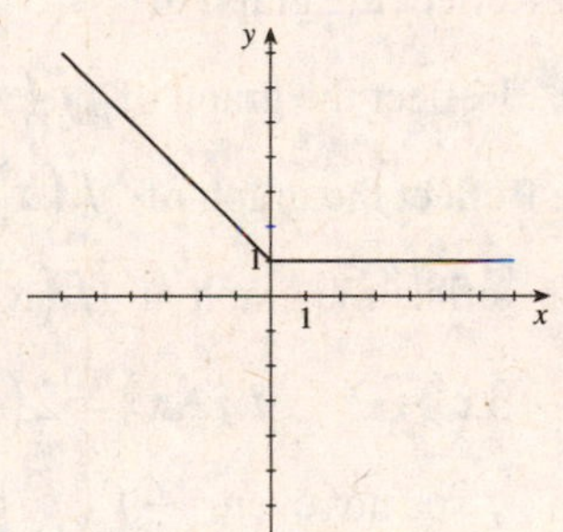

69. $f(x) = \begin{cases} x + 6 & \text{if } x < -2 \\ x^2 & \text{if } x \geq -2 \end{cases}$

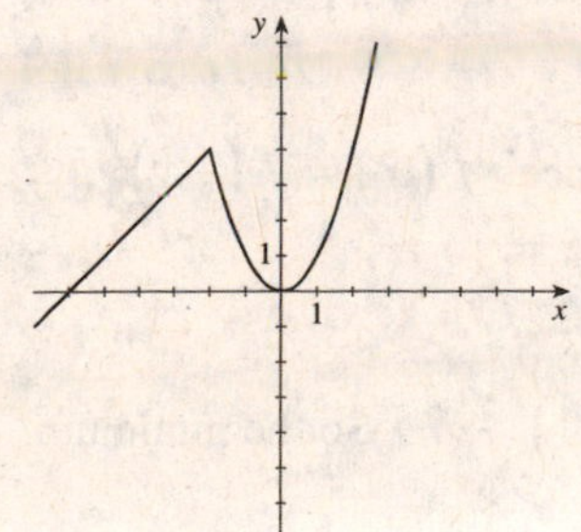

71. $x+y^2=14 \;\Rightarrow\; y^2=14-x \;\Rightarrow\; y=\pm\sqrt{14-x}$, so the original equation does not define y as a function of x .

73. $x^3-y^3=27 \;\Leftrightarrow\; y^3=x^3-27 \;\Leftrightarrow\; y=\left(x^3-27\right)^{1/3}$, so the original equation defines y as a function of x (since the cube root function is one-to-one).

75. $f(x)=\sqrt{x^3-4x+1}$. The domain consists of all x where $x^3-4x+1\ge 0$. Using a graphing device, we see that the domain is approximately $[-2.1,0.2]\cup[1.9,\infty)$.

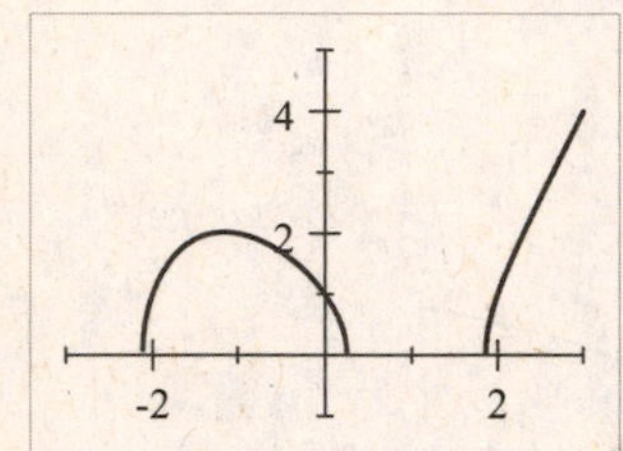

77. $f(x)=x^3-4x^2$ is graphed in the viewing rectangle $[-5,5]$ by $[-20,10]$. $f(x)$ is increasing on $(-\infty,0]$ and $[2.67,\infty)$. It is decreasing on $[0,2.67]$.

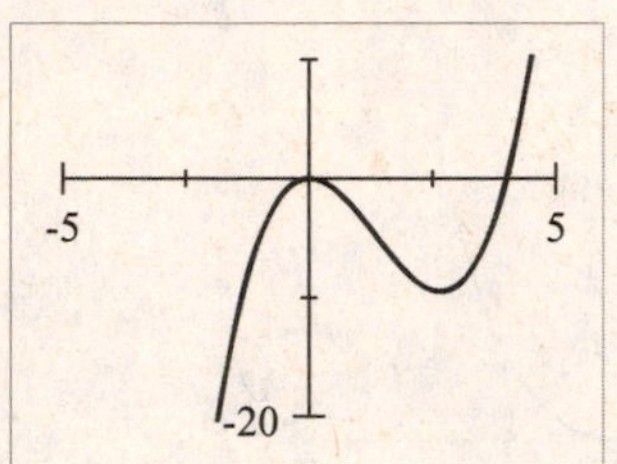

79. (a) $y=f(x)+8$. Shift the graph of $f(x)$ upward 8 units.

(b) $y=f(x+8)$. Shift the graph of $f(x)$ to the left 8 units.

(c) $y=1+2f(x)$. Stretch the graph of $f(x)$ vertically by a factor of 2 , then shift it upward 1 unit.

(d) $y=f(x-2)-2$. Shift the graph of $f(x)$ to the right 2 units, then downward 2 units.

(e) $y=f(-x)$. Reflect the graph of $f(x)$ about the y -axis.

(f) $y=-f(-x)$. Reflect the graph of $f(x)$ first about the y -axis, then reflect about the x -axis.

(g) $y=-f(x)$. Reflect the graph of $f(x)$ about the x -axis.

(h) $y=f^{-1}(x)$. Reflect the graph of $f(x)$ about the line $y=x$.

81. (a) $f(x)=2x^5-3x^2+2$. $f(-x)=2(-x)^5-3(-x)^2+2=-2x^5-3x^2+2$. Since $f(x)\ne f(-x)$, f is not even. $-f(x)=-2x^5+3x^2-2$. Since $-f(x)\ne f(-x)$, f is not odd.

(b) $f(x)=x^3-x^7$. $f(-x)=(-x)^3-(-x)^7=-\left(x^3-x^7\right)=-f(x)$, hence f is odd.

(c) $f(x)=\dfrac{1-x^2}{1+x^2}$. $f(-x)=\dfrac{1-(-x)^2}{1+(-x)^2}=\dfrac{1-x^2}{1+x^2}=f(x)$. Since $f(x)=f(-x)$, f is even.

(d) $f(x)=\dfrac{1}{x+2}$. $f(-x)=\dfrac{1}{(-x)+2}=\dfrac{1}{2-x}$. $-f(x)=-\dfrac{1}{x+2}$. Since $f(x)\ne f(-x)$, f is not even, and since $f(-x)\ne -f(x)$, f is not odd.

83. $g(x)=2x^2+4x-5=2\left(x^2+2x\right)-5=2\left(x^2+2x+1\right)-5-2=2(x+1)^2-7$. So the minimum value is $g(-1)=-7$.

85. $h(t) = -16t^2 + 48t + 32 = -16(t^2 - 3t) + 32 = -16(t^2 - 3t + \frac{9}{4}) + 32 + 36$ The stone reaches a

$= -16(t^2 - 3t + \frac{9}{4}) + 68 = -16(t - \frac{3}{2})^2 + 68$

maximum height of 68 feet.

87. $f(x) = 3.3 + 1.6x - 2.5x^3$. In the first viewing rectangle, $[-2,2]$ by $[-4,8]$, we see that $f(x)$ has a local maximum and a local minimum. In the next viewing rectangle, $[0.4,0.5]$ by $[3.78,3.80]$, we isolate the local maximum value as approximately 3.79 when $x \approx 0.46$. In the last viewing rectangle, $[-0.5,-0.4]$ by $[2.80,2.82]$, we isolate the local minimum value as 2.81 when $x \approx -0.46$.

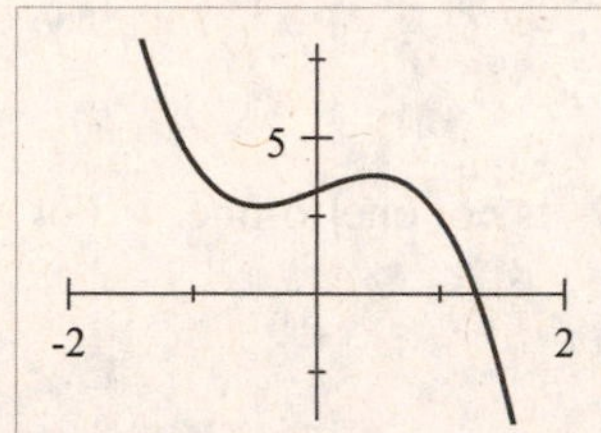

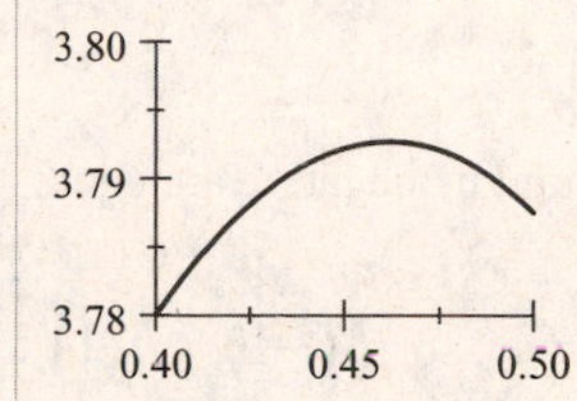

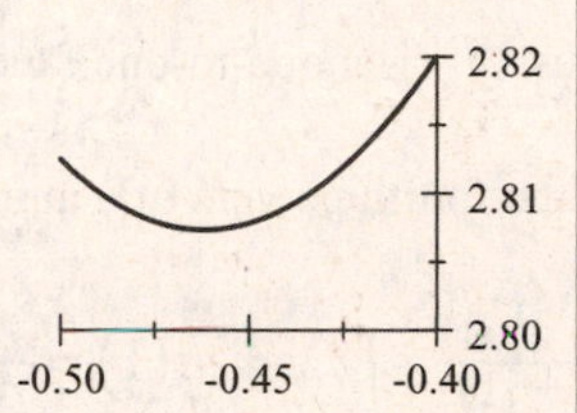

89. $f(x) = x + 2$, $g(x) = x^2$

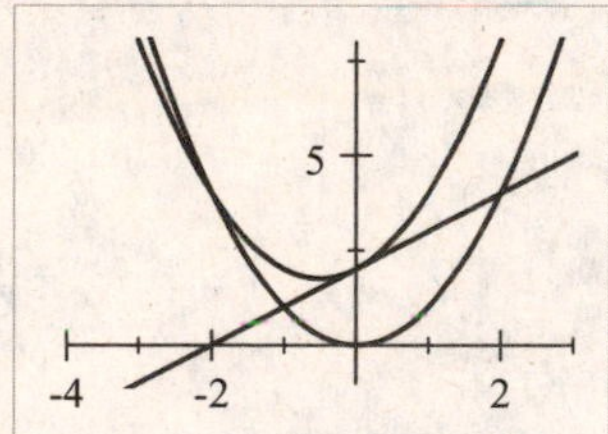

91. $f(x) = x^2 - 3x + 2$ and $g(x) = 4 - 3x$.

(a) $(f+g)(x) = (x^2 - 3x + 2) + (4 - 3x) = x^2 - 6x + 6$

(b) $(f-g)(x) = (x^2 - 3x + 2) - (4 - 3x) = x^2 - 2$

(c) $(fg)(x) = (x^2 - 3x + 2)(4 - 3x) = 4x^2 - 12x + 8 - 3x^3 + 9x^2 - 6x \quad = -3x^3 + 13x^2 - 18x + 8$

(d) $\left(\frac{f}{g}\right)(x) = \frac{x^2 - 3x + 2}{4 - 3x}$, $x \neq \frac{4}{3}$

(e) $(f \circ g)(x) = f(4 - 3x) = (4 - 3x)^2 - 3(4 - 3x) + 2 = 16 - 24x + 9x^2 - 12 + 9x + 2$

$= 9x^2 - 15x + 6$

(f) $(g \circ f)(x) = g(x^2 - 3x + 2) = 4 - 3(x^2 - 3x + 2) = -3x^2 + 9x - 2$

93. $f(x) = 3x - 1$ and $g(x) = 2x - x^2$.

$(f \circ g)(x) = f(2x - x^2) = 3(2x - x^2) - 1 = -3x^2 + 6x - 1$, and the domain is $(-\infty, \infty)$.

$(g \circ f)(x) = g(3x - 1) = 2(3x - 1) - (3x - 1)^2 = 6x - 2 - 9x^2 + 6x - 1 = -9x^2 + 12x - 3$, and the domain is $(-\infty, \infty)$

$(f \circ f)(x) = f(3x - 1) = 3(3x - 1) - 1 = 9x - 4$, and the domain is $(-\infty, \infty)$.

$(g \circ g)(x) = g(2x - x^2) = 2(2x - x^2) - (2x - x^2)^2 = 4x - 2x^2 - 4x^2 + 4x^3 - x^4 = -x^4 + 4x^3 - 6x^2 + 4x$ and domain is $(-\infty, \infty)$.

95. $f(x) = \sqrt{1-x}$, $g(x) = 1 - x^2$ and $h(x) = 1 + \sqrt{x}$.

$$\begin{aligned}(f \circ g \circ h)(x) &= f(g(h(x))) = f\left(g\left(1+\sqrt{x}\right)\right) = f\left(1-\left(1+\sqrt{x}\right)^2\right) = f\left(1-\left(1+2\sqrt{x}+x\right)\right) \\ &= f\left(-x-2\sqrt{x}\right) = \sqrt{1-\left(-x-2\sqrt{x}\right)} = \sqrt{1+2\sqrt{x}+x} = \sqrt{\left(1+\sqrt{x}\right)^2} = 1+\sqrt{x}\end{aligned}$$

97. $f(x) = 3 + x^3$. If $x_1 \neq x_2$, then $x_1^3 \neq x_2^3$ (unequal numbers have unequal cubes), and therefore $3 + x_1^3 \neq 3 + x_2^3$. Thus f is a one-to-one function.

99. $h(x) = \dfrac{1}{x^4}$. Since the fourth powers of a number and its negative are equal, h is not one-to-one. For example, $h(-1) = \dfrac{1}{(-1)^4} = 1$ and $h(1) = \dfrac{1}{(1)^4} = 1$, so $h(-1) = h(1)$.

101. $p(x) = 3.3 + 1.6x - 2.5x^3$. Using a graphing device and the Horizontal Line Test, we see that p is not a one-to-one function.

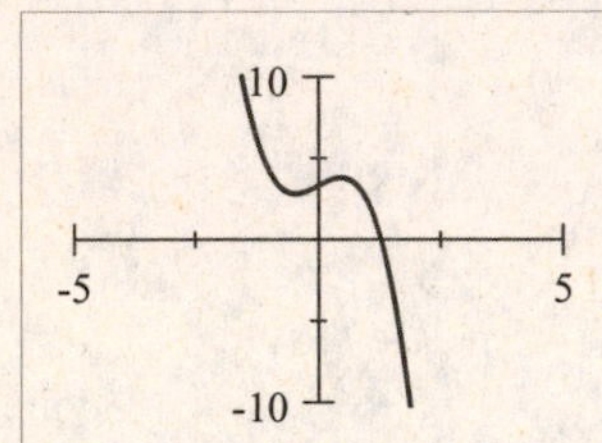

103. $f(x) = 3x - 2 \Leftrightarrow y = 3x - 2 \Leftrightarrow 3x = y + 2 \Leftrightarrow x = \frac{1}{3}(y+2)$. So $f^{-1}(x) = \frac{1}{3}(x+2)$.

105. $f(x) = (x+1)^3 \Leftrightarrow y = (x+1)^3 \Leftrightarrow x + 1 = \sqrt[3]{y} \Leftrightarrow x = \sqrt[3]{y} - 1$. So $f^{-1}(x) = \sqrt[3]{x} - 1$.

.

107. (a), (b) $f(x) = x^2 - 4$, $x \geq 0$

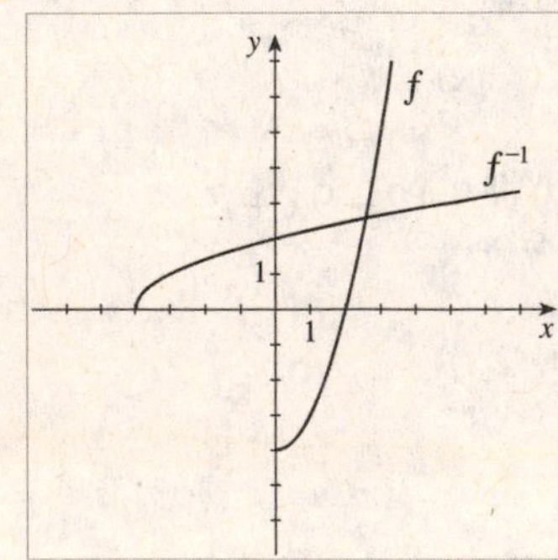

(c) $f(x) = x^2 - 4$, $x \geq 0 \Leftrightarrow y = x^2 - 4$, $y \geq -4 \Leftrightarrow x^2 = y + 4 \Leftrightarrow x = \sqrt{y+4}$. So $f^{-1}(x) = \sqrt{x+4}$, $x \geq -4$.

108. (a) If $x_1 \neq x_2$, then $\sqrt[4]{x_1} \neq \sqrt[4]{x_2}$, and so $1+\sqrt[4]{x_1} \neq 1+\sqrt[4]{x_2}$. Therefore, f is a one-to-one function.

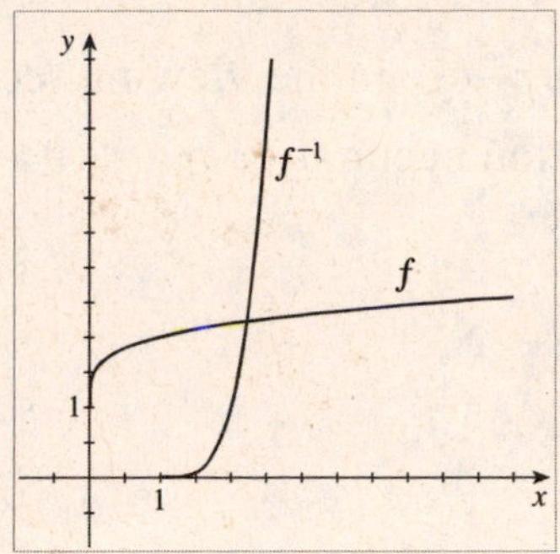

(b), (c)

(d) $f(x)=1+\sqrt[4]{x}$. $y=1+\sqrt[4]{x} \iff \sqrt[4]{x}=y-1 \iff x=(y-1)^4$. So $f^{-1}(x)=(x-1)^4$, $x \geq 1$. Note that the domain of f is $[0,\infty)$, so $y=1+\sqrt[4]{x} \geq 1$. Hence, the domain of f^{-1} is $[1,\infty)$.

Chapter 1 Test

1. (a) $x^3 - 9x - 1 = 0$. We graph the equation $y = x^3 - 9x - 1$ in the viewing rectangle $[-5,5]$ by $[-10,10]$. We find that the points of intersection occur at $x \approx -2.94$, -0.11 , 3.05 .

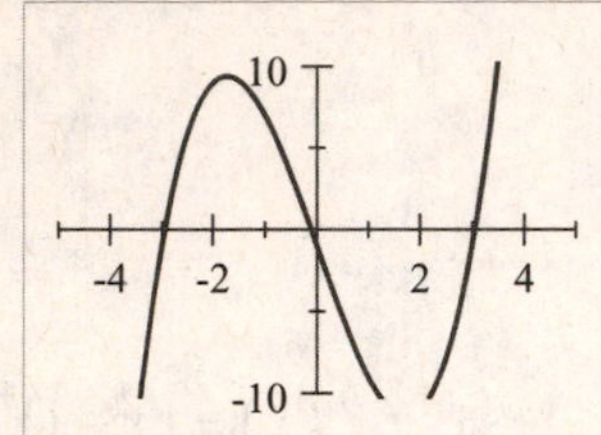

(b) $x^2 - 1 \le |x+1|$. We graph the equations $y_1 = x^2 - 1$ and $y_2 = |x+1|$ in the viewing rectangle $[-5,5]$ by $[-5,10]$. We find that the points of intersection occur at $x = -1$ and $x = 2$. Since we want $x^2 - 1 \le |x+1|$, the solution is the interval $[-1,2]$.

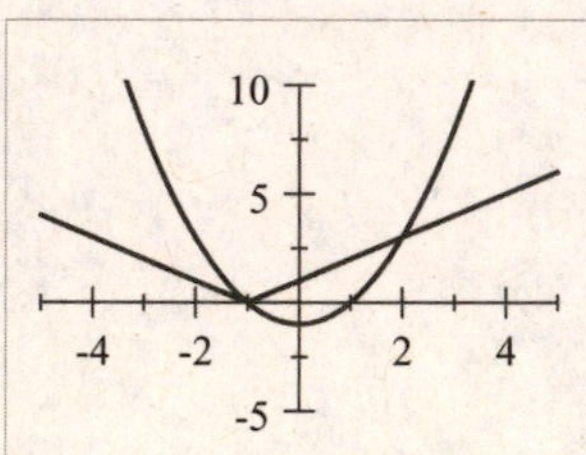

2. (a)

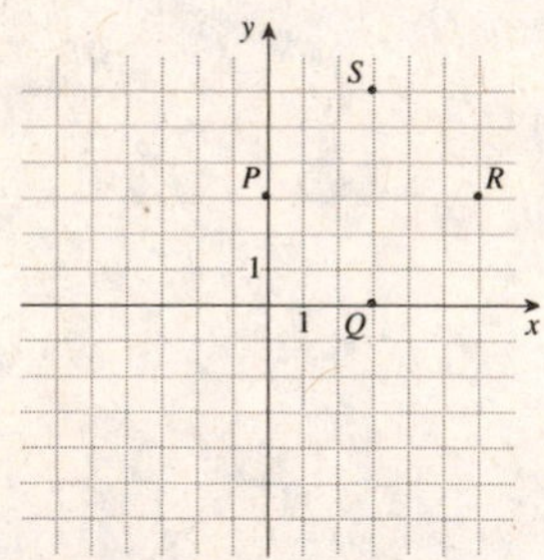

There are several ways to determine the coordinates of S . The diagonals of a square have equal length and are perpendicular. The diagonal PR is horizontal and has length is 6 units. So the diagonal QS is vertical and also has length 6 . Thus, the coordinates of S are $(3,6)$.

(b) The length of PQ is $\sqrt{(0-3)^2 + (3-0)^2} = \sqrt{18} = 3\sqrt{2}$. So the area of $PQRS$ is $(3\sqrt{2})^2 = 18$.

3. (a)

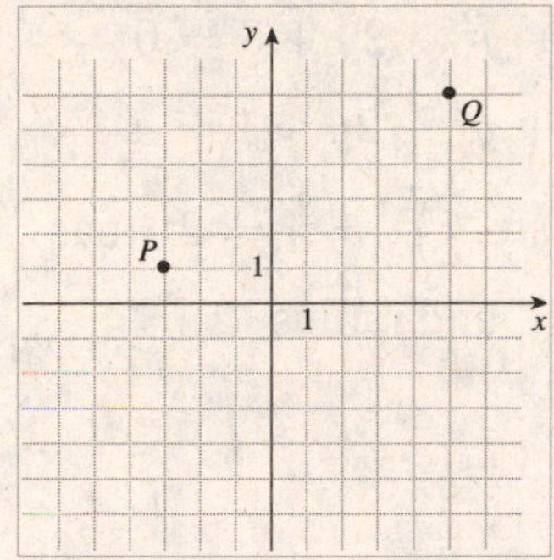

(b) The distance between P and Q is

$d(P,Q) = \sqrt{(-3-5)^2 + (1-6)^2} = \sqrt{64+25} = \sqrt{89}$.

(c) The midpoint is $\left(\dfrac{-3+5}{2}, \dfrac{1+6}{2}\right) = \left(1, \frac{7}{2}\right)$.

(d) The slope of the line is $\dfrac{1-6}{-3-5} = \dfrac{-5}{-8} = \dfrac{5}{8}$.

(e) The perpendicular bisector of PQ contains the midpoint, $\left(1, \frac{7}{2}\right)$, and it slope is the negative reciprocal of $\frac{5}{8}$. Thus the slope is $-\frac{1}{5/8} = -\frac{8}{5}$. Hence the equation is $y - \frac{7}{2} = -\frac{8}{5}(x-1) \Leftrightarrow y = -\frac{8}{5}x + \frac{8}{5} + \frac{7}{2} = -\frac{8}{5}x + \frac{51}{10}$. That is, $y = -\frac{8}{5}x + \frac{51}{10}$.

(f) The center of the circle is the midpoint, $\left(1, \frac{7}{2}\right)$, and the length of the radius is $\frac{1}{2}\sqrt{89}$. Thus the equation of the circle whose diameter is PQ is $(x-1)^2 + \left(y - \frac{7}{2}\right)^2 = \left(\frac{1}{2}\sqrt{89}\right)^2$ $\Leftrightarrow (x-1)^2 + \left(y - \frac{7}{2}\right)^2 = \frac{89}{4}$.

4. (a) $x^2 + y^2 = 25 = 5^2$ has center $(0,0)$ and radius 5.

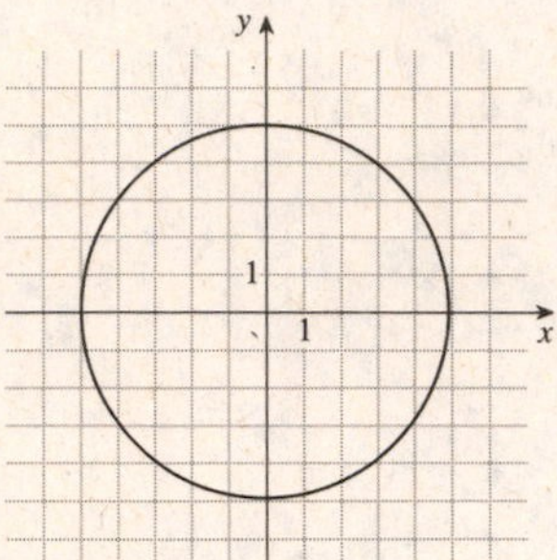

(b) $(x-2)^2 + (y+1)^2 = 9 = 3^2$ has center $(2,-1)$ and radius 3.

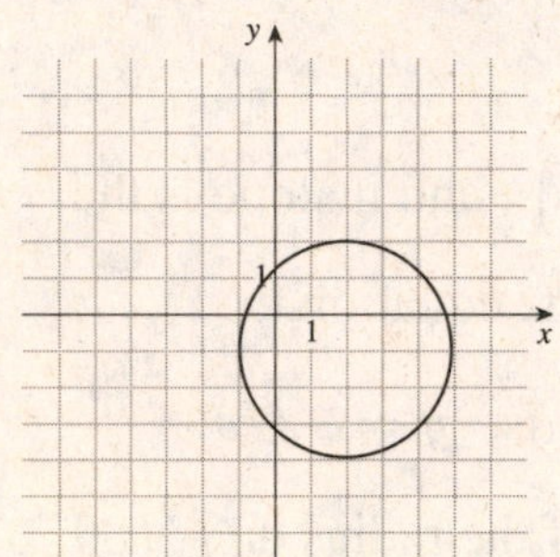

(c) $x^2 + 6x + y^2 - 2y + 6 = 0 \iff x^2 + 6x + 9 + y^2 - 2y + 1 = 4 \iff (x+3)^2 + (y-1)^2 = 4 = 2^2$ has center $(-3,1)$ and radius 2.

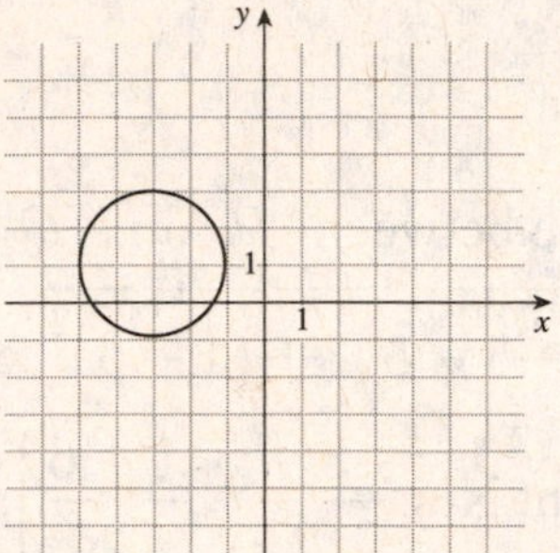

5. $2x - 3y = 15 \iff -3y = -2x + 15 \iff y = \frac{2}{3}x - 5$. The slope is $\frac{2}{3}$ and the y-intercept is -5.

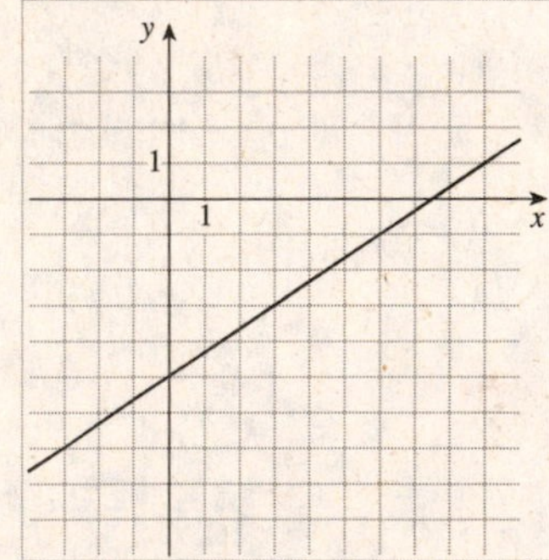

6. (a) $3x + y - 10 = 0 \iff y = -3x + 10$, so the slope of the line we seek is -3. Using the point-slope, $y - (-6) = -3(x-3) \iff y + 6 = -3x + 9 \iff 3x + y - 3 = 0$.

(b) Using the intercept form we get $\dfrac{x}{6} + \dfrac{y}{4} = 1 \iff 2x + 3y = 12 \iff 2x + 3y - 12 = 0$.

7. By the Vertical Line Test, figures (a) and (b) are graphs of functions. By the Horizontal Line Test, only figure (a) is the graph of a one-to-one function.

8. (a) $f(3)=\dfrac{\sqrt{3+1}}{3}=\dfrac{\sqrt{4}}{3}=\frac{2}{3}$; $f(5)=\dfrac{\sqrt{5+1}}{5}=\dfrac{\sqrt{6}}{5}$; $f(a-+1)=\dfrac{\sqrt{(a-1)+1}}{a-1}=\dfrac{\sqrt{a}}{a-1}$.

(b) $f(x)=\dfrac{\sqrt{x+1}}{x}$. Our restrictions are that the input to the radical is nonnegative, and the denominator must not be equal to zero. Thus $x+1\geq 0 \quad \Leftrightarrow \quad x\geq -1$ and $x\neq 0$. In interval notation, the domain is $[-1,0)\cup(0,\infty)$.

9. (a) Subtract 2, then cube the result can be expressed algebraically as $f(x)=(x-2)^3$.

(b)

x	$f(x)$
−1	−27
0	−8
1	−1
2	0
3	1
4	8

(c)

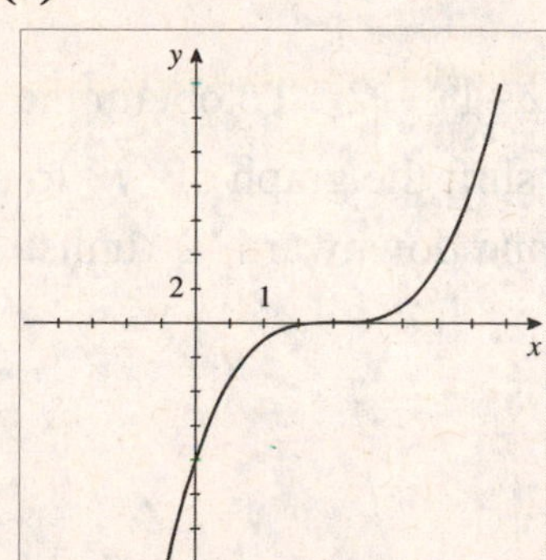

(d) We know that f has an inverse because it passes the Horizontal Line Test. A verbal description for f^{-1} is, Take the cube root, then add 2 .

(e) $y=(x-2)^3 \quad \Leftrightarrow \quad \sqrt[3]{y}=x-2 \quad \Leftrightarrow \quad x=\sqrt[3]{y}+2$. Thus, a formula for f^{-1} is $f^{-1}(x)=\sqrt[3]{x}+2$.

10. $R(x) = -500x^2 + 3000x$

(a) $R(2) = -500(2)^2 + 3000(2) = \$4000$ represents their total sales revenue when their price is $\$2$ per bar and $R(4) = -500(4)^2 + 3000(4) = \$4000$ represents their total sales revenue when their price is $\$4$ per bar

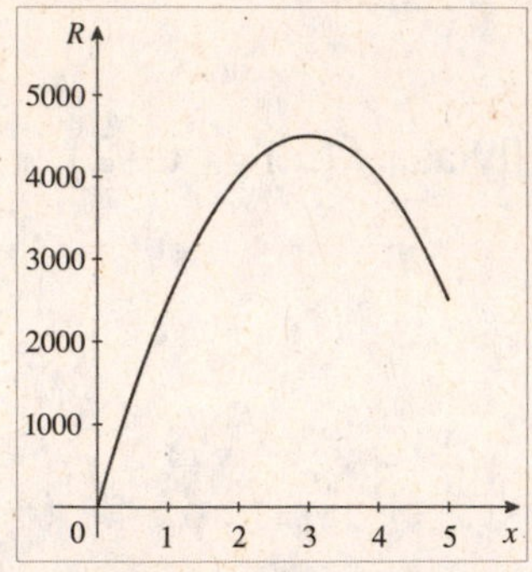

(b)

(c) The maximum revenue is $\$4500$, and it is achieved at a price of $x = \$3$.

11. The average rate of change is

$$\frac{f(2)-f(5)}{2-5} = \frac{\left[2^2-2(2)\right]-\left[5^2-2(5)\right]}{-3} = \frac{4-4-(25-10)}{-3} = \frac{-15}{-3} = 5.$$

12. (a) $f(x) = x^3$

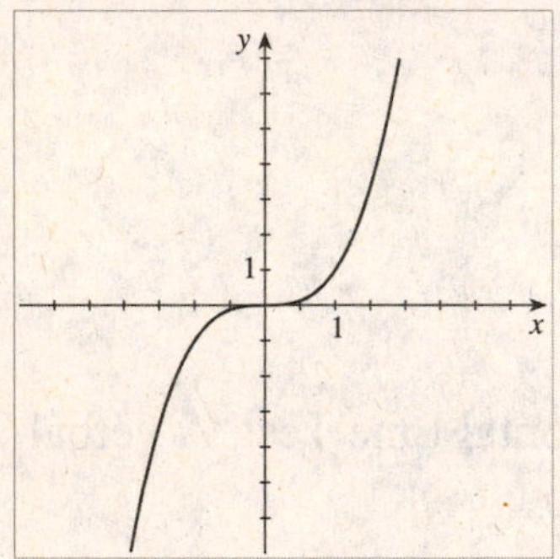

(b) $g(x) = (x-1)^3 - 2$. To obtain the graph of g, shift the graph of f to the right 1 unit and downward 2 units.

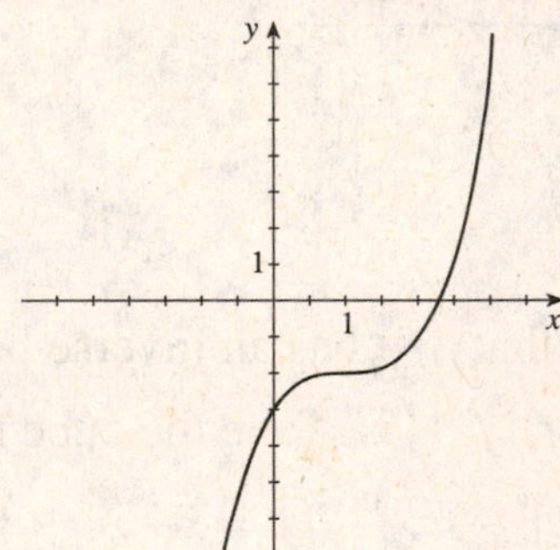

13. (a) $y = f(x-3)+2$. Shift the graph of $f(x)$ to the right 3 units, then shift the graph upward 2 units.

(b) $y = f(-x)$. Reflect the graph of $f(x)$ about the y-axis.

14. (a) $f(-2)=1-(-2)=1+2=3$ (since $-2\le 1$).

$f(1)=1-1=0$ (since $1\le 1$).

(b)

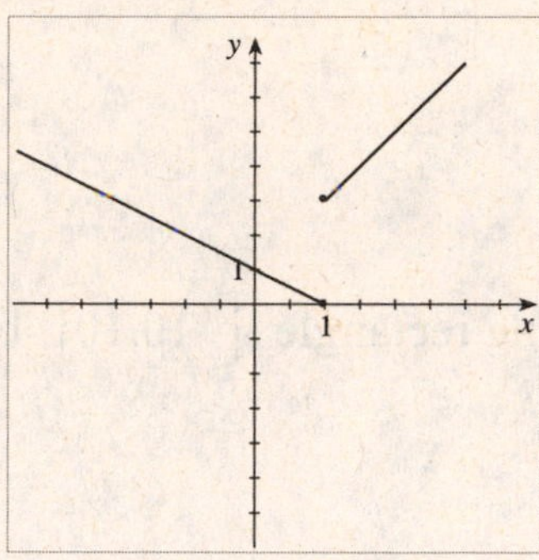

15. $f(x)=x^2+1$; $g(x)=x-3$.

(a) $(f\circ g)(x)=f(g(x))=f(x-3)=(x-3)^2+1=x^2-6x+9+1=x^2-6x+10$

(b) $(g\circ f)(x)=g(f(x))=g(x^2+1)=(x^2+1)-3=x^2-2$

(c) $f(g(2))=f(-1)=(-1)^2+1=2$. (We have used the fact that $g(2)=(2)-3=-1$.)

(d) $g(f(2))=g(5)=5-3=2$. (We have used the fact that $f(2)=2^2+1=5$.)

(e) $(g\circ g\circ g)(x)=g(g(g(x)))=g(g(x-3))=g(x-6)=(x-6)-3=x-9$. (We have used the fact that $g(x-3)=(x-3)-3=x-6$.)

16. (a) $f(x)=\sqrt{3-x}$, $x\le 3$ $\Leftrightarrow$ $y=\sqrt{3-x}$ $\Leftrightarrow$ $y^2=3-x$ $\Leftrightarrow$ $x=3-y^2$. Thus $f^{-1}(x)=3-x^2$, $x\ge 0$.

(b) $f(x)=\sqrt{3-x}$, $x\le 3$ and $f^{-1}(x)=3-x^2$, $x\ge 0$

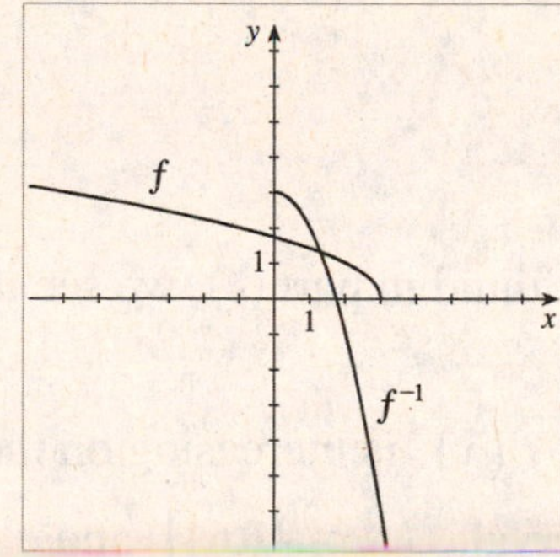

17. (a) The domain of f is $[0,6]$, and the range of f is $[1,7]$.

(b)

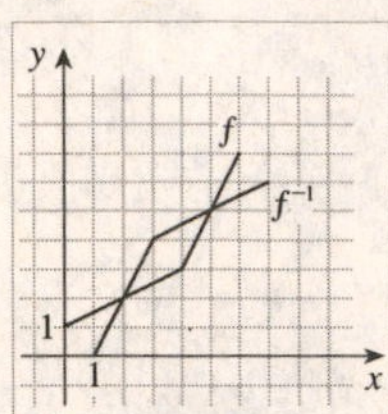

18. (a) $f(x)=3x^4-14x^2+5x-3$. The graph is shown in the viewing rectangle $[-10,10]$ by $[-30,10]$.

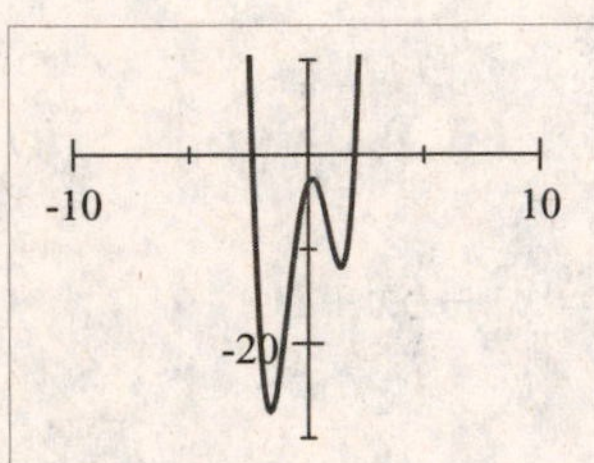

(b) No, by the Horizontal Line Test.

(c) The local maximum is approximately -2.55 when $x \approx 0.18$, as shown in the first viewing rectangle $[0.15,0.25]$ by $[-2.6,-2.5]$. One local minimum is approximately -27.18 when $x \approx -1.61$, as shown in the second viewing rectangle $[-1.65,-1.55]$ by $[-27.5,-27]$. The other local minimum is approximately -11.93 when $x \approx 1.43$, as shown is the viewing rectangle $[1.4,1.5]$ by $[-12,-11.9]$.

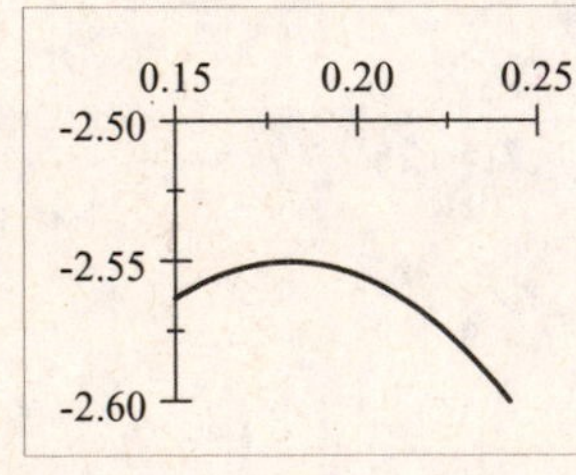

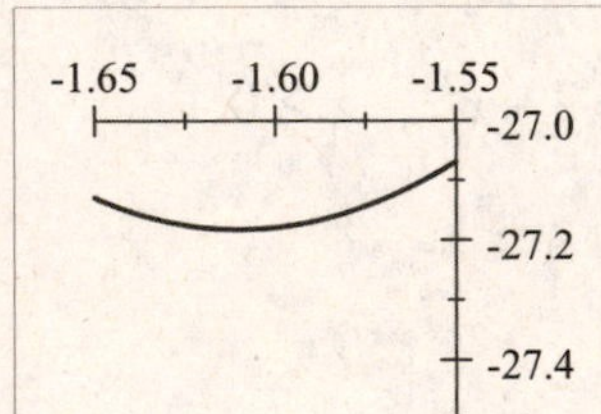

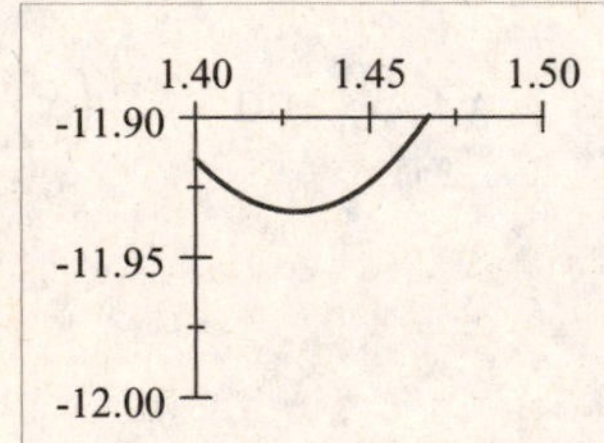

(d) Using the graph in part (a) and the local minimum, -27.18, found in part (c), we see that the range is $[-27.18,\infty)$.

(e) Using the information from part (c) and the graph in part (a), $f(x)$ is increasing on the intervals $[-1.61,0.18]$ and $[1.43,\infty)$ and decreasing on the intervals $(-\infty,-1.61]$ and $[0.18,1.43]$.

1. Let w be the width of the building lot. Then the length of the lot is $3w$. So the area of the building lot is $A(w) = 3w^2$, $w > 0$.

3. Let w be the width of the base of the rectangle. Then the height of the rectangle is $\frac{1}{2}w$. Thus the volume of the box is given by the function $V(w) = \frac{1}{2}w^3$, $w > 0$.

5. Let P be the perimeter of the rectangle and y be the length of the other side. Since $P = 2x + 2y$ and the perimeter is 20, we have $2x + 2y = 20 \Leftrightarrow x + y = 10 \Leftrightarrow y = 10 - x$. Since area is $A = xy$, substituting gives $A(x) = x(10 - x) = 10x - x^2$, and since A must be positive, the domain is $0 < x < 10$.

7.

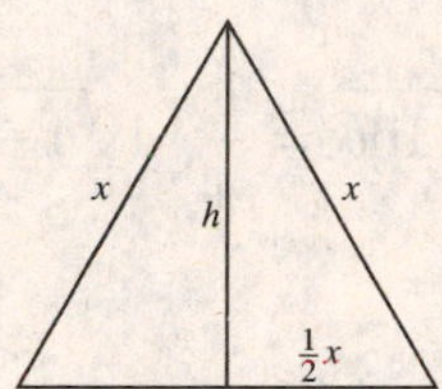

Let h be the height of an altitude of the equilateral triangle whose side has length x, as shown in the diagram. Thus the area is given by $A = \frac{1}{2}xh$. By the Pythagorean Theorem, $h^2 + \left(\frac{1}{2}x\right)^2 = x^2$ $\Leftrightarrow h^2 + \frac{1}{4}x^2 = x^2 \Leftrightarrow h^2 = \frac{3}{4}x^2 \Leftrightarrow h = \frac{\sqrt{3}}{2}x$. Substituting into the area of a triangle, we get $A(x) = \frac{1}{2}xh = \frac{1}{2}x\left(\frac{\sqrt{3}}{2}x\right) = \frac{\sqrt{3}}{4}x^2$, $x > 0$.

9. We solve for r in the formula for the area of a circle. This gives $A = \pi r^2 \Leftrightarrow r^2 = \dfrac{A}{\pi} \Rightarrow r = \sqrt{\dfrac{A}{\pi}}$, so the model is $r(A) = \sqrt{\dfrac{A}{\pi}}$, $A > 0$.

11. Let h be the height of the box in feet. The volume of the box is $V = 60$. Then $x^2h = 60 \Leftrightarrow h = \dfrac{60}{x^2}$. The surface area, S, of the box is the sum of the area of the 4 sides and the area of the base and top. Thus $S = 4xh + 2x^2 = 4x\left(\dfrac{60}{x^2}\right) + 2x^2 = \dfrac{240}{x} + 2x^2$, so the model is $S(x) = \dfrac{240}{x} + 2x^2$, $x > 0$.

13.

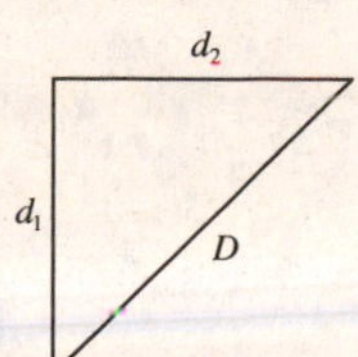

Let d_1 be the distance traveled south by the first ship and d_2 be the distance traveled east by the second ship. The first ship travels south for t hours at 5 mi/h, so l and, similarly, $d_2 = 20t$. Since the ships are traveling at right angles to each other, we can apply the Pythagorean Theorem to get

$$D(t) = \sqrt{d_1^2 + d_2^2} = \sqrt{(15t)^2 + (20t)^2} = \sqrt{225t^2 + 400t^2} = 25t$$

15.

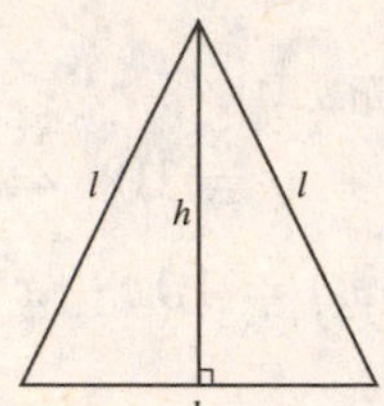

Let b be the length of the base, l be the length of the equal sides, and h be the height in centimeters. Since the perimeter is 8 , $2l + b = 8 \Leftrightarrow 2l = 8 - b \Leftrightarrow l = \frac{1}{2}(8 - b)$. By the Pythagorean Theorem, $h^2 + \left(\frac{1}{2}b\right)^2 = l^2 \Leftrightarrow h = \sqrt{l^2 - \frac{1}{4}b^2}$. Therefore the area of the triangle is

$$\begin{aligned} A &= \tfrac{1}{2} \cdot b \cdot h = \tfrac{1}{2} \cdot b\sqrt{l^2 - \tfrac{1}{4}b^2} = \frac{b}{2}\sqrt{\tfrac{1}{4}(8 - b)^2 - \tfrac{1}{4}b^2} \\ &= \frac{b}{4}\sqrt{64 - 16b + b^2 - b^2} = \frac{b}{4}\sqrt{64 - 16b} = \frac{b}{4} \cdot 4\sqrt{4 - b} = b\sqrt{4 - b} \end{aligned}$$

so the model is $A(b) = b\sqrt{4 - b}$, $0 < b < 4$.

17. Let w be the length of the rectangle. By the Pythagorean Theorem, $\left(\frac{1}{2}w\right)^2 + h^2 = 10^2 \Leftrightarrow$

$$\frac{w^2}{4} + h^2 = 10^2 \Leftrightarrow w^2 = 4\left(100 - h^2\right) \Leftrightarrow w = 2\sqrt{100 - h^2} \text{ (since } w > 0\text{).}$$

Therefore, the area of the rectangle is $A = wh = 2h\sqrt{100 - h^2}$, so the model is $A(h) = 2h\sqrt{100 - h^2}$, $0 < h < 10$.

19. (a) We complete the table.

First number	Second number	Product
1	18	18
2	17	34
3	16	48
4	15	60
5	14	70
6	13	78
7	12	84
8	11	88
9	10	90
10	9	90
11	8	88

From the table we conclude that the numbers is still increasing, the numbers whose product is a maximum should both be 9.5 .

(b) Let x be one number: then 14 is the other number, and so the product, p , is

$$p(x) = x(19 - x) = 19x - x^2 .$$

(c) $p(x) = 19x - x^2 = -(x^2 - 19x)$

$$= -\left[x^2 - 19x + \left(\tfrac{19}{2}\right)^2\right] + \left(\tfrac{19}{2}\right)^2$$

$$= -(x - 9.5)^2 + 90.25$$

So the product is maximized when the numbers are both 9.5 .

21. (a) Let x be the width of the field (in feet) and l be the length of the field (in feet). Since the farmer has 2400 ft of fencing we must have $2x + l = 2400$.

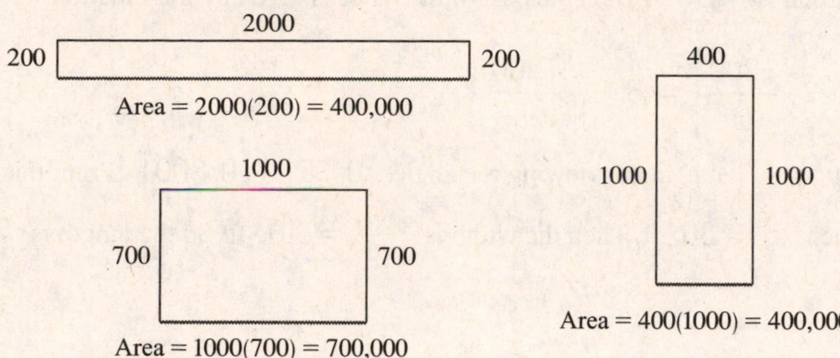

Width	Length	Area
200	2000	400,000
300	1800	540,000
400	1600	640,000
500	1400	700,000
600	1200	720,000
700	1000	700,000
800	800	640,000

It appears that the field of largest area is about 600 ft $\times 1200$ ft.

(b) Let x be the width of the field (in feet) and l be the length of the field (in feet). Since the farmer has 2400 ft of fencing we must have $2x + l = 2400 \quad \Leftrightarrow \quad .l = 2400 - 2x$ .. The area of the fenced-in field is given by

$$A(x) = l \cdot x = (2400 - 2x)x = -2x^2 + 2400x = -2(x^2 - 1200x) .$$

(c) The area is $A(x) = -2(x^2 - 1200x + 600^2) + 2(600^2) = -2(x - 600)^2 + 720$, 000 . So the maximum area occurs when $x = 600$ feet and $l = 2400 - 2(600) = 1200$ feet.

23. (a) Let x be the length of the fence along the road. If the area is 1200, we have $1200 = x \cdot$ width, so the width of the garden is $\dfrac{1200}{x}$. Then the cost of the fence is given by the function $C(x) = 5(x) + 3\left[x + 2 \cdot \dfrac{1200}{x}\right] = 8x + \dfrac{7200}{x}$.

(b) We graph the function $y = C(x)$ in the viewing rectangle $[0,75] \times [0,800]$. From this we get the cost is minimized when $x = 30$ ft. Then the width is $\frac{1200}{30} = 40$ ft. So the length is 30 ft and the width is 40 ft.

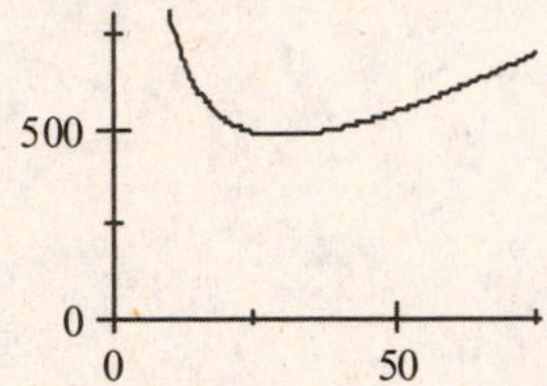

(c) We graph the function $y = C(x)$ and $y = 600$ in the viewing rectangle $[10,65] \times [450,650]$. From this we get that the cost is at most $\$600$ when $15 \le x \le 60$. So the range of lengths he can fence along the road is 15 feet to 60 feet.

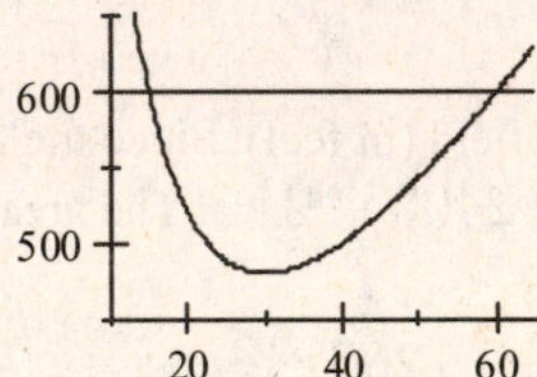

25. (a) Let h be the height in feet of the straight portion of the window. The circumference of the semicircle is $C = \frac{1}{2}\pi x$. Since the perimeter of the window is 30 feet, we have

$x + 2h + \frac{1}{2}\pi x = 30$. Solving for h , we get $2h = 30 - x - \frac{1}{2}\pi x \Leftrightarrow$

$h = 15 - \frac{1}{2}x - \frac{1}{4}\pi x$. The area of the window is

$A(x) = xh + \frac{1}{2}\pi\left(\frac{1}{2}x\right)^2 = x\left(15 - \frac{1}{2}x - \frac{1}{4}\pi x\right) + \frac{1}{8}\pi x^2 = 15x - \frac{1}{2}x^2 - \frac{1}{8}\pi x^2$.

(b)

$$A(x) = -\frac{1}{8}(\pi + 4)\left[x^2 - \frac{120}{\pi + 4}x + \left(\frac{60}{\pi + 4}\right)^2\right] + \frac{450}{\pi + 4} = -\frac{1}{8}(\pi + 4)\left(x - \frac{60}{\pi + 4}\right)^2 + \frac{450}{\pi + 4}$$

$$= 15x - \frac{1}{8}(\pi + 4)x^2 = -\frac{1}{8}(\pi + 4)\left[x^2 - \frac{120}{\pi + 4}x\right]$$

The area is maximized when $x = \dfrac{60}{\pi + 4} \approx 8.40$, and hence

$h \approx 15 - \frac{1}{2}(8.40) - \frac{1}{4}\pi(8.40) \approx 4.20$.

27. (a) Let x be the length of one side of the base and let h be the height of the box in feet. Since the volume of the box is $V = x^2h = 12$, we have $x^2h = 12 \Leftrightarrow h = \dfrac{12}{x^2}$. The surface area, A , of the box is sum of the area of the four sides and the area of the base. Thus the surface area of the box is given by the formula $A(x) = 4xh + x^2 = 4x\left(\dfrac{12}{x^2}\right) + x^2 = \dfrac{48}{x} + x^2$, $x > 0$.

(b) The function $y = A(x)$ is shown in the first viewing rectangle below. In the second viewing rectangle, we isolate the minimum, and we see that the amount of material is minimized when x (the length and width) is 2.88 ft. Then the height is $h = \dfrac{12}{x^2} \approx 1.44$ ft.

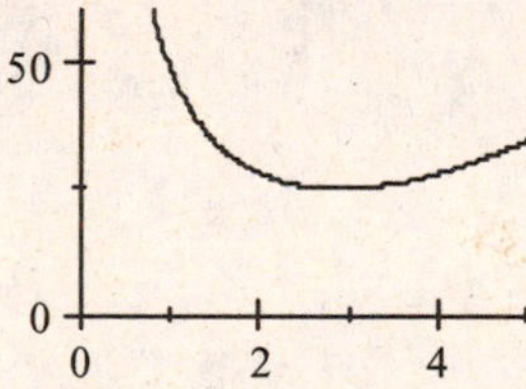

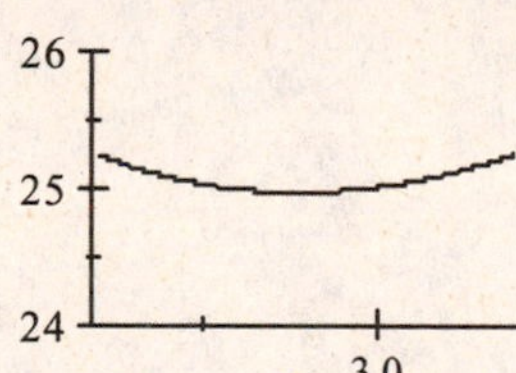

29. (a) Let w be the width of the pen and l be the length in meters. We use the area to establish a relationship between w and l. Since the area is $100\ \text{m}^2$, we have $l \cdot w = 100 \Leftrightarrow l = \frac{100}{w}$. So the amount of fencing used is $F = 2l + 2w = 2\left(\frac{100}{w}\right) + 2w = \frac{200 + 2w^2}{w}$.

(b) Using a graphing device, we first graph F in the viewing rectangle $[0, 40]$ by $[0, 100]$, and locate the approximate location of the minimum value. In the second viewing rectangle, $[8, 12]$ by $[39, 41]$, we see that the minimum value of F occurs when $w = 10$. Therefore the pen should be a square with side 10 m.

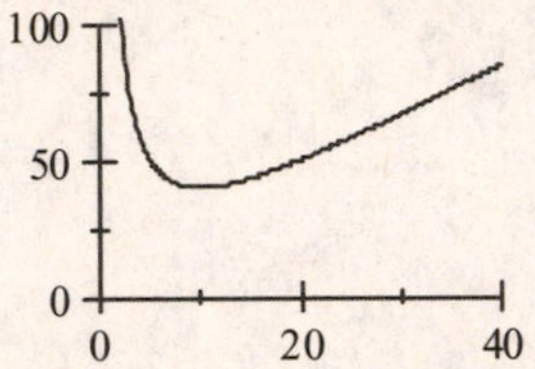

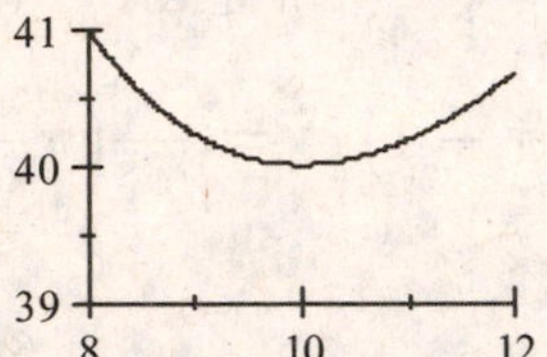

31. (a) Let x be the distance from point B to C, in miles. Then the distance from A to C is $\sqrt{x^2 + 25}$, and the energy used in flying from A to C then C to D is $f(x) = 14\sqrt{x^2 + 25} + 10(12 - x)$.

(b) By using a graphing device, the energy expenditure is minimized when the distance from B to C is about 5.1 miles.

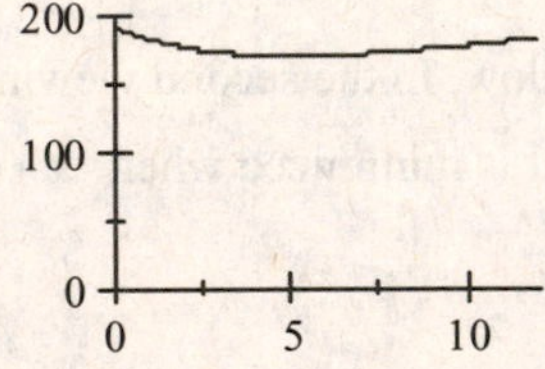

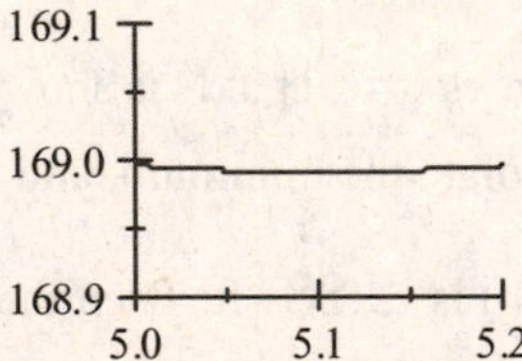

Chapter 2: Trigonometric Functions: Unit Circle Approach

2.1 The Unit Circle

1. (a) The unit circle is the circle centered at $(0,0)$ with radius 1.

(b) The equation of the unit circle is $x^2+y^2=1$.

(c) Since $1^2+0^2=1$, the point is $P(1,0)$.

(ii) $P(0,1)$

(iii) $P(-1,0)$

(iv) $P(0,-1)$

3. Since $\left(\frac{4}{5}\right)^2+\left(-\frac{3}{5}\right)^2=\frac{16}{25}+\frac{9}{25}=1$, $P\left(\frac{4}{5},-\frac{3}{5}\right)$ lies on the unit circle.

5. Since $\left(\frac{7}{25}\right)^2+\left(\frac{24}{25}\right)^2=\frac{49}{625}+\frac{576}{625}=1$, $P\left(\frac{7}{25},\frac{24}{25}\right)$ lies on the unit circle.

7. Since $\left(-\frac{\sqrt{5}}{3}\right)^2+\left(\frac{2}{3}\right)^2=\frac{5}{9}+\frac{4}{9}=1$, $P\left(-\frac{\sqrt{5}}{3},\frac{2}{3}\right)$ lies on the unit circle.

9. $\left(-\frac{3}{5}\right)^2+y^2=1 \Leftrightarrow y^2=1-\frac{9}{25} \Leftrightarrow y^2=\frac{16}{25} \Leftrightarrow y=\pm\frac{4}{5}$. Since $P(x,y)$ is in quadrant III, y is negative, so the point is $P\left(-\frac{3}{5},-\frac{4}{5}\right)$.

11. $x^2+\left(\frac{1}{3}\right)^2=1 \Leftrightarrow x^2=1-\frac{1}{9} \Leftrightarrow x^2=\frac{8}{9} \Leftrightarrow x=\pm\frac{2\sqrt{2}}{3}$. Since P is in quadrant II, x is negative, so the point is $P\left(-\frac{2\sqrt{2}}{3},\frac{1}{3}\right)$.

13. $x^2+\left(-\frac{2}{7}\right)^2=1 \Leftrightarrow x^2=1-\frac{4}{49} \Leftrightarrow x^2=\frac{45}{49} \Leftrightarrow x=\pm\frac{3\sqrt{5}}{7}$. Since $P(x,y)$ is in quadrant IV, x is positive, so the point is $P\left(\frac{3\sqrt{5}}{7},-\frac{2}{7}\right)$.

15. $\left(\frac{4}{5}\right)^2+y^2=1 \Leftrightarrow y^2=1-\frac{16}{25} \Leftrightarrow y^2=\frac{9}{25} \Leftrightarrow y=\pm\frac{3}{5}$. Since its y-coordinate is positive, the point is $P\left(\frac{4}{5},\frac{3}{5}\right)$.

17. $x^2+\left(\frac{2}{3}\right)^2=1 \Leftrightarrow x^2=1-\frac{4}{9} \Leftrightarrow x^2=\frac{5}{9} \Leftrightarrow x=\pm\frac{\sqrt{5}}{3}$. Since its x-coordinate is negative, the point is $P\left(-\frac{\sqrt{5}}{3},\frac{2}{3}\right)$.

19. $\left(-\frac{\sqrt{2}}{3}\right)^2+y^2=1 \Leftrightarrow y^2=1-\frac{2}{9} \Leftrightarrow y^2=\frac{7}{9} \Leftrightarrow y=\pm\frac{\sqrt{7}}{3}$. Since P lies below the x-axis, its y-coordinate is negative, so the point is $P\left(-\frac{\sqrt{2}}{3},-\frac{\sqrt{7}}{3}\right)$.

21.

t	Terminal Point
0	$(1,0)$
$\frac{\pi}{4}$	$\left(\frac{\sqrt{2}}{2},\frac{\sqrt{2}}{2}\right)$
$\frac{\pi}{2}$	$(0,1)$
$\frac{3\pi}{4}$	$\left(-\frac{\sqrt{2}}{2},\frac{\sqrt{2}}{2}\right)$
π	$(-1,0)$

t	Terminal Point
π	$(-1,0)$
$\frac{5\pi}{4}$	$\left(-\frac{\sqrt{2}}{2},-\frac{\sqrt{2}}{2}\right)$
$\frac{3\pi}{2}$	$(0,-1)$
$\frac{7\pi}{4}$	$\left(\frac{\sqrt{2}}{2},-\frac{\sqrt{2}}{2}\right)$
2π	$(1,0)$

23. $P(x,y)=(0,1)$

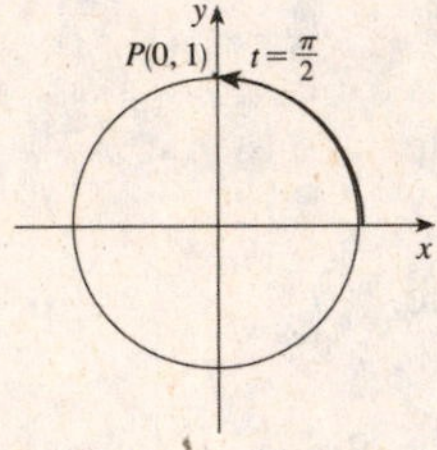

25. $P(x,y)=\left(-\frac{\sqrt{3}}{2},\frac{1}{2}\right)$

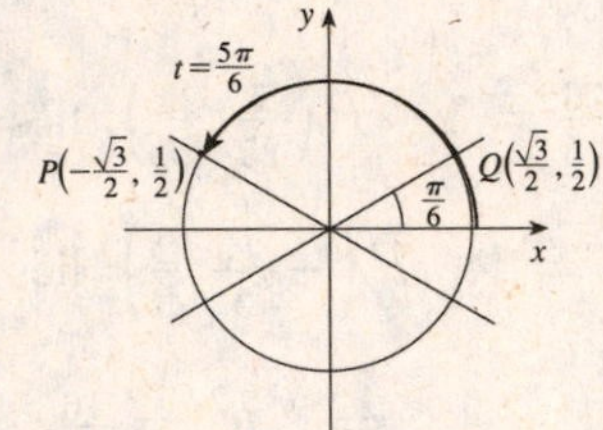

27. $P(x,y)=\left(\frac{1}{2},-\frac{\sqrt{3}}{2}\right)$

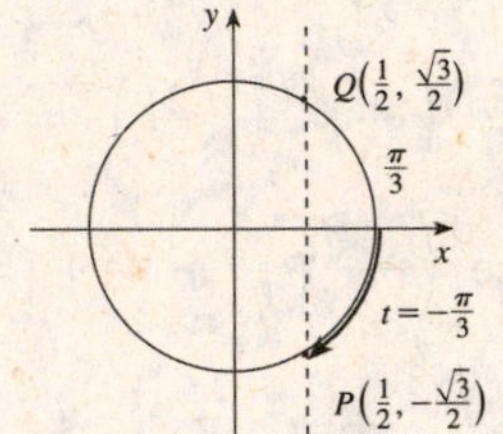

29. $P(x,y)=\left(-\frac{1}{2},\frac{\sqrt{3}}{2}\right)$

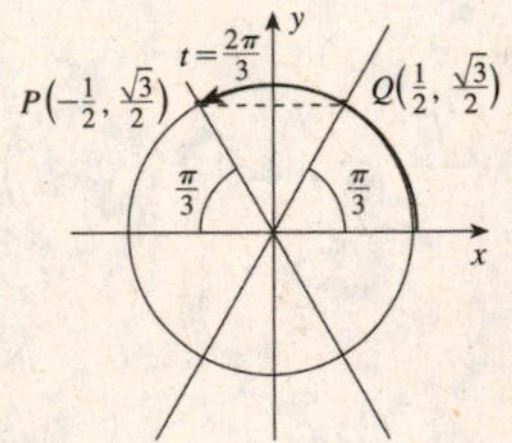

31. $P(x,y)=\left(-\frac{\sqrt{2}}{2},-\frac{\sqrt{2}}{2}\right)$

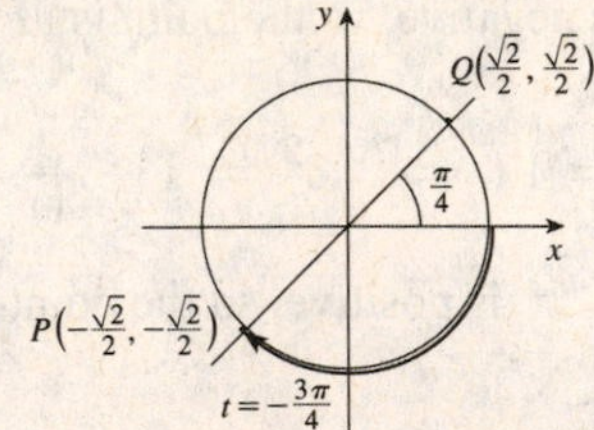

33. Let $Q\left(x,y\right)=\left(\frac{3}{5},\frac{4}{5}\right)$ be the terminal point determined by t.

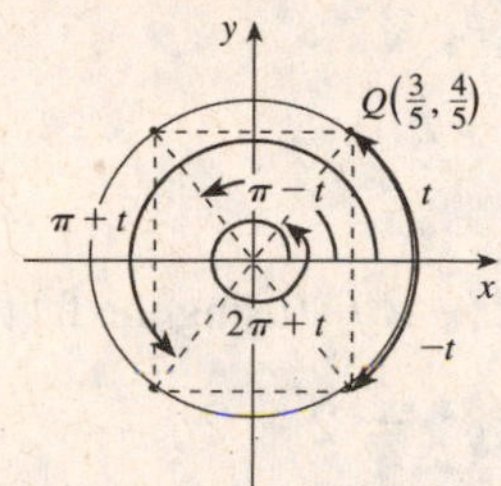

(a) $\pi-t$ determines the point $P\left(-x,y\right)=\left(-\frac{3}{5},\frac{4}{5}\right)$.

(b) $-t$ determines the point $P\left(x,-y\right)=\left(\frac{3}{5},-\frac{4}{5}\right)$.

(c) $\pi+t$ determines the point $P\left(-x,-y\right)=\left(-\frac{3}{5},-\frac{4}{5}\right)$.

(d) $2\pi+t$ determines the point $P\left(x,y\right)=\left(\frac{3}{5},\frac{4}{5}\right)$.

35. (a) $\overline{t}=\frac{5\pi}{4}-\pi=\frac{\pi}{4}$ **(b)** $\overline{t}=\frac{7\pi}{3}-2\pi=\frac{\pi}{3}$

(c) $\overline{t}=\frac{4\pi}{3}-\pi=\frac{\pi}{3}$ **(d)** $\overline{t}=\frac{\pi}{6}$

37. (a) $\overline{t}=\pi-\frac{5\pi}{7}=\frac{2\pi}{7}$ **(b)** $\overline{t}=\pi-\frac{7\pi}{9}=\frac{2\pi}{9}$

(c) $\overline{t}=\pi-3\approx 0.142$ **(d)** $\overline{t}=2\pi-5\approx 1.283$

39. (a) $\overline{t}=\pi-\frac{2\pi}{3}=\frac{\pi}{3}$ **(b)** $P\left(-\frac{1}{2},\frac{\sqrt{3}}{2}\right)$

41. (a) $\overline{t}=\pi-\frac{3\pi}{4}=\frac{\pi}{4}$ **(b)** $P\left(-\frac{\sqrt{2}}{2},\frac{\sqrt{2}}{2}\right)$

43. (a) $\overline{t}=\pi-\frac{2\pi}{3}=\frac{\pi}{3}$ **(b)** $P\left(-\frac{1}{2},-\frac{\sqrt{3}}{2}\right)$

45. (a) $\overline{t}=\frac{13\pi}{4}-3\pi=\frac{\pi}{4}$ **(b)** $P\left(-\frac{\sqrt{2}}{2},-\frac{\sqrt{2}}{2}\right)$

47. (a) $\overline{t}=\frac{7\pi}{6}-\pi=\frac{\pi}{6}$ **(b)** $P\left(-\frac{\sqrt{3}}{2},-\frac{1}{2}\right)$

49. (a) $\overline{t}=4\pi-\frac{11\pi}{3}=\frac{\pi}{3}$ **(b)** $P\left(\frac{1}{2},\frac{\sqrt{3}}{2}\right)$

51. (a) $\overline{t}=\frac{16\pi}{3}-5\pi=\frac{\pi}{3}$ **(b)** $P\left(-\frac{1}{2},-\frac{\sqrt{3}}{2}\right)$

53. $t=1\ \Rightarrow\ \left(0.5,0.8\right)$

55. $t=-1.1\ \Rightarrow\ \left(0.5,-0.9\right)$

57. The distances PQ and PR are equal because they both subtend arcs of length $\frac{\pi}{3}$. Since $P(x,y)$ is a point on the unit circle, $x^2 + y^2 = 1$. Now $d(P,Q) = \sqrt{(x-x)^2 + (y-(-y))^2} = 2y$ and $d(R,S) = \sqrt{(x-0)^2 + (y-1)^2} = \sqrt{x^2 + y^2 - 2y + 1} = \sqrt{2-2y}$ (using the fact that $x^2 + y^2 = 1$). Setting these equal gives $2y = \sqrt{2-2y} \Rightarrow 4y^2 = 2 - 2y \Leftrightarrow 4y^2 + 2y - 2 = 0 \Leftrightarrow 2(2y-1)(y+1) = 0$. So $y = -1$ or $y = \frac{1}{2}$. Since P is in quadrant I, $y = \frac{1}{2}$ is the only viable solution. Again using $x^2 + y^2 = 1$ we have $x^2 + \left(\frac{1}{2}\right)^2 = 1 \Leftrightarrow x^2 = \frac{3}{4} \Rightarrow x = \pm\frac{\sqrt{3}}{2}$. Again, since P is in quadrant I the coordinates must be $\left(\frac{\sqrt{3}}{2}, \frac{1}{2}\right)$.

2.2 Trigonometric Functions of Real Numbers

1. If $P(x,y)$ is the terminal point on the unit circle determined by t, then $\sin t = y$, $\cos t = x$, and $\tan t = y / x$.

3.

t	$\sin t$	$\cos t$
0	0	1
$\frac{\pi}{4}$	$\frac{\sqrt{2}}{2}$	$\frac{\sqrt{2}}{2}$
$\frac{\pi}{2}$	1	0
$\frac{3\pi}{4}$	$\frac{\sqrt{2}}{2}$	$-\frac{\sqrt{2}}{2}$
π	0	-1
$\frac{5\pi}{4}$	$-\frac{\sqrt{2}}{2}$	$-\frac{\sqrt{2}}{2}$
$\frac{3\pi}{2}$	-1	0
$\frac{7\pi}{4}$	$-\frac{\sqrt{2}}{2}$	$\frac{\sqrt{2}}{2}$
2π	0	1

5.(a) $\sin\frac{2\pi}{3} = \frac{\sqrt{3}}{2}$

(b) $\cos\frac{2\pi}{3} = -\frac{1}{2}$

(c) $\tan\frac{2\pi}{3} = -\sqrt{3}$

7. (a) $\sin\frac{7\pi}{6} = -\frac{1}{2}$

(b) $\sin(-\frac{\pi}{6}) = -\frac{1}{2}$

(c) $\sin\frac{11\pi}{6} = -\frac{1}{2}$

9.(a) $\cos\frac{3\pi}{4} = -\frac{\sqrt{2}}{2}$

(b) $\cos\frac{5\pi}{4} = -\frac{\sqrt{2}}{2}$

(c) $\cos\frac{7\pi}{4} = \frac{\sqrt{2}}{2}$

11. (a) $\sin\frac{7\pi}{3} = \frac{\sqrt{3}}{2}$

(b) $\csc\frac{7\pi}{3} = \frac{2\sqrt{3}}{3}$

(c) $\cot\frac{7\pi}{3} = \frac{\sqrt{3}}{3}$

13. (a) $\sin(-\frac{\pi}{2}) = -1$

(b) $\cos(-\frac{\pi}{2}) = 0$

(c) $\cot(-\frac{\pi}{2}) = 0$

15. (a) $\sec\frac{11\pi}{3} = 2$

(b) $\csc\frac{11\pi}{3} = -\frac{2\sqrt{3}}{3}$

(c) $\sec(-\frac{\pi}{3}) = 2$

17. (a) $\tan\frac{5\pi}{6} = -\frac{\sqrt{3}}{3}$

(b) $\tan\frac{7\pi}{6} = \frac{\sqrt{3}}{3}$

(c) $\tan\frac{11\pi}{6} = -\frac{\sqrt{3}}{3}$

19. (a) $\cos\left(-\frac{\pi}{4}\right) = \frac{\sqrt{2}}{2}$

(b) $\csc\left(-\frac{\pi}{4}\right) = -\sqrt{2}$

(c) $\cot\left(-\frac{\pi}{4}\right) = -1$

21. (a) $\csc\left(-\frac{\pi}{2}\right) = -1$

(b) $\csc\frac{\pi}{2} = 1$

(c) $\csc\frac{3\pi}{2} = -1$

23. (a) $\sin 13\pi = 0$

(b) $\cos 14\pi = 1$

(c) $\tan 15\pi = 0$

25. $t = 0 \Rightarrow \sin t = 0$, $\cos t = 1$, $\tan t = 0$, $\sec t = 1$, $\csc t$ and $\cot t$ are undefined.

27. $t = \pi \Rightarrow \sin t = 0$, $\cos t = -1$, $\tan t = 0$, $\sec t = -1$, $\csc t$ and $\cot t$ are undefined.

29. $\left(\frac{3}{5}\right)^2 + \left(\frac{4}{5}\right)^2 = \frac{9}{25} + \frac{1}{2} = 1$. So $\sin t = \frac{4}{5}$, $\cos t = \frac{3}{5}$, and

$$\tan t = \frac{\frac{4}{5}}{\frac{3}{5}} = \frac{4}{3} .$$

31. $\left(\frac{\sqrt{5}}{4}\right)^2 + \left(-\frac{\sqrt{11}}{4}\right)^2 = \frac{5}{16} + \frac{11}{16} = 1$. So $\sin t = -\frac{\sqrt{11}}{4}$, $\cos t = \frac{\sqrt{5}}{4}$, and

$$\tan t = \frac{-\frac{\sqrt{11}}{4}}{\frac{\sqrt{5}}{4}} = -\frac{\sqrt{11}}{\sqrt{5}} = -\frac{\sqrt{55}}{5} .$$

33. $\left(-\frac{6}{7}\right)^2 + \left(\frac{\sqrt{13}}{7}\right)^2 = \frac{36}{49} + \frac{13}{49} = 1$. So $\sin t = \frac{\sqrt{13}}{7}$, $\cos t = -\frac{6}{7}$, and

$$\tan t = \frac{\frac{\sqrt{13}}{7}}{-\frac{6}{7}} = -\frac{\sqrt{13}}{6} .$$

35. $\left(-\frac{5}{13}\right)^2 + \left(-\frac{12}{13}\right)^2 = \frac{25}{169} + \frac{144}{169} = 1$. So $\sin t = -\frac{12}{13}$, $\cos t = -\frac{5}{13}$, and

$$\tan t \frac{-\frac{12}{13}}{-\frac{5}{13}} = \frac{12}{5} .$$

37. $\left(-\frac{20}{29}\right)^2 + \left(\frac{21}{29}\right)^2 = \frac{400}{841} + \frac{441}{841} = 1$. So $\sin t = \frac{21}{29}$, $\cos t = -\frac{20}{29}$, and

$$\tan t = \frac{\frac{21}{29}}{-\frac{20}{29}} = -\frac{21}{20} .$$

39. (a) 0.8
(b) 0.84147

41. (a) 0.9
(b) 0.93204

43. (a) 1.0
(b) 1.02964

45. (a) -0.6
(b) -0.57482

47. $\sin t \cdot \cos t$. Since $\sin t$ is positive in quadrant II and $\cos t$ is negative in quadrant II, their product is negative.

49.

$$\frac{\tan t \cdot \sin t}{\cot t} = \tan t \cdot \frac{1}{\cot t} \cdot \sin t = \tan t \cdot \tan t \cdot \sin t = \tan^2 t \cdot \sin t$$

. Since $\tan^2 t$ is always positive and $\sin t$ is negative in quadrant III, the expression is negative in quadrant III.

51. Quadrant II

53. Quadrant II

55. $\sin t = \sqrt{1 - \cos^2 t}$

57. $\tan t = \frac{\sin t}{\cos t} = \frac{\sin t}{\sqrt{1 - \sin^2 t}}$

59. $\sec t = -\sqrt{1 + \tan^2 t}$

61. $\tan t = \sqrt{\sec^2 t - 1}$

63. $\tan^2 t = \dfrac{\sin^2 t}{\cos^2 t} = \dfrac{\sin^2 t}{1-\sin^2 t}$

65. $\sin t = \frac{3}{5}$ and t is in quadrant II, so the terminal point determined by t is $P\left(x, \frac{3}{5}\right)$. Since P is on the unit circle $x^2 + \left(\frac{3}{5}\right)^2 = 1$. Solving for x gives $x = \pm\sqrt{1 - \frac{9}{25}} = \pm\sqrt{\frac{1}{2}} = \pm\frac{4}{5}$. Since t is in quadrant III, $x = -\frac{4}{5}$. Thus the terminal point is $P\left(-\frac{4}{5}, \frac{3}{5}\right)$. Thus, $\cos t = -\dfrac{4}{5}$, $\tan t = -\frac{3}{4}$, $\csc t = \frac{5}{3}$, $\sec t = -\frac{5}{4}$, $\cot t = -\frac{4}{3}$.

67. $\sec t = 3$ and t lies in quadrant IV. Thus, $\cos t = \frac{1}{3}$ and the terminal point determined by t is $P\left(\frac{1}{3}, y\right)$. Since P is on the unit circle $\left(\frac{1}{3}\right)^2 + y^2 = 1$. Solving for y gives $y = \pm\sqrt{1 - \frac{1}{9}} = \pm\sqrt{\frac{8}{9}} = \pm\frac{2\sqrt{2}}{3}$. Since t is in quadrant IV, $y = -\frac{2\sqrt{2}}{3}$. Thus the terminal point is $P\left(\frac{1}{3}, -\frac{2\sqrt{2}}{3}\right)$. Therefore, $\sin t = -\frac{2\sqrt{2}}{3}$, $\cos t = \frac{1}{3}$, $\tan t = -2\sqrt{2}$, $\csc t = -\frac{3}{2\sqrt{2}} = -\frac{3\sqrt{2}}{4}$, $\cot t = -\frac{1}{2\sqrt{2}} = -\frac{\sqrt{2}}{4}$.

69. $\tan t = -\frac{3}{4}$ and $\cos t > 0$, so t is in quadrant IV. Since $\sec^2 t = \tan^2 t + 1$ we have $\sec^2 t = \left(-\frac{3}{4}\right)^2 + 1 = \frac{9}{16} + 1 = \frac{25}{16}$. Thus $\sec t = \pm\sqrt{\frac{25}{16}} = \pm\frac{5}{4}$. Since $\cos t > 0$, we have $\cos t = \dfrac{1}{\sec t} = \dfrac{1}{\frac{5}{4}} = \frac{4}{5}$. Let $P\left(\frac{4}{5}, y\right)$. Since $\tan t \cdot \cos t = \sin t$ we have $\sin t = \left(-\frac{3}{4}\right)\left(\frac{4}{5}\right) = -\frac{3}{5}$. Thus, the terminal point determined by t is $P\left(\frac{4}{5}, -\frac{3}{5}\right)$, and so $\sin t = -\frac{3}{5}$, $\cos t = \frac{4}{5}$, $\csc t = -\frac{5}{3}$, $\sec t = \frac{5}{4}$, $\cot t = -\frac{4}{3}$.

71. $\sin t = -\frac{1}{4}$, $\sec t < 0$, so t is in quadrant III. So the terminal point determined by t is $P\left(x, -\frac{1}{4}\right)$. Since P is on the unit circle $x^2 + \left(-\frac{1}{4}\right)^2 = 1$. Solving for x gives $x = \pm\sqrt{1 - \frac{1}{16}} = \pm\sqrt{\frac{15}{16}} = \pm\frac{\sqrt{15}}{4}$. Since t is in quadrant III, $x = -\dfrac{\sqrt{15}}{4}$. Thus, the terminal point determined by t is $P\left(-\frac{\sqrt{15}}{4}, -\frac{1}{4}\right)$, and so $\cos t = -\frac{\sqrt{15}}{4}$, $\tan t = \frac{1}{\sqrt{15}} = \frac{\sqrt{15}}{15}$, $\csc t = -4$, $\sec t = -\frac{4}{\sqrt{15}} = -\frac{4\sqrt{15}}{15}$, $\cot t = \sqrt{15}$.

73. $f(-x) = (-x)^2 \sin(-x) = -x^2 \sin x = -f(x)$, so f is odd.

75. $f(-x) = \sin(-x)\cos(-x) = -\sin x \cos x = -f(x)$, so f is odd.

77. $f(-x) = |-x|\cos(-x) = |x|\cos x = f(x)$, so f is even.

79. $f(-x)=(-x)^3+\cos(-x)=-x^3+\cos\ x$ which is neither $f(x)$ nor $-f(x)$, so f is neither even nor odd.

81.

t	0	0.25	0.50	0.75	1.00	1.25
$y(t)$	4	-2.83	0	2.83	-4	2.83

83. **(a)** $I(0.1)=0.8e^{-0.3}\sin 1\approx 0.499\ \text{A}$

(b) $I(0.5)=0.8e^{-1.5}\sin 5\approx -0.171\ \text{A}$

85. Notice that if $P(t)=(x,y)$, then $P(t+\pi)=(-x,-y)$. Thus,

(a) $\sin(t+\pi)=-y$ and $\sin t=y$. Therefore, $\sin(t+\pi)=-\sin t$.

(b) $\cos(t+\pi)=-x$ and $\cos t=x$. Therefore, $\cos(t+\pi)=-\cos t$.

(c) $\tan(t+\pi)=\dfrac{\sin(t+\pi)}{\cos(t+\pi)}=\dfrac{-y}{-x}=\dfrac{y}{x}=\dfrac{\sin t}{\cos t}=\tan t$.

2.3 Trigonometric Graphs

1. The trigonometric functions $y = \sin x$ and $y = \cos x$ have amplitude 1 and period 2π .

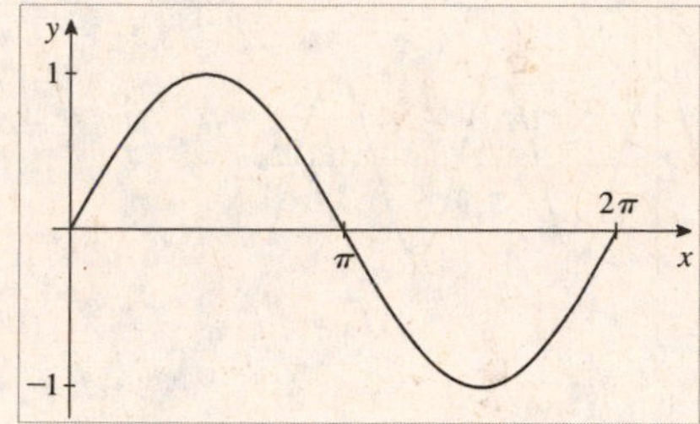

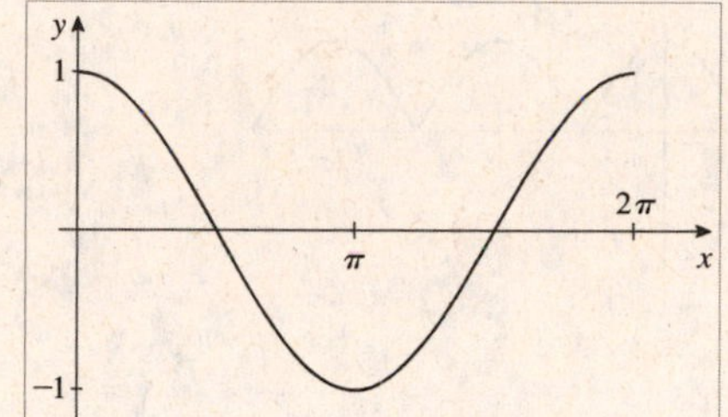

3. $f(x) = 1 + \cos x$

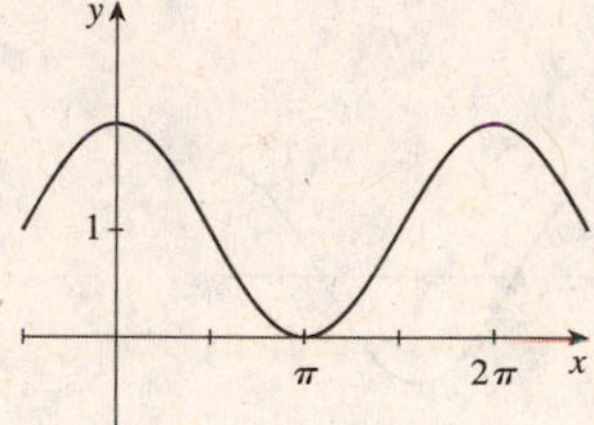

5. $f(x) = -\sin x$

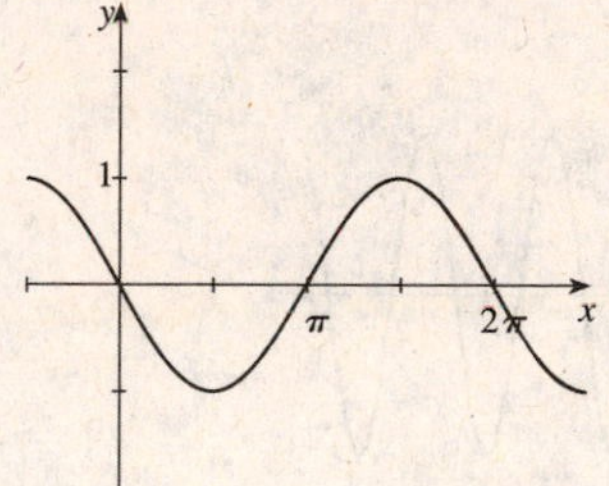

7. $f(x) = -2 + \sin x$

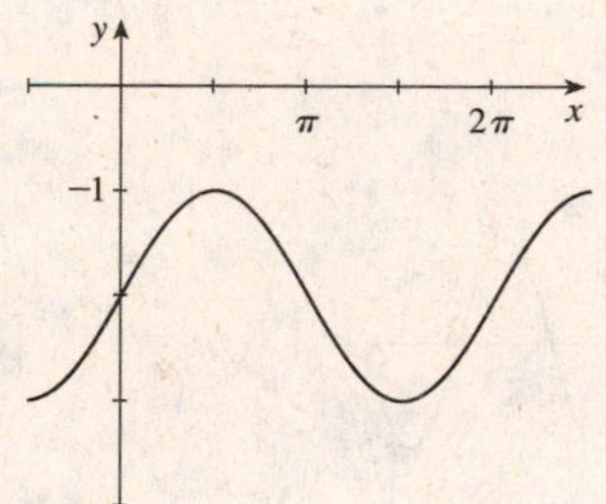

9. $g(x) = 3\cos x$

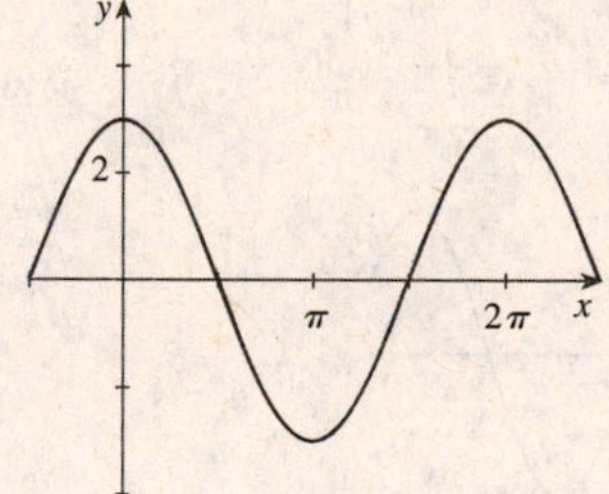

11. $g(x) = -\frac{1}{2}\sin x$

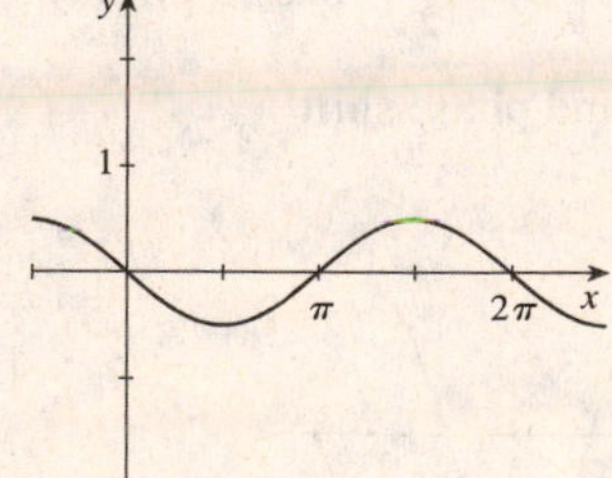

13. $g(x) = 3 + 3\cos x$

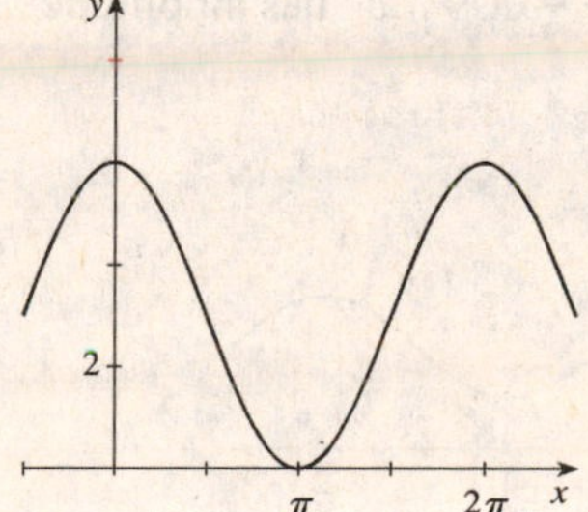

15. $h(x) = |\cos x|$

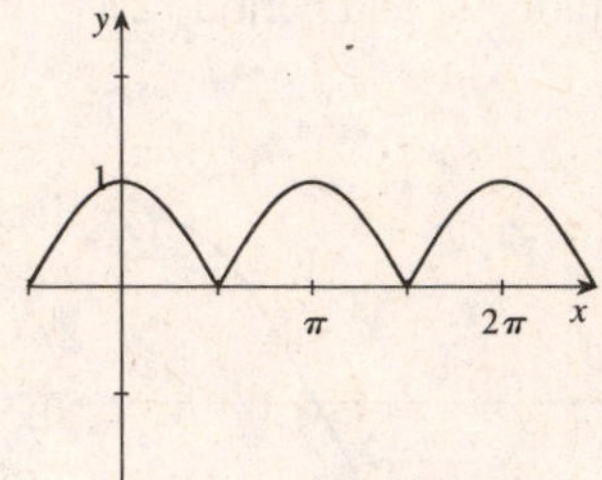

17. $y = \cos 2x$ has amplitude 1 and period π .

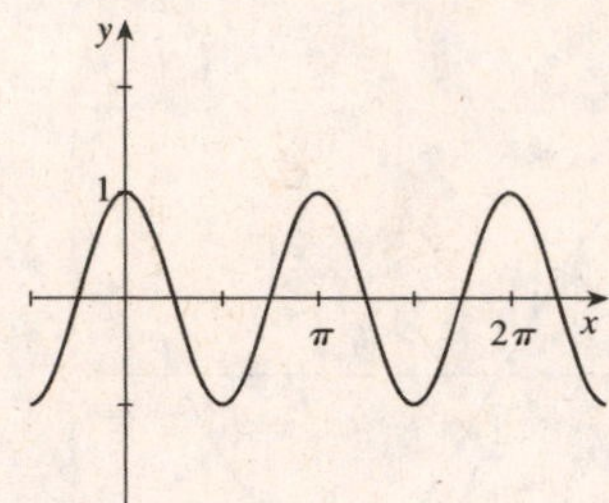

19. $y = -3\sin 3x$ has amplitude 3 and period $\frac{2\pi}{3}$.

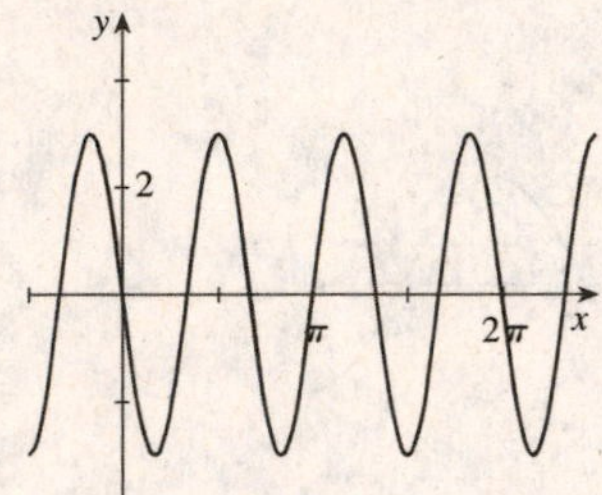

21. $y = 10\sin \frac{1}{2}x$ has amplitude 10 and period 4π .

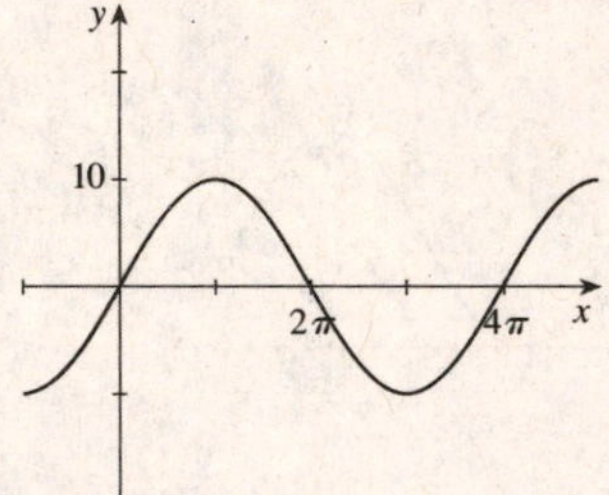

23. $y = -\frac{1}{3}\cos \frac{1}{3}x$ has amplitude $\frac{1}{3}$ and period 6π .

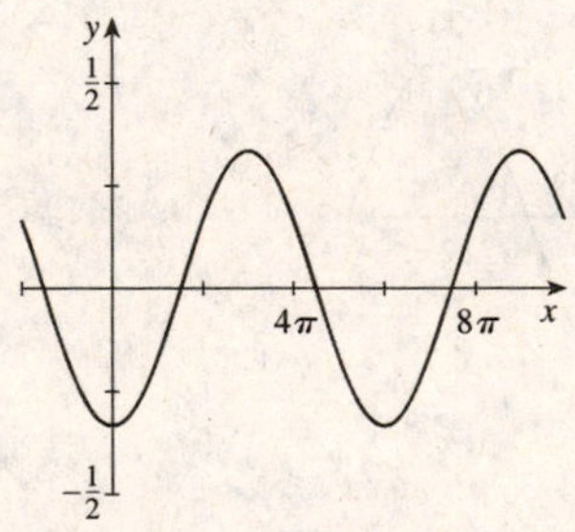

25. $y = -2\sin 2\pi x$ has amplitude 2 and period 1 .

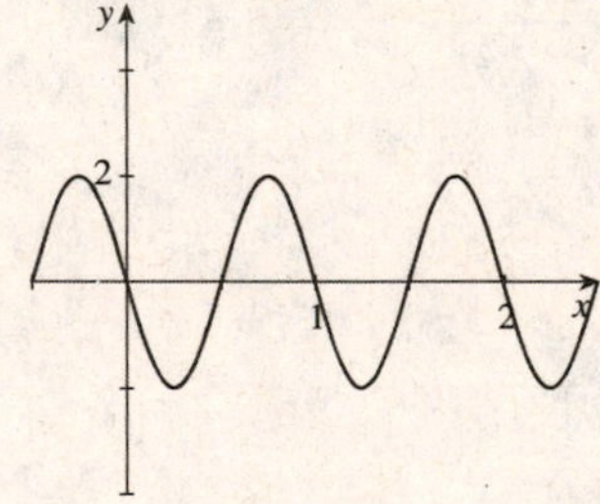

27. $y = 1 + \frac{1}{2}\cos \pi x$ has amplitude $\frac{1}{2}$ and period 2 .

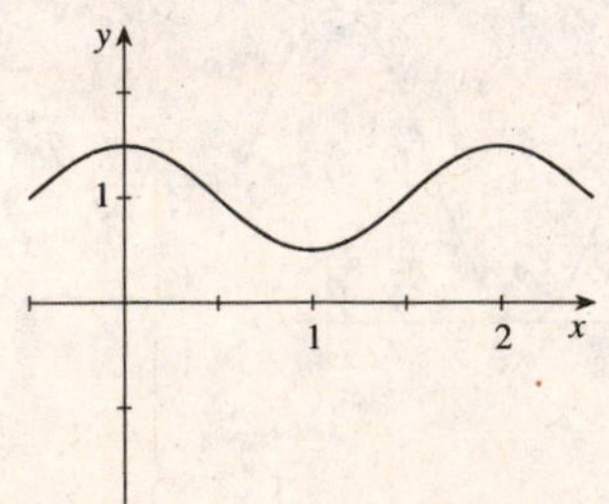

29. $y = \cos\left(x - \frac{\pi}{2}\right)$ has amplitude 1 , period 2π , and phase shift $\frac{\pi}{2}$.

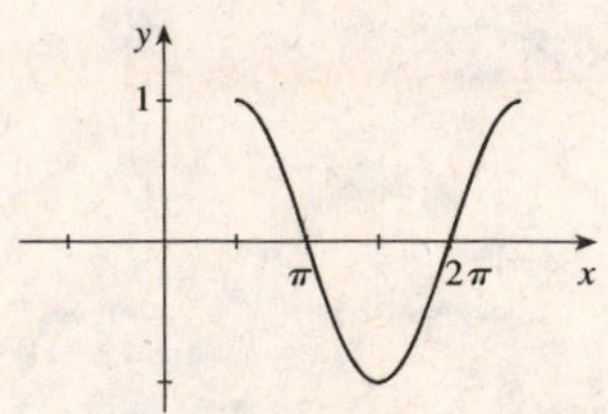

31. $y = -2\sin\left(x - \frac{\pi}{6}\right)$ has amplitude 2, period 2π, and phase shift $\frac{\pi}{6}$.

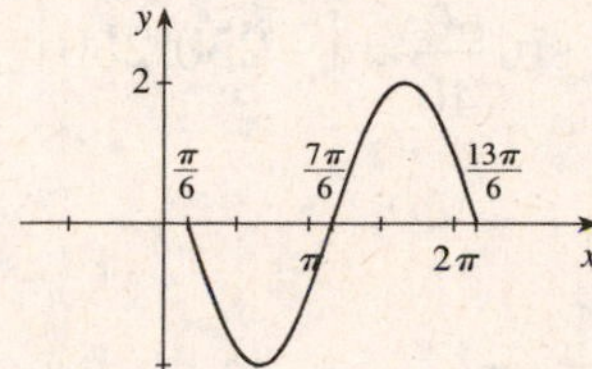

33. $y = -4\sin 2\left(x + \frac{\pi}{2}\right)$ has amplitude 4, period π, and phase shift $-\frac{\pi}{2}$.

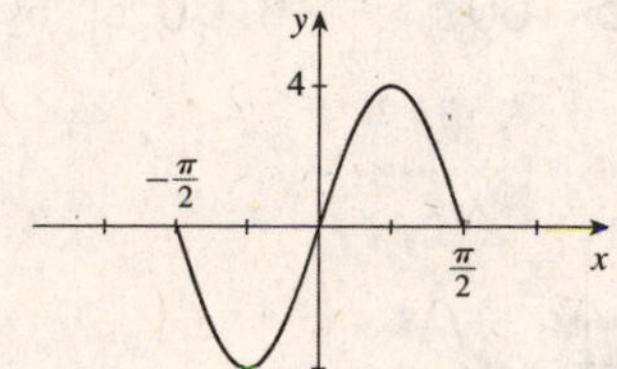

35.
$y = 5\cos\left(3x - \frac{\pi}{4}\right) = 5\cos 3\left(x - \frac{\pi}{12}\right)$ has amplitude 5, period $\frac{2\pi}{3}$, and phase shift $\frac{\pi}{12}$.

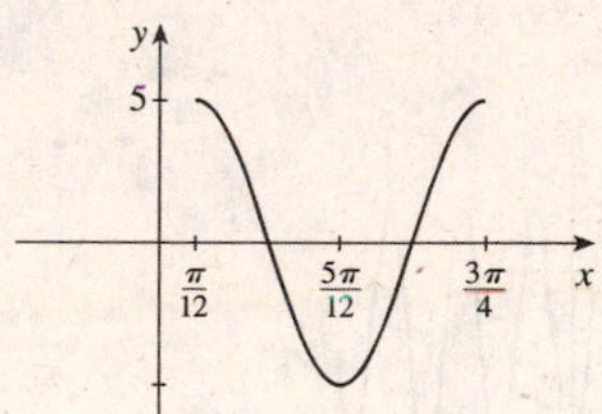

37.
$y = \frac{1}{2} - \frac{1}{2}\cos\left(2x - \frac{\pi}{3}\right) = \frac{1}{2} - \frac{1}{2}\cos 2\left(x - \frac{\pi}{6}\right)$ has amplitude $\frac{1}{2}$, period π, and phase shift $\frac{\pi}{6}$.

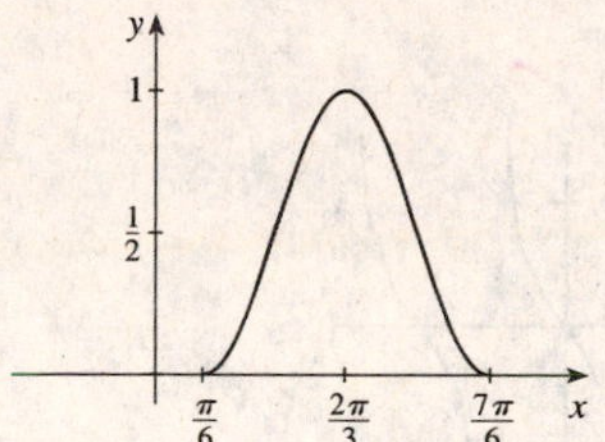

39. $y = 3\cos\pi\left(x + \frac{1}{2}\right)$ has amplitude 3, period 2, and phase shift $-\frac{1}{2}$.

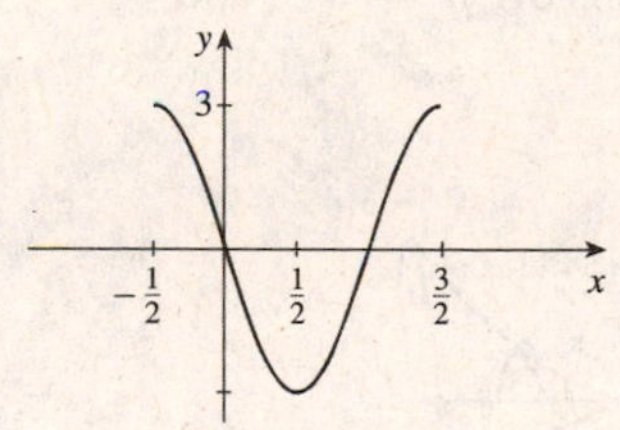

41. $y = \sin\left(3x + \pi\right) = \sin\ 3\left(x + \frac{\pi}{3}\right)$ has amplitude 1, period $\frac{2\pi}{3}$, and phase shift $-\frac{\pi}{3}$

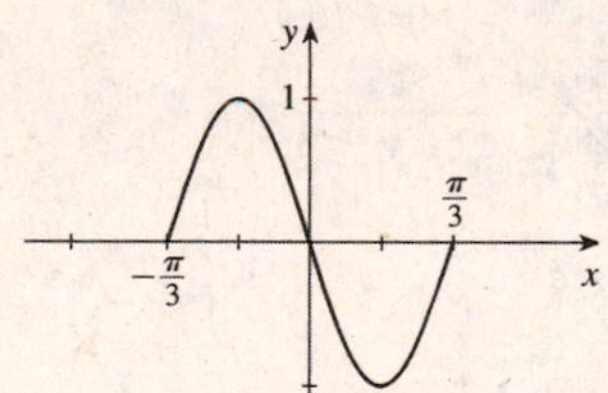

43. (a) This function has amplitude $a = 4$, period $\frac{2\pi}{k} = 2\pi$, and phase shift $b = 0$ as a sine curve.

(b) $y = a\sin k\left(x - b\right) = 4\sin x$

45. (a) This curve has amplitude $a = \frac{3}{2}$, period $\frac{2\pi}{k} = \frac{2\pi}{3}$, and phase shift $b = 0$ as a cosine curve.

(b) $y = a\cos k\left(x - b\right) = \frac{3}{2}\cos 3x$

47. (a) This curve has amplitude $a = \frac{1}{2}$, period $\frac{2\pi}{k} = \pi$, and phase shift $b = -\frac{\pi}{3}$ as a cosine curve.

(b) $y = -\frac{1}{2}\cos 2\left(x + \frac{\pi}{3}\right)$

49. (a) This curve has amplitude $a = 4$, period $\frac{2\pi}{k} = \frac{3}{2}$, and phase shift $b = -\frac{1}{2}$ as a sine curve.

(b) $y = 4\sin\frac{4\pi}{3}\left(x + \frac{1}{2}\right)$

51. $f(x) = \cos 100x$, $[-0.1, 0.1]$ by $[-1.5, 1.5]$

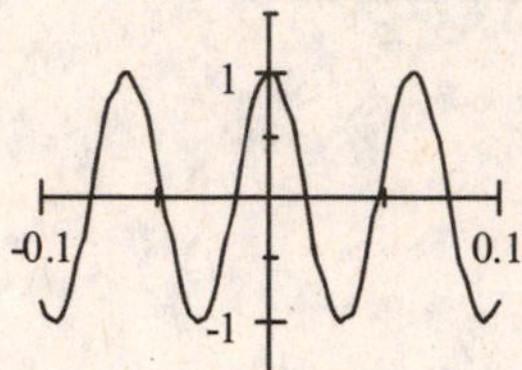

53. $f(x) = \sin\dfrac{x}{40}$, $[-250, 250]$ by $[-1.5, 1.5]$

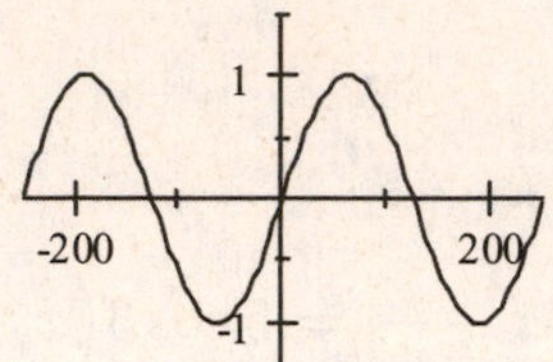

55. $y = \tan\ 25x$, $[-0.2, 0.2]$ by $[-3, 3]$

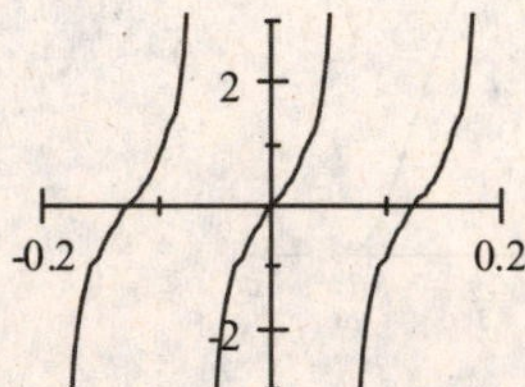

57. $y = \sin^2 20x$, $[-0.5, 0.5]$ by $[-0.2, 1.2]$

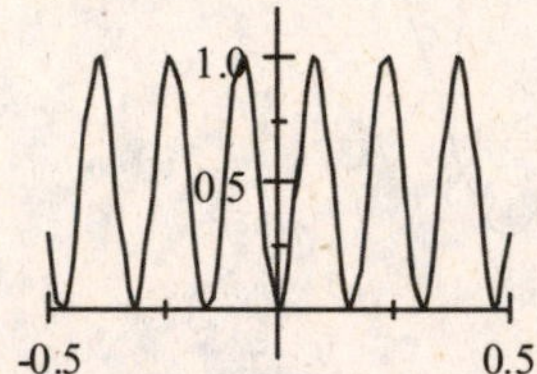

59. $f(x) = x$, $g(x) = \sin x$

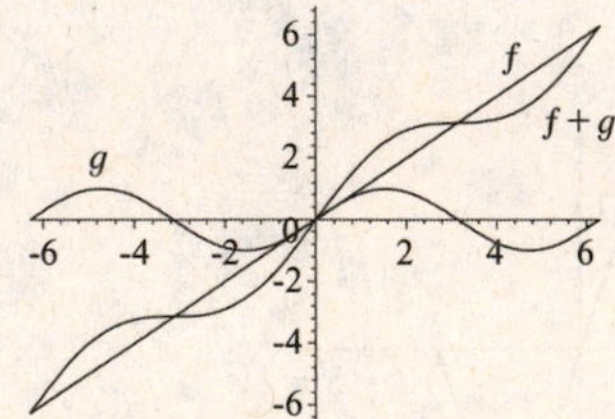

61. $y = x^2 \sin x$ is a sine curve that lies between the graphs of $y = x^2$ and $y = -x^2$.

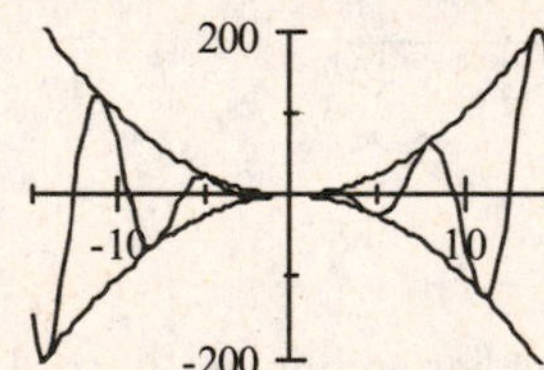

63. $y = \sqrt{x} \sin 5\pi x$ is a sine curve that lies between the graphs of $y = \sqrt{x}$ and $y = -\sqrt{x}$.

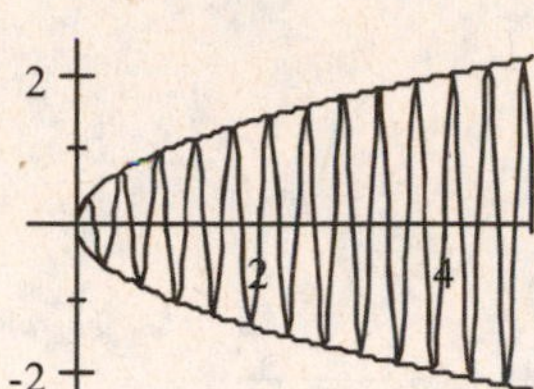

65. $y = \cos 3\pi x \cos 21\pi x$ is a cosine curve that lies between the graphs of $y = \cos 3\pi x$ and $y = -\cos 3\pi x$.

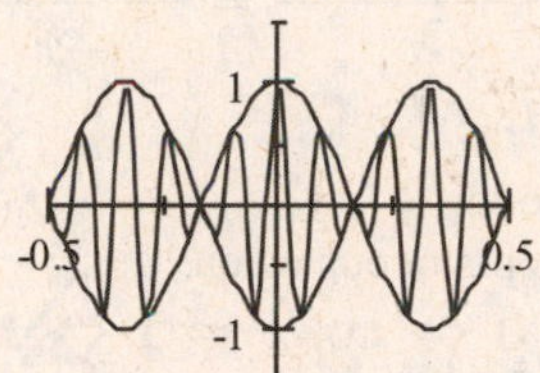

67. $y = \sin x + \sin 2x$. The period is 2π, so we graph the function over one period, $(-\pi, \pi)$. Maximum value 1.76 when $x \approx 0.94 + 2n\pi$, minimum value -1.76 when $x \approx -0.94 + 2n\pi$, n any integer.

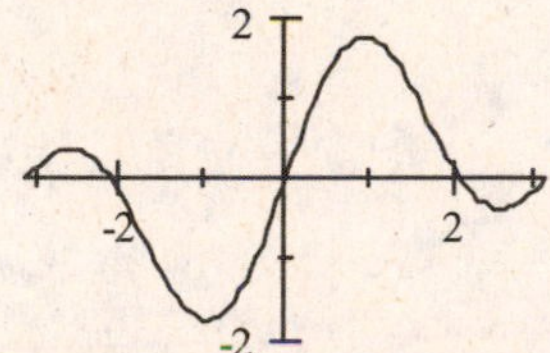

69. $y = 2\sin x + \sin^2 x$. The period is 2π, so we graph the function over one period, $(-\pi, \pi)$. Maximum value 3.00 when $x \approx 1.57 + 2n\pi$, minimum value -1.00 when $x \approx -1.57 + 2n\pi$, n any integer.

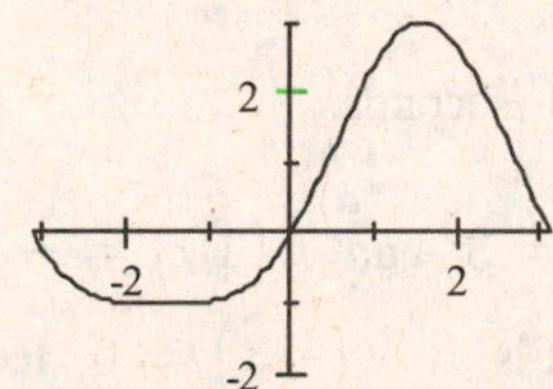

71. $\cos x = 0.4$, $x \in [0, \pi]$. The solution is $x \approx 1.16$.

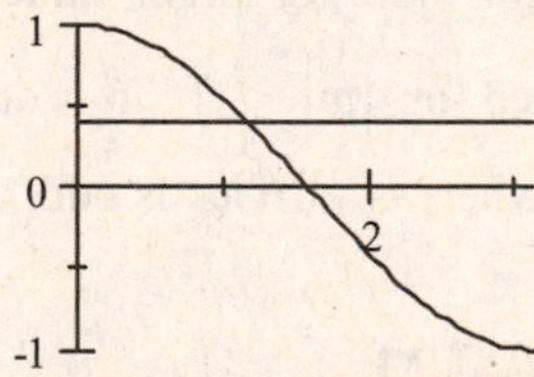

73. $\csc x = 3$, $x \in [0, \pi]$. The solutions are $x \approx 0.34$, 2.80.

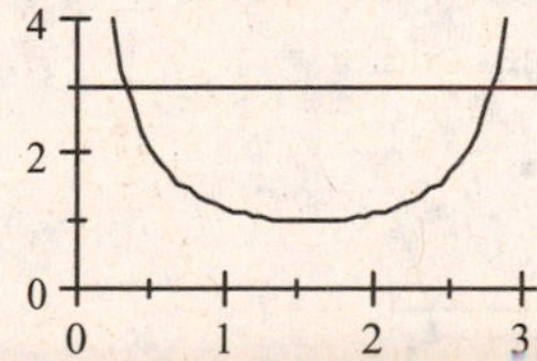

75. $f(x) = \dfrac{1-\cos x}{x}$

(a) Since

$$f(-x) = \frac{1-\cos(-x)}{-x} = \frac{1-\cos x}{-x} = -f(x),$$

the function is odd.

(b) The x -intercepts occur when $1-\cos x = 0 \Leftrightarrow \cos x = 1 \Leftrightarrow x = 0$, $\pm 2\pi$, $\pm 4\pi$, $\pm 6\pi$, ...

(c)

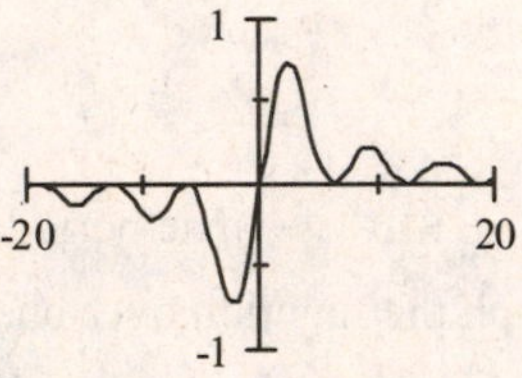

(d) As $x \to \pm\infty$, $f(x) \to 0$.

(e) As $x \to 0$, $f(x) \to 0$.

77. (a) The period of the wave is $\dfrac{2\pi}{\pi / 10} = 20$ seconds.

(b) Since $h(0) = 3$ and $h(10) = -3$, the wave height is $3-(-3) = 6$ feet.

79. (a) The period of p is $\dfrac{2\pi}{160\pi} = \dfrac{1}{80}$ minute.

(b) Since each period represents a heart beat, there are 80 heart beats per minute.

(c)

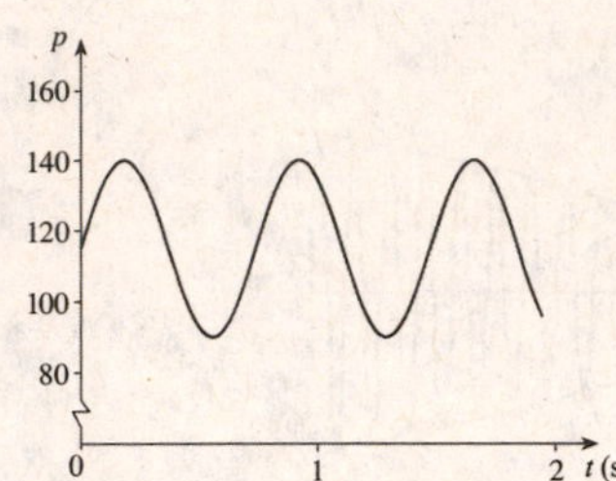

(d) The maximum (or systolic) is $115+25 = 140$ and the minimum (or diastolic) is $115-25 = 90$. The read would be $140 / 90$ which higher than normal.

81. (a) $y = \sin(\sqrt{x})$. This graph looks like a sine function which has been stretched horizontally (stretched more for larger values of x). It is defined only for $x \geq 0$, so it is neither even nor odd.

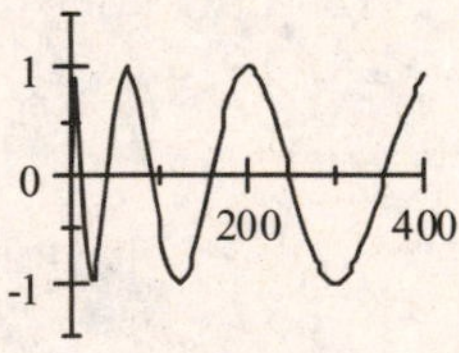

(b) $y = \sin(x^2)$. This graph looks like a graph of $\sin|x|$ which has been shrunk for $|x| > 1$ (shrunk more for larger values of x) and stretched for $|x| < 1$. It is an even function, whereas $\sin x$ is odd.

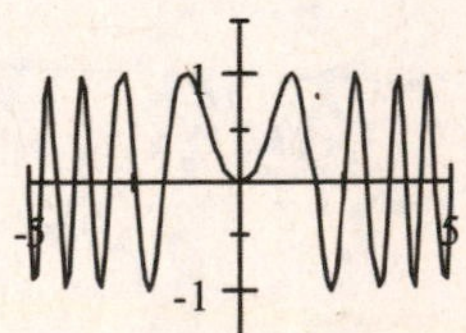

83. (a) The graph of $y = \left|\sin x\right|$ is shown in the viewing rectangle $\left[-6.28, 6.28\right]$ by $\left[-0.5, 1.5\right]$. This function is periodic with period π.

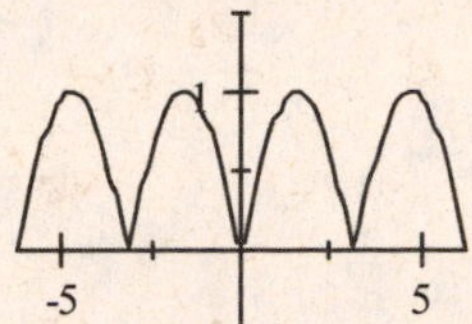

(b) The graph of $y = \sin\left|x\right|$ is shown in the viewing rectangle $\left[-10, 10\right]$ by $\left[-1.5, 1.5\right]$. The function is not periodic. Note that while $\sin\left|x + 2\pi\right| = \sin\left|x\right|$ for many values of x, it is false for $x \in \left(-2\pi, 0\right)$. For example $\sin\left|-\frac{\pi}{2}\right| = \sin\frac{\pi}{2} = 1$ while $\sin\left|-\frac{\pi}{2} + 2\pi\right| = \sin\frac{3\pi}{2} = -1$.

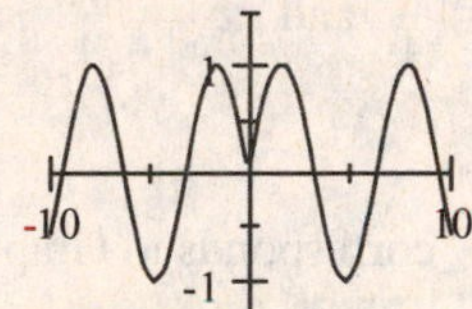

(c) The graph of $y = 2^{\cos x}$ is shown in the viewing rectangle $\left[-10, 10\right]$ by $\left[-1, 3\right]$. This function is periodic with period $= 2\pi$.

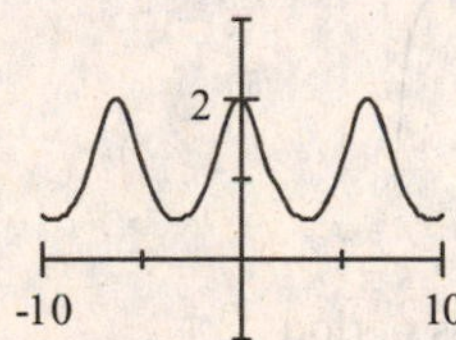

(d) The graph of $y = x - \left[?\left[x\right]?\right]$ is shown in the viewing rectangle $\left[-7.5, 7.5\right]$ by $\left[-0.5, 1.5\right]$. This function is periodic with period 1. Be sure to turn off connected mode when graphing functions with gaps in their graph.

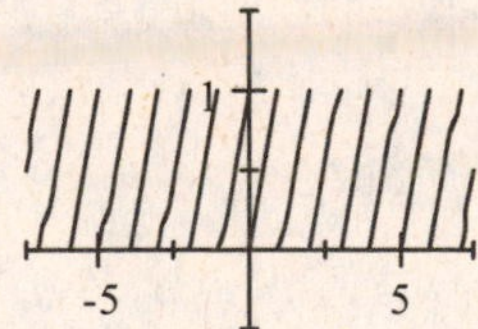

2.4 More Trigonometric Graphs

1. The trigonometric function $y = \tan x$ has period π and asymptotes $x = \frac{\pi}{2} + n\pi$, n an integer.

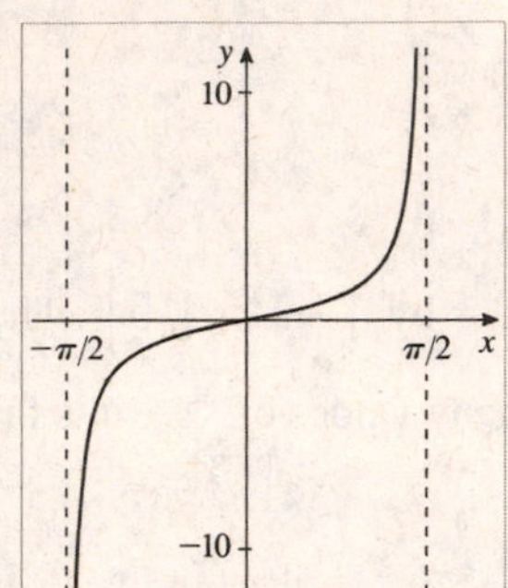

3. $f(x) = \tan\left(x + \frac{\pi}{4}\right)$ corresponds to Graph II. f is undefined at $x = \frac{\pi}{4}$ and $x = \frac{3\pi}{4}$, and Graph II has the shape of a graph of a tangent function.

5. $f(x) = \cot 2x$ corresponds to Graph VI.

7. $f(x) = 2\sec x$ corresponds to Graph IV.

9. $y = 4\tan x$ has period π.

11. $y = -\frac{1}{2}\tan x$ has period π.

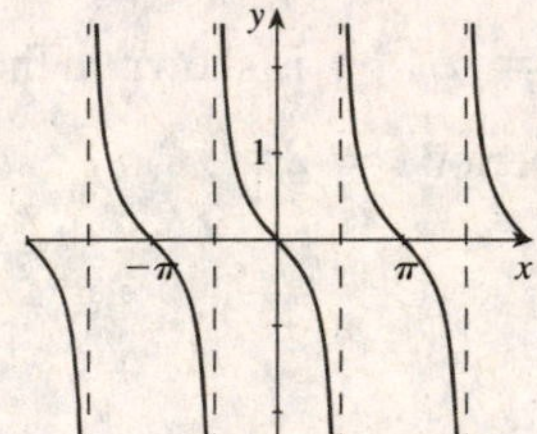

13. $y = -\cot x$ has period π.

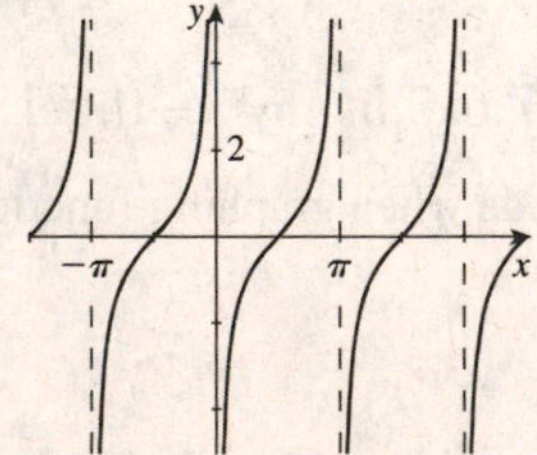

15. $y = 2\csc x$ has period 2π.

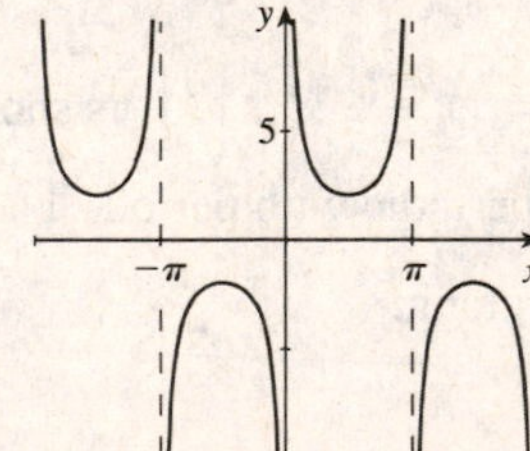

17. $y = 3\sec x$ has period 2π .

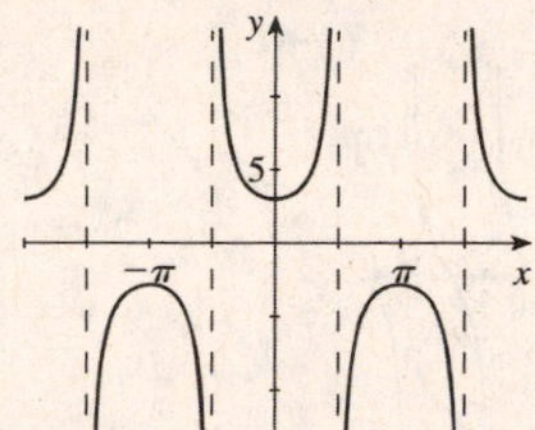

19. $y = \tan\left(x + \frac{\pi}{2}\right)$ has period π .

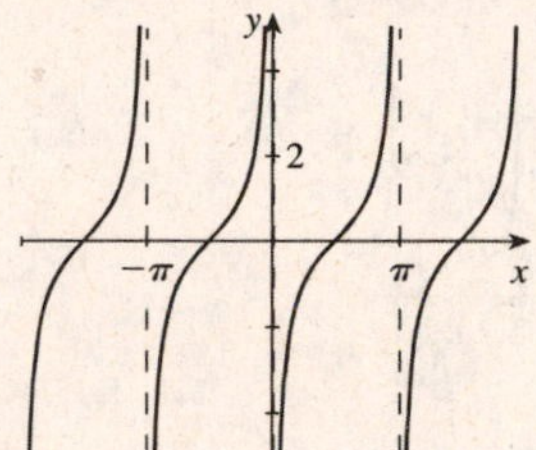

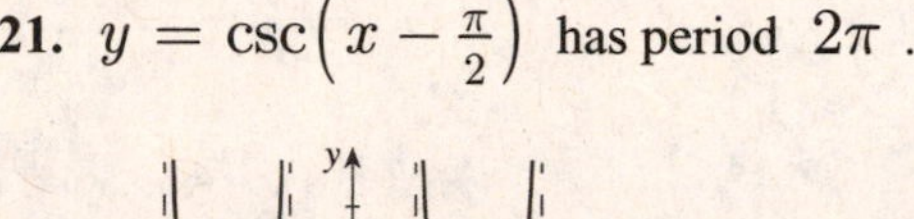

21. $y = \csc\left(x - \frac{\pi}{2}\right)$ has period 2π .

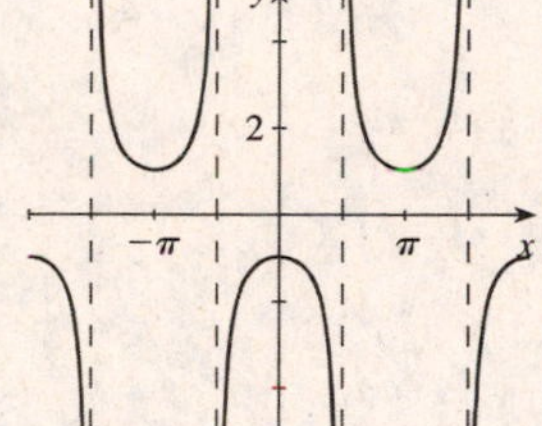

23. $y = \cot\left(x + \frac{\pi}{4}\right)$ has period π .

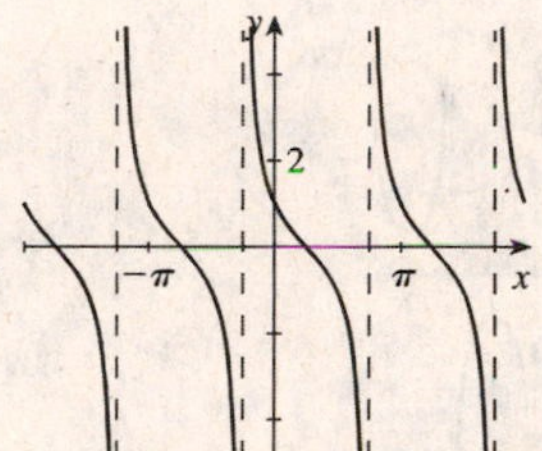

25. $y = \frac{1}{2}\sec\left(x - \frac{\pi}{6}\right)$ has period 2π .

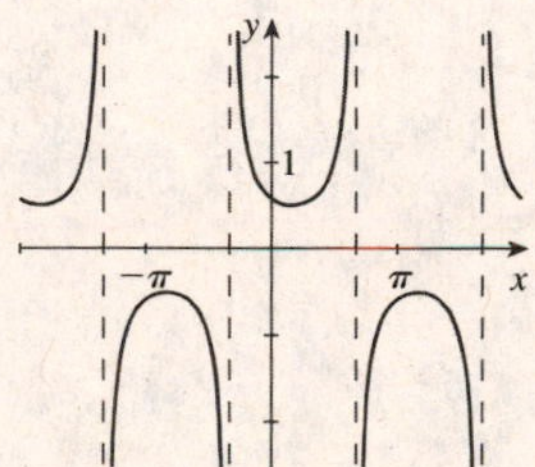

27. $y = \tan 4x$ has period $\frac{\pi}{4}$.

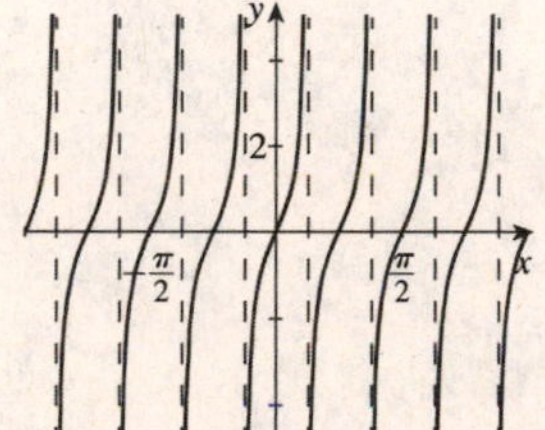

29. $y = \tan\left(\frac{\pi}{4}x\right)$ has period $\dfrac{\pi}{\frac{\pi}{4}} = 4$.

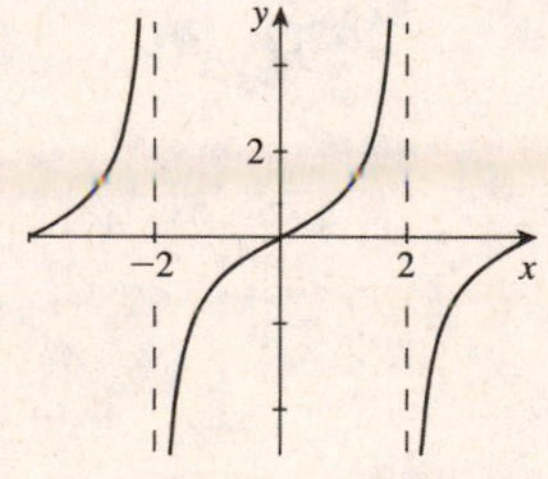

31. $y = \sec 2x$ has period $\dfrac{2\pi}{2} = \pi$.

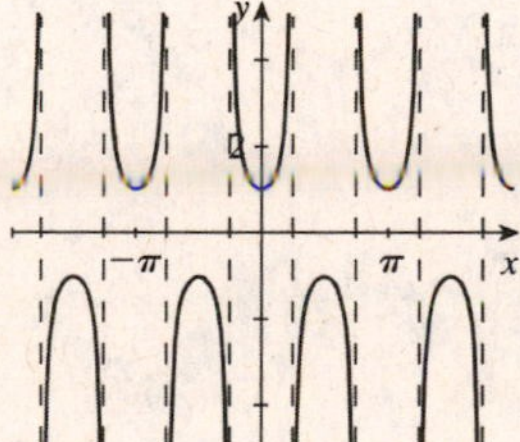

33. $y = \csc 4x$ has period $\dfrac{2\pi}{4} = \dfrac{\pi}{2}$.

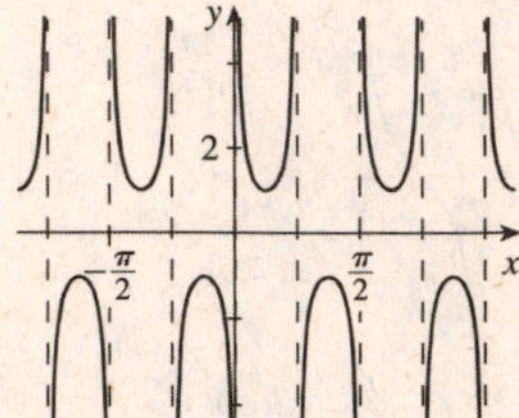

35. $y = 2\tan 3\pi x$ has period $\frac{1}{3}$.

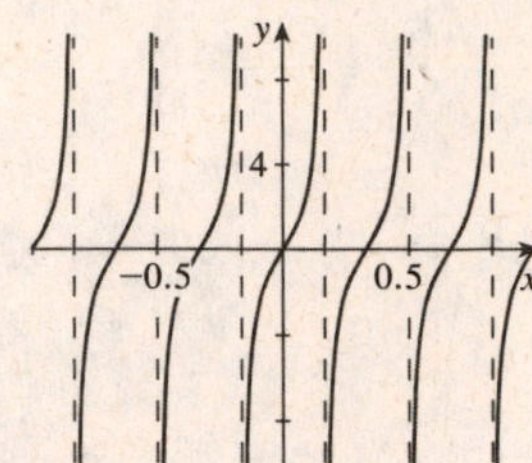

37. $y = 5\csc \frac{3\pi}{2} x$ has period $\dfrac{2\pi}{\frac{3\pi}{2}} = \dfrac{4}{3}$.

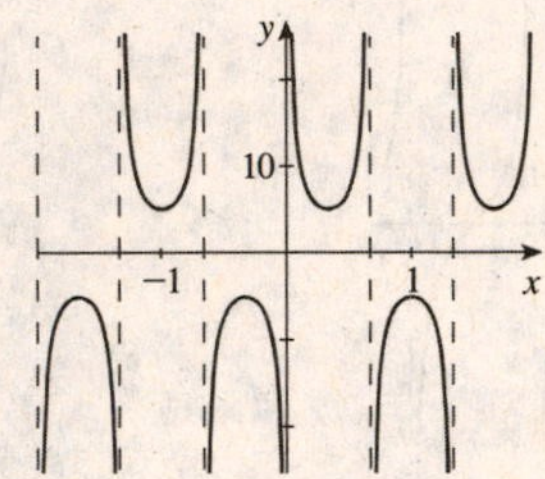

39. $y = \tan 2\left(x + \frac{\pi}{2}\right)$ has period $\frac{\pi}{2}$.

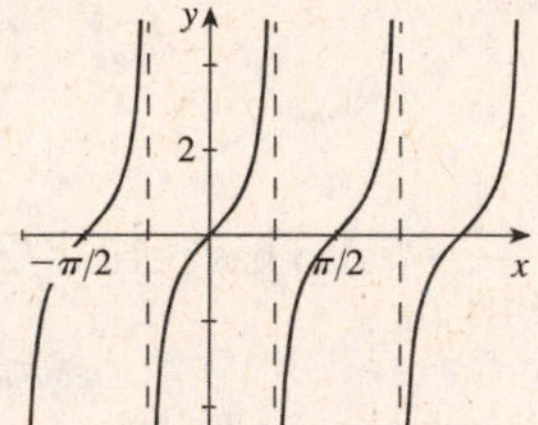

41. $y = \tan 2\left(x - \pi\right) = \tan 2x$ has period $\frac{\pi}{2}$.

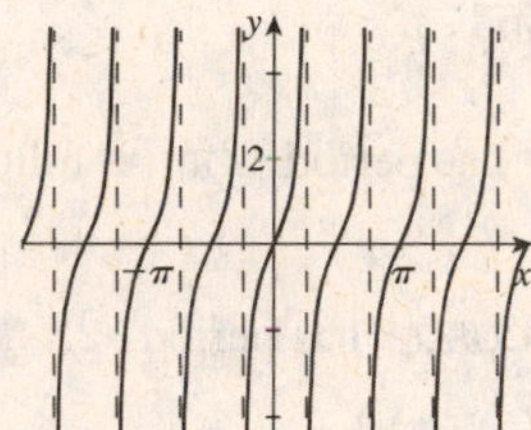

43. $y = \cot\left(2x - \frac{\pi}{2}\right) = \cot 2\left(x - \frac{\pi}{4}\right)$ has period $\frac{\pi}{2}$.

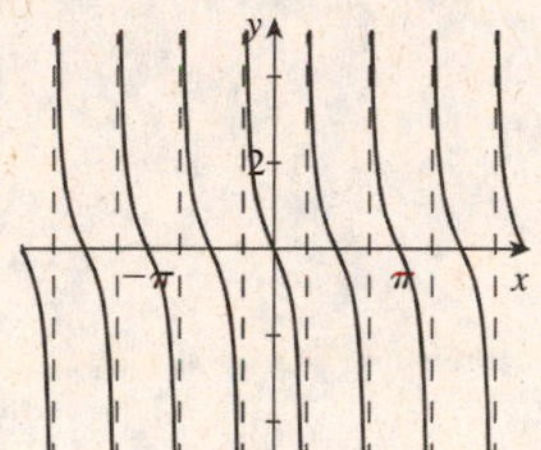

45. $y = 2\csc\left(\pi x - \frac{\pi}{3}\right) = 2\csc \pi\left(x - \frac{1}{3}\right)$ has period $\frac{2\pi}{\pi} = 2$.

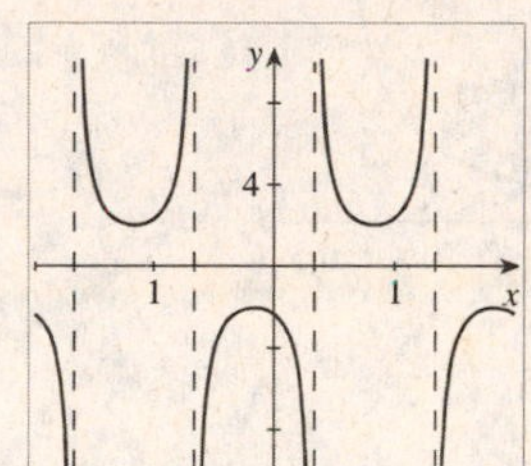

47. $y = 5\sec\left(3x - \frac{\pi}{2}\right) = 5\sec 3\left(x - \frac{\pi}{6}\right)$ has period $\frac{2\pi}{3}$.

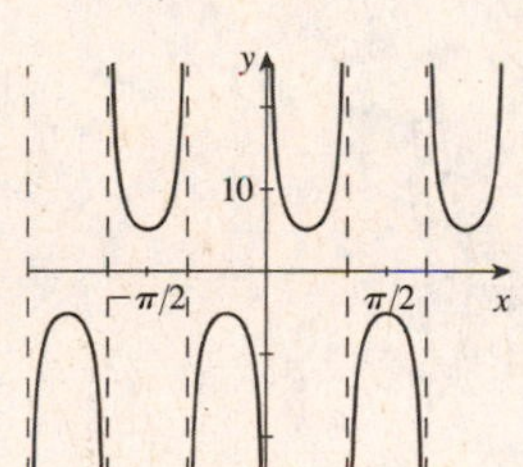

49. $y = \tan\left(\frac{2}{3}x - \frac{\pi}{6}\right) = \tan\frac{2}{3}\left(x - \frac{\pi}{4}\right)$ has period $\pi \,/ \left(\frac{2}{3}\right) = \frac{3\pi}{2}$.

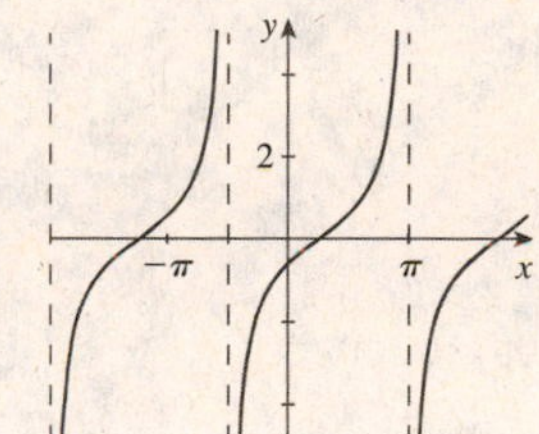

51. $y = 3\sec \pi\left(x + \frac{1}{2}\right)$ has period $\frac{2\pi}{\pi} = 2$.

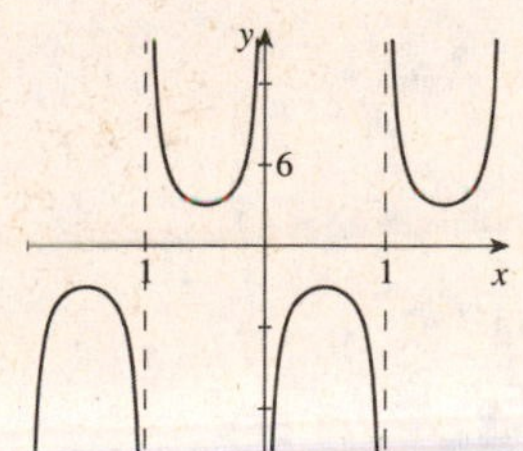

53. $y = -2\tan\left(2x - \frac{\pi}{3}\right) = -2\tan 2\left(x - \frac{\pi}{6}\right)$ has period $\frac{\pi}{2}$.

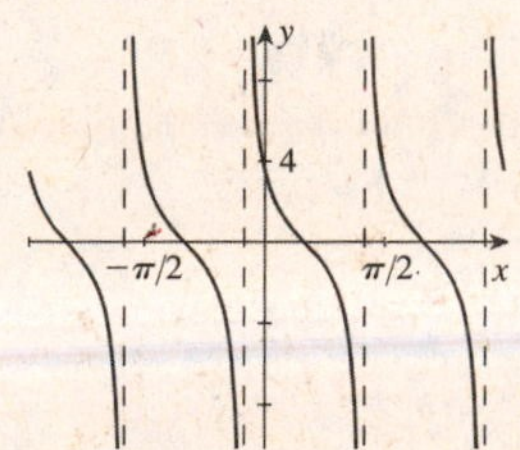

55. (a) If f is periodic with period p, then by the definition of a period, $f(x+p)=f(x)$ for all x in the domain of f. Therefore, $\dfrac{1}{f(x+p)}=\dfrac{1}{f(x)}$ for all $f(x)\neq 0$. Thus, $\dfrac{1}{f}$ is also periodic with period p.

(b) Since $\sin x$ has period 2π, it follows from part (a) that $\csc x=\dfrac{1}{\sin x}$ also has period 2π. Similarly, since $\cos x$ has period 2π, we conclude $\sec x=\dfrac{1}{\cos x}$ also has period 2π.

57. (a) $d(t)=3\tan \pi t$, so $d(0.15)\approx 1.53$, $d(0.25)\approx 3.00$, and $d(0.45)\approx 18.94$.

(b)

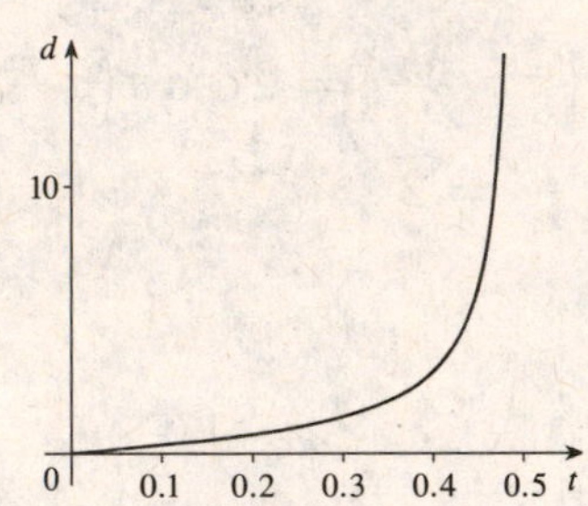

(c) $d\to\infty$ as $t\to\frac{1}{2}$.

59. The graph of $y=-\cot x$ is the same as the graph of $y=\tan x$ shifted $\frac{\pi}{2}$ units to the right, and the graph of $y=\csc x$ is the same as the graph of $y=\sec x$ shifted $\frac{\pi}{2}$ units to the right.

2.5 Inverse Trigonometric Functions and Their Graphs

1. (a) To define the inverse sine function we restrict the domain of sine to the interval $\left[-\frac{\pi}{2},\frac{\pi}{2}\right]$. On this interval the sine function is one-to-one and its inverse function $\sin^{-1}$ is defined by $\sin^{-1}x=y \Leftrightarrow \sin y=x$. For example, $\sin^{-1}\frac{1}{2}=\frac{\pi}{6}$ because $\sin\frac{\pi}{6}=\frac{1}{2}$.

(b) To define the inverse cosine function we restrict the domain of cosine to the interval $\left[0,\pi\right]$. On this interval the cosine function is one-to-one and its inverse function $\cos^{-1}$ is defined by $\cos^{-1}x=y \Leftrightarrow \cos y=x$. For example, $\cos^{-1}\frac{1}{2}=\frac{\pi}{3}$ because $\cos\frac{\pi}{3}=\frac{1}{2}$.

3. (a) $\sin^{-1}1=\frac{\pi}{2}$ because $\sin\frac{\pi}{2}=1$ and $\frac{\pi}{2}$ lies in $\left[-\frac{\pi}{2},\frac{\pi}{2}\right]$.

(b) $\sin^{-1}\frac{\sqrt{3}}{2}=\frac{\pi}{3}$ because $\sin\frac{\pi}{3}=\frac{\sqrt{3}}{2}$ and $\frac{\pi}{3}$ lies in $\left[-\frac{\pi}{2},\frac{\pi}{2}\right]$.

(c) $\sin^{-1}2$ is undefined because there is no real number x such that $\sin x=2$.

5. (a) $\cos^{-1}(-1)=\pi$

(b) $\cos^{-1}\frac{1}{2}=\frac{\pi}{3}$

(c) $\cos^{-1}\left(-\frac{\sqrt{3}}{2}\right)=\frac{5\pi}{6}$

7. (a) $\tan^{-1}(-1)=-\frac{\pi}{4}$

(b) $\tan^{-1}\sqrt{3}=\frac{\pi}{3}$

(c) $\tan^{-1}\frac{\sqrt{3}}{3}=\frac{\pi}{6}$

9. (a) $\cos^{-1}\left(-\frac{1}{2}\right)=\frac{2\pi}{3}$

(b) $\sin^{-1}\left(-\frac{\sqrt{2}}{2}\right)=-\frac{\pi}{4}$

(c) $\tan^{-1}1=\frac{\pi}{4}$

11. $\sin^{-1}\frac{2}{3}=0.72973$

13. $\cos^{-1}\left(-\frac{3}{7}\right)=2.01371$

15. $\cos^{-1}(-0.92761)=2.75876$

17. $\tan^{-1}10=1.47113$

19. $\tan^{-1}(1.23456)=0.88998$

21. $\sin^{-1}(-0.25713)=-0.26005$

23. $\sin\left(\sin^{-1}\frac{1}{4}\right)=\frac{1}{4}$

25. $\tan\left(\tan^{-1}5\right)=5$

27. $\sin\left(\sin^{-1}\left(\frac{3}{2}\right)\right)$ is undefined because $\frac{3}{2}>1$.

29. $\cos^{-1}\left(\cos\frac{5\pi}{6}\right)=\frac{5\pi}{6}$ because $\frac{5\pi}{6}$ lies in $\left[0,2\pi\right]$.

31. $\sin^{-1}\left(\sin\left(-\frac{\pi}{6}\right)\right)=-\frac{\pi}{6}$ because $-\frac{\pi}{6}$ lies in $\left[-\pi,\pi\right]$.

33. $\sin^{-1}\left(\sin\left(\frac{5\pi}{6}\right)\right)=\frac{\pi}{6}$ because $\sin\frac{\pi}{6}=\sin\frac{5\pi}{6}$ and $\frac{\pi}{6}$ lies in $\left[-\frac{\pi}{2},\frac{\pi}{2}\right]$.

35. $\cos^{-1}\left(\cos\left(\frac{17\pi}{6}\right)\right)=\frac{\pi}{6}$ because $\cos\frac{\pi}{6}=\cos\frac{17\pi}{6}$ and $\frac{\pi}{6}$ lies in $\left[0,2\pi\right]$.

37. $\tan^{-1}\left(\tan\frac{2\pi}{3}\right)=-\frac{\pi}{3}$ because $\tan\left(-\frac{\pi}{3}\right)=\tan\frac{2\pi}{3}$ and $-\frac{\pi}{3}$ lies in $\left(-\frac{\pi}{2},\frac{\pi}{2}\right)$.

39. $\tan\left(\sin^{-1}\frac{1}{2}\right)=\tan\frac{\pi}{6}=\frac{\sqrt{3}}{3}$

41. $\cos\left(\sin^{-1}\frac{\sqrt{3}}{2}\right)=\cos\frac{\pi}{3}=\frac{1}{2}$

43.

$\sin\left(\tan^{-1}(-1)\right) = \sin\left(-\frac{\pi}{4}\right) = -\frac{\sqrt{2}}{2}$

45. The domain of $f(x) = \sin\left(\sin^{-1} x\right)$ is the same as that of $\sin^{-1} x$, $\left[-1,1\right]$, and the graph of f is the same as that of $y = x$ on $\left[1,1\right]$.

The domain of $g(x) = \sin^{-1}(\sin x)$ is the same as that of $\sin x$, $\left(-\infty, \infty\right)$, because for all x , the value of $\sin x$ lies within the domain of $\sin^{-1} x$. $g(x) = \sin^{-1}(\sin x) = x$ for $-\frac{\pi}{2} \leq x \leq \frac{\pi}{2}$. Because the graph of $y = \sin x$ is symmetric about the line $x = \frac{\pi}{2}$, we can obtain the part of the graph of g for $\frac{\pi}{2} \leq x \leq \frac{3\pi}{2}$ by reflecting the graph of $y = x$ about this vertical line. The graph of g is periodic with period 2π .

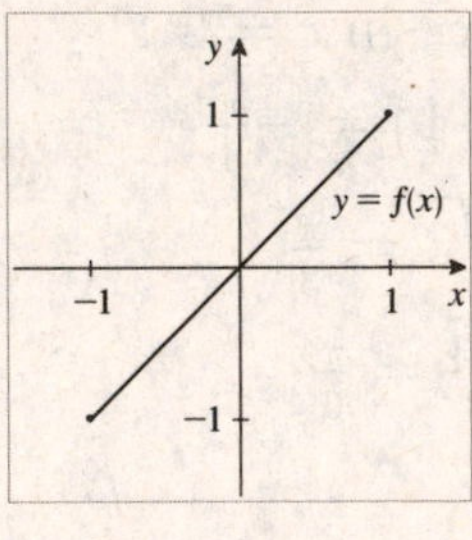

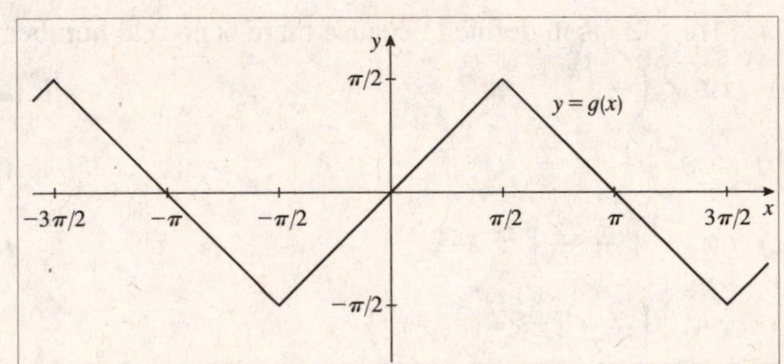

47. (a)

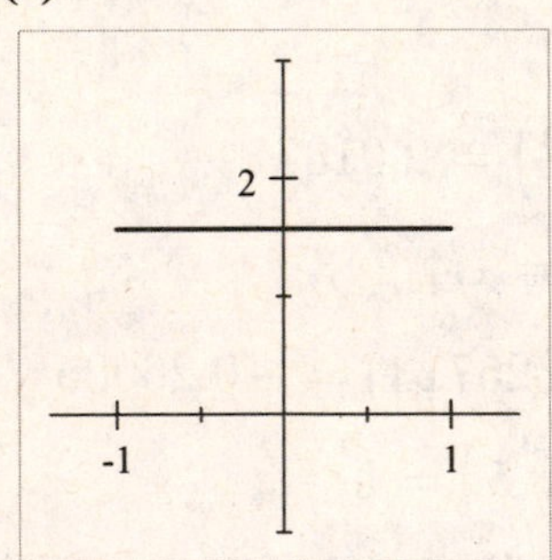

From the graph of $y = \sin^{-1} x + \cos^{-1} x$, it appears that $y \approx 1.57$. We suspect that the actual value is $\frac{\pi}{2}$.

(b) To show that $\sin^{-1} x + \cos^{-1} x = \frac{\pi}{2}$, start with the identity $\sin\left(a - \frac{\pi}{2}\right) = -\cos a$ and take $\arcsin$ of both sides to obtain $a - \frac{\pi}{2} = \sin^{-1}\left(-\cos a\right)$. Now let $a = \cos^{-1} x$. Then $\cos^{-1} x - \frac{\pi}{2} = \sin^{-1}\left(-\cos\left(\cos^{-1} x\right)\right) = \sin^{-1}(-x) = -\sin^{-1} x$, so $\sin^{-1} x + \cos^{-1} x = \frac{\pi}{2}$.

2.6 Modeling Harmonic Motion

1. $a \sin \omega t$

3. $y = 2\sin 3t$

(a) Amplitude 2 , period $\frac{2\pi}{3}$, frequency $\frac{1}{\text{period}} = \frac{3}{2\pi}$.

(b)

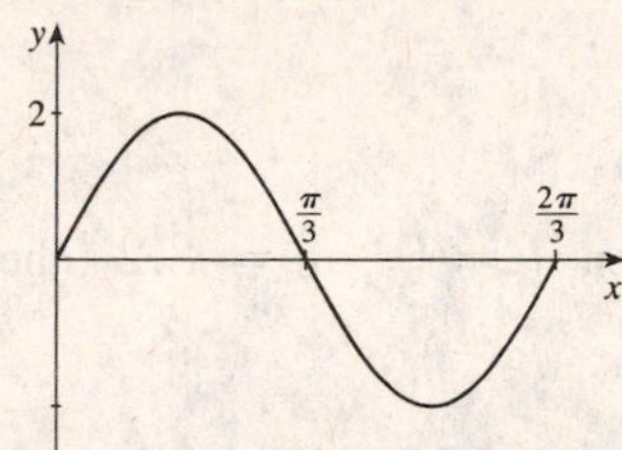

5. $y = -\cos 0.3t$

(a) Amplitude 1 , period $\frac{2\pi}{0.3} = \frac{20\pi}{3}$, frequency $\frac{3}{20\pi}$.

(b)

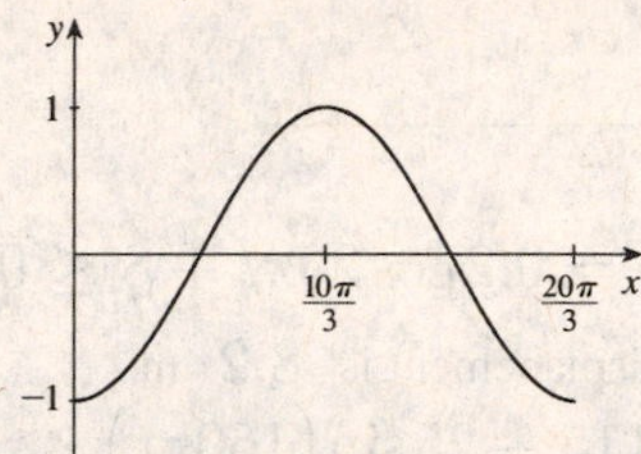

7.

$$y = -0.25\cos\left(1.5t - \frac{\pi}{3}\right) = -0.25\cos\left(\frac{3}{2}t - \frac{\pi}{3}\right) = -0.25\cos\frac{3}{2}\left(t - \frac{2\pi}{9}\right)$$

(a) Amplitude 0.25 , period $\frac{2\pi}{3/2} = \frac{4\pi}{3}$, frequency $\frac{3}{4\pi}$.

(b)

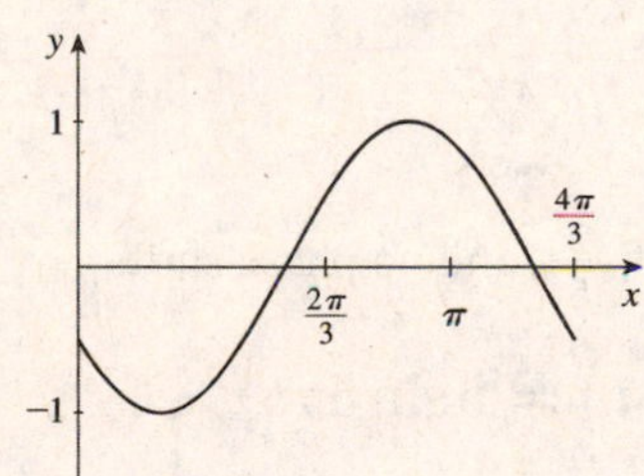

9. $y = 5\cos\left(\frac{2}{3}t + \frac{3}{4}\right) = 5\cos\frac{2}{3}\left(t + \frac{9}{8}\right)$

(a) Amplitude 5 , period $\frac{2\pi}{2/3} = 3\pi$, frequency $\frac{1}{3\pi}$.

(b)

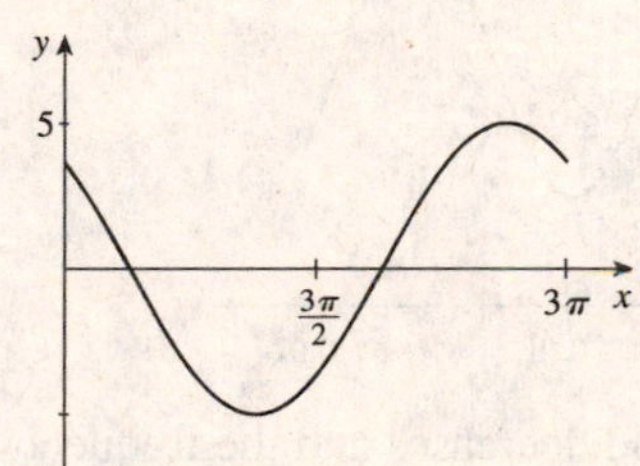

11. The amplitude is $a = 10$ cm, the period is $\frac{2\pi}{k} = 3$ s, and $f(0) = 0$, so $f(x) = 10\sin\frac{2\pi}{3}t$.

13. The amplitude is 6 in., the frequency is $\frac{k}{2\pi} = \frac{5}{\pi}$ Hz, and $f(0) = 0$, so $f(x) = 6\sin 10t$.

15. The amplitude is 60 ft, the period is $\frac{2\pi}{k} = 0.5$ min, and $f(0) = 60$, so $f(x) = 60\cos 4\pi t$.

17. The amplitude is 2.4 m, the frequency is $\frac{k}{2\pi} = 750$ Hz, and $f(0) = 2.4$, so $f(x) = 2.4\cos 1500\pi t$.

19. $y = 0.2 \cos 20\pi t + 8$

(a) The frequency is $\frac{20\pi}{2\pi} = 10$ cycles/min.

(b)

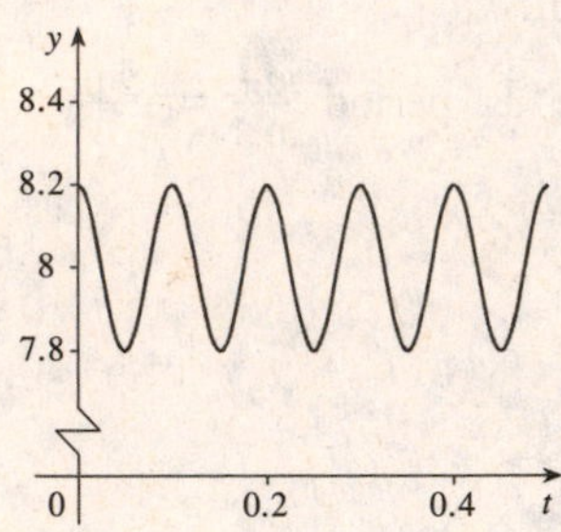

(c) Since $y = 0.2 \cos 20\pi t + 8 \le 0.2(1) + 8 = 8.2$ and when $t = 0$, $y = 8.2$, the maximum displacement is 8.2 m.

21. $p(t) = 115 + 25 \sin(160\pi t)$

(a) Amplitude 25, period $\frac{2\pi}{160\pi} = \frac{1}{80} = 0.0125$, frequency $\dfrac{1}{\text{period}} = 80$.

(b)

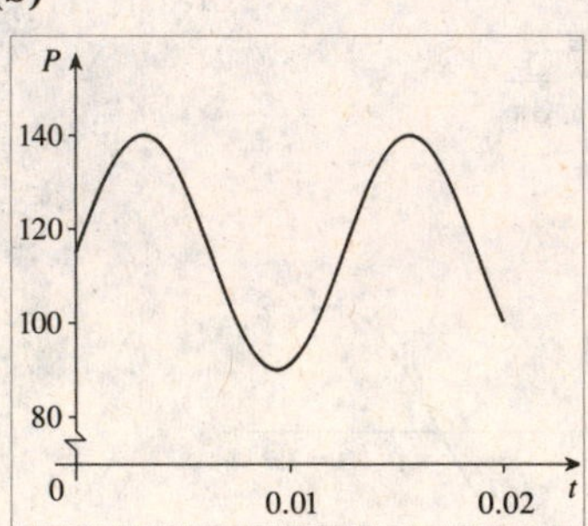

(c) The period decreases and the frequency increases.

23. The graph resembles a sine wave with an amplitude of 5, a period of $\frac{2}{5}$, and no phase shift. Therefore, $a = 5$, $\dfrac{2\pi}{\omega} = \frac{2}{5} \Leftrightarrow \omega = 5\pi$, and a formula is $d(t) = 5 \sin 5\pi t$.

25. $a = 21$, $f = \frac{1}{12}$ cycle/hour $\Rightarrow$ $\dfrac{\omega}{2\pi} = \frac{1}{12} \Leftrightarrow \omega = \frac{\pi}{6}$. So, $y = 21 \sin\left(\frac{\pi}{6} t\right)$ (assuming the tide is at mean level and rising when $t = 0$).

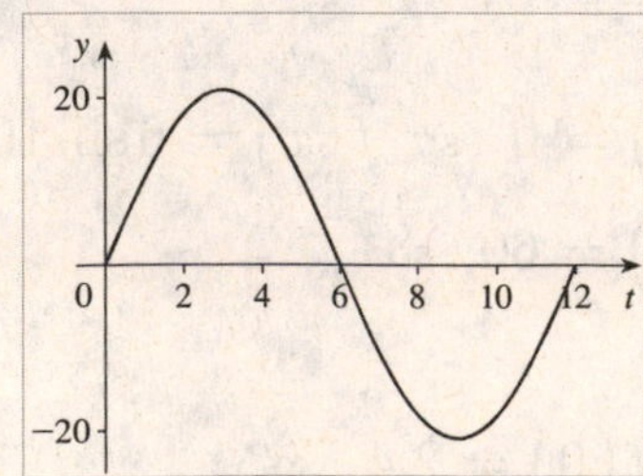

27. Since the mass travels from its highest point (compressed spring) to its lowest point in $\frac{1}{2}$ s, it completes half a period in $\frac{1}{2}$ s. So, $\frac{1}{2}\left(\text{one period}\right) = \frac{1}{2}$ s $\Rightarrow \frac{1}{2} \cdot \frac{2\pi}{\omega} = \frac{1}{2} \Leftrightarrow \omega = 2\pi$. Also, $a = 5$. So $y = 5\cos 2\pi t$.

29. Since the Ferris wheel has a radius of 10 m and the bottom of the wheel is 1 m above the ground, the minimum height is 1 m and the maximum height is 21 m. Then $a = 10$ and $\frac{2\pi}{\omega} = 20$ s $\Leftrightarrow \quad \omega = \frac{\pi}{10}$, and so $y = 11 + 10\sin\left(\frac{\pi}{10}t\right)$, where t is in seconds.

31. $a = 0.2$, $\frac{2\pi}{\omega} = 10 \Leftrightarrow \omega = \frac{\pi}{5}$. Then $y = 3.8 + 0.2\sin\left(\frac{\pi}{5}t\right)$.

33. The amplitude is $\frac{1}{2}\left(100 - 80\right) = 10$ mmHG, the period is 24 hours, and the phase shift is 8 hours, so $f\left(t\right) = 10\sin\left(\frac{\pi}{12}\left(t - 8\right)\right) + 90$.

35. (a) The maximum voltage is the amplitude, that is, $V_{\text{max}} = a = 45$ V.

(b) From the graph we see that 4 cycles are completed every 0.1 seconds, or equivalently, 40 cycles are completed every second, so $f = 40$.

(c) The number of revolutions per second of the armature is the frequency, that is, $\frac{\omega}{2\pi} = f = 40$.

(d) $a = 45$, $f = \frac{\omega}{2\pi} = 40 \Leftrightarrow \omega = 80\pi$. Then $V\left(t\right) = 45\cos 80\pi t$.

36. (a) As the car approaches, the perceived frequency is $f = 500\left(\frac{1130}{1130-110}\right) \approx 553.9$ Hz. As it moves away, the perceived frequency is $f = 500\left(\frac{1130}{1130+110}\right) \approx 455.6$ Hz.

(b) The frequency is $\frac{\omega}{2\pi} = 500\left(\frac{1130}{1130\pm110}\right)$, so $\omega = 1000\pi\left(\frac{1130}{1130\pm110}\right) \approx 1107.8\pi$ (approaching) or 911.3π (receding). Thus, models are $y = A\sin 1107.8\pi t$ and $A\sin 911.3\pi t$.

Chapter 2 Review

1. (a) Since $\left(-\frac{\sqrt{3}}{2}\right)^2+\left(\frac{1}{2}\right)^2=\frac{3}{4}+\frac{1}{4}=1$, the point $P\left(-\frac{\sqrt{3}}{2},\frac{1}{2}\right)$ lies on the unit circle.

(b) $\sin t=\frac{1}{2}$, $\cos t=-\frac{\sqrt{3}}{2}$,

$$\tan t=\frac{\frac{1}{2}}{-\frac{\sqrt{3}}{2}}=-\frac{\sqrt{3}}{3}.$$

3. $t=\frac{2\pi}{3}$

(a) $\overline{t}=\pi-\frac{2\pi}{3}=\frac{\pi}{3}$

(b) $P\left(-\frac{1}{2},\frac{\sqrt{3}}{2}\right)$

(c) $\sin t=\frac{\sqrt{3}}{2}$, $\cos t=-\frac{1}{2}$, $\tan t=-\sqrt{3}$, $\csc t=\frac{2\sqrt{3}}{3}$, $\sec t=-2$, and $\cot t=-\frac{\sqrt{3}}{3}$.

5. $t=-\frac{11\pi}{4}$

(a) $\overline{t}=3\pi+\left(-\frac{11\pi}{4}\right)=\frac{\pi}{4}$

(b) $P\left(-\frac{\sqrt{2}}{2},-\frac{\sqrt{2}}{2}\right)$

(c) $\sin t=-\frac{\sqrt{2}}{2}$, $\cos t=-\frac{\sqrt{2}}{2}$, $\tan t=1$, $\csc t=-\sqrt{2}$, $\sec t=-\sqrt{2}$, and $\cot t=1$.

7. (a) $\sin\frac{3\pi}{4}=\sin\frac{\pi}{4}=\frac{\sqrt{2}}{2}$

(b) $\cos\frac{3\pi}{4}=-\cos\frac{\pi}{4}=-\frac{\sqrt{2}}{2}$

9. (a) $\sin 1.1\approx 0.89121$

(b) $\cos 1.1\approx 0.45360$

11. (a) $\cos\frac{9\pi}{2}=\cos\frac{\pi}{2}=0$

(b) $\sec\frac{9\pi}{2}$ is undefined

13. (a) $\tan\frac{5\pi}{2}$ is undefined

(b) $\cot\frac{5\pi}{2}=\cot\frac{\pi}{2}=0$

15. (a) $\tan\frac{5\pi}{6}=-\frac{\sqrt{3}}{3}$

(b) $\cot\frac{5\pi}{6}=-\sqrt{3}$

17. $$\frac{\tan t}{\cos t}=\frac{\frac{\sin t}{\cos t}}{\cos t}=\frac{\sin t}{\cos^2 t}=\frac{\sin t}{1-\sin^2 t}$$

19. $\tan t=\frac{\sin t}{\cos t}=\frac{\sin t}{\pm\sqrt{1-\sin^2 t}}=\frac{\sin t}{\sqrt{1-\sin^2 t}}$ (because t is in quadrant IV, $\cos t$ is positive).

21. $\sin t=\frac{5}{13}$, $\cos t=-\frac{12}{13}$. Then $\tan t=\frac{\frac{5}{13}}{-\frac{12}{13}}=-\frac{5}{12}$, $\csc t=\frac{13}{5}$, $\sec t=-\frac{13}{12}$, and $\cot t=-\frac{12}{5}$.

23. $\cot t = -\frac{1}{2}$, $\csc t = \frac{\sqrt{5}}{2}$. Since $\csc t = \dfrac{1}{\sin t}$, we know $\sin t = \frac{2}{\sqrt{5}} = \frac{2\sqrt{5}}{5}$. Now

$\cot t = \dfrac{\cos t}{\sin t}$, so $\cos t = \sin t \cdot \cot t = \frac{2\sqrt{5}}{5} \cdot \left(-\frac{1}{2}\right) = -\frac{\sqrt{5}}{5}$, and

$\tan t = \dfrac{1}{\left(-\frac{1}{2}\right)} = -2$ while $\sec t = \dfrac{1}{\cos t} = \dfrac{1}{\left(-\frac{\sqrt{5}}{5}\right)} = -\frac{5}{\sqrt{5}} = -\sqrt{5}$.

25. $\tan t = \frac{1}{4}$, t is in quadrant III $\Rightarrow$

$$\sec t + \cot t = -\sqrt{\tan^2 t + 1} + \frac{1}{\tan t} = -\sqrt{\left(\tfrac{1}{4}\right)^2 + 1} + 4 = -\sqrt{\tfrac{17}{16}} + 4 = 4 - \tfrac{\sqrt{17}}{4} = \frac{16 - \sqrt{17}}{4}$$

27. $\cos t = \frac{3}{5}$, t is in quadrant I

$$\Rightarrow \tan t + \sec t = \frac{\sin t}{\cos t} + \frac{1}{\cos t} = \frac{\sqrt{1 - \cos^2 t}}{\cos t} + \frac{1}{\cos t} = \frac{\sqrt{1 - \left(\frac{3}{5}\right)^2}}{\frac{3}{5}} + \frac{5}{3}$$

$$= \frac{\sqrt{\frac{1}{2}}}{\frac{3}{5}} + \frac{5}{3} = \frac{4}{5} \cdot \frac{5}{3} + \frac{5}{3} = \frac{5}{3} = \frac{9}{3} = 3$$

29. $y = 10\cos\frac{1}{2}x$

(a) This function has amplitude 10 , period

$\dfrac{2\pi}{\frac{1}{2}} = 4\pi$, and phase shift 0 .

(b)

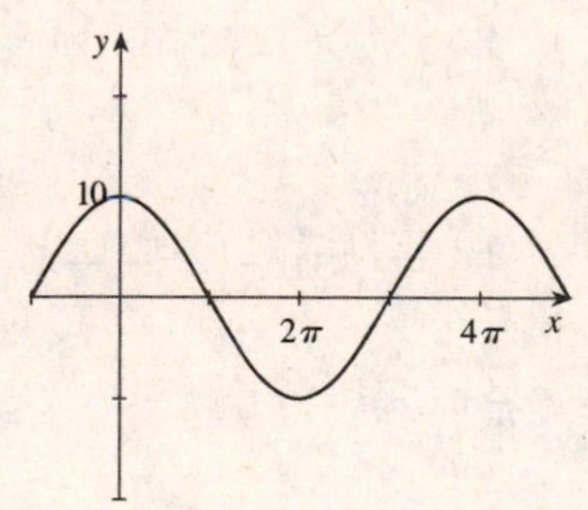

31. $y = -\sin\frac{1}{2}x$

(a) This function has amplitude 1 , period

$\frac{2\pi}{1/2} = 4\pi$, and phase shift 0 .

(b)

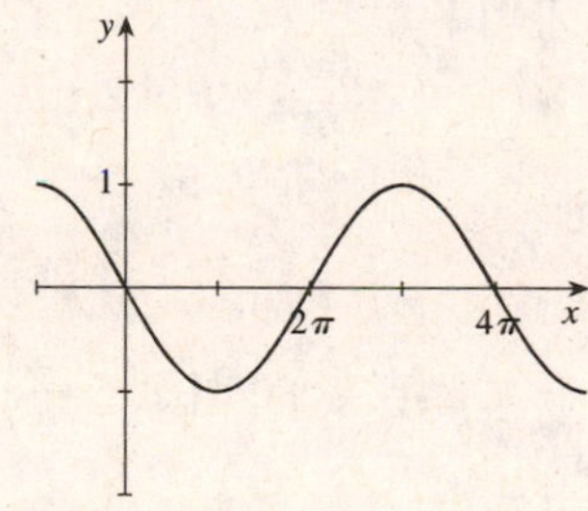

33. $y = 3\sin\left(2x - 2\right) = 3\sin 2\left(x - 1\right)$

(a) This function has amplitude 3, period $\frac{2\pi}{2} = \pi$, and phase shift 1.

(b)

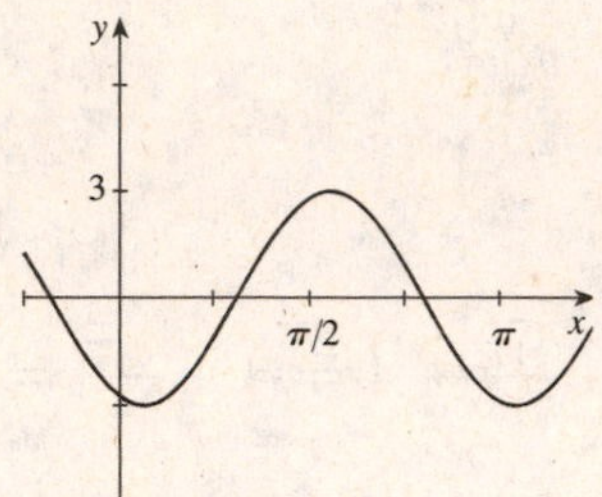

35. $y = -\cos\left(\frac{\pi}{2}x + \frac{\pi}{6}\right) = -\cos\frac{\pi}{2}\left(x + \frac{1}{3}\right)$

(a) This function has amplitude 1, period $\frac{2\pi}{\pi/2} = 4$, and phase shift $-\frac{1}{3}$.

(b)

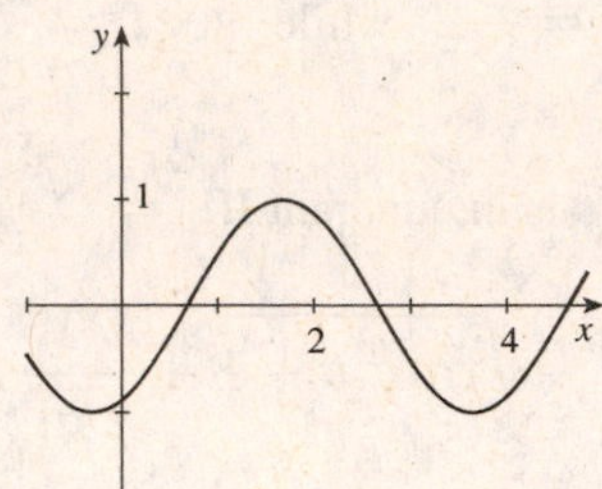

37. From the graph we see that the amplitude is 5, the period is $\frac{\pi}{2}$, and there is no phase shift. Therefore, the function is $y = 5\sin 4x$.

39. From the graph we see that the amplitude is $\frac{1}{2}$, the period is 1, and there is a phase shift of $-\frac{1}{3}$. Therefore, the function is $y = \frac{1}{2}\sin 2\pi\left(x + \frac{1}{3}\right)$.

41. $y = 3\tan x$ has period π.

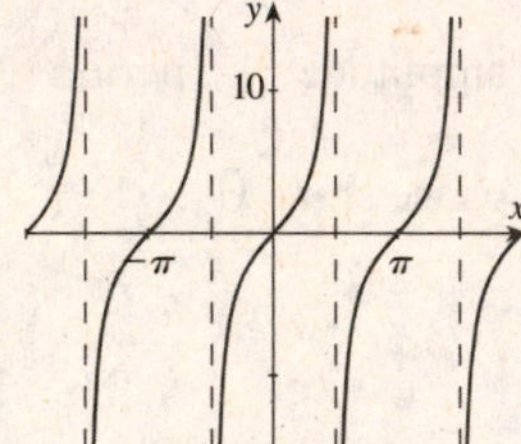

43. $y = 2\cot\left(x - \frac{\pi}{2}\right)$ has period π.

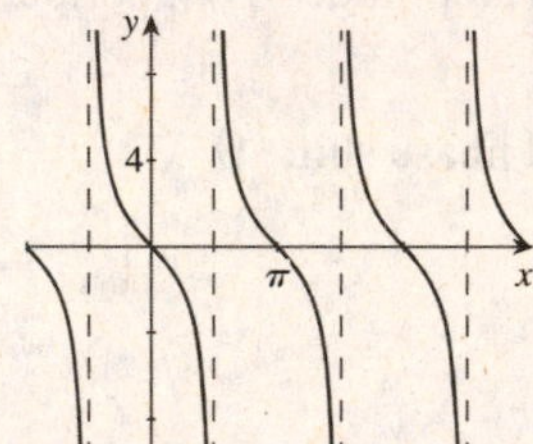

45.

$y = 4\csc\left(2x + \pi\right) = 4\csc 2\left(x + \frac{\pi}{2}\right)$

has period $\frac{2\pi}{2} = \pi$.

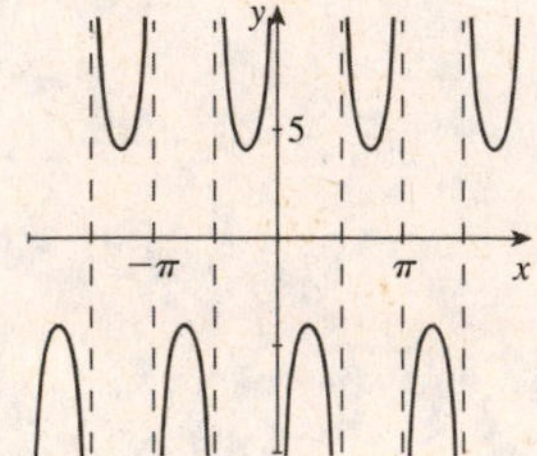

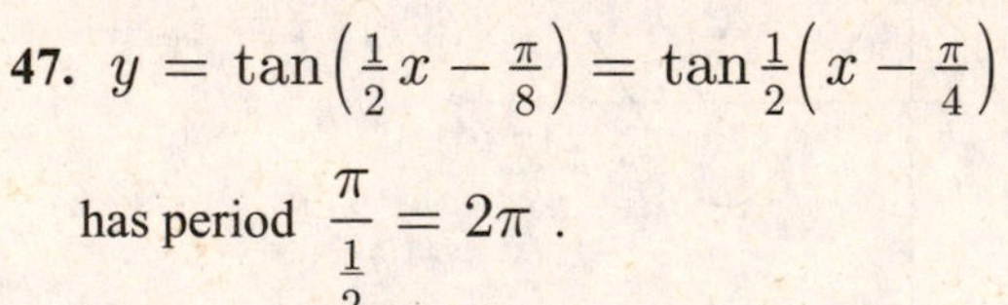

47. $y = \tan\left(\frac{1}{2}x - \frac{\pi}{8}\right) = \tan\frac{1}{2}\left(x - \frac{\pi}{4}\right)$

has period $\dfrac{\pi}{\frac{1}{2}} = 2\pi$.

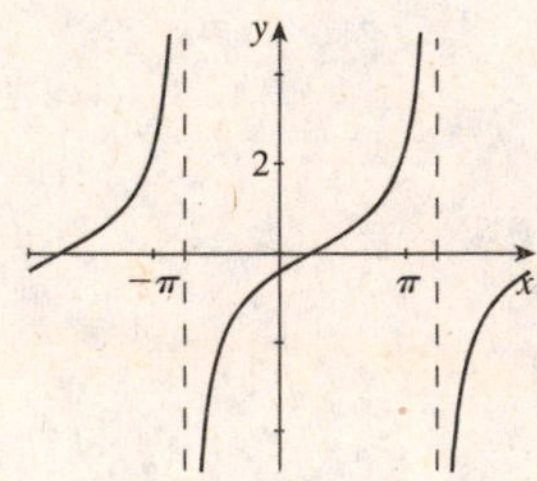

49. $\sin^{-1} 1 = \frac{\pi}{2}$

51. $\sin^{-1}\left(\sin \frac{13\pi}{6}\right) = \frac{\pi}{6}$

53. (a) $y = \left|\cos x\right|$

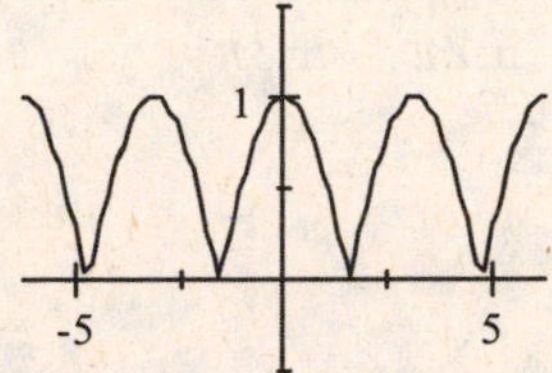

(b) This function has period π .
(c) This function is even.

55. (a) $y = \cos\left(2^{0.1x}\right)$

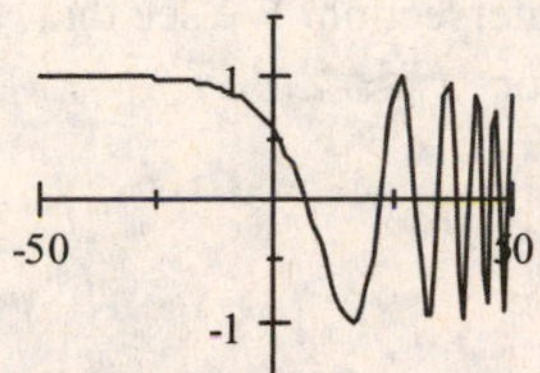

(b) This function is not periodic.
(c) This function is neither even nor odd.

57. (a) $y = \left|x\right| \cos 3x$

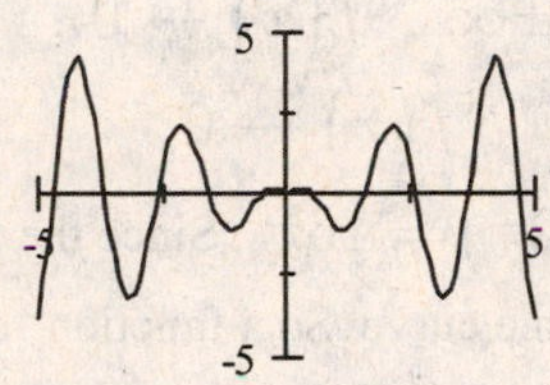

(b) This function is not periodic.
(c) This function is even.

59. $y = x \sin x$ is a sine function whose graph lies between those of $y = x$ and $y = -x$.

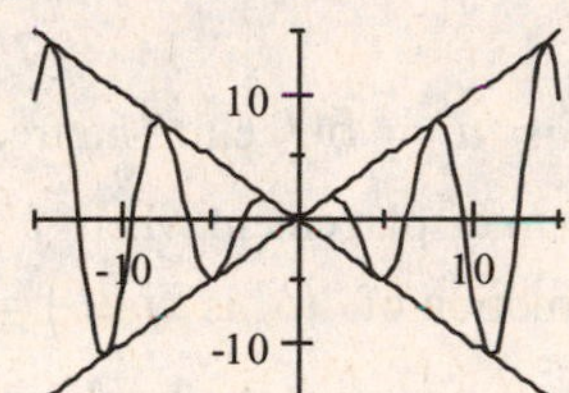

61. $y = x + \sin 4x$ is the sum of the two functions $y = x$ and $y = \sin 4x$.

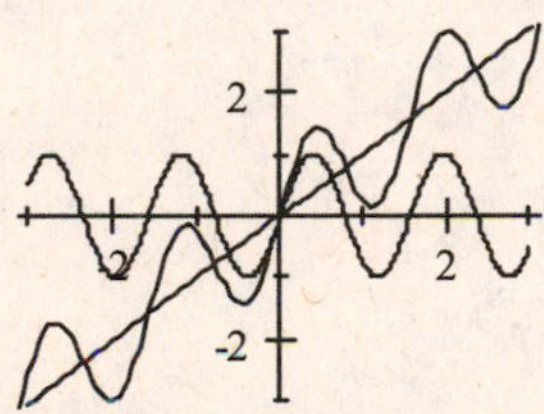

63. $y = \cos x + \sin 2x$. Since the period is 2π , we graph over the interval $\left[-\pi, \pi\right]$. The maximum value is 1.76 when $x \approx 0.63 \pm 2n\pi$, the minimum value is -1.76 when $x \approx 2.51 \pm 2n\pi$, n an integer.

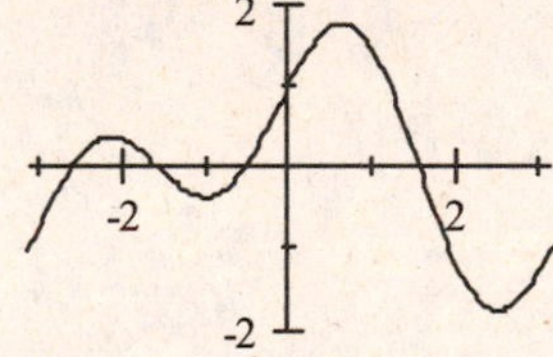

65. We want to find solutions to $\sin x = 0.3$ in the interval $\left[0, 2\pi\right]$, so we plot the functions $y = \sin x$ and $y = 0.3$ and look for their intersection. We see that $x \approx 0.305$ or $x \approx 2.837$.

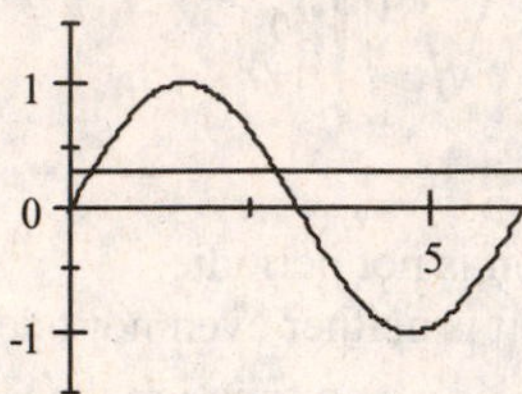

67. $f\left(x\right) = \dfrac{\sin^2 x}{x}$

(a) The function is odd.

(b) The graph intersects the x-axis at $x = 0, \pm\pi$, $\pm 2\pi$, $\pm 3\pi$, ...

(c)

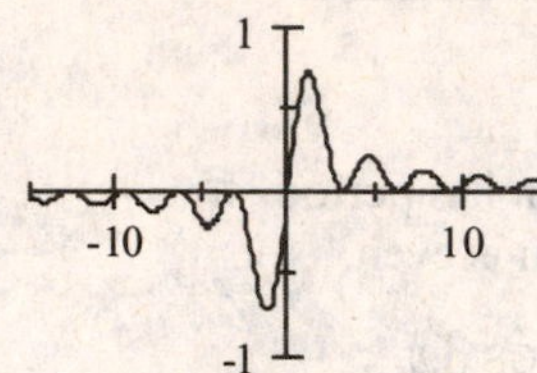

(d) As $x \to \pm\infty$, $f\left(x\right) \to 0$.

(e) As $x \to 0$, $f\left(x\right) \to 0$.

69. The amplitude is $a = 50$ cm. The frequency is 8 Hz, so $\omega = 8\left(2\pi\right) = 16\pi$. Since the mass is at its maximum displacement when $t = 0$, the motion follows a cosine curve. So a function describing the motion of P is $f\left(t\right) = 50\cos 16\pi t$.

71. From the graph, we see that the amplitude is 4 ft, the period is 12 hours, and there is no phase shift. Thus, the variation in water level is described by $y = 4\cos\frac{\pi}{6}t$.

Chapter 2 Review

1. (a) Since $\left(-\frac{\sqrt{3}}{2}\right)^2 + \left(\frac{1}{2}\right)^2 = \frac{3}{4} + \frac{1}{4} = 1$, the point $P\left(-\frac{\sqrt{3}}{2}, \frac{1}{2}\right)$ lies on the unit circle.

(b) $\sin t = \frac{1}{2}$, $\cos t = -\frac{\sqrt{3}}{2}$, $\tan t = \dfrac{\frac{1}{2}}{-\frac{\sqrt{3}}{2}} = -\frac{\sqrt{3}}{3}$.

3. $t = \frac{2\pi}{3}$

(a) $\bar{t} = \pi - \frac{2\pi}{3} = \frac{\pi}{3}$

(b) $P\left(-\frac{1}{2}, \frac{\sqrt{3}}{2}\right)$

(c) $\sin t = \frac{\sqrt{3}}{2}$, $\cos t = -\frac{1}{2}$, $\tan t = -\sqrt{3}$, $\csc t = \frac{2\sqrt{3}}{3}$, $\sec t = -2$, and $\cot t = -\frac{\sqrt{3}}{3}$.

5. $t = -\frac{11\pi}{4}$

(a) $\bar{t} = 3\pi + \left(-\frac{11\pi}{4}\right) = \frac{\pi}{4}$

(b) $P\left(-\frac{\sqrt{2}}{2}, -\frac{\sqrt{2}}{2}\right)$

(c) $\sin t = -\frac{\sqrt{2}}{2}$, $\cos t = -\frac{\sqrt{2}}{2}$, $\tan t = 1$, $\csc t = -\sqrt{2}$, $\sec t = -\sqrt{2}$, and $\cot t = 1$.

7. (a) $\sin\frac{3\pi}{4} = \sin\frac{\pi}{4} = \frac{\sqrt{2}}{2}$

(b) $\cos\frac{3\pi}{4} = -\cos\frac{\pi}{4} = -\frac{\sqrt{2}}{2}$.

9. (a) $\sin 1.1 \approx 0.89121$

(b) $\cos 1.1 \approx 0.45360$

11. (a) $\cos\frac{9\pi}{2} = \cos\frac{\pi}{2} = 0$

(b) $\sec\frac{9\pi}{2}$ is undefined

13. (a) $\tan\frac{5\pi}{2}$ is undefined

(b) $\cot\frac{5\pi}{2} = \cot\frac{\pi}{2} = 0$

15. (a) $\tan\frac{5\pi}{6} = -\frac{\sqrt{3}}{3}$

(b) $\cot\frac{5\pi}{6} = -\sqrt{3}$

17. $\dfrac{\tan t}{\cos t} = \dfrac{\frac{\sin t}{\cos t}}{\cos t} = \dfrac{\sin t}{\cos^2 t} = \dfrac{\sin t}{1 - \sin^2 t}$

19. $\tan t = \dfrac{\sin t}{\cos t} = \dfrac{\sin t}{\pm\sqrt{1 - \sin^2 t}} = \dfrac{\sin t}{\sqrt{1 - \sin^2 t}}$ (because t is in quadrant IV, $\cos t$ is positive).

21. $\sin t = \frac{5}{13}$, $\cos t = -\frac{12}{13}$. Then $\tan t = \dfrac{\frac{5}{13}}{-\frac{12}{13}} = -\frac{5}{12}$, $\csc t = \frac{13}{5}$, $\sec t = -\frac{13}{12}$, and $\cot t = -\frac{12}{5}$.

23. $\cot t = -\frac{1}{2}$, $\csc t = \frac{\sqrt{5}}{2}$. Since $\csc t = \dfrac{1}{\sin t}$, we know $\sin t = \frac{2}{\sqrt{5}} = \frac{2\sqrt{5}}{5}$. Now

$\cot t = \dfrac{\cos t}{\sin t}$, so $\cos t = \sin t \cdot \cot t = \frac{2\sqrt{5}}{5} \cdot \left(-\frac{1}{2}\right) = -\frac{\sqrt{5}}{5}$, and

$\tan t = \dfrac{1}{\left(-\frac{1}{2}\right)} = -2$ while $\sec t = \dfrac{1}{\cos t} = \dfrac{1}{\left(-\frac{\sqrt{5}}{5}\right)} = -\frac{5}{\sqrt{5}} = -\sqrt{5}$.

25. $\tan t = \frac{1}{4}$, t is in quadrant III $\Rightarrow$

$$\sec t + \cot t = -\sqrt{\tan^2 t + 1} + \frac{1}{\tan t} = -\sqrt{\left(\frac{1}{4}\right)^2 + 1} + 4 = -\sqrt{\frac{17}{16}} + 4 = 4 - \frac{\sqrt{17}}{4} = \frac{16 - \sqrt{17}}{4}$$

27. $\cos t = \frac{3}{5}$, t is in quadrant I

$$\Rightarrow \tan t + \sec t = \frac{\sin t}{\cos t} + \frac{1}{\cos t} = \frac{\sqrt{1 - \cos^2 t}}{\cos t} + \frac{1}{\cos t} = \frac{\sqrt{1 - \left(\frac{3}{5}\right)^2}}{\frac{3}{5}} + \frac{5}{3}$$

$$= \frac{\sqrt{\frac{1}{2}}}{\frac{3}{5}} + \frac{5}{3} = \frac{4}{5} \cdot \frac{5}{3} + \frac{5}{3} = \frac{9}{3} = 3$$

29. $y = 10 \cos \frac{1}{2} x$

(a) This function has amplitude 10, period

$\dfrac{2\pi}{\frac{1}{2}} = 4\pi$, and phase shift 0.

(b)

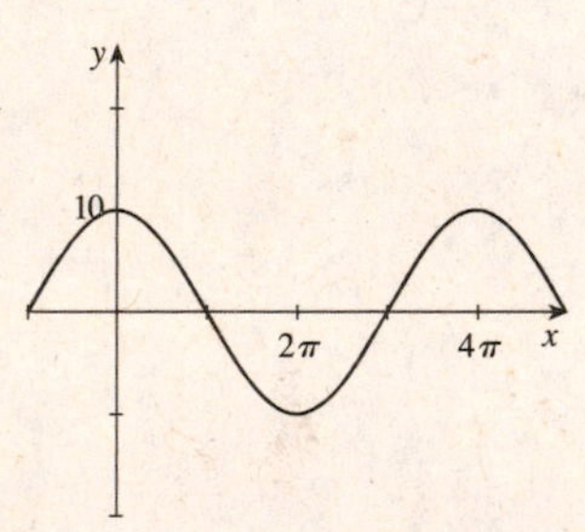

31. $y = -\sin \frac{1}{2} x$

(a) This function has amplitude 1, period

$\frac{2\pi}{1/2} = 4\pi$, and phase shift 0.

(b)

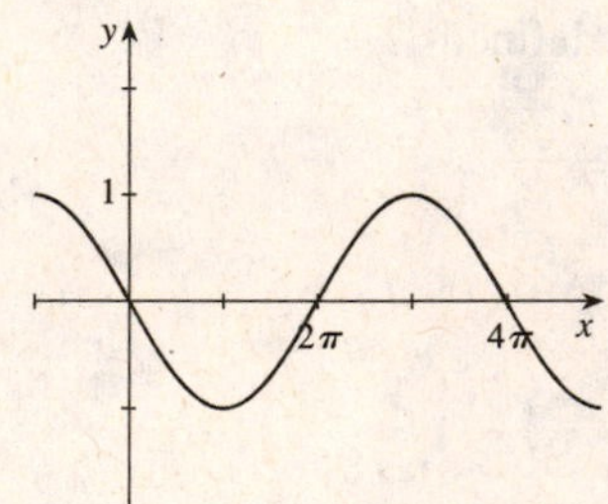

33. $y = 3\sin\left(2x - 2\right) = 3\sin 2\left(x - 1\right)$

(a) This function has amplitude 3 , period $\frac{2\pi}{2} = \pi$, and phase shift 1 .

(b)

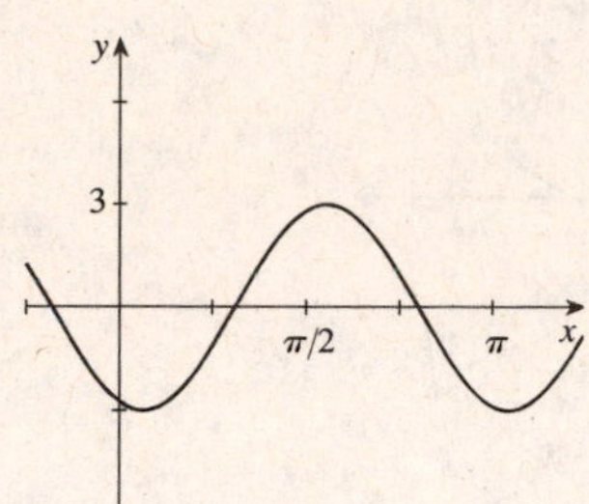

35. $y = -\cos\left(\frac{\pi}{2}x + \frac{\pi}{6}\right) = -\cos\frac{\pi}{2}\left(x + \frac{1}{3}\right)$

(a) This function has amplitude 1 , period $\frac{2\pi}{\pi/2} = 4$, and phase shift $-\frac{1}{3}$.

(b)

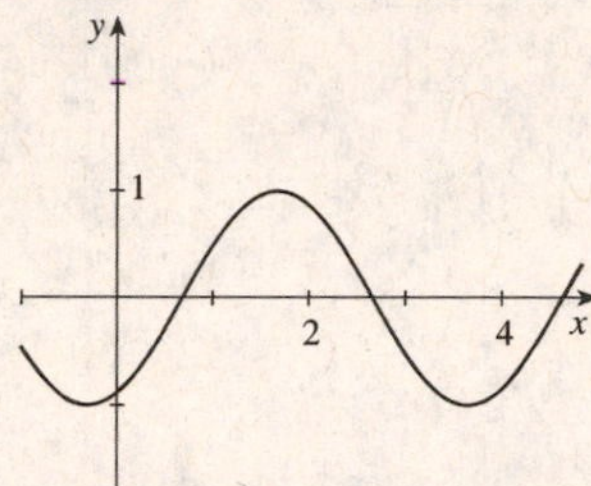

36. $y = 10\sin\left(2x - \frac{\pi}{2}\right) = 10\sin 2\left(x - \frac{\pi}{4}\right)$

(a) This function has amplitude 10 , period $\frac{2\pi}{2} = \pi$, and phase shift $\frac{\pi}{4}$.

(b)

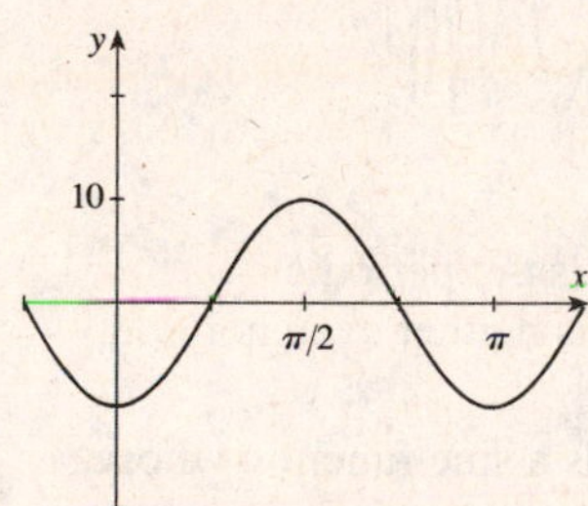

37. From the graph we see that the amplitude is 5 , the period is $\frac{\pi}{2}$, and there is no phase shift. Therefore, the function is $y = 5\sin 4x$.

39. From the graph we see that the amplitude is $\frac{1}{2}$, the period is 1 , and there is a phase shift of $-\frac{1}{3}$. Therefore, the function is $y = \frac{1}{2}\sin 2\pi\left(x + \frac{1}{3}\right)$.

41. $y = 3\tan x$ has period π .

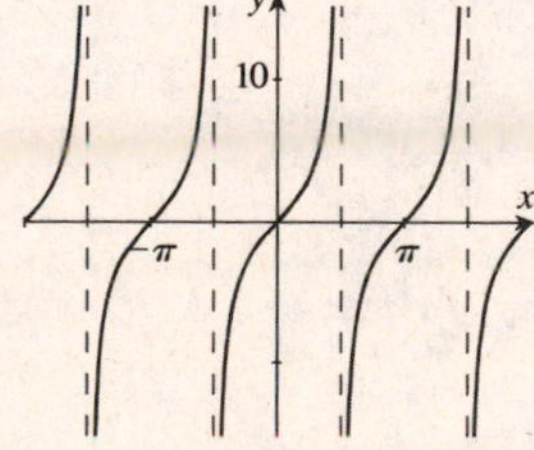

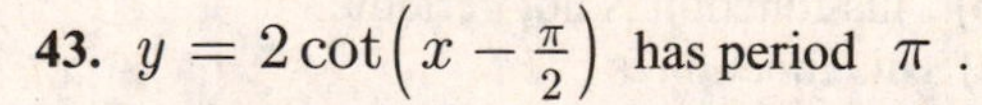

43. $y = 2\cot\left(x - \frac{\pi}{2}\right)$ has period π .

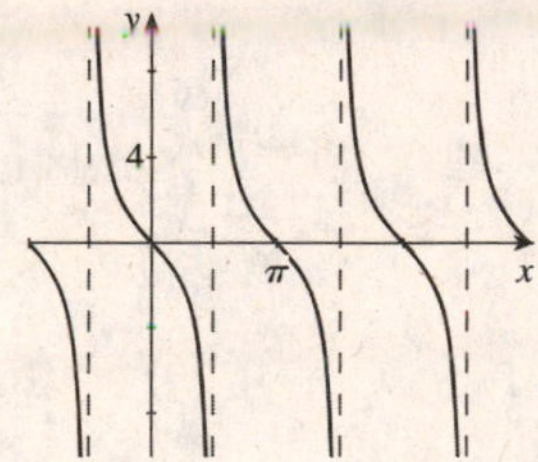

45.
$y = 4\csc\left(2x + \pi\right) = 4\csc 2\left(x + \frac{\pi}{2}\right)$
has period $\frac{2\pi}{2} = \pi$.

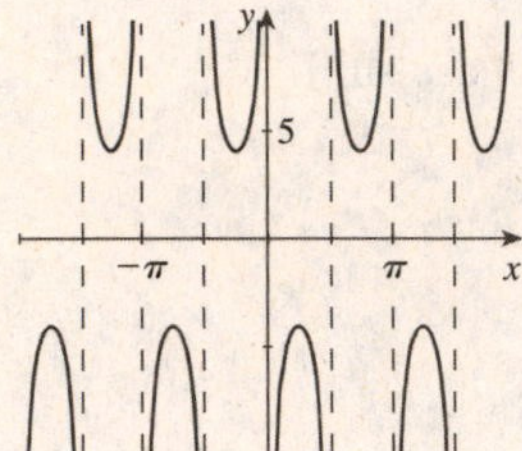

47. $y = \tan\left(\frac{1}{2}x - \frac{\pi}{8}\right) = \tan\frac{1}{2}\left(x - \frac{\pi}{4}\right)$
has period $\dfrac{\pi}{\frac{1}{2}} = 2\pi$.

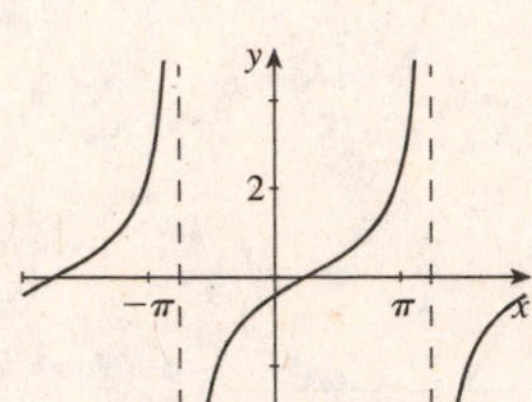

49. $\sin^{-1} 1 = \frac{\pi}{2}$

51. $\sin^{-1}\left(\sin\frac{13\pi}{6}\right) = \frac{\pi}{6}$

53. (a) $y = \left|\cos x\right|$

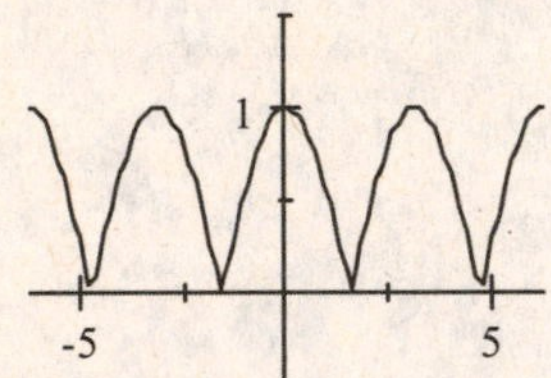

(b) This function has period π .
(c) This function is even.

55. (a) $y = \cos\left(2^{0.1x}\right)$

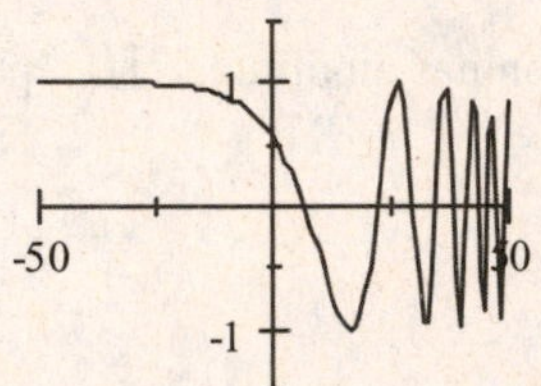

(b) This function is not periodic.
(c) This function is neither even nor odd.

57. (a) $y = \left|x\right| \cos 3x$

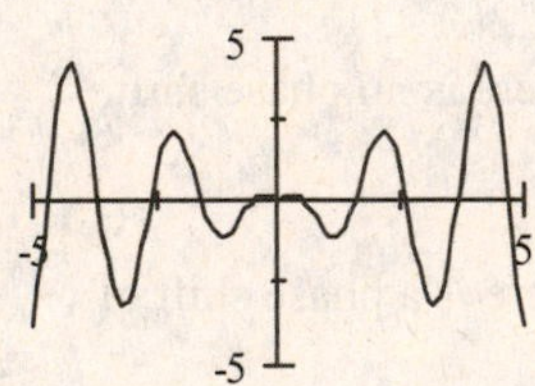

(b) This function is not periodic.
(c) This function is even.

59. $y = x\sin x$ is a sine function whose graph lies between those of $y = x$ and $y = -x$.

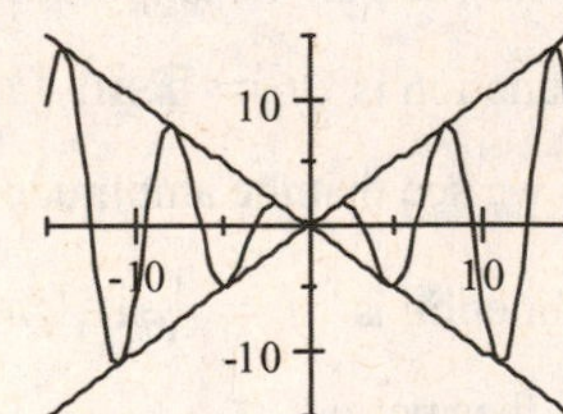

61. $y = x + \sin 4x$ is the sum of the two functions $y = x$ and $y = \sin 4x$.

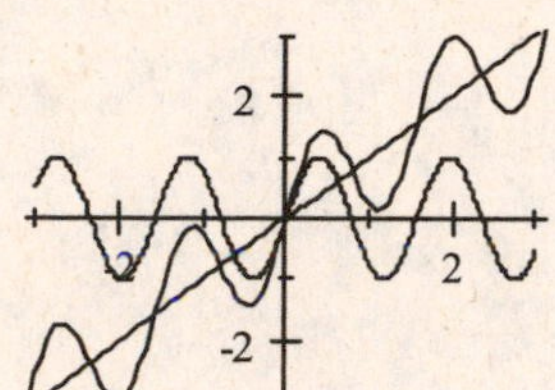

63. $y = \cos x + \sin 2x$. Since the period is 2π, we graph over the interval $[-\pi, \pi]$. The maximum value is 1.76 when $x \approx 0.63 \pm 2n\pi$, the minimum value is -1.76 when $x \approx 2.51 \pm 2n\pi$, n an integer.

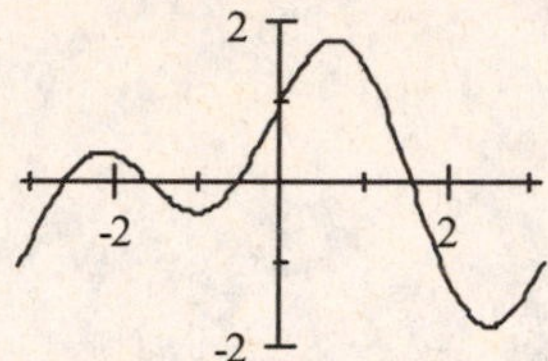

65. We want to find solutions to $\sin x = 0.3$ in the interval $[0, 2\pi]$, so we plot the functions $y = \sin x$ and $y = 0.3$ and look for their intersection. We see that $x \approx 0.305$ or $x \approx 2.837$.

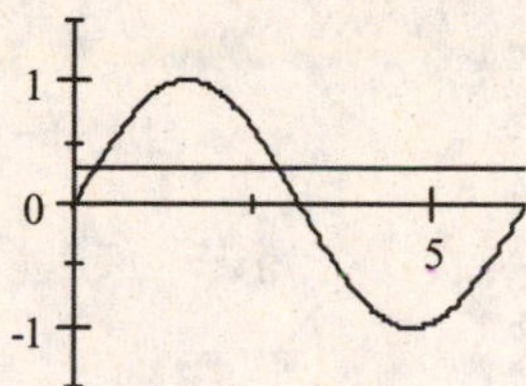

67. $f(x) = \dfrac{\sin^2 x}{x}$

(a) The function is odd.

(b) The graph intersects the x-axis at $x = 0, \pm\pi$, $\pm 2\pi$, $\pm 3\pi$, ….

(c)

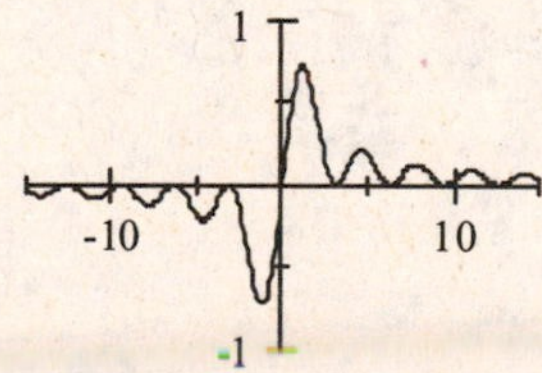

(d) As $x \to \pm\infty$, $f(x) \to 0$.

(e) As $x \to 0$, $f(x) \to 0$.

69. The amplitude is $a = 50$ cm. The frequency is 8 Hz, so $\omega = 8(2\pi) = 16\pi$. Since the mass is at its maximum displacement when $t = 0$, the motion follows a cosine curve. So a function describing the motion of P is $f(t) = 50 \cos 16\pi t$.

71. From the graph, we see that the amplitude is 4 ft, the period is 12 hours, and there is no phase shift. Thus, the variation in water level is described by $y = 4\cos\frac{\pi}{6}t$.

Chapter 2 Test

1. Since $P(x,y)$ lies on the unit circle, $x^2+y^2=1$

$\Rightarrow y=\pm\sqrt{1-\left(\frac{\sqrt{11}}{6}\right)^2}=\pm\sqrt{\frac{25}{36}}=\pm\frac{5}{6}$. But $P(x,y)$ lies in the fourth quadrant. Therefore y is negative $\Rightarrow y=-\frac{5}{6}$.

2. Since P is on the unit circle, $x^2+y^2=1 \quad\Leftrightarrow\quad x^2=1-y^2$. Thus, $x^2=1-\left(\frac{4}{5}\right)^2=\frac{9}{25}$, and so $x=\pm\frac{3}{5}$. From the diagram, x is clearly negative, so $x=-\frac{3}{5}$. Therefore, P is the point $\left(-\frac{3}{5},\frac{4}{5}\right)$.

(a) $\sin t=\frac{4}{5}$

(b) $\cos t=-\frac{3}{5}$

(c) $\tan t=\dfrac{\frac{4}{5}}{-\frac{3}{5}}=-\frac{4}{3}$

(d) $\sec(t)=-\frac{5}{3}$

3. (a) $\sin\frac{7\pi}{6}=-0.5$

(b) $\cos\dfrac{13\pi}{4}=-\frac{\sqrt{2}}{2}$

(c) $\tan\left(-\frac{5\pi}{3}\right)=\sqrt{3}$

(d) $\csc\left(\frac{3\pi}{2}\right)=-1$

4. $\tan t=\dfrac{\sin t}{\cos t}=\dfrac{\sin t}{\pm\sqrt{1-\sin^2 t}}$. But t is in quadrant II $\Rightarrow \cos t$ is negative, so we choose the negative square root. Thus,

$$\tan t=\frac{\sin t}{-\sqrt{1-\sin^2 t}}.$$

5. $\cos t=-\frac{8}{17}$, t in quadrant III

$$\Rightarrow \tan t\cdot\cot t+\csc t=1+\frac{1}{-\sqrt{1-\cos^2 t}}$$

(since t is in quadrant III)

$$=1-\frac{1}{\sqrt{1-\frac{64}{289}}}=1-\frac{1}{\frac{15}{17}}=-\frac{2}{15}.$$

6. $y = -5\cos 4x$

(a) This function has amplitude 5, period $\frac{2\pi}{4} = \frac{\pi}{2}$, and phase shift 0.

(b)

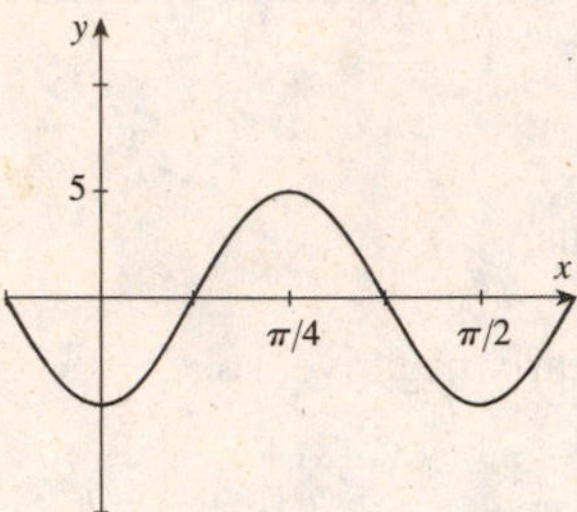

7. $y = 2\sin\left(\frac{1}{2}x - \frac{\pi}{6}\right) = \sin\frac{1}{2}\left(x - \frac{\pi}{3}\right)$

(a) This function has amplitude 2, period $\dfrac{2\pi}{\frac{1}{2}} = 4\pi$, and phase shift $\frac{\pi}{3}$.

(b)

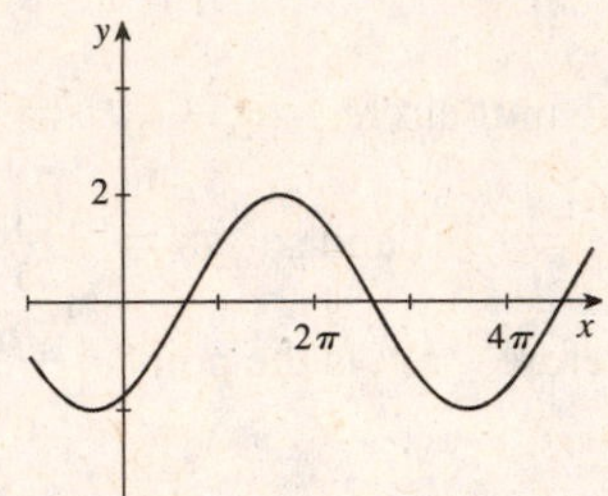

8. $y = -\csc 2x$ has period $\frac{2\pi}{2} = \pi$.

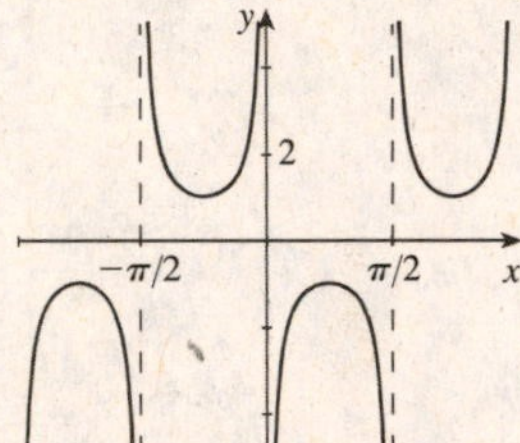

9. $y = \tan 2\left(x - \frac{\pi}{4}\right)$ has period $\frac{\pi}{2}$.

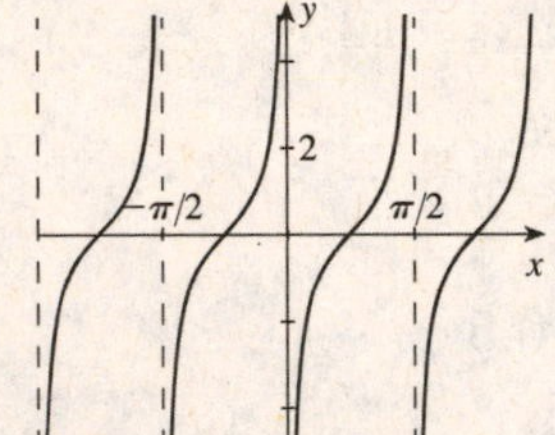

10. (a) $\tan^{-1} 1 = \frac{\pi}{4}$

(b) $\cos^{-1}\left(-\frac{\sqrt{3}}{2}\right) = \frac{5\pi}{6}$

(c) $\tan^{-1}\left(\tan 3\pi\right) = 0$

(d) $\sin\left(\sin^{-1}\frac{1}{2}\right) = \frac{1}{2}$

11. From the graph, we see that the amplitude is 2 and the phase shift is $-\frac{\pi}{3}$. Also, the period is π, so $\frac{2\pi}{k} = \pi \Rightarrow k = \frac{2\pi}{\pi} = 2$. Thus, the function is $y = 2\sin 2\left(x + \frac{\pi}{3}\right)$.

12. $y = \dfrac{\cos x}{1 + x^2}$

(a)

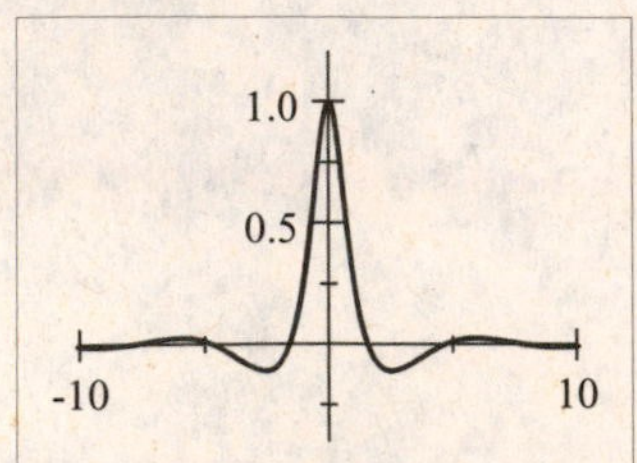

(b) The function is even.

(c) The function has a minimum value of approximately -0.11 when $x \approx \pm 2.54$ and a maximum value of 1 when $x = 0$.

13. The amplitude is $\frac{1}{2}\left(10\right)=5$ cm and the frequency is 2 Hz. Assuming that the mass is at its rest position and moving upward when $t=0$, a function describing the distance of the mass from its rest position is $f\left(t\right)=5\sin 4\pi t$.

1. (a) See the graph in part (c).

(b) Using the method of Example 1, we find the vertical shift

$b=\frac{1}{2}(\text{maximum value}+\text{minimum value})$
$=\frac{1}{2}(2.1-2.1)=0$,

the amplitude

$a=\frac{1}{2}(\text{maximum value}-\text{minimum value})$
$=\frac{1}{2}(2.1-(-2.1))=2.1$,

the period $\frac{2\pi}{\omega}=2(6-0)=12$ (so $\omega\approx 0.5236$), and the phase shift $c=0$. Thus, our model is $y=2.1\cos\frac{\pi}{6}t$.

(c)

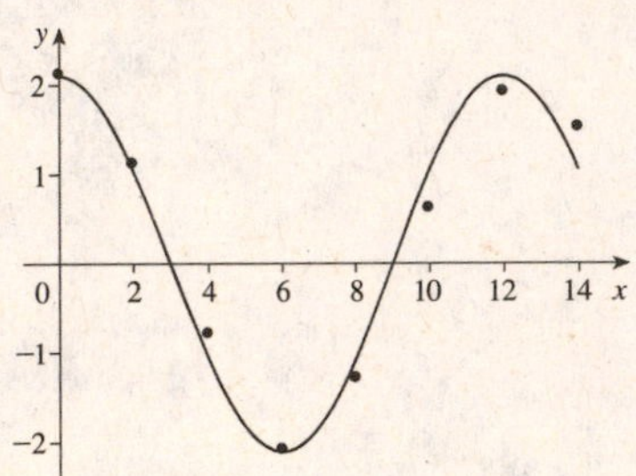

The curve fits the data quite well.

(d) Using the SinReg command on the TI-83, we find

$$y=2.048714222\sin\begin{pmatrix}0.5030795477t+\\1.551856108\end{pmatrix}-0.0089616507$$

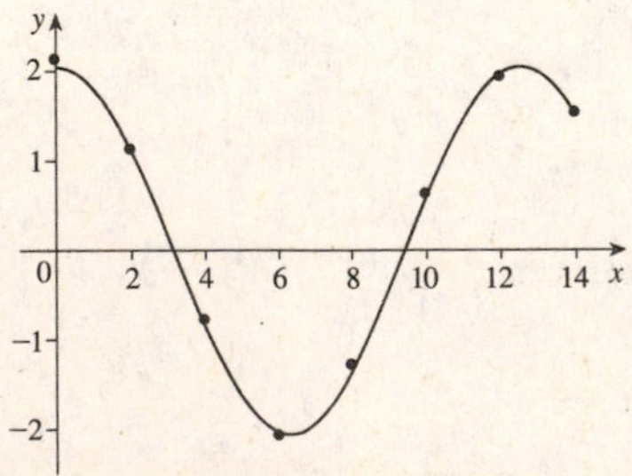

(e) Our model from part (d) is equivalent to

$$y=2.05\cos\left(0.50t+1.55-\tfrac{\pi}{2}\right)-0.01$$
$$\approx 2.05\cos(0.50t-0.02)-0.01$$

This is the same as the function in part (b), correct to one decimal place.

3. (a) See the graph in part (c).

(b) Using the method of Example 1, we find the vertical shift

$b=\frac{1}{2}(\text{maximum value}+\text{minimum value})$
$=\frac{1}{2}(25.1+1.0)=13.05$

the amplitude

$a=\frac{1}{2}(\text{maximum value}-\text{minimum value})$
$=\frac{1}{2}(25.1-1.0)=12.05$

the period $\frac{2\pi}{\omega}=2(1.5-0.9)=1.2$ (so $\omega\approx 5.236$), and the phase shift $c=0.3$. Thus, our model is

$y=12.05\cos(5.236(t-0.3))+13.05$.

(c)

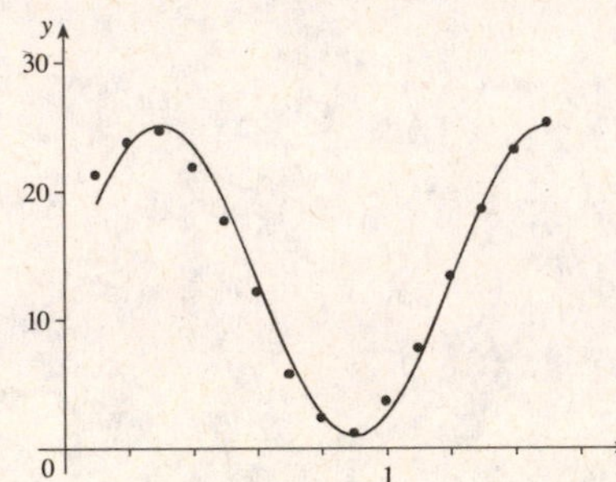

The curve fits the data fairly well.

(d) Using the SinReg command on the TI-83, we find .

$$y=11.71905062\sin\begin{pmatrix}5.048853286t\\+0.2388957877\end{pmatrix}+12.96070536$$

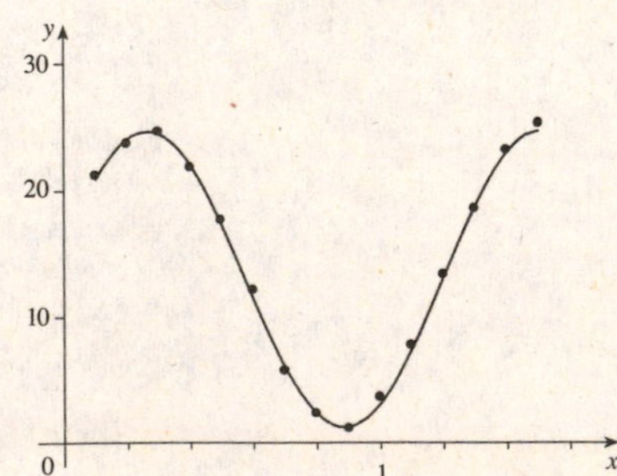

(e) Our model from part (d) is equivalent to

$$y=11.72\cos\left(5.05t+0.24-\tfrac{\pi}{2}\right)+12.96$$
$$\approx 11.72\cos(5.05t-1.33)+12.96$$

This is close but not identical to the function in part (b).

5. (a) See the graph in part (c).

(b) Let t be the time (in months) from January. We find a function of the form $y = a\cos\omega(t-c)+b$, where y is the temperature in $^\circ$F. $a = \frac{1}{2}(85.8-40) = 22.9$. The period is $2(\text{Jul}-\text{Jan}) = 2(6-0) = 12$, and so $\omega = \frac{2\pi}{12} \approx 0.52$. $b = \frac{1}{2}(85.8+40) = 62.9$. Because the maximum value occurs in July, $c = 6$. Thus the function is $y = 22.9\cos 0.52(t-6)+62.9$.

(c)

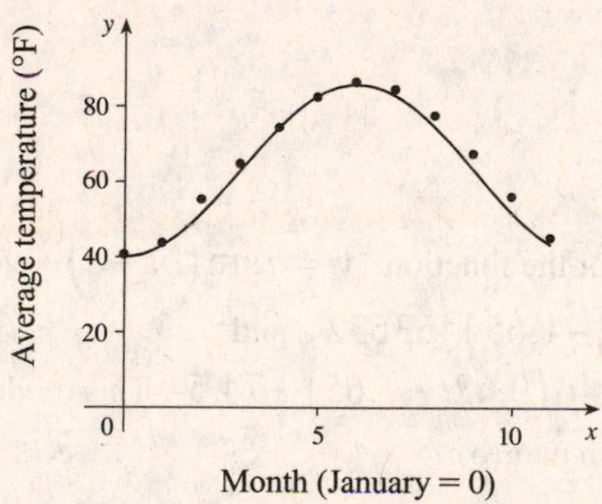

(d) Using the SinReg command on the TI-83 we find that for the function $y = a\sin(bt+c)+d$, where $a = 23.4$, $b = 0.48$, $c = -1.36$, and $d = 62.2$. Thus we get the model $y = 23.4\sin(0.48t-1.36)+62.2$.

7. (a) See the graph in part (c).

(b) Let t be the time years. We find a function of the form $y = a\sin\omega(t-c)+b$, where y is the owl population. $a = \frac{1}{2}(80-20) = 30$. The period is $2(9-3) = 12$ and so $\omega = \frac{2\pi}{12} \approx 0.52$. $b = \frac{1}{2}(80+20) = 50$. Because the values start at the middle we have $c = 0$. Thus the function is $y = 30\sin 0.52t + 50$.

(c)

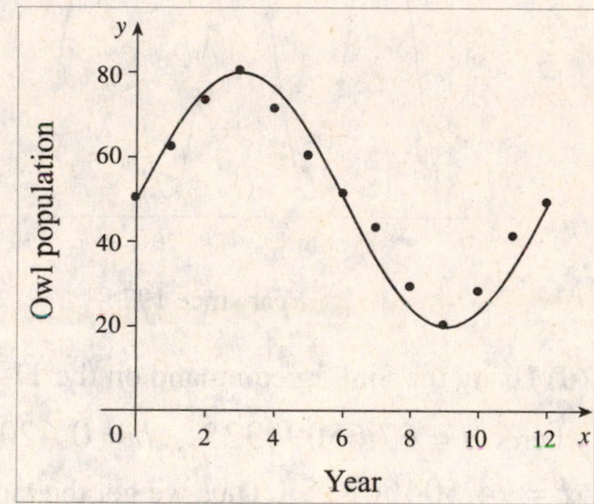

(d) Using the SinReg command on the TI--83 we find that for the function $y = a\sin(bt+c)+d$, where $a = 25.8$, $b = 0.52$, $c = -0.02$, and $d = 50.6$. Thus we get the model $y = 25.8\sin(0.52t-0.02)+50.6$.

9. (a) See the graph in part (c).

(b) Let t be the number of years since 1975. We find a function of the form $y = a\cos\omega(t-c)+b$. We have $a = \frac{1}{2}(85.8-40) = 22.9$. The number of sunspots varies over a cycle of approximately 11 years (as the problem indicates), so $\omega = \frac{2\pi}{11} \approx 0.57$. Also, we have $b = \frac{1}{2}(158+9) = 83.5$. The first maximum value occurs at approximately $c = 4.5$. Thus, our model is $y = 74.5\cos(0.57(t-4.5))+83.5$, where y is the average daily sunspot count and t is the number of years since 1975.

(c)

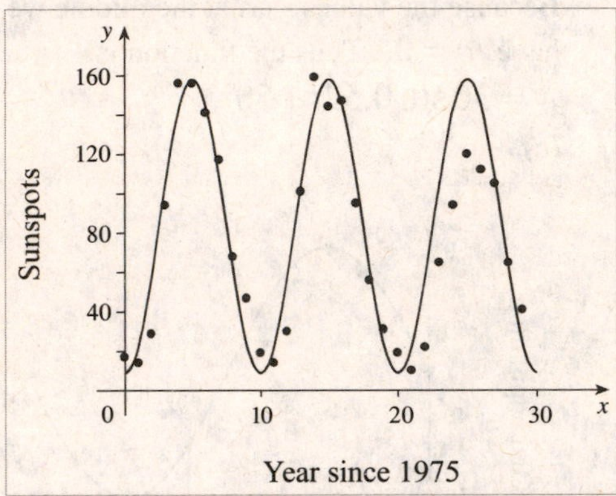

(d) Using the SinReg command on the TI-83, we find that for the function $y = a\sin(bt+c)+d$, where $a = 67.65094323$, $b = 0.6205550572$, $c = -1.654463632$, and $d = 74.50460325$. Thus we get the model $y = 67.65\sin(0.62t-1.65)+74.5$. This model is more accurate for more recent years than the one we found in part (b).

3 Trigonometric Functions: Right Triangle Approach

3.1 Angle Measure

1. (a) The radian measure of an angle θ is the length of the *arc* that subtends the angle in a circle of radius 1.

(b) To convert degrees to radians we multiply by $\frac{\pi}{180}$.

(c) To convert radians to degrees we multiply by $\frac{180}{\pi}$.

3. $72^\circ = 72^\circ \cdot \frac{\pi}{180^\circ}$ rad $= \frac{2\pi}{5}$ rad ≈ 1.257 rad

5. $-45^\circ = -45^\circ \cdot \frac{\pi}{180^\circ}$ rad $= -\frac{\pi}{4}$ rad ≈ -0.785 rad

7. $-75^\circ = -75^\circ \cdot \frac{\pi}{180^\circ}$ rad $= -\frac{5\pi}{12}$ rad ≈ -1.309 rad

9. $1080^\circ = 1080^\circ \cdot \frac{\pi}{180^\circ}$ rad $= 6\pi$ rad ≈ 18.850 rad

11. $96^\circ = 96^\circ \cdot \frac{\pi}{180^\circ}$ rad $= \frac{8\pi}{15}$ rad ≈ 1.676 rad

13. $7.5^\circ = 7.5^\circ \cdot \frac{\pi}{180^\circ}$ rad $= \frac{\pi}{24}$ rad ≈ 0.131 rad

15. $\frac{7\pi}{6} = \frac{7\pi}{6} \cdot \frac{180^\circ}{\pi} = 210^\circ$

17. $-\frac{5\pi}{4} = -\frac{5\pi}{4} \cdot \frac{180^\circ}{\pi} = -225^\circ$

19. $3 = 3 \cdot \frac{180^\circ}{\pi} = \frac{540^\circ}{\pi} \approx 171.9^\circ$

21. $-1.2 = -1.2 \cdot \frac{180^\circ}{\pi} = -\frac{216^\circ}{\pi} = -68.8^\circ$

23. $\frac{\pi}{10} = \frac{\pi}{10} \cdot \frac{180^\circ}{\pi} = 18^\circ$

25. $-\frac{2\pi}{15} = -\frac{2\pi}{15} \cdot \frac{180^\circ}{\pi} = -24^\circ$

27. 50° is coterminal with $50^\circ + 360^\circ = 410^\circ$, $50^\circ + 720^\circ = 770^\circ$, $50^\circ - 360^\circ = -310^\circ$, and $50^\circ - 720^\circ = -670^\circ$. (Other answers are possible.)

29. $\frac{3\pi}{4}$ is coterminal with $\frac{3\pi}{4} + 2\pi = \frac{11\pi}{4}$, $\frac{3\pi}{4} + 4\pi = \frac{19\pi}{4}$, $\frac{3\pi}{4} - 2\pi = -\frac{5\pi}{4}$, and $\frac{3\pi}{4} - 4\pi = -\frac{13\pi}{4}$. (Other answers are possible.)

31. $-\frac{\pi}{4}$ is coterminal with $-\frac{\pi}{4} + 2\pi = \frac{7\pi}{4}$, $-\frac{\pi}{4} + 4\pi = \frac{15\pi}{4}$, $-\frac{\pi}{4} - 2\pi = -\frac{9\pi}{4}$, and $-\frac{\pi}{4} - 4\pi = -\frac{17\pi}{4}$. (Other answers are possible.)

33. Since $430^\circ - 70^\circ = 360^\circ$, the angles are coterminal.

35. Since $\frac{17\pi}{6} - \frac{5\pi}{6} = \frac{12\pi}{6} = 2\pi$; the angles are coterminal.

37. Since $875^\circ - 155^\circ = 720^\circ = 2 \cdot 360^\circ$, the angles are coterminal.

39. Since $733^\circ - 2 \cdot 360^\circ = 13^\circ$, the angles 733° and 13° are coterminal.

41. Since $1110^\circ - 3 \cdot 360^\circ = 30^\circ$, the angles 1110° and 30° are coterminal.

43. Since $-800^\circ + 3 \cdot 360^\circ = 280^\circ$, the angles -800° and 280° are coterminal.

45. Since $\frac{17\pi}{6} - 2\pi = \frac{5\pi}{6}$, the angles $\frac{17\pi}{6}$ and $\frac{5\pi}{6}$ are coterminal.

47. Since $87\pi - 43 \cdot 2\pi = \pi$, the angles 87π and π are coterminal.

49. Since $\frac{17\pi}{4} - 2 \cdot 2\pi = \frac{\pi}{4}$, the angles $\frac{17\pi}{4}$ and $\frac{\pi}{4}$ are coterminal.

51. Using the formula $s = \theta r$, the length of the arc is $s = \left(220° \cdot \frac{\pi}{180°}\right) \cdot 5 = \frac{55\pi}{9} \approx 19.2$.

53. Solving for r we have $r = \frac{s}{\theta}$, so the radius of the circle is $r = \frac{8}{2} = 4$.

55. Using the formula $s = \theta r$, the length of the arc is $s = 2 \cdot 2 = 4$ mi.

57. Solving for θ , we have $\theta = \frac{s}{r}$, so the measure of the central angle is $\theta = \frac{100}{50} = 2$ rad.

Converting to degrees we have $\theta = 2 \cdot \frac{180°}{\pi} \approx 114.6°$

59. Solving for r , we have $r = \frac{s}{\theta}$, so the radius of the circle is $r = \frac{6}{\pi/6} = \frac{36}{\pi} \approx 11.46$ m.

61. (a) $A = \frac{1}{2}r^2\theta = \frac{1}{2} \cdot 8^2 \cdot 80° \cdot \frac{\pi}{180°} = 32 \cdot \frac{4\pi}{9} = \frac{128\pi}{9} \approx 44.68$

(b) $A = \frac{1}{2}r^2\theta = \frac{1}{2} \cdot 10^2 \cdot 0.5 = 25$

63. $A = \frac{1}{2}r^2\theta = \frac{1}{2} \cdot 10^2 \cdot 1 = 50$ m ^{2}

65. $\theta = 2$ rad, $A = 16$ m ^{2} . Since $A = \frac{1}{2}r^2\theta$, we have

$r = \sqrt{2A / \theta} = \sqrt{2 \cdot 16 / 2} = \sqrt{16} = 4$ m.

67. Since the area of the circle is 72 cm ^{2} , the radius of the circle is $r = \sqrt{A / \pi} = \sqrt{72 / \pi}$.

Then the area of the sector is $A = \frac{1}{2}r^2\theta = \frac{1}{2} \cdot \frac{72}{\pi} \cdot \frac{\pi}{6} = 6$ cm ^{2} .

69. The circumference of each wheel is $\pi d = 28\pi$ in. If the wheels revolve $10,000$ times, the distance traveled is $10,000 \cdot 28\pi$ in. $\cdot \frac{1 \text{ ft}}{12 \text{ in.}} \cdot \frac{1 \text{ mi}}{5280 \text{ ft}} \approx 13.88$ mi.

71. We find the measure of the angle in degrees and then convert to radians.

$\theta = 40.5° - 25.5° = 15°$ and $15 \cdot \frac{\pi}{180°}$ rad $= \frac{\pi}{12}$ rad. Then using the formula $s = \theta r$,

we have $s = \frac{\pi}{12} \cdot 3960 = 330\pi \approx 1036.725$ and so the distance between the two cities is

roughly 1037 mi.

73. In one day, the earth travels $\frac{1}{365}$ of its orbit which is $\frac{2\pi}{365}$ rad. Then $s = \theta r = \frac{2\pi}{365} \cdot 93,000,000 \approx 1,600,911.3$, so the distance traveled is approximately 1.6 million miles.

75. The central angle is 1 minute $= \left(\frac{1}{60}\right)° = \frac{1}{60} \cdot \frac{\pi}{180°}$ rad $= \frac{\pi}{10{,}800}$ rad. Then

$s = \theta r = \frac{\pi}{10{,}800} \cdot 3960 \approx 1.152$, and so a nautical mile is approximately 1.152 mi.

77. The area is equal to the area of the large sector (with radius 34 in.) minus the area of the small sector (with radius 14 in.) Thus, $A = \frac{1}{2}r_1^2\theta - \frac{1}{2}r_2^2\theta = \frac{1}{2}\left(34^2 - 14^2\right)\left(135° \cdot \frac{\pi}{180°}\right) \approx 1131$ in. ^{2}

.

79. (a) The angular speed is $\omega = \frac{45 \cdot 2\pi \text{ rad}}{1 \text{ min}} = 90\pi$ rad/min.

(b) The linear speed is $v = \frac{45 \cdot 2\pi \cdot 16}{1} = 1440\pi$ in./min ≈ 4523.9 in./min.

81. $v = \dfrac{8 \cdot 2\pi \cdot 2}{15} = \dfrac{32\pi}{15} \approx 6.702$ ft/s.

83. 23 h 56 min 4 s $= 23.9344$ hr. So the linear speed is

$$\frac{1 \cdot 2\pi \cdot (3960)}{1 \text{ day}} \cdot \frac{1 \text{ day}}{23.9344 \text{ hr}} \approx 1039.57 \text{ mi/h.}$$

85. $v = \dfrac{100 \cdot 2\pi \cdot 0.20 \text{ m}}{60 \text{ s}} = \dfrac{2\pi}{3} \approx 2.09$ m/s.

87. (a) The circumference of the opening is the length of the arc subtended by the angle θ on the flat piece of paper, that is, $C = s = r\theta = 6 \cdot \frac{5\pi}{3} = 10\pi \approx 31.4$ cm.

(b) Solving for r, we find $r = \dfrac{C}{2\pi} = \dfrac{10\pi}{2\pi} = 5$ cm.

(c) By the Pythagorean Theorem, $h^2 = 6^2 - 5^2 = 11$, so $h = \sqrt{11} \approx 3.3$ cm.

(d) The volume of a cone is $V = \frac{1}{3}\pi r^2 h$. In this case $V = \frac{1}{3}\pi \cdot 5^2 \cdot \sqrt{11} \approx 86.8$ cm ^{3}.

89. Answers will vary, although of course everyone prefers radians.

3.2 Trigonometry of Right Triangles

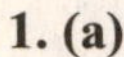

1. (a)

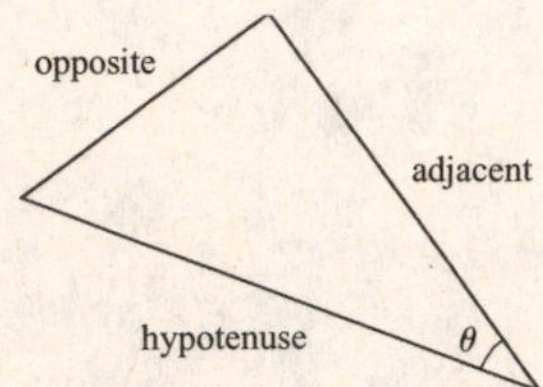

(b) $\sin\theta = \dfrac{\text{opposite}}{\text{hypotenuse}}$, $\cos\theta = \dfrac{\text{adjacent}}{\text{hypotenuse}}$, and $\tan\theta = \dfrac{\text{opposite}}{\text{adjacent}}$.

(c) The trigonometric ratios do not depend on the size of the triangle because all right triangles with angle θ are *similar*.

3. $\sin\theta = \frac{4}{5}$, $\cos\theta = \frac{3}{5}$, $\tan\theta = \frac{4}{3}$, $\csc\theta = \frac{5}{4}$, $\sec\theta = \frac{5}{3}$, $\cot\theta = \frac{3}{4}$

5. The remaining side is obtained by the Pythagorean Theorem: $\sqrt{41^2 - 40^2} = \sqrt{81} = 9$. Then $\sin\theta = \frac{40}{41}$, $\cos\theta = \frac{9}{41}$, $\tan\theta = \frac{40}{9}$, $\csc\theta = \frac{41}{40}$, $\sec\theta = \frac{41}{9}$, $\cot\theta = \frac{9}{40}$

7. The remaining side is obtained by the Pythagorean Theorem: $\sqrt{3^2 + 2^2} = \sqrt{13}$. Then $\sin\theta = \frac{2}{\sqrt{13}} = \frac{2\sqrt{13}}{13}$, $\cos\theta = \frac{3}{\sqrt{13}} = \frac{3\sqrt{13}}{13}$, $\tan\theta = \frac{2}{3}$, $\csc\theta = \frac{\sqrt{13}}{2}$, $\sec\theta = \frac{\sqrt{13}}{3}$, $\cot\theta = \frac{3}{2}$

9. $c = \sqrt{5^2 + 3^2} = \sqrt{34}$

(a) $\sin\alpha = \cos\beta = \frac{3}{\sqrt{34}} = \frac{3\sqrt{34}}{34}$

(b) $\tan\alpha = \cot\beta = \frac{3}{5}$

(c) $\sec\alpha = \csc\beta = \frac{\sqrt{34}}{5}$

11. Since $\sin 30^\circ = \dfrac{x}{25}$, we have $x = 25\sin 30^\circ = 25 \cdot \frac{1}{2} = \frac{25}{2}$.

13. Since $\sin 60^\circ = \dfrac{x}{13}$, we have $x = 13\sin 60^\circ = 13 \cdot \frac{\sqrt{3}}{2} = \frac{13\sqrt{3}}{2}$.

15. Since $\tan 36^\circ = \dfrac{12}{x}$, we have $x = \dfrac{12}{\tan 36^\circ} \approx 16.51658$.

17. $\dfrac{x}{28} = \cos\theta \;\Leftrightarrow\; x = 28\cos\theta$, and $\dfrac{y}{28} = \sin\theta \;\Leftrightarrow\; y = 28\sin\theta$.

19. $\sin\theta = \frac{3}{5}$. Then the third side is $x = \sqrt{5^2 - 3^2} = 4$. The other five ratios are $\cos\theta = \frac{4}{5}$, $\tan\theta = \frac{3}{4}$, $\csc\theta = \frac{5}{3}$, $\sec\theta = \frac{5}{4}$, and $\cot\theta = \frac{4}{3}$.

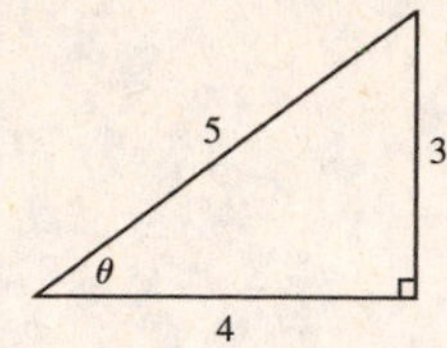

21. $\cot\theta = 1$. Then the third side is $r = \sqrt{1^2 + 1^2} = \sqrt{2}$. The other five ratios are $\sin\theta = \frac{1}{\sqrt{2}} = \frac{\sqrt{2}}{2}$, $\cos\theta = \frac{1}{\sqrt{2}} = \frac{\sqrt{2}}{2}$, $\tan\theta = 1$, $\csc\theta = \sqrt{2}$, and $\sec\theta = \sqrt{2}$.

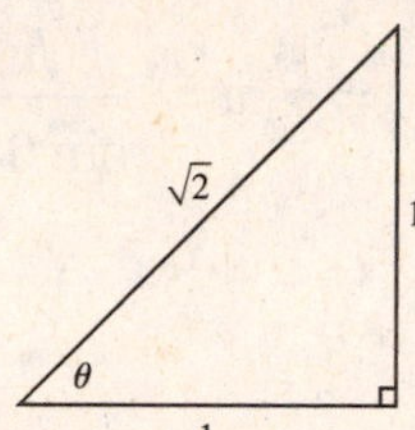

23. $\sec\theta = 7$. The third side is $y = \sqrt{7^2 - 2^2} = \sqrt{45} = 3\sqrt{5}$ The other five ratios are $\sin\theta = \frac{3\sqrt{5}}{7}$, $\cos\theta = \frac{2}{7}$, $\tan\theta = \frac{3\sqrt{5}}{2}$, $\csc\theta = \frac{7}{3\sqrt{5}} = \frac{7\sqrt{5}}{15}$, and $\cot\theta = \frac{2}{3\sqrt{5}} = \frac{2\sqrt{5}}{15}$.

25. $\sin\frac{\pi}{6} + \cos\frac{\pi}{6} = \frac{1}{2} + \frac{\sqrt{3}}{2} = \frac{1+\sqrt{3}}{2}$

27. $\sin 30^\circ \cos 60^\circ + \sin 60^\circ \cos 30^\circ = \frac{1}{2}\cdot\frac{1}{2} + \frac{\sqrt{3}}{2}\cdot\frac{\sqrt{3}}{2} = \frac{1}{4} + \frac{3}{4} = 1$

29. $\left(\cos 30^\circ\right)^2 - \left(\sin 30^\circ\right)^2 = \left(\frac{\sqrt{3}}{2}\right)^2 - \left(\frac{1}{2}\right)^2 = \frac{3}{4} - \frac{1}{4} = \frac{1}{2}$

31. This is an isosceles right triangle, so the other leg has length $16\tan 45^\circ = 16$, the hypotenuse has length $\dfrac{16}{\sin 45^\circ} = 16\sqrt{2} \approx 22.63$, and the other angle is $90^\circ - 45^\circ = 45^\circ$.

33. The other leg has length $35\tan 52^\circ \approx 44.79$, the hypotenuse has length $\dfrac{35}{\cos 52^\circ} \approx 56.85$, and the other angle is $90^\circ - 52^\circ = 38^\circ$.

35. The adjacent leg has length $33.5\cos\frac{\pi}{8} \approx 30.95$, the opposite leg has length $33.5\sin\frac{\pi}{8} \approx 12.82$, and the other angle is $\frac{\pi}{2} - \frac{\pi}{8} = \frac{3\pi}{8}$.

37. The adjacent leg has length $\dfrac{106}{\tan\frac{\pi}{5}} \approx 145.90$, the hypotenuse has length $\dfrac{106}{\sin\frac{\pi}{5}} \approx 180.34$, and the other angle is $\frac{\pi}{2} - \frac{\pi}{5} = \frac{3\pi}{10}$.

39. $\sin\theta \approx \frac{1}{2.24} \approx 0.45. \cos\theta \approx \frac{2}{2.24} \approx 0.89$, $\tan\theta = \frac{1}{2}$, csc $\theta \approx 2.24$, sec $\theta \approx \frac{2.24}{2} \approx 1.12$, $\cot\theta \approx 2.00$.

41. $x = \dfrac{100}{\tan 60°} + \dfrac{100}{\tan 30°} \approx 230.9$

43. Let h be the length of the shared side. Then $\sin 60° = \dfrac{50}{h} \Leftrightarrow h = \dfrac{50}{\sin 60°} \approx 57.735$ $\Leftrightarrow \sin 65° = \dfrac{h}{x} \Leftrightarrow x = \dfrac{h}{\sin 65°} \approx 63.7$

45.

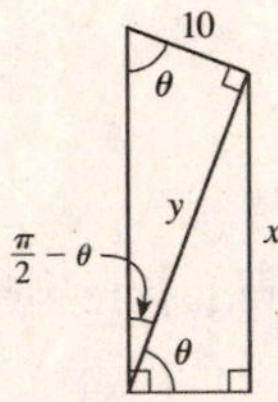

From the diagram, $\sin\theta = \dfrac{x}{y}$ and $\tan\theta = \dfrac{y}{10}$, so $x = y\sin\theta = 10\sin\theta\tan\theta$.

47. Let h be the height, in feet, of the Empire State Building. Then $\tan 11° = \dfrac{h}{5280} \Leftrightarrow$ $h = 5280 \cdot \tan 11° \approx 1026$ ft.

49. **(a)** Let h be the distance, in miles, that the beam has diverged. Then $\tan 0.5° = \dfrac{h}{240{,}000} \Leftrightarrow$ $h = 240\ ,\ 000 \cdot \tan 0.5° \approx 2100$ mi.

(b) Since the deflection is about 2100 mi whereas the radius of the moon is about 1000 mi, the beam will not strike the moon.

51. Let h represent the height, in feet, that the ladder reaches on the building. Then $\sin 72° = \dfrac{h}{20}$ $\Leftrightarrow h = 20\sin 72° \approx 19$ ft.

53. Let h be the height, in feet, of the kite above the ground. Then $\sin 50° = \dfrac{h}{450} \Leftrightarrow$ $h = 450\sin 50° \approx 345$ ft.

55. Let h_1 be the height of the window in feet and h_2 be the height from the window to the top of the tower. Then $\tan 25° = \dfrac{h_1}{325} \Leftrightarrow h_1 = 325 \cdot \tan 25° \approx 152$ ft. Also, $\tan 39° = \dfrac{h_2}{325}$ $\Leftrightarrow h_2 = 325 \cdot \tan 39° \approx 263$ ft. Therefore, the height of the window is approximately 152 ft and the height of the tower is approximately $152 + 263 = 415$ ft.

57. Let d_1 be the distance, in feet, between a point directly below the plane and one car, and d_2 be the distance, in feet, between the same point and the other car. Then $\tan 52° = \dfrac{d_1}{5150} \Leftrightarrow$

$d_1 = 5150 \cdot \tan 52° \approx 6591.7$ ft. Also, $\tan 38° = \dfrac{d_2}{5150} \Leftrightarrow$

$d_2 = 5150 \cdot \tan 38° \approx 4023.6$ ft. So in this case, the distance between the two cars is about 2570 ft.

59. Let x be the horizontal distance, in feet, between a point on the ground directly below the top of the mountain and the point on the plain closest to the mountain. Let h be the height, in feet, of the mountain. Then $\tan 35° = \dfrac{h}{x}$ and $\tan 32° = \dfrac{h}{x+1000}$. So

$h = x \tan 35° = (x+1000)\tan 32° \Leftrightarrow x = \dfrac{1000 \cdot \tan 32°}{\tan 35° - \tan 32°} \approx 8294.2$. Thus

$h \approx 8294.2 \cdot \tan 35° \approx 5808$ ft.

61. Let d be the distance, in miles, from the earth to the sun. Then $\sec 89.85° = \dfrac{d}{240{,}000} \Leftrightarrow$

$d = 240{,}000 \cdot \sec 89.85° \approx 91.7$ million miles.

63. Let r represent the radius, in miles, of the earth. Then $\sin 60.276° = \frac{r}{r+600} \Leftrightarrow$

$(r+600)\sin 60.276° = r \Leftrightarrow 600 \sin 60.276° = r(1 - \sin 60.276°) \Leftrightarrow$

$r = \frac{600 \sin 60.276°}{1 - \sin 60.276°} \approx 3960.099$. So the earth's radius is about 3960 mi.

65. Let d be the distance, in AU, between Venus and the sun. Then $\sin 46.3° = \dfrac{d}{1} = d$, so

$d = \sin 46.3° \approx 0.723$ AU.

3.3 Trigonometric Functions of Angles

1. If the angle θ is in standard position and $P\left(x,y\right)$ is a point on the terminal side of θ, and r is the distance from the origin to P, then $\sin\theta = \dfrac{y}{r}$, $\cos\theta = \dfrac{x}{r}$, and $\tan\theta = \dfrac{y}{x}$.

3. (a) The reference angle for 150° is $180^\circ - 150^\circ = 30^\circ$.

(b) The reference angle for 330° is $360^\circ - 330^\circ = 30^\circ$.

(c) The reference angle for -30° is $-\left(-30^\circ\right) = 30^\circ$.

5. (a) The reference angle for 225° is $225^\circ - 180^\circ = 45^\circ$.

(b) The reference angle for 810° is $810^\circ - 720^\circ = 90^\circ$.

(c) The reference angle for -105° is $180^\circ - 105^\circ = 75^\circ$.

7. (a) The reference angle for $\frac{11\pi}{4}$ is $3\pi - \frac{11\pi}{4} = \frac{\pi}{4}$.

(b) The reference angle for $-\frac{11\pi}{6}$ is $2\pi - \frac{11\pi}{6} = \frac{\pi}{6}$.

(c) The reference angle for $\frac{11\pi}{3}$ is $4\pi - \frac{11\pi}{3} = \frac{\pi}{3}$.

9. (a) The reference angle for $\frac{5\pi}{7}$ is $\pi - \frac{5\pi}{7} = \frac{2\pi}{7}$.

(b) The reference angle for -1.4π is $1.4\pi - \pi = 0.4\pi$.

(c) The reference angle for 1.4 is 1.4 because $1.4 < \frac{\pi}{2}$.

11. $\sin 150^\circ = \sin 30^\circ = \frac{1}{2}$

13. $\cos 210^\circ = -\cos 30^\circ = -\frac{\sqrt{3}}{2}$

15. $\tan\left(-60^\circ\right) = -\tan 60^\circ = -\sqrt{3}$

17. $\csc\left(-630^\circ\right) = \csc 90^\circ = \frac{1}{\sin 90^\circ} = 1$

19. $\cos 570^\circ = -\cos 30^\circ = -\frac{\sqrt{3}}{2}$

21. $\tan 750^\circ = \tan 30^\circ = \frac{1}{\sqrt{3}} = \frac{\sqrt{3}}{3}$

23. $\sin\frac{2\pi}{3} = \sin\frac{\pi}{3} = \frac{\sqrt{3}}{2}$

25. $\sin\frac{3\pi}{2} = -\sin\frac{\pi}{2} = -1$

27. $\cos\left(-\frac{7\pi}{3}\right) = \cos\frac{\pi}{3} = \frac{1}{2}$

29. $\sec\frac{17\pi}{3} = \sec\frac{\pi}{3} = \dfrac{1}{\cos\frac{\pi}{3}} = 2$

31. $\cot\left(-\frac{\pi}{4}\right) = -\cot\frac{\pi}{4} = \dfrac{-1}{\tan\frac{\pi}{4}} = -1$

33. $\tan\frac{5\pi}{2} = \tan\frac{\pi}{2}$ which is undefined.

35. Since $\sin\theta < 0$ and $\cos\theta < 0$, θ is in quadrant III.

37. $\sec\theta > 0 \;\Rightarrow\; \cos\theta > 0$. Also $\tan\theta < 0 \;\Rightarrow\; \dfrac{\sin\theta}{\cos\theta} < 0 \;\Leftrightarrow\; \sin\theta < 0$ (since $\cos\theta > 0$). Since $\sin\theta < 0$ and $\cos\theta > 0$, θ is in quadrant IV.

39. $\sec^2\theta = 1 + \tan^2\theta \quad \Leftrightarrow \quad \tan^2\theta = \dfrac{1}{\cos^2\theta} - 1 \quad \Leftrightarrow$

$\tan\theta = \sqrt{\dfrac{1}{\cos^2\theta} - 1} = \sqrt{\dfrac{1-\cos^2\theta}{\cos^2\theta}} = \dfrac{\sqrt{1-\cos^2\theta}}{|\cos\theta|} = \dfrac{\sqrt{1-\cos^2\theta}}{-\cos\theta}$ (since

$\cos\theta < 0$ in quadrant III, $|\cos\theta| = -\cos\theta$). Thus $\tan\theta = -\dfrac{\sqrt{1-\cos^2\theta}}{\cos\theta}$.

41. $\cos^2\theta + \sin^2\theta = 1 \quad \Leftrightarrow \quad \cos\theta = \sqrt{1-\sin^2\theta}$ because $\cos\theta > 0$ in quadrant IV.

43. $\sec^2\theta = 1 + \tan^2\theta \quad \Leftrightarrow \quad \sec\theta = -\sqrt{1+\tan^2\theta}$ because $\sec\theta < 0$ in quadrant II.

45. $\sin\theta = \frac{3}{5}$. Then $x = -\sqrt{5^2 - 3^2} = -\sqrt{16} = -4$, since θ is in quadrant II. Thus,

$\cos\theta = -\frac{4}{5}$, $\tan\theta = -\frac{3}{4}$, $\csc\theta = \frac{5}{3}$, $\sec\theta = -\frac{5}{4}$, and $\cot\theta = -\frac{4}{3}$.

47. $\tan\theta = -\frac{3}{4}$. Then $r = \sqrt{3^2 + 4^2} = 5$, and so $\sin\theta = -\frac{3}{5}$, $\cos\theta = \frac{4}{5}$,

$\csc\theta = -\frac{5}{3}$, $\sec\theta = \frac{5}{4}$, and $\cot\theta = -\frac{4}{3}$.

49. $\csc\theta = 2$. Then $\sin\theta = \frac{1}{2}$ and $x = \sqrt{2^2 - 1^2} = \sqrt{3}$. So $\sin\theta = \frac{1}{2}$, $\cos\theta = \frac{\sqrt{3}}{2}$,

$\tan\theta = \frac{1}{\sqrt{3}} = \frac{\sqrt{3}}{3}$, $\sec\theta = \frac{2}{\sqrt{3}} = \frac{2\sqrt{3}}{3}$, and $\cot\theta = \sqrt{3}$.

51. $\cos\theta = -\frac{2}{7}$. Then $y = \sqrt{7^2 - 2^2} = \sqrt{45} = 3\sqrt{5}$, and so $\sin\theta = \frac{3\sqrt{5}}{7}$,

$\tan\theta = -\frac{3\sqrt{5}}{2}$, $\csc\theta = \frac{7}{3\sqrt{5}} = \frac{7\sqrt{5}}{15}$, $\sec\theta = -\frac{7}{2}$, and $\cot\theta = -\frac{2}{3\sqrt{5}} = -\frac{2\sqrt{5}}{15}$.

53. (a) $\sin 2\theta = \sin\left(2 \cdot \frac{\pi}{3}\right) = \sin\frac{2\pi}{3} = \sin\frac{\pi}{3} = \frac{\sqrt{3}}{2}$, while

$2\sin\theta = 2\sin\frac{\pi}{3} = 2 \cdot \frac{\sqrt{3}}{2} = \sqrt{3}$.

(b) $\sin\frac{1}{2}\theta = \sin\left(\frac{1}{2} \cdot \frac{\pi}{3}\right) = \sin\frac{\pi}{6} = \frac{1}{2}$, while $\frac{1}{2}\sin\theta = \frac{1}{2}\sin\frac{\pi}{3} = \frac{1}{2} \cdot \frac{\sqrt{3}}{2} = \frac{\sqrt{3}}{4}$.

(c) $\sin^2\theta = \left(\sin\frac{\pi}{3}\right)^2 = \left(\frac{\sqrt{3}}{2}\right)^2 = \frac{3}{4}$, while $\sin\left(\theta^2\right) = \sin\left(\frac{\pi}{3}\right)^2 = \sin\frac{\pi^2}{9} \approx 0.88967$.

55. $a = 10$, $b = 22$, and $\theta = 10^\circ$. Thus, the area of the triangle is

$A = \frac{1}{2}(10)(22)\sin 10^\circ = 110\sin 10^\circ \approx 19.1$.

57. $A = 16$, $a = 5$, and $b = 7$. So $\sin\theta = \dfrac{2A}{ab} = \dfrac{2 \cdot 16}{5 \cdot 7} = \dfrac{32}{35} \quad \Leftrightarrow$

$\theta = \sin^{-1}\frac{32}{35} \approx 66.1^\circ$.

59. For the sector defined by the two sides, $A_1 = \frac{1}{2}r^2\theta = \frac{1}{2} \cdot 2^2 \cdot 120^\circ \cdot \frac{\pi}{180^\circ} = \frac{4\pi}{3}$. For the triangle defined by the two sides, $A_2 = \frac{1}{2}ab\sin\theta = \frac{1}{2} \cdot 2 \cdot 2 \cdot \sin 120^\circ = 2\sin 60^\circ = \sqrt{3}$. Thus the area of the region is $A_1 - A_2 = \frac{4\pi}{3} - \sqrt{3} \approx 2.46$.

61. $\sin^2\theta + \cos^2\theta = 1 \iff \left(\sin^2\theta + \cos^2\theta\right)\cdot\dfrac{1}{\cos^2\theta} = 1\cdot\dfrac{1}{\cos^2\theta} \iff$

$\tan^2\theta + 1 = \sec^2\theta$

63. (a) $\tan\theta = \dfrac{h}{1\text{ mile}}$, so $h = \tan\theta\cdot 1\text{ mile}\cdot\dfrac{5280\text{ ft}}{1\text{ mile}} = 5280\tan\theta$ ft.

(b)

θ	20°	60°	80°	85°
h	1922	9145	29 , 944	60 , 351

65. (a) From the figure in the text, we express depth and width in terms of θ. Since $\sin\theta = \dfrac{\text{depth}}{20}$ and $\cos\theta = \dfrac{\text{width}}{20}$, we have depth $= 20\sin\theta$ and width $= 20\cos\theta$. Thus, the cross-section area of the beam is

$A(\theta) = (\text{depth})(\text{width}) = (20\cos\theta)(20\sin\theta) = 400\cos\theta\sin\theta$.

(c) The beam with the largest cross-sectional area is the square beam, $10\sqrt{2}$ by $10\sqrt{2}$ (about 14.14 by 14.14).

(b)

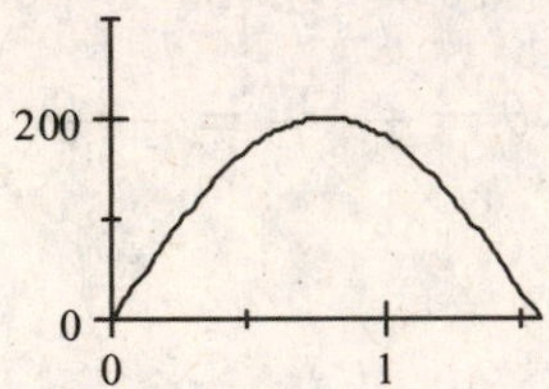

67. (a) On Earth, the range is $R = \dfrac{v_0^2\sin(2\theta)}{g} = \dfrac{12^2\sin\frac{\pi}{3}}{32} = \dfrac{9\sqrt{3}}{4} \approx 3.897$ ft and the height is

$H = \dfrac{v_0^2\sin^2\theta}{2g} = \dfrac{12^2\sin^2\frac{\pi}{6}}{2\cdot 32} = \dfrac{9}{16} = 0.5625$ ft.

(b) On the moon, $R = \dfrac{12^2\sin\frac{\pi}{3}}{5.2} \approx 23.982$ ft and $H = \dfrac{12^2\sin^2\frac{\pi}{6}}{2\cdot 5.2} \approx 3.462$ ft

69. (a) $W = 3.02 - 0.38\cot\theta + 0.65\csc\theta$

(b) From the graph, it appears that W has its minimum value at about $\theta = 0.946 \approx 54.2°$.

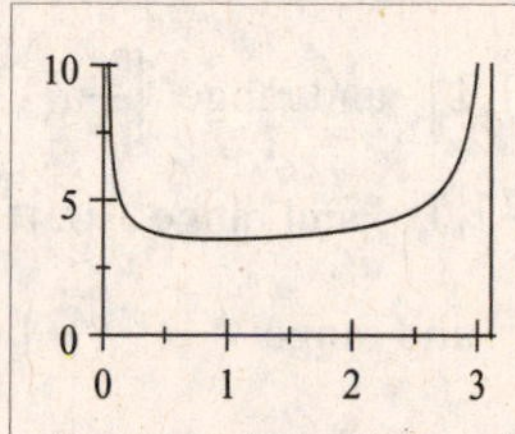

71. We have $\sin\alpha = k\sin\beta$, where $\alpha = 59.4°$ and $k = 1.33$. Substituting,

$\sin 59.4° = 1.33\sin\beta \;\Rightarrow\; \sin\beta = \dfrac{\sin 59.4°}{1.33} \approx 0.6472$. Using a calculator, we find that

$\beta \approx \sin^{-1} 0.6472 \approx 40.3°$, so $\theta = 4\beta - 2\alpha \approx 4(40.3°) - 2(59.4°) = 42.4°$.

73. $\cos\theta = \dfrac{\text{adj}}{\text{hyp}} = \dfrac{|OP|}{|OR|} = \dfrac{|OP|}{1} = |OP|$. Since QS is tangent to the circle at R ,

ΔORQ is a right triangle. Then $\tan\theta = \dfrac{\text{opp}}{\text{adj}} = \dfrac{|RQ|}{|OR|} = |RQ|$ and

$\sec\theta = \dfrac{\text{hyp}}{\text{adj}} = \dfrac{|OQ|}{|OR|} = |OQ|$. Since $\angle SOQ$ is a right angle ΔSOQ is a right triangle

and $\angle OSR = \theta$. Then $\csc\theta = \dfrac{\text{hyp}}{\text{opp}} = \dfrac{|OS|}{|OR|} = |OS|$ and

$\cot\theta = \dfrac{\text{adj}}{\text{opp}} = \dfrac{|SR|}{|OR|} = |SR|$. Summarizing, we have $\sin\theta = |PR|$, $\cos\theta = |OP|$,

$\tan\theta = |RQ|$, $\sec\theta = |OQ|$, $\csc\theta = |OS|$, and $\cot\theta = |SR|$.

3.4 Inverse Trigonometric Functions and Triangles

1. (a) The function $\sin^{-1}$ has domain $[-1,1]$ and range $\left[-\frac{\pi}{2},\frac{\pi}{2}\right]$.

(b) The function $\cos^{-1}$ has domain $[-1,1]$ and range $[0,\pi]$.

(c) The function $\tan^{-1}$ has domain $\mathbb{R}$ and range $\left(-\frac{\pi}{2},\frac{\pi}{2}\right)$.

3. (a) $\sin^{-1}\frac{1}{2}=\frac{\pi}{6}$ because $\sin\frac{\pi}{6}=\frac{1}{2}$ and $\frac{\pi}{6}$ lies in the range of $\sin^{-1}$, $\left[-\frac{\pi}{2},\frac{\pi}{2}\right]$.

(b) $\cos^{-1}\left(-\frac{\sqrt{3}}{2}\right)=\frac{5\pi}{6}$ because $\cos\frac{5\pi}{6}=-\frac{\sqrt{3}}{2}$ and $\frac{5\pi}{6}$ lies in the range of $\cos^{-1}$, $[0,\pi]$

(c) $\tan^{-1}(-1)=-\frac{\pi}{4}$ because $\tan\left(-\frac{\pi}{4}\right)=-1$ and $-\frac{\pi}{4}$ lies in the range of $\tan^{-1}$, $\left(-\frac{\pi}{2},\frac{\pi}{2}\right)$.

5. (a) $\sin^{-1}\left(-\frac{1}{2}\right)=-\frac{\pi}{6}$

(b) $\cos^{-1}\frac{1}{2}=\frac{\pi}{3}$

(c) $\tan^{-1}\left(\frac{\sqrt{3}}{3}\right)=\frac{\pi}{6}$

7. $\sin^{-1}(0.45)\approx 0.46677$

9. $\cos^{-1}\left(-\frac{1}{4}\right)\approx 1.82348$

11. $\tan^{-1}3\approx 1.24905$

13. $\cos^{-1}3$ is undefined.

15. $\sin\theta=\frac{6}{10}=\frac{3}{5}$, so

$\theta=\sin^{-1}\frac{3}{5}\approx 36.9^\circ$.

17. $\tan\theta=\frac{9}{13}$, so

$\theta=\tan^{-1}\frac{9}{13}\approx 34.7^\circ$.

19. $\sin\theta=\frac{4}{7}$, so $\theta=\sin^{-1}\frac{4}{7}\approx 34.8^\circ$.

21. We use $\sin^{-1}$ to find one solution in the interval $\left[-\frac{\pi}{2},\frac{\pi}{2}\right]$. $\sin\theta=\frac{1}{2}$ $\Rightarrow$ $\theta=\sin^{-1}\frac{1}{2}=30^\circ$. Another solution with θ between 0° and 180° is obtained by taking the supplement of the angle: $180^\circ-30^\circ=150^\circ$. So the solutions of the equation with θ between 0° and 180° are $\theta=30^\circ$ and $\theta=150^\circ$.

23. $\sin\theta=0.7$, so the solutions with $0^\circ<\theta<180^\circ$ are $\theta=\sin^{-1}0.7\approx 44.4^\circ$ and $\theta=180^\circ-\sin^{-1}0.7\approx 135.6^\circ$.

25. The cosine function is one-to-one on the interval $[0,\pi]$, so there is only one solution of the equation with θ between 0° and 180°. We find that solution by taking $\cos^{-1}$ of each side: $\cos\theta=0.7$ $\Leftrightarrow$ $\theta=\cos^{-1}0.7\approx 45.6^\circ$.

27. To find $\sin\left(\cos^{-1}\frac{3}{5}\right)$, first let $\theta = \cos^{-1}\frac{3}{5}$. Then θ is the number in the interval $\left[-\frac{\pi}{2},\frac{\pi}{2}\right]$ whose cosine is $\frac{3}{5}$. We draw a right triangle with θ as one of its acute angles, with adjacent side 3 and hypotenuse 5. The remaining leg of the triangle is found by the Pythagorean Theorem to be 4. From the figure we get $\sin\left(\cos^{-1}\frac{3}{5}\right) = \sin\theta = \frac{4}{5}$.

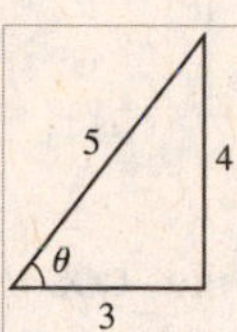

Another method: By the cancellation properties of inverse functions, $\cos\left(\cos^{-1}\frac{3}{5}\right)$ is exactly $\frac{3}{5}$. To find $\sin\left(\cos^{-1}\frac{3}{5}\right)$, we first write the sine function in terms of the cosine function. Let $u = \cos^{-1}\frac{3}{5}$. Since $0 \leq u \leq \pi$, $\sin u$ is positive, and since $\cos^2 u + \sin^2 u = 1$, we can write

$$\sin u = \sqrt{1-\cos^2 u} = \sqrt{1-\cos^2\left(\cos^{-1}\frac{3}{5}\right)} = \sqrt{1-\left(\frac{3}{5}\right)^2} = \sqrt{1-\frac{9}{25}} = \sqrt{\frac{16}{25}} = \frac{4}{5}.$$

Therefore, $\sin\left(\cos^{-1}\frac{3}{5}\right) = \frac{4}{5}$.

29. To find $\sec\left(\sin^{-1}\frac{12}{13}\right)$, we draw a right triangle with angle θ, opposite side 12, and hypotenuse 13. From the figure we see that $\sec\left(\sin^{-1}\frac{12}{13}\right) = \sec\theta = \frac{13}{5}$.

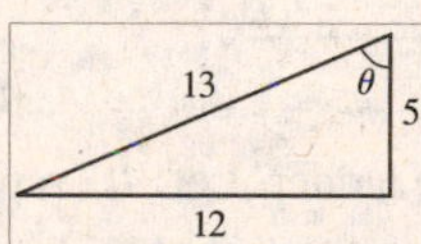

31. To find $\tan\left(\sin^{-1}\frac{12}{13}\right)$, we draw a right triangle with angle θ, opposite side 12, and hypotenuse 13. From the figure we see that $\tan\left(\sin^{-1}\frac{12}{13}\right) = \tan\theta = \frac{12}{5}$.

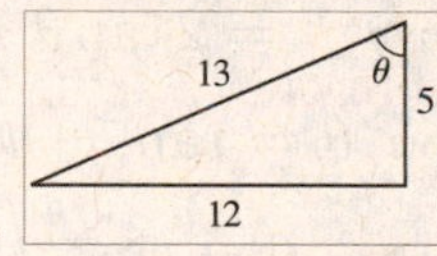

33. We want to find $\cos\left(\sin^{-1} x\right)$. Let $\theta = \sin^{-1} x$, so $\sin\theta = x$. We sketch a right triangle with an acute angle θ, opposite side x, and hypotenuse 1. By the Pythagorean Theorem, the remaining leg is $\sqrt{1 - x^2}$. From the figure we have $\cos\left(\sin^{-1} x\right) = \cos\theta = \sqrt{1 - x^2}$.

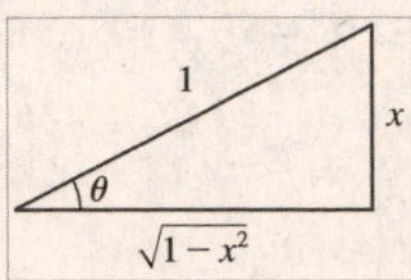

Another method: Let $u = \sin^{-1} x$. We need to find $\cos u$ in terms of x. To do so, we write cosine in terms of sine. Note that $-\frac{\pi}{2} \le u \le \frac{\pi}{2}$ because $u = \sin^{-1} x$. Now $\cos u = \sqrt{1 - \sin^2 u}$ is positive because u lies in the interval $\left[-\frac{\pi}{2}, \frac{\pi}{2}\right]$. Substituting $u = \sin^{-1} x$ and using the cancellation property $\sin(\sin^{-1} x) = x$ gives $\cos(\sin^{-1} x) = \sqrt{1 - x^2}$.

35. We want to find $\tan\left(\sin^{-1} x\right)$. Let $\theta = \sin^{-1} x$, so $\sin\theta = x$. We sketch a right triangle with an acute angle θ, opposite side x, and hypotenuse 1. By the Pythagorean Theorem, the remaining leg is $\sqrt{1 - x^2}$. From the figure we have $\tan\left(\sin^{-1} x\right) = \tan\theta = \dfrac{x}{\sqrt{1 - x^2}}$.

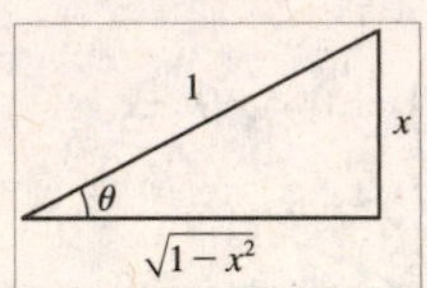

37. Let θ represent the angle of elevation of the ladder. Let h represent the height, in feet, that the ladder reaches on the building. Then $\cos\theta = \frac{6}{20} = 0.3 \Leftrightarrow \theta = \cos^{-1} 0.3 \approx 1.266$ rad $\approx 72.5^\circ$. By the Pythagorean Theorem, $h^2 + 6^2 = 20^2 \Leftrightarrow$ $h = \sqrt{400 - 36} = \sqrt{364} \approx 19$ ft.

39. (a) Solving $\tan\theta = h / 2$ for h, we have $h = 2\tan\theta$.

(b) Solving $\tan\theta = h / 2$ for θ we have $\theta = \tan^{-1}\left(h / 2\right)$.

41. (a) Solving $\sin\theta = h / 680$ for θ we have $\theta = \sin^{-1}\left(h / 680\right)$.

(b) Set $h = 500$ to get $\theta = \sin^{-1}\left(\frac{500}{680}\right) \approx 0.826$ rad.

43. (a) $\theta = \sin^{-1}\left(\dfrac{1}{(2\cdot 3+1)\tan 10^\circ}\right) = \sin^{-1}\left(\dfrac{1}{7\tan 10^\circ}\right) \approx \sin^{-1} 0.8102 \approx 54.1^\circ$

(b) For $n = 2$, $\theta = \sin^{-1}\left(\dfrac{1}{5\tan 15^\circ}\right) \approx 48.3^\circ$. For $n = 3$,

$\theta = \sin^{-1}\left(\dfrac{1}{7\tan 15^\circ}\right) \approx 32.2^\circ$. For $n = 4$, $\theta = \sin^{-1}\left(\dfrac{1}{9\tan 15^\circ}\right) \approx 24.5^\circ$.

$n = 0$ and $n = 1$ are outside of the domain for $\beta = 15^\circ$, because $\dfrac{1}{\tan 15^\circ} \approx 3.732$ and

$\dfrac{1}{3\tan 15^\circ} \approx 1.244$, neither of which is in the domain of $\sin^{-1}$.

45. Let $\beta = \csc^{-1} x$. Then $\csc \gamma = x$, as shown in the figure. Then $\sin \beta = \dfrac{1}{x}$, so

$\beta = \sin^{-1}\left(\dfrac{1}{x}\right)$. Thus, $\csc^{-1} x = \sin^{-1}\left(\dfrac{1}{x}\right)$, $x \ge 1$.

In particular, $\csc^{-1} 3 = \sin^{-1} \frac{1}{3} \approx 0.340$.

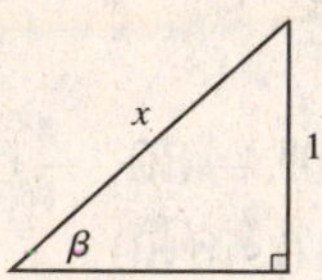

3.5 The Law of Sines

1. In triangle ABC with sides a, b, and c the Law of Sines states that $\dfrac{\sin A}{a} = \dfrac{\sin B}{b} = \dfrac{\sin C}{c}$.

3. $\angle C = 180° - 98.4° - 24.6° = 57°$.

$x = \dfrac{376 \sin 57°}{\sin 98.4°} \approx 318.75$.

5. $\angle C = 180° - 52° - 70° = 58°$.

$x = \dfrac{26.7 \sin 52°}{\sin 58°} \approx 24.8$.

7. $\sin C = \dfrac{36 \sin 120°}{45} \approx 0.693 \Leftrightarrow$

$\angle C \approx \sin^{-1} 0.693 \approx 44°$.

9. $\angle C = 180° - 46° - 20° = 114°$.

Then $a = \dfrac{65 \sin 46°}{\sin 114°} \approx 51$ and

$b = \dfrac{65 \sin 20°}{\sin 114°} \approx 24$.

11. $\angle B = 68°$, so

$\angle A = 180° - 68° - 68° = 44°$ and

$a = \dfrac{12 \sin 44°}{\sin 68°} \approx 8.99$.

13. $\angle C = 180° - 50° - 68° = 62°$.

Then $a = \dfrac{230 \sin 50°}{\sin 62°} \approx 200$ and

$b = \dfrac{230 \sin 68°}{\sin 62°} \approx 242$.

15. $\angle B = 180° - 30° - 65° = 85°$.

Then $a = \dfrac{10 \sin 30°}{\sin 85°} \approx 5.0$ and

$c = \dfrac{10 \sin 65°}{\sin 85°} \approx 9$.

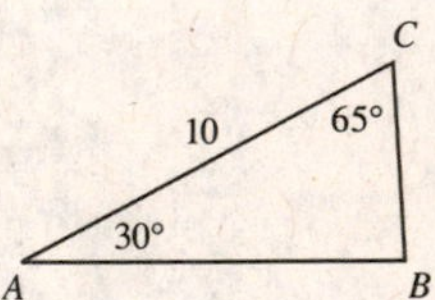

17. $\angle A = 180° - 51° - 29° = 100°$. Then $a = \dfrac{44 \sin 100°}{\sin 29°} \approx 89$ and

$c = \dfrac{44 \sin 51°}{\sin 29°} \approx 71$.

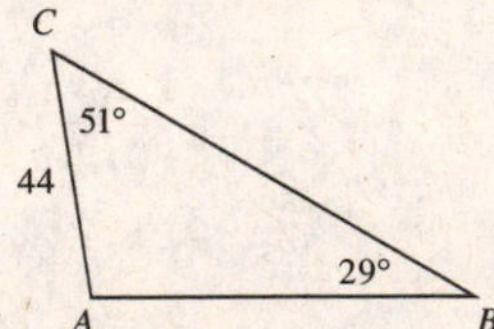

19. Since $\angle A > 90°$ there is only one triangle. $\sin B = \dfrac{15\sin 110°}{28} \approx 0.503 \quad \Leftrightarrow$

$\angle B \approx \sin^{-1} 0.503 \approx 30°$. Then $\angle C \approx 180° - 110° - 30° = 40°$, and so

$c = \dfrac{28\sin 40°}{\sin 110°} \approx 19$. Thus $\angle B \approx 30°$, $\angle C \approx 40°$, and $c \approx 19$.

21. $\angle A = 125°$ is the largest angle, but since side a is not the longest side, there can be no such triangle.

23. $\sin C = \dfrac{30\sin 25°}{25} \approx 0.507 \quad \Leftrightarrow \quad \angle C_1 \approx \sin^{-1} 0.507 \approx 30.47°$ or

$\angle C_2 \approx 180° - 39.47° = 149.53°$.

If $\angle C_1 = 30.47°$, then $\angle A_1 \approx 180° - 25° - 30.47° = 124.53°$ and

$a_1 = \dfrac{25\sin 124.53°}{\sin 25°} \approx 48.73$.

If $\angle C_2 = 149.53°$, then $\angle A_2 \approx 180° - 25° - 149.53° = 5.47°$ and

$a_2 = \dfrac{25\sin 5.47°}{\sin 25°} \approx 5.64$.

Thus, one triangle has $\angle A_1 \approx 125°$, $\angle C_1 \approx 30°$, and $a_1 \approx 49$; the other has $\angle A_2 \approx 5°$, $\angle C_2 \approx 150°$, and $a_2 \approx 5.6$.

25. $\sin B = \dfrac{100\sin 50°}{50} \approx 1.532$. Since $\left|\sin\theta\right| \leq 1$ for all θ , there can be no such angle B , and thus no such triangle.

27. $\sin A = \dfrac{26\sin 29°}{15} \approx 0.840 \quad \Leftrightarrow \quad \angle A_1 \approx \sin^{-1} 0.840 \approx 57.2°$ or

$\angle A_2 \approx 180° - 57.2° = 122.8°$.

If $\angle A_1 \approx 57.2°$, then $\angle B_1 = 180° - 29° - 57.2° = 93.8°$ and

$b_1 \approx \dfrac{15\sin 93.8°}{\sin 29°} \approx 30.9$.

If $\angle A_2 \approx 122.8°$, then $\angle B_2 = 180 - 29° - 122.8° = 28.2°$ and

$b_2 \approx \dfrac{15\sin 28.1°}{\sin 29°} \approx 14.6$.

Thus, one triangle has $\angle A_1 \approx 57.2°$, $\angle B_1 \approx 93.8°$, and $b_1 \approx 30.9$; the other has $\angle A_2 \approx 122.8°$, $\angle B_2 \approx 28.2°$, and $b_2 \approx 14.6$.

29. (a) From $\Delta\ ABC$ and the Law of Sines we get $\dfrac{\sin 30^\circ}{20} = \dfrac{\sin B}{28} \Leftrightarrow$

$\sin B = \dfrac{28 \sin 30^\circ}{20} = 0.7$, so $\angle B \approx \sin^{-1} 0.7 \approx 44.427^\circ$. Since $\Delta\ BCD$ is isosceles,

$\angle B = \angle BDC \approx 44.427^\circ$. Thus, $\angle BCD = 180^\circ - 2\angle B \approx 91.146^\circ \approx 91.1^\circ$.

(b) From $\Delta\ ABC$ we get

$\angle BCA = 180^\circ - \angle A - \angle B \approx 180^\circ - 30^\circ - 44.427^\circ = 105.573^\circ$. Hence
$\angle DCA = \angle BCA - \angle BCD \approx 105.573^\circ - 91.146^\circ = 14.4^\circ$.

31. (a) $\sin B = \dfrac{20 \sin 40^\circ}{15} \approx 0.857 \Leftrightarrow \angle B_1 \approx \sin^{-1} 0.857 \approx 58.99^\circ$ or

$\angle B_2 \approx 180^\circ - 58.99^\circ \approx 121.01^\circ$.

If $\angle B_1 = 58.99^\circ$, then $\angle C_1 \approx 180^\circ - 40^\circ - 58.99^\circ \approx 81.01^\circ$ and

$c_1 = \dfrac{15 \sin 81.01^\circ}{\sin 40^\circ} \approx 23.05$.

If $\angle B_2 = 121.01^\circ$, then $\angle C_2 \approx 180^\circ - 40^\circ - 121.01^\circ = 18.99^\circ$ and

$c_2 = \dfrac{15 \sin 18.99^\circ}{\sin 40^\circ} \approx 7.59$. Thus there are two triangles.

(b) By the area formula given in Section 6.3, $\dfrac{\text{Area of } \Delta\ ABC}{\text{Area of } \Delta\ A'B'C'} = \dfrac{\frac{1}{2} ab \sin C}{\frac{1}{2} ab \sin C'} = \dfrac{\sin C}{\sin C'}$,

because a and b are the same in both triangles.

33. (a) Let a be the distance from satellite to the tracking station A in miles. Then the subtended angle at the satellite is $\angle C = 180^\circ - 93^\circ - 84.2^\circ = 2.8^\circ$, and so $a = \dfrac{50 \sin 84.2^\circ}{\sin 2.8^\circ} \approx 1018$ mi.

(b) Let d be the distance above the ground in miles. Then $d = 1018.3 \sin 87^\circ \approx 1017$ mi.

35. $\angle C = 180^\circ - 82^\circ - 52^\circ = 46^\circ$, so by the Law of Sines, $\dfrac{|AC|}{\sin 52^\circ} = \dfrac{|AB|}{\sin 46^\circ} \Leftrightarrow$

$|AC| = \dfrac{|AB| \sin 52^\circ}{\sin 46^\circ}$, so substituting we have $|AC| = \dfrac{200 \sin 52^\circ}{\sin 46^\circ} \approx 219$ ft.

37.

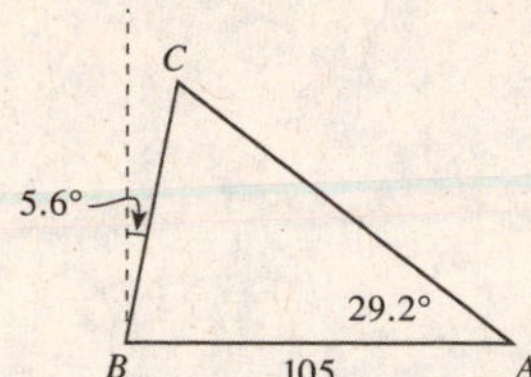

We draw a diagram. A is the position of the tourist and C is the top of the tower. $\angle B = 90° - 5.6° = 84.4°$ and so $\angle C = 180° - 29.2° - 84.4° = 66.4°$. Thus, by the Law of Sines, the length of the tower is $\left|BC\right| = \dfrac{105 \sin 29.2°}{\sin 66.4°} \approx 55.9$ m.

39. The angle subtended by the top of the tree and the sun's rays is $\angle A = 180° - 90° - 52° = 38°$ Thus the height of the tree is $h = \dfrac{215 \sin 30°}{\sin 38°} \approx 175$ ft.

41. Call the balloon's position R. Then in ΔPQR, we see that $\angle P = 62° - 32° = 30°$, and $\angle Q = 180° - 71° + 32° = 141°$. Therefore, $\angle R = 180° - 30° - 141° = 9°$. So by the Law of Sines, $\dfrac{\left|QR\right|}{\sin 30°} = \dfrac{\left|PQ\right|}{\sin 9°} \Leftrightarrow \left|QR\right| = 60 \cdot \dfrac{\sin 30°}{\sin 9°} \approx 192$ m.

43. Let d be the distance from the earth to Venus, and let β be the angle formed by sun, Venus, and earth. By the Law of Sines, $\dfrac{\sin \beta}{1} = \dfrac{\sin 39.4°}{0.723} \approx 0.878$, so either $\beta \approx \sin^{-1} 0.878 \approx 61.4°$ or $\beta \approx 180° - \sin^{-1} 0.878 \approx 118.6°$. In the first case,

$$\frac{d}{\sin\left(180° - 39.4° - 61.4°\right)} = \frac{0.723}{\sin 39.4°} \Leftrightarrow d \approx 1.119 \text{ AU; in the second case,}$$

$$\frac{d}{\sin\left(180° - 39.4° - 118.6°\right)} = \frac{0.723}{\sin 39.4°} \Leftrightarrow d \approx 0.427 \text{ AU.}$$

45.

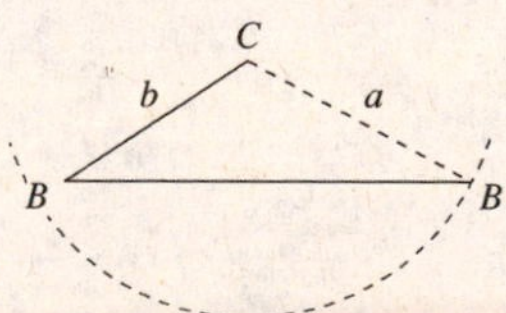

$a \geq b$: One solution

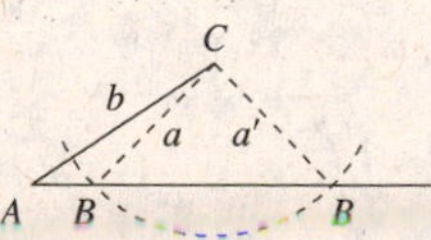

$b > a > b \sin A$: Two solutions

$a = b \sin A$: One solution

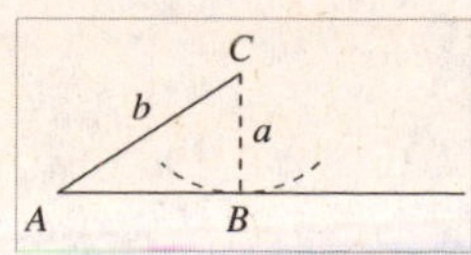

$a < b\sin A$: No solution

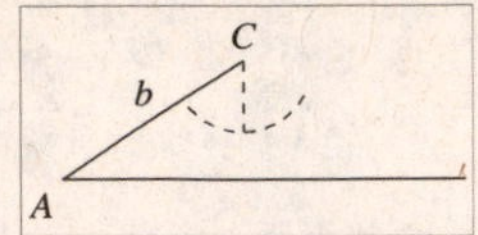

$\angle A = 30°$, $b = 100$, $\sin A = \frac{1}{2}$. If $a \geq b = 100$ then there is one triangle. If $100 > a > 100\sin 30° = 50$, then there are two possible triangles. If $a = 50$, then there is one (right) triangle. And if $a < 50$, then no triangle is possible.

3.6 The Law of Cosines

1. For triangle ABC with sides a, b, and c the Law of Cosines states $c^2 = a^2 + b^2 - 2ab\cos C$.

3. $x^2 = 21^2 + 42^2 - 2 \cdot 21 \cdot 42 \cdot \cos 39° = 441 + 1764 - 1764\cos 39° \approx 834.115$ and so $x \approx \sqrt{834.115} \approx 28.9$.

5. $x^2 = 25^2 + 25^2 - 2 \cdot 25 \cdot 25 \cdot \cos 140° = 625 + 625 - 1250\cos 140° \approx 2207.556$ and so $x \approx \sqrt{2207.556} \approx 47$.

7. $37.83^2 = 68.01^2 + 42.15^2 - 2 \cdot 68.01 \cdot 42.15 \cdot \cos\theta$. Then
$$\cos\theta = \frac{37.83^2 - 68.01^2 - 42.15^2}{-2 \cdot 68.01 \cdot 42.15} \approx 0.867 \quad\Leftrightarrow\quad \theta \approx \cos^{-1} 0.867 \approx 29.89°.$$

9. $x^2 = 24^2 + 30^2 - 2 \cdot 24 \cdot 30 \cdot \cos 30° = 576 + 900 - 1440\cos 30° \approx 228.923$ and so $x \approx \sqrt{228.923} \approx 15$.

11. $c^2 = 10^2 + 18^2 - 2 \cdot 10 \cdot 18 \cdot \cos 120° = 100 + 324 - 360\cos 120° = 604$ and so $c \approx \sqrt{604} \approx 24.576$. Then $\sin A \approx \dfrac{18\sin 120°}{24.576} \approx 0.634295 \Leftrightarrow$ $\angle A \approx \sin^{-1} 0.634295 \approx 39.4°$, and $\angle B \approx 180° - 120° - 39.4° = 20.6°$.

13. $c^2 = 3^2 + 4^2 - 2 \cdot 3 \cdot 4 \cdot \cos 53° = 9 + 16 - 24\cos 53° \approx 10.556 \Leftrightarrow$ $c \approx \sqrt{10.556} \approx 3.2$. Then $\sin B = \dfrac{4\sin 53°}{3.25} \approx 0.983 \Leftrightarrow$ $\angle B \approx \sin^{-1} 0.983 \approx 79°$ and $\angle A \approx 180° - 53° - 79° = 48°$.

15. $20^2 = 25^2 + 22^2 - 2 \cdot 25 \cdot 22 \cdot \cos A \Leftrightarrow \cos A = \frac{20^2 - 25^2 - 22^2}{-2 \cdot 25 \cdot 22} \approx 0.644 \Leftrightarrow$ $\angle A \approx \cos^{-1} 0.644 \approx 50°$. Then $\sin B \approx \frac{25\sin 49.9°}{20} \approx 0.956 \Leftrightarrow$ $\angle B \approx \sin^{-1} 0.956 \approx 73°$, and so $\angle C \approx 180° - 50° - 73° = 57°$.

17. $\sin C = \frac{162\sin 40°}{125} \approx 0.833 \Leftrightarrow \angle C_1 \approx \sin^{-1} 0.833 \approx 56.4°$ or $\angle C_2 \approx 180° - 56.4° \approx 123.6°$.

If $\angle C_1 \approx 56.4°$, then $\angle A_1 \approx 180° - 40° - 56.4° = 83.6°$ and $a_1 = \frac{125\sin 83.6°}{\sin 40°} \approx 193$.

If $\angle C_2 \approx 123.6°$, then $\angle A_2 \approx 180° - 40° - 123.6° = 16.4°$ and $a_2 = \frac{125\sin 16.4°}{\sin 40°} \approx 54.9$.

Thus, one triangle has $\angle A \approx 83.6°$, $\angle C \approx 56.4°$, and $a \approx 193$; the other has $\angle A \approx 16.4°$, $\angle C \approx 123.6°$, and $a \approx 54.9$.

19. $\sin B = \frac{65\sin 55°}{50} \approx 1.065$. Since $|\sin\theta| \le 1$ for all θ, there is no such $\angle B$, and hence there is no such triangle.

21. $\angle B = 180° - 35° - 85° = 60°$. Then $x = \frac{3\sin 35°}{\sin 60°} \approx 2$.

23. $x = \frac{50\sin 30°}{\sin 100°} \approx 25.4$

25. $b^2 = 110^2 + 138^2 - 2(110)(138)\cdot\cos 38° = 12,100 + 19,044 - 30,360\cos 38° \approx 7220.0$ and so $b \approx 85.0$. Therefore, using the Law of Cosines again, we have $\cos\theta = \frac{110^2+85^2-128^2}{2(110)(138)} \Leftrightarrow \theta \approx 89.15°$.

27. $x^2 = 38^2 + 48^2 - 2\cdot 38\cdot 48\cdot\cos 30° = 1444 + 2304 - 3648\cos 30° \approx 588.739$ and so $x \approx 24.3$.

29. The semiperimeter is $s = \frac{9+12+15}{2} = 18$, so by Heron's Formula the area is

$A = \sqrt{18(18-9)(18-12)(18-15)} = \sqrt{2916} = 54$.

31. The semiperimeter is $s = \frac{7+8+9}{2} = 12$, so by Heron's Formula the area is

$A = \sqrt{12(12-7)(12-8)(12-9)} = \sqrt{720} = 12\sqrt{5} \approx 26.8$.

33. The semiperimeter is $s = \frac{3+4+6}{2} = \frac{13}{2}$, so by Heron's Formula the area is

$A = \sqrt{\frac{13}{2}\left(\frac{13}{2}-3\right)\left(\frac{13}{2}-4\right)\left(\frac{13}{2}-6\right)} = \sqrt{\frac{455}{16}} = \frac{\sqrt{455}}{4} \approx 5.33$.

35. We draw a diagonal connecting the vertices adjacent to the $100°$ angle. This forms two triangles. Consider the triangle with sides of length 5 and 6 containing the $100°$ angle. The area of this triangle is $A_1 = \frac{1}{2}(5)(6)\sin 100° \approx 14.77$. To use Heron's Formula to find the area of the second triangle, we need to find the length of the diagonal using the Law of Cosines:

$c^2 = a^2 + b^2 - 2ab\cos C = 5^2 + 6^2 - 2\cdot 5\cdot 6\cos 100° \approx 71.419 \Rightarrow c \approx 8.45$.

Thus the second triangle has semiperimeter $s = \dfrac{8+7+8.45}{2} \approx 11.7255$ and area

$A_2 = \sqrt{11.7255(11.7255-8)(11.7255-7)(11.7255-8.45)} \approx 26.00$. The area of the quadrilateral is the sum of the areas of the two triangles:

$A = A_1 + A_2 \approx 14.77 + 26.00 = 40.77$.

37.

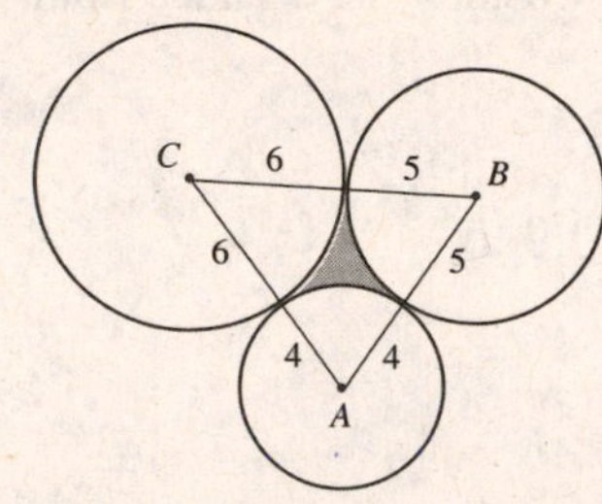

Label the centers of the circles A , B , and C , as in the figure. By the Law of Cosines,

$\cos A = \dfrac{AB^2 + AC^2 - BC^2}{2(AB)(AC)} = \dfrac{9^2 + 10^2 - 11^2}{2(9)(10)} = \frac{1}{3} \Rightarrow \angle A \approx 70.53°$. Now, by the Law of Sines, $\dfrac{\sin 70.53°}{11} = \dfrac{\sin B}{AC} = \dfrac{\sin C}{AB}$. So $\sin B = \frac{10}{11} \sin 70.53° \approx 0.85710 \Rightarrow$ $B \approx \sin^{-1} 0.85710 \approx 58.99°$ and $\sin C = \frac{9}{11} \sin 70.53° \approx 0.77139 \Rightarrow$ $C \approx \sin^{-1} 0.77139 \approx 50.48°$. The area of ΔABC is $\frac{1}{2}(AB)(AC)\sin A = \frac{1}{2}(9)(10)(\sin 70.53°) \approx 42.426$.

The area of sector A is given by $S_A = \pi R^2 \cdot \dfrac{\theta}{360°} = \pi(4)^2 \cdot \dfrac{70.53°}{360°} \approx 9.848$. Similarly, the areas of sectors B and C are $S_B \approx 12.870$ and $S_C \approx 15.859$. Thus, the area enclosed between the circles is $A = \Delta ABC - S_A - S_B - S_C \Rightarrow$

$A \approx 42.426 - 9.848 - 12.870 - 15.859 \approx 3.85 \text{ cm}^2$.

39. Let c be the distance across the lake, in miles. Then

$c^2 = 2.82^2 + 3.56^2 - 2(2.82)(3.56) \cdot \cos 40.3° \approx 5.313 \Leftrightarrow c \approx 2.30$ mi.

41. In half an hour, the faster car travels 25 miles while the slower car travels 15 miles. The distance between them is given by the Law of Cosines: $d^2 = 25^2 + 15^2 - 2(25)(15) \cdot \cos 65° \Rightarrow$

$d = \sqrt{25^2 + 15^2 - 2(25)(15) \cdot \cos 65°} = 5\sqrt{25 + 9 - 30 \cdot \cos 65°} \approx 23.1$ mi.

43. The pilot travels a distance of $625 \cdot 1.5 = 937.5$ miles in her original direction and $625 \cdot 2 = 1250$ miles in the new direction. Since she makes a course correction of $10°$ to the right, the included angle is $180° - 10° = 170°$. From the figure, we use the Law of Cosines to get the expression $d^2 = 937.5^2 + 1250^2 - 2(937.5)(1250) \cdot \cos 170° \approx 4,749,549.42$, so $d \approx 2179$ miles. Thus, the pilot's distance from her original position is approximately 2179 miles.

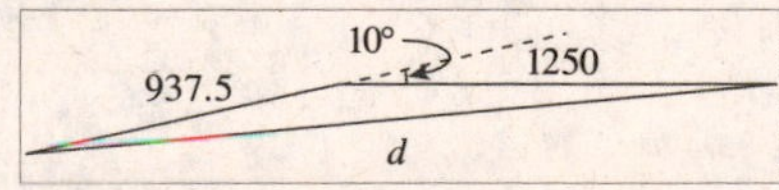

45. (a) The angle subtended at Egg Island is $100°$. Thus using the Law of Cosines, the distance from Forrest Island to the fisherman's home port is

$$x^2 = 30^2 + 50^2 - 2 \cdot 30 \cdot 50 \cdot \cos 100°$$
$$= 900 + 2500 - 3000 \cos 100° \approx 3920.945$$

and so $x \approx \sqrt{3920.945} \approx 62.62$ miles.

(b) Let θ be the angle shown in the figure. Using the Law of Sines,

$\sin \theta = \dfrac{50 \sin 100°}{62.62} \approx 0.7863 \quad \Leftrightarrow \quad \theta \approx \sin^{-1} 0.7863 \approx 51.8°$. Then

$\gamma = 90° - 20° - 51.8° = 18.2°$. Thus the bearing to his home port is S $18.2°$ E.

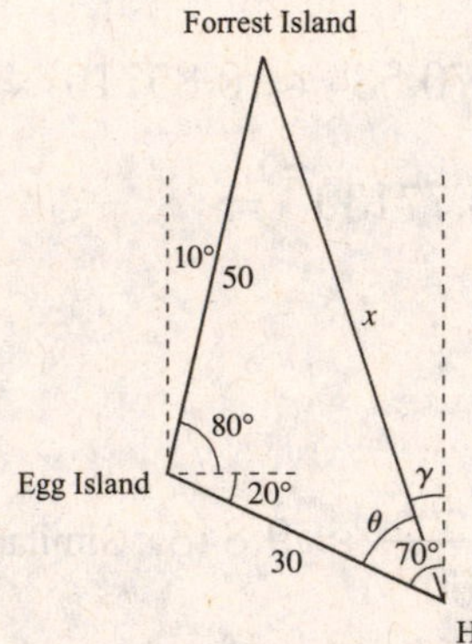

47. The largest angle is the one opposite the longest side; call this angle θ. Then by the Law of Cosines,

$$44^2 = 36^2 + 22^2 - 2(36)(22) \cdot \cos \theta \quad \Leftrightarrow \quad \cos \theta = \frac{36^2 + 22^2 - 44^2}{2(36)(22)} = -0.09848$$

$\Rightarrow \quad \theta \approx \cos^{-1}(-0.09848) \approx 96°$.

49. Let d be the distance between the kites. Then $d^2 \approx 380^2 + 420^2 - 2(380)(420) \cdot \cos 30°$

$\Rightarrow \quad d \approx \sqrt{380^2 + 420^2 - 2(380)(420) \cdot \cos 30°} \approx 211$ ft.

51. *Solution 1:* From the figure, we see that $\gamma = 106°$ and $\sin 74° = \dfrac{3400}{b} \Leftrightarrow$

$b = \dfrac{3400}{\sin 74°} \approx 3537$. Thus, $x^2 = 800^2 + 3537^2 - 2(800)(3537)\cos 106° \Rightarrow$

$x = \sqrt{800^2 + 3537^2 - 2(800)(3537)\cos 106°} \Rightarrow x \approx 3835$ ft.

Solution 2: Notice that $\tan 74° = \dfrac{3400}{a} \Leftrightarrow a = \dfrac{3400}{\tan 74°} \approx 974.9$. By the Pythagorean

Theorem, $x^2 = (a + 800)^2 + 3400^2$. So $x = \sqrt{(974.9 + 800)^2 + 3400^2} \approx 3835$ ft.

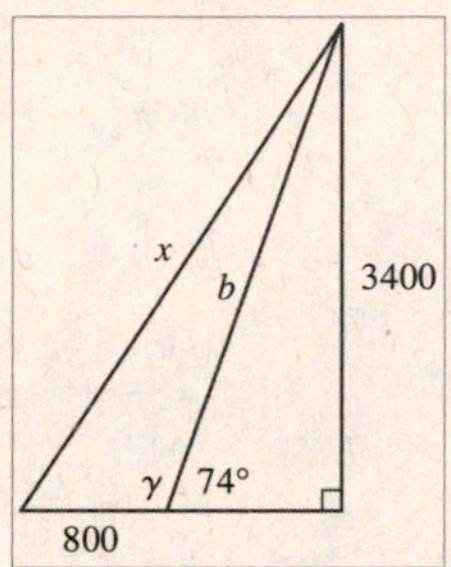

53. By Heron's formula, $A = \sqrt{s(s-a)(s-b)(s-c)}$, where

$s = \dfrac{a+b+c}{2} = \dfrac{112 + 148 + 190}{2} = 225$. Thus,

$A = \sqrt{225(225-112)(225-148)(225-190)} \approx 8277.7$ ft ^{2} . Since the land value is $\$20$ per square foot, the value of the lot is approximately $8277.7 \cdot 20 = \$165,554$.

Chapter 3 Review

1. (a) $60^\circ = 60 \cdot \frac{\pi}{180} = \frac{\pi}{3} \approx 1.05$ rad

(b) $330^\circ = 330 \cdot \frac{\pi}{180} = \frac{11\pi}{6} \approx 5.76$ rad

(c) $-135^\circ = -135 \cdot \frac{\pi}{180} = -\frac{3\pi}{4} \approx -3.36$ rad

(d) $-90^\circ = -90 \cdot \frac{\pi}{180} = -\frac{\pi}{2} \approx -1.57$ rad

3. (a) $\frac{5\pi}{2}$ rad $= \frac{5\pi}{2} \cdot \frac{180}{\pi} = 450^\circ$

(b) $-\frac{\pi}{6}$ rad $= -\frac{\pi}{6} \cdot \frac{180}{\pi} = -30^\circ$

(c) $\frac{9\pi}{4}$ rad $= \frac{9\pi}{4} \cdot \frac{180}{\pi} = 405^\circ$

(d) 3.1 rad $= 3.1 \cdot \frac{180}{\pi} = \frac{558}{\pi} \approx 177.6^\circ$

5. $r = 8$ m, $\theta = 1$ rad. Then $s = r\theta = 8 \cdot 1 = 8$ m.

7. $s = 100$ ft, $\theta = 70^\circ = 70 \cdot \frac{\pi}{180} = \frac{7\pi}{18}$ rad. Then $r = \frac{s}{\theta} = 100 \cdot \frac{18}{7\pi} = \frac{1800}{7\pi} \approx 82$ ft.

9. $r = 3960$ miles, $s = 2450$ miles. Then $\theta = \dfrac{s}{r} = \dfrac{2450}{3960} \approx 0.619$ rad

$= 0.619 \cdot \frac{\pi}{180} \approx 35.448^\circ$ and so the angle is approximately 35.4° .

11. $A = \frac{1}{2}r^2\theta = \frac{1}{2}(200)^2\left(52^\circ \cdot \frac{\pi}{180^\circ}\right) \approx 18\,,151$ ft ^{2}

13. The angular speed is $\omega = \dfrac{150 \cdot 2\pi \text{ rad}}{1 \text{ min}} = 300\pi$ rad/min ≈ 942.5 rad/min. The linear speed

is $v = \dfrac{150 \cdot 2\pi \cdot 8}{1} = 2400\pi$ in./min ≈ 7539.8 in./min.

15. $r = \sqrt{5^2 + 7^2} = \sqrt{74}$. Then $\sin\theta = \frac{5}{\sqrt{74}}$, $\cos\theta = \frac{7}{\sqrt{74}}$, $\tan\theta = \frac{5}{7}$, $\csc\theta = \frac{\sqrt{74}}{5}$

, $\sec\theta = \frac{\sqrt{74}}{7}$, and $\cot\theta = \frac{7}{5}$.

17. $\dfrac{x}{5} = \cos 40^\circ \quad \Leftrightarrow \quad x = 5\cos 40^\circ \approx 3.83$, and $\frac{y}{5} = \sin 40^\circ$

$\Leftrightarrow \quad y = 5\sin 40^\circ \approx 3.21$.

19. $\dfrac{1}{x} = \sin 20^\circ \quad \Leftrightarrow \quad x = \frac{1}{\sin 20^\circ} \approx 2.92$, and $\frac{x}{y} = \cos 20^\circ$

$\Leftrightarrow \quad y = \frac{x}{\cos 20^\circ} \approx \frac{2.924}{0.9397} \approx 3.11$.

21. $A = 90° - 20° = 70°$, $a = 3\cos 20° \approx 2.819$, and $b = 3\sin 20° \approx 1.026$.

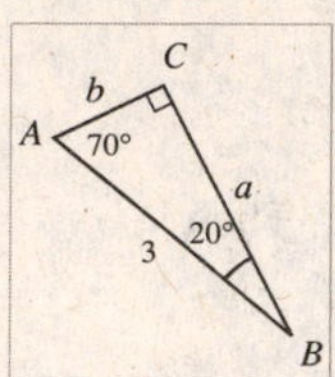

23. $c = \sqrt{25^2 - 7^2} = 24$, $A = \sin^{-1}\frac{7}{24} \approx 0.2960 \approx 17.0°$, and

$C = \sin^{-1}\frac{24}{25} \approx 1.2870 \approx 73.7°$.

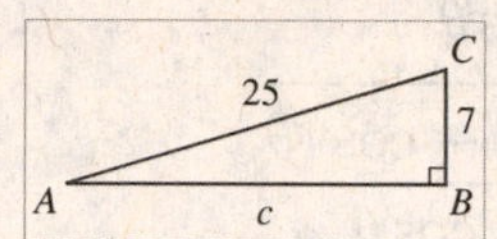

25. $\tan\theta = \dfrac{1}{a} \quad \Leftrightarrow \quad a = \dfrac{1}{\tan\theta} = \cot\theta$, $\sin\theta = \dfrac{1}{b} \quad \Leftrightarrow \quad b = \dfrac{1}{\sin\theta} = \csc\theta$

27.

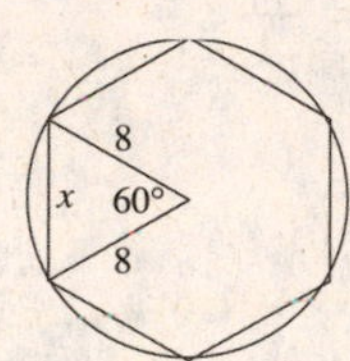

One side of the hexagon together with radial line segments through its endpoints forms a triangle with two sides of length 8 m and subtended angle $60°$. Let x be the length of one such side (in meters). By the Law of Cosines, $x^2 = 8^2 + 8^2 - 2 \cdot 8 \cdot 8 \cdot \cos 60° = 64 \quad \Leftrightarrow \quad x = 8$. Thus the perimeter of the hexagon is $6x = 6 \cdot 8 = 48$ m.

29. Let r represent the radius, in miles, of the moon. Then $\tan\dfrac{\theta}{2} = \dfrac{r}{r + |AB|}$, $\theta = 0.518°$

$\Leftrightarrow \quad r = (r + 236{,}900) \cdot \tan 0.259° \quad \Leftrightarrow \quad r(1 - \tan 0.259°) = 236$,

$900 \cdot \tan 0.259° \quad \Leftrightarrow \quad r = \dfrac{236{,}900 \cdot \tan 0.259°}{1 - \tan 0.259°} \approx 1076$ and so the radius of the moon is roughly 1076 miles.

31. $\sin 315° = -\sin 45° = -\frac{1}{\sqrt{2}} = -\frac{\sqrt{2}}{2}$

33. $\tan(-135°) = \tan 45° = 1$

35. $\cot\left(-\frac{22\pi}{3}\right) = \cot\frac{2\pi}{3} = \cot\frac{\pi}{3} = -\frac{1}{\sqrt{3}} = -\frac{\sqrt{3}}{3}$

37. $\cos 585° = \cos 225° = -\cos 45° = -\frac{1}{\sqrt{2}} = -\frac{\sqrt{2}}{2}$

39. $\csc\frac{8\pi}{3} = \csc\frac{2\pi}{3} = \csc\frac{\pi}{3} = \frac{2}{\sqrt{3}} = \frac{2\sqrt{3}}{3}$

41. $\cot(-390°) = \cot(-30°) = -\cot 30° = -\sqrt{3}$

43. $r=\sqrt{(-5)^2+12^2}=\sqrt{169}=13$. Then $\sin\theta=\frac{12}{13}$, $\cos\theta=-\frac{5}{13}$, $\tan\theta=-\frac{12}{5}$, $\csc\theta=\frac{13}{12}$, $\sec\theta=-\frac{13}{5}$, and $\cot\theta=-\frac{5}{12}$.

45. $y-\sqrt{3}x+1=0 \Leftrightarrow y=\sqrt{3}x-1$, so the slope of the line is $m=\sqrt{3}$. Then $\tan\theta=m=\sqrt{3} \Leftrightarrow \theta=60°$.

47. $\sec^2\theta=1+\tan^2\theta \quad\Leftrightarrow\quad \tan^2\theta=\dfrac{1}{\cos^2\theta}-1 \quad\Leftrightarrow$

$\tan\theta=\sqrt{\dfrac{1}{\cos^2\theta}-1}=\sqrt{\dfrac{1-\cos^2\theta}{\cos^2\theta}}=\dfrac{\sqrt{1-\cos^2\theta}}{|\cos\theta|}=\dfrac{\sqrt{1-\cos^2\theta}}{-\cos\theta}$ (because $\cos\theta<0$ in quadrant III, $|\cos\theta|=-\cos\theta$). Thus $\tan\theta=-\dfrac{\sqrt{1-\cos^2\theta}}{\cos\theta}$.

49. $\tan^2\theta=\dfrac{\sin^2\theta}{\cos^2\theta}=\dfrac{\sin^2\theta}{1-\sin^2\theta}$

51. $\tan\theta=\frac{\sqrt{7}}{3}$, $\sec\theta=\frac{4}{3}$. Then $\cos\theta=\frac{3}{4}$ and $\sin\theta=\tan\theta\cdot\cos\theta=\frac{\sqrt{7}}{4}$, $\csc\theta=\frac{4}{\sqrt{7}}=\frac{4\sqrt{7}}{7}$, and $\cot\theta=\frac{3}{\sqrt{7}}=\frac{3\sqrt{7}}{7}$.

53. $\sin\theta=\frac{3}{5}$. Since $\cos\theta<0$, θ is in quadrant II. Thus,

$x=-\sqrt{5^2-3^2}=-\sqrt{16}=-4$ and so $\cos\theta=-\frac{4}{5}$, $\tan\theta=-\frac{3}{4}$, $\csc\theta=\frac{5}{3}$, $\sec\theta=-\frac{5}{4}$, $\cot\theta=-\frac{4}{3}$.

55. $\tan\theta=-\frac{1}{2}$. $\sec^2\theta=1+\tan^2\theta=1+\frac{1}{4}=\frac{5}{4} \quad\Leftrightarrow\quad \cos^2\theta=\dfrac{4}{5} \quad\Rightarrow$

$\cos\theta=-\sqrt{\frac{4}{5}}=-\frac{2}{\sqrt{5}}$ since $\cos\theta<0$ in quadrant II. But $\tan\theta=\dfrac{\sin\theta}{\cos\theta}=-\frac{1}{2} \quad\Leftrightarrow$

$\sin\theta=-\frac{1}{2}\cos\theta=-\frac{1}{2}\left(-\frac{2}{\sqrt{5}}\right)=\frac{1}{\sqrt{5}}$. Therefore,

$\sin\theta+\cos\theta=\frac{1}{\sqrt{5}}+\left(-\frac{2}{\sqrt{5}}\right)=-\frac{1}{\sqrt{5}}=-\frac{\sqrt{5}}{5}$.

57. By the Pythagorean Theorem, $\sin^2\theta+\cos^2\theta=1$ for any angle θ.

59. $\sin^{-1}\frac{\sqrt{3}}{2}=\frac{\pi}{3}$

61. Let $u=\sin^{-1}\frac{2}{5}$ and so $\sin u=\frac{2}{5}$. Then from the triangle, $\tan\left(\sin^{-1}\frac{2}{5}\right)=\tan u=\frac{2}{\sqrt{21}}$

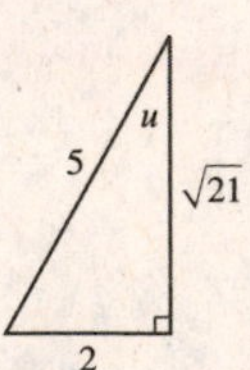

63. Let $\theta=\tan^{-1}x \quad\Leftrightarrow\quad \tan\theta=x$. Then from the triangle, we have

$\sin\left(\tan^{-1} x\right) = \sin\theta = \dfrac{x}{\sqrt{1+x^2}}$.

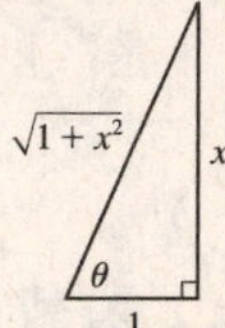

65. $\cos\theta = \dfrac{x}{3} \Rightarrow \theta = \cos^{-1}\left(\dfrac{x}{3}\right)$

67. $\angle B = 180° - 30° - 80° = 70°$, and so by the Law of Sines, $x = \dfrac{10\sin 30°}{\sin 70°} \approx 5.32$.

69. $x^2 = 100^2 + 210^2 - 2\cdot 100\cdot 210\cdot \cos 40° \approx 21{,}926.133 \Leftrightarrow x \approx 148.07$

71. $x^2 = 2^2 + 8^2 - 2(2)(8)\cos 120° = 84 \Leftrightarrow x \approx \sqrt{84} \approx 9.17$

73. By the Law of Sines, $\dfrac{\sin\theta}{23} = \dfrac{\sin 25°}{12} \Rightarrow \sin\theta = \dfrac{23\sin 25°}{12} \Rightarrow$

$\theta = \sin^{-1}\left(\dfrac{23\sin 25°}{12}\right) \approx 54.1°$ or $\theta \approx 180° - 54.1° = 125.9°$.

75. By the Law of Cosines, $120^2 = 100^2 + 85^2 - 2(100)(85)\cos\theta$, so

$\cos\theta = \dfrac{120^2 - 100^2 - 85^2}{-2(100)(85)} \approx 0.16618$. Thus, $\theta \approx \cos^{-1}(0.16618) \approx 80.4°$.

77. After 2 hours the ships have traveled distances $d_1 = 40$ mi and $d_2 = 56$ mi. The subtended angle is $180° - 32° - 42° = 106°$. Let d be the distance between the two ships in miles. Then by the Law of Cosines, $d^2 = 40^2 + 56^2 - 2(40)(56)\cos 106° \approx 5970.855 \Leftrightarrow$ $d \approx 77.3$ miles.

79. Let d be the distance, in miles, between the points A and B . Then by the Law of Cosines, $d^2 = 3.2^2 + 5.6^2 - 2(3.2)(5.6)\cos 42° \approx 14.966 \Leftrightarrow d \approx 3.9$ mi.

81. $A = \frac{1}{2}ab\sin\theta = \frac{1}{2}(8)(14)\sin 35° \approx 32.12$

Chapter 3 Test

1. $330° = 330 \cdot \frac{\pi}{180} = \frac{11\pi}{6}$ rad. $-135° = -135 \cdot \frac{\pi}{180} = -\frac{3\pi}{4}$ rad.

2. $\frac{4\pi}{3}$ rad $= \frac{4\pi}{3} \cdot \frac{180}{\pi} = 240°$. -1.3 rad $= -1.3 \cdot \frac{180}{\pi} = -\frac{234}{\pi} \approx -74.5°$

3. (a) The angular speed is $\omega = \dfrac{120 \cdot 2\pi \text{ rad}}{1 \text{ min}} = 240\pi$ rad/min ≈ 753.98 rad/min.**(b)** The linear speed is $v = \dfrac{120 \cdot 2\pi \cdot 16}{1} = 3840\pi$ ft/min $\approx 12{,}063.7$ ft/min ≈ 137 mi/h .

4. (a) $\sin 405° = \sin 45° = \frac{1}{\sqrt{2}} = \frac{\sqrt{2}}{2}$

(b) $\tan\left(-150°\right) = \tan 30° = \frac{1}{\sqrt{3}} = \frac{\sqrt{3}}{3}$

(c) $\sec \frac{5\pi}{3} = \sec \frac{\pi}{3} = 2$

(d) $\csc \frac{5\pi}{2} = \csc \frac{\pi}{2} = 1$

5. $r = \sqrt{3^2 + 2^2} = \sqrt{13}$. Then

$$\tan\theta + \sin\theta = \frac{2}{3} + \frac{2}{\sqrt{13}} = \frac{2\left(\sqrt{13}+3\right)}{3\sqrt{13}} = \frac{26 + 6\sqrt{13}}{39}.$$

6. $\sin\theta = \dfrac{a}{24} \Leftrightarrow a = 24\sin\theta$. Also, $\cos\theta = \dfrac{b}{24} \Leftrightarrow b = 24\cos\theta$.

7. $\cos\theta = -\frac{1}{3}$ and θ is in quadrant III, so $r = 3$, $x = -1$, and $y = -\sqrt{3^2 - 1^2} = -2\sqrt{2}$. Then

$$\tan\theta\cot\theta + \csc\theta = \tan\theta \cdot \frac{1}{\tan\theta} + \csc\theta = 1 - \frac{3}{2\sqrt{2}} = \frac{2\sqrt{2} - 3}{2\sqrt{2}} = \frac{4 - 3\sqrt{2}}{4}.$$

8. $\sin\theta = \frac{5}{13}$, $\tan\theta = -\frac{5}{12}$. Then

$$\sec\theta = \frac{1}{\cos\theta} = \frac{1}{\cos\theta} \cdot \frac{\sin\theta}{\sin\theta} = \tan\theta \cdot \frac{1}{\sin\theta} = -\frac{5}{12} \cdot \frac{13}{5} = -\frac{13}{12}.$$

9. $\sec^2\theta = 1 + \tan^2\theta \Leftrightarrow \tan\theta = \pm\sqrt{\sec^2\theta - 1}$. Thus, $\tan\theta = -\sqrt{\sec^2\theta - 1}$ since $\tan\theta < 0$ in quadrant II.

10. $\tan 73° = \dfrac{h}{6} \Rightarrow h = 6\tan 73° \approx 19.6$ ft.

11. (a) $\tan\theta = \dfrac{x}{4} \Rightarrow \theta = \tan^{-1}\left(\dfrac{x}{4}\right)$

(b) $\cos\theta = \dfrac{3}{x} \Rightarrow \theta = \cos^{-1}\left(\dfrac{3}{x}\right)$

12. Let $u = \tan^{-1}\frac{9}{40}$ so $\tan u = \frac{9}{40}$. From the triangle, $r = \sqrt{9^2 + 40^2} = 41$. So $\cos\left(\tan^{-1}\frac{9}{40}\right) = \cos u = \frac{40}{41}$.

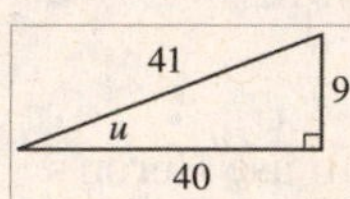

13. By the Law of Cosines, $x^2 = 10^2 + 12^2 - 2(10)(12)\cdot\cos 48^\circ \approx 8.409 \Leftrightarrow x \approx 9.1$.

14. $\angle C = 180^\circ - 52^\circ - 69^\circ = 59^\circ$. Then by the Law of Sines, $x = \dfrac{230\sin 69^\circ}{\sin 59^\circ} \approx 250.5$.

15. Let h be the height of the shorter altitude. Then $\tan 20^\circ = \dfrac{h}{50} \Leftrightarrow h = 50\tan 20^\circ$ and

$\tan 28^\circ = \dfrac{x+h}{50} \Leftrightarrow x + h = 50\tan 28^\circ$

$\Leftrightarrow x = 50\tan 28^\circ - h = 50\tan 28^\circ - 50\tan 20^\circ \approx 8.4$.

16. Let $\angle A$ and $\angle X$ be the other angles in the triangle. Then $\sin A = \dfrac{15\sin 108^\circ}{28} \approx 0.509$

$\Leftrightarrow \angle A \approx 30.63^\circ$. Then $\angle X \approx 180^\circ - 108^\circ - 30.63^\circ \approx 41.37^\circ$, and so

$x \approx \dfrac{28\sin 41.37^\circ}{\sin 108^\circ} \approx 19.5$.

17. By the Law of Cosines, $9^2 = 8^2 + 6^2 - 2(8)(6)\cos\theta \Rightarrow$

$\cos\theta = \dfrac{8^2 + 6^2 - 9^2}{2(8)(6)} \approx 0.1979$, so $\theta \approx \cos^{-1}(0.1979) \approx 78.6^\circ$.

18. We find the length of the third side x using the Law of Cosines:

$x^2 = 5^2 + 7^2 - 2(5)(7)\cos 75^\circ \approx 55.88 \Rightarrow x \approx 7.475$. Therefore, by the Law of Sines, $\dfrac{\sin\theta}{5} \approx \dfrac{\sin 75^\circ}{7.475} \Rightarrow \sin\theta \approx \dfrac{5\sin 75^\circ}{7.475} \approx 0.6461$, so

$\theta \approx \sin^{-1}(0.6461) \approx 40.2^\circ$.

19. (a) $A(\text{sector}) = \frac{1}{2}r^2\theta = \frac{1}{2}\cdot 10^2\cdot 72\cdot\frac{\pi}{180} = 50\cdot\frac{72\pi}{180}$.

$A(\text{triangle}) = \frac{1}{2}r\cdot r\sin\theta = \frac{1}{2}\cdot 10^2\sin 72^\circ$. Thus, the area of the shaded region is

$A(\text{shaded}) = A(\text{sector}) - A(\text{triangle}) = 50\left(\frac{72\pi}{180} - \sin 72^\circ\right) \approx 15.3\ \text{m}^2$.

(b) The shaded region is bounded by two pieces: one piece is part of the triangle, the other is part of the circle. The first part has length

$l = \sqrt{10^2 + 10^2 - 2(10)(10)\cdot\cos 72^\circ} = 10\sqrt{2 - 2\cdot\cos 72^\circ}$. The second has length

$s = 10\cdot 72\cdot\frac{\pi}{180} = 4\pi$. Thus, the perimeter of the shaded region is

$p = l + s = 10\sqrt{2 - 2\cos 72^\circ} + 4\pi \approx 24.3$ m.

20. (a) If θ is the angle opposite the longest side, then by the Law of Cosines

$$\cos\theta = \frac{9^2 + 13^2 - 20^2}{2(9)(20)} = -0.6410$$. Therefore, $\theta = \cos^{-1}(-0.6410) \approx 129.9°$.

(b) From part (a), $\theta \approx 129.9°$, so the area of the triangle is

$A = \frac{1}{2}(9)(13)\sin 129.9° \approx 44.9$ units ^{2} . Another way to find the area is to use Heron's

Formula: $A = \sqrt{s(s-a)(s-b)(s-c)}$, where $s = \dfrac{a+b+c}{2} = \dfrac{9+13+20}{2} = 21$

. Thus, $A = \sqrt{21(21-20)(21-13)(21-9)} = \sqrt{2016} \approx 44.9$ units ^{2} .

21. Label the figure as shown. Now $\angle\beta = 85° - 75° = 10°$, so by the Law of Sines,

$$\frac{x}{\sin 75°} = \frac{100}{\sin 10°} \quad\Leftrightarrow\quad x = 100 \cdot \frac{\sin 75°}{\sin 10°}$$. Now $\sin 85° = \dfrac{h}{x} \Leftrightarrow$

$$h = x\sin 85° = 100 \cdot \frac{\sin 75°}{\sin 10°}\sin 85° \approx 554 .$$

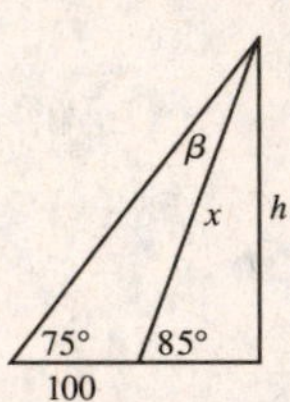

FOCUS ON MODELING Surveying

1. Let x be the distance between the church and City Hall. To apply the Law of Sines to the triangle with vertices at City Hall, the church, and the first bridge, we first need the measure of the angle at the first bridge, which is $180° - 25° - 30° = 125°$. Then $\dfrac{x}{\sin 125°} = \dfrac{0.86}{\sin 30°} \Leftrightarrow$ $x = \dfrac{0.86 \sin 125°}{\sin 30°} \approx 1.4089$. So the distance between the church and City Hall is about 1.41 miles.

3. First notice that $\angle DBC = 180° - 20° - 95° - 65°$ and $\angle DAC = 180° - 60° - 45° = 75°$. From $\Delta\, ACD$ we get $\dfrac{|AC|}{\sin 45°} = \dfrac{20}{\sin 75°} \Leftrightarrow$ $|AC| = \dfrac{20 \sin 45°}{\sin 75°} \approx 14.6°$. From $\Delta\, BCD$ we get $\dfrac{|BC|}{\sin 95°} = \dfrac{20}{\sin 65°} \Leftrightarrow$ $|BC| = \dfrac{20 \sin 95°}{\sin 65°} \approx 22.0$. By applying the Law of Cosines to $\Delta\, ABC$ we get $|AB|^2 = |AC|^2 + |BC|^2 - 2|AC||BC|\cos 40° \approx 14.6^2 + 22.0^2 - 2 \cdot 14.6 \cdot 22.0 \cdot \cos 40° \approx 205$, so $|AB| \approx \sqrt{205} \approx 14.3$ m. Therefore, the distance between A and B is approximately 14.3 m.

5. (a) In $\Delta\, ABC$, $\angle B = 180° - \beta$, so $\angle C = 180° - \alpha - (180° - \beta) = \beta - \alpha$. By the Law of Sines, $\dfrac{|BC|}{\sin\alpha} = \dfrac{|AB|}{\sin(\beta - \alpha)} \Rightarrow |BC| = |AB|\dfrac{\sin\alpha}{\sin(\beta - \alpha)} = \dfrac{d \sin\alpha}{\sin(\beta - \alpha)}$

(b) From part (a) we know that $|BC| = \dfrac{d \sin\alpha}{\sin(\beta - \alpha)}$. But $\sin\beta = \dfrac{h}{|BC|} \Leftrightarrow$ $|BC| = \dfrac{h}{\sin\beta}$. Therefore, $|BC| = \dfrac{d \sin\alpha}{\sin(\beta - \alpha)} = \dfrac{h}{\sin\beta} \Rightarrow h = \dfrac{d \sin\alpha \sin\beta}{\sin(\beta - \alpha)}$.

(c) $h = \dfrac{d \sin\alpha \sin\beta}{\sin(\beta - \alpha)} = \dfrac{800 \sin 25° \sin 29°}{\sin 4°} \approx 2350$ ft

7. We start by labeling the edges and calculating the remaining angles, as shown in the first figure. Using the Law of Sines, we find the following: $\dfrac{a}{\sin 29°} = \dfrac{150}{\sin 60°} \Leftrightarrow a = \dfrac{150 \sin 29°}{\sin 60°} \approx 83.97$ $\dfrac{b}{\sin 91°} = \dfrac{150}{\sin 60°} \Leftrightarrow b = \dfrac{150 \sin 91°}{\sin 60°} \approx 173.18$, $\dfrac{c}{\sin 32°} = \dfrac{173.18}{\sin 87°} \Leftrightarrow$ $c = \dfrac{173.18 \sin 32°}{\sin 87°} \approx 91.90$, $\dfrac{d}{\sin 61°} = \dfrac{173.18}{\sin 87°} \Leftrightarrow$

$$e = \frac{173.18\sin 61^\circ}{\sin 87^\circ} \approx 151.67\ ,\ \frac{e}{\sin 41^\circ} = \frac{151.67}{\sin 51^\circ} \Leftrightarrow$$

$$e = \frac{151.67\sin 41^\circ}{\sin 51^\circ} \approx 128.04\ ,\ \frac{f}{\sin 88^\circ} = \frac{151.67}{\sin 51^\circ} \Leftrightarrow$$

$$f = \frac{151.67\sin 88^\circ}{\sin 51^\circ} \approx 195.04\ ,\ \frac{g}{\sin 50^\circ} = \frac{195.04}{\sin 92^\circ} \Leftrightarrow$$

$$g = \frac{195.04\sin 50^\circ}{\sin 92^\circ} \approx 149.50\ \text{, and}\ \frac{h}{\sin 38^\circ} = \frac{195.04}{\sin 92^\circ} \Leftrightarrow$$

$h = \dfrac{195.04\sin 38^\circ}{\sin 92^\circ} \approx 120.15$. Note that we used two decimal places throughout our calculations. Our results are shown (to one decimal place) in the second figure.

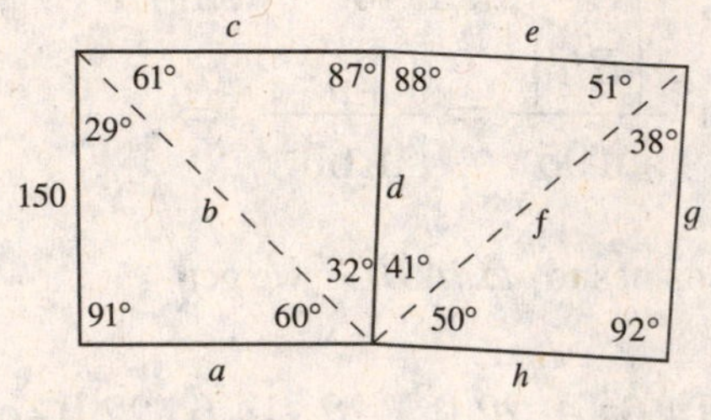

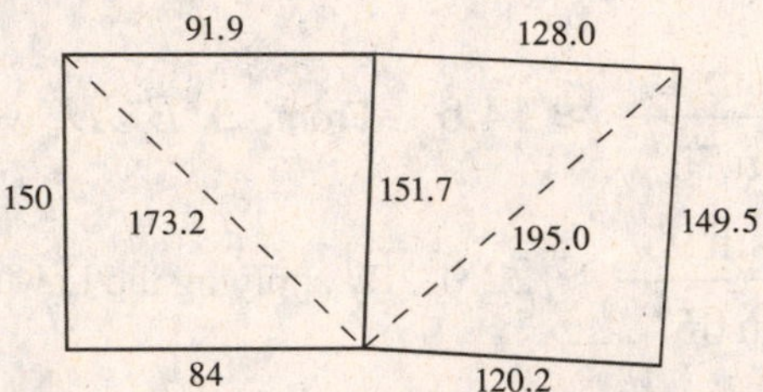

4 Analytic Trigonometry

4.1 Trigonometric Identities

1. An equation is called an identity if it is valid for *all* values of the variable. The equation $2x = x + x$ is an algebraic identity and the equation $\sin^2 x + \cos^2 x = 1$ is a trigonometric identity.

3. $\cos t \tan t = \cos t \cdot \dfrac{\sin t}{\cos t} = \sin t$

5. $\sin\theta \sec\theta = \sin\theta \cdot \dfrac{1}{\cos\theta} = \tan\theta$

7. $\tan^2 x - \sec^2 x = \dfrac{\sin^2 x}{\cos^2 x} - \dfrac{1}{\cos^2 x} = \dfrac{\sin^2 x - 1}{\cos^2 x} = \dfrac{-\cos^2 x}{\cos^2 x} = -1$

9. $\sin u + \cot u \cos u = \sin u + \dfrac{\cos u}{\sin u} \cdot \cos u = \dfrac{\sin^2 u + \cos^2 u}{\sin u} = \dfrac{1}{\sin u} = \csc u$

11. $\dfrac{\sec\theta - \cos\theta}{\sin\theta} = \dfrac{\dfrac{1}{\cos\theta} - \cos\theta}{\sin\theta} = \dfrac{1 - \cos^2\theta}{\sin\theta\cos\theta} = \dfrac{\sin^2\theta}{\sin\theta\cos\theta} = \dfrac{\sin\theta}{\cos\theta} = \tan\theta$

13. $\dfrac{\sin x \sec x}{\tan x} = \dfrac{\sin x \cdot \dfrac{1}{\cos x}}{\dfrac{\cos x}{\sin x}} = 1$

15. $\dfrac{1 + \cos y}{1 + \sec y} = \dfrac{1 + \cos y}{1 + \dfrac{1}{\cos y}} = \dfrac{1 + \cos y}{\dfrac{\cos y + 1}{\cos y}} = \dfrac{1 + \cos y}{1} \cdot \dfrac{\cos y}{\cos y + 1} = \cos y$

17. $\dfrac{\sec^2 x - 1}{\sec^2 x} = \dfrac{\tan^2 x}{\sec^2 x} = \dfrac{\sin^2 x}{\cos^2 x} \cdot \cos^2 x = \sin^2 x$. *Another method:*

$\dfrac{\sec^2 x - 1}{\sec^2 x} = 1 - \dfrac{1}{\sec^2 x} = 1 - \cos^2 x = \sin^2 x$

19. $\dfrac{1 + \csc x}{\cos x + \cot x} = \dfrac{1 + \dfrac{1}{\sin x}}{\cos x + \dfrac{\cos x}{\sin x}} = \dfrac{1 + \dfrac{1}{\sin x}}{\cos x + \dfrac{\cos x}{\sin x}} \cdot \dfrac{\sin x}{\sin x} = \dfrac{\sin x + 1}{\cos x(\sin x + 1)} = \dfrac{1}{\cos x} = \sec x$

21.

$$\dfrac{1 + \sin u}{\cos u} + \dfrac{\cos u}{1 + \sin u} = \dfrac{(1 + \sin u)^2 + \cos^2 u}{\cos u(1 + \sin u)} = \dfrac{1 + 2\sin u + \sin^2 u + \cos^2 u}{\cos u(1 + \sin u)} = \dfrac{1 + 2\sin u + 1}{\cos u(1 + \sin u)}$$

$$= \dfrac{2 + 2\sin u}{\cos u(1 + \sin u)} = \dfrac{2(1 + \sin u)}{\cos u(1 + \sin u)} = \dfrac{2}{\cos u} = 2\sec u$$

23.

$$\frac{2+\tan^2 x}{\sec^2 x}-1 = \frac{1+1+\tan^2 x}{\sec^2 x}-1=\frac{1}{\sec^2 x}+\frac{1+\tan^2 x}{\sec^2 x}-1=\frac{1}{\sec^2 x}+\frac{\sec^2 x}{\sec^2 x}-1$$

$$=\frac{1}{\sec^2 x}+1-1=\frac{1}{\sec^2 x}=\cos^2 x$$

25. $\tan\theta+\cos(-\theta)+\tan(-\theta)=\tan\theta+\cos\theta-\tan\theta=\cos\theta$

27. (a)

$$\frac{\cos x}{\sec x\sin x}=\frac{\cos x}{\frac{1}{\cos x}\cdot\sin x}=\frac{\cos^2 x}{\sin x}=\frac{1-\sin^2 x}{\sin x}=\frac{1}{\sin x}-\frac{\sin^2 x}{\sin x}=\csc x-\sin x$$

(b) We graph each side of the equation and see that the graphs of $y=\dfrac{\cos x}{\sec x\sin x}$ and $y=\csc x-\sin x$ are identical, confirming that the equation is an identity.

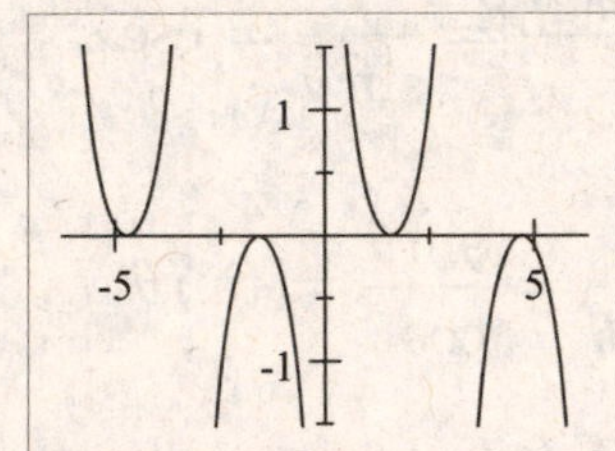

29. $\dfrac{\sin\theta}{\tan\theta}=\dfrac{\sin\theta}{\frac{\sin\theta}{\cos\theta}}=\sin\theta\cdot\dfrac{\cos\theta}{\sin\theta}=\cos\theta$

31. $\dfrac{\cos u\sec u}{\tan u}=\cos u\dfrac{1}{\cos u}\cot u=\cot u$

33. $\sin B+\cos B\cot B=\sin B+\cos B\dfrac{\cos B}{\sin B}=\dfrac{\sin^2 B+\cos^2 B}{\sin B}=\dfrac{1}{\sin B}=\csc B$

35.

$$\cot(-\alpha)\cos(-\alpha)+\sin(-\alpha)=-\frac{\cos\alpha}{\sin\alpha}\cos\alpha-\sin\alpha=\frac{-\cos^2\alpha-\sin^2\alpha}{\sin\alpha}=\frac{-1}{\sin\alpha}=-\csc\alpha$$

37. $\tan\theta+\cot\theta=\dfrac{\sin\theta}{\cos\theta}+\dfrac{\cos\theta}{\sin\theta}=\dfrac{\sin^2\theta+\cos^2\theta}{\cos\theta\sin\theta}=\dfrac{1}{\cos\theta\sin\theta}=\sec\theta\csc\theta$

39. $(1-\cos\beta)(1+\cos\beta)=1-\cos^2\beta=\sin^2\beta=\dfrac{1}{\csc^2\beta}$

41.

$$\begin{aligned}\frac{(\sin x+\cos x)^2}{\sin^2 x-\cos^2 x} &= \frac{(\sin x+\cos x)^2}{(\sin x+\cos x)(\sin x-\cos x)}\\ &= \frac{\sin x+\cos x}{\sin x-\cos x}\\ &= \frac{(\sin x+\cos x)(\sin x-\cos x)}{(\sin x-\cos x)(\sin x-\cos x)}\\ &= \frac{\sin^2 x-\cos^2 x}{(\sin x-\cos x)^2}\end{aligned}$$

43. $\dfrac{\sec t-\cos t}{\sec t}=\dfrac{\dfrac{1}{\cos t}-\cos t}{\dfrac{1}{\cos t}}=\dfrac{\dfrac{1}{\cos t}-\cos t}{\dfrac{1}{\cos t}}\cdot\dfrac{\cos t}{\cos t}=\dfrac{1-\cos^2 t}{1}=\sin^2 t$

45. $\dfrac{1}{1-\sin^2 y}=\dfrac{1}{\cos^2 y}=\sec^2 y=1+\tan^2 y$

47.

$$\begin{aligned}(\cot x-\csc x)(\cos x+1) &= \cot x\cos x+\cot x-\csc x\cos x-\csc x=\frac{\cos^2 x}{\sin x}+\frac{\cos x}{\sin x}-\frac{\cos x}{\sin x}-\frac{1}{\sin x}\\ &= \frac{\cos^2 x-1}{\sin x}=\frac{-\sin^2 x}{\sin x}=-\sin x\end{aligned}$$

49. $(1-\cos^2 x)(1+\cot^2 x)=\sin^2 x\left(1+\dfrac{\cos^2 x}{\sin^2 x}\right)=\sin^2 x+\cos^2 x=1$

51. $2\cos^2 x-1=2(1-\sin^2 x)-1=2-2\sin^2 x-1=1-2\sin^2 x$

53.

$$\frac{1-\cos\alpha}{\sin\alpha}=\frac{1-\cos\alpha}{\sin\alpha}\cdot\frac{1+\cos\alpha}{1+\cos\alpha}=\frac{1-\cos^2\alpha}{\sin\alpha(1+\cos\alpha)}=\frac{\sin^2\alpha}{\sin\alpha(1+\cos\alpha)}=\frac{\sin\alpha}{1+\cos\alpha}$$

55. $\tan^2\theta\sin^2\theta=\tan^2\theta(1-\cos^2\theta)=\tan^2\theta-\dfrac{\sin^2\theta}{\cos^2\theta}\cos^2\theta=\tan^2\theta-\sin^2\theta$

57. $\dfrac{\sin x-1}{\sin x+1}=\dfrac{\sin x-1}{\sin x+1}\cdot\dfrac{\sin x+1}{\sin x+1}=\dfrac{\sin^2 x-1}{(\sin x+1)^2}=\dfrac{-\cos^2 x}{(\sin x+1)^2}$

59.

$$\begin{aligned}\frac{(\sin t+\cos t)^2}{\sin t\cos t} &= \frac{\sin^2 t+2\sin t\cos t+\cos^2 t}{\sin t\cos t}=\frac{\sin^2 t+\cos^2 t}{\sin t\cos t}+\frac{2\sin t\cos t}{\sin t\cos t}\\ &= \frac{1}{\sin t\cos t}+2\\ &= 2+\sec t\cos t\end{aligned}$$

61.

$$\frac{1+\tan^2 u}{1-\tan^2 u}=\frac{1+\dfrac{\sin^2 u}{\cos^2 u}}{1-\dfrac{\sin^2 u}{\cos^2 u}}=\frac{1+\dfrac{\sin^2 u}{\cos^2 u}}{1-\dfrac{\sin^2 u}{\cos^2 u}}\cdot\frac{\cos^2 u}{\cos^2 u}=\frac{\cos^2 u+\sin^2 u}{\cos^2 u-\sin^2 u}=\frac{1}{\cos^2 u-\sin^2 u}$$

63.

$$\begin{aligned}\frac{\sec x}{\sec x-\tan x}&=\frac{\sec x}{\sec x-\tan x}\cdot\frac{\sec x+\tan x}{\sec x+\tan x}=\frac{\sec x\left(\sec x+\tan x\right)}{\sec^2 x-\tan^2 x}\\&=\frac{\sec x\left(\sec x+\tan x\right)}{1}\\&=\sec x\left(\sec x+\tan x\right)\end{aligned}$$

65. $\sec v-\tan v=\left(\sec v-\tan v\right)\cdot\dfrac{\sec v+\tan v}{\sec v+\tan v}=\dfrac{\sec^2 v-\tan^2 v}{\sec v+\tan v}=\dfrac{1}{\sec v+\tan v}$

67.

$$\frac{\sin x+\cos x}{\sec x+\csc x}=\frac{\sin x+\cos x}{\dfrac{1}{\cos x}+\dfrac{1}{\sin x}}=\frac{\sin x+\cos x}{\dfrac{\sin x+\cos x}{\cos x\sin x}}=\left(\sin x+\cos x\right)\frac{\cos x\sin x}{\sin x+\cos x}=\cos x\sin x$$

69.

$$\begin{aligned}\frac{\csc x-\cot x}{\sec x-1}&=\frac{\dfrac{1}{\sin x}-\dfrac{\cos x}{\sin x}}{\dfrac{1}{\cos x}-1}=\frac{\dfrac{1}{\sin x}-\dfrac{\cos x}{\sin x}}{\dfrac{1}{\cos x}-1}\cdot\frac{\sin x\cos x}{\sin x\cos x}\\&=\frac{\cos x\left(1-\cos x\right)}{\sin x\left(1-\cos x\right)}=\frac{\cos x}{\sin x}=\cot x\end{aligned}$$

71. $\tan^2 u-\sin^2 u=\dfrac{\sin^2 u}{\cos^2 u}-\dfrac{\sin^2 u\cos^2 u}{\cos^2 u}=\dfrac{\sin^2 u}{\cos^2 u}\left(1-\cos^2 u\right)=\tan^2 u\sin^2 u$

73.

$$\begin{aligned}\sec^4 x-\tan^4 x&=\left(\sec^2 x-\tan^2 x\right)\left(\sec^2 x+\tan^2 x\right)\\&=1\left(\sec^2 x+\tan^2 x\right)\\&=\sec^2 x+\tan^2 x\end{aligned}$$

75.

$$\frac{\sin\theta-\csc\theta}{\cos\theta-\cot\theta}=\frac{\sin\theta-\dfrac{1}{\sin\theta}}{\cos\theta-\dfrac{\cos\theta}{\sin\theta}}=\frac{\dfrac{\sin^2\theta-1}{\sin\theta}}{\dfrac{\cos\theta\sin\theta-\cos\theta}{\sin\theta}}=\frac{\cos^2\theta}{\cos\theta\left(\sin\theta-1\right)}=\frac{\cos\theta}{\sin\theta-1}$$

77.

$$\frac{\cos^2 t + \tan^2 t - 1}{\sin^2 t} = \frac{-\sin^2 t + \tan^2 t}{\sin^2 t} = -1 + \frac{\sin^2 t}{\cos^2 t} \cdot \frac{1}{\sin^2 t} = -1 + \sec^2 t = \tan^2 t$$

79.

$$\frac{1}{\sec x + \tan x} + \frac{1}{\sec x - \tan x} = \frac{\sec x - \tan x + \sec x + \tan x}{(\sec x + \tan x)(\sec x - \tan x)} = \frac{2\sec x}{\sec^2 x - \tan^2 x}$$
$$= \frac{2\sec x}{1}$$
$$= 2\sec x$$

81.

$$\left(\tan x + \cot x\right)^2 = \tan^2 x + 2\tan x \cot x + \cot^2 x = \tan^2 x + 2 + \cot^2 x$$
$$= \left(\tan^2 x + 1\right) + \left(\cot^2 x + 1\right)$$
$$= \sec^2 x + \csc^2 x$$

83. $$\frac{\sec u - 1}{\sec u + 1} = \frac{\frac{1}{\cos u} - 1}{\frac{1}{\cos u} + 1} \cdot \frac{\cos u}{\cos u} = \frac{1 - \cos u}{1 + \cos u}$$

85. $$\frac{\sin^3 x + \cos^3 x}{\sin x + \cos x} = \frac{(\sin x + \cos x)\left(\sin^2 x - \sin x \cos x + \cos^2 x\right)}{\sin x + \cos x} = \sin^2 - \sin x$$

$\cos x + \cos^2 x = 1 - \sin x \quad \cos x$

87.

$$\frac{1 + \sin x}{1 - \sin x} = \frac{1 + \sin x}{1 - \sin x} \cdot \frac{1 + \sin x}{1 + \sin x} = \frac{(1 + \sin x)^2}{1 - \sin^2 x} = \frac{(1 + \sin x)^2}{\cos^2 x}$$
$$= \left(\frac{1 + \sin x}{\cos x}\right)^2$$
$$= \left(\tan x + \sec x\right)^2$$

89.

$$\left(\tan x + \cot x\right)^4 = \left(\frac{\sin x}{\cos x} + \frac{\cos x}{\sin x}\right)^4 = \left(\frac{\sin^2 x + \cos^2 x}{\sin x \cos x}\right)^4$$
$$= \left(\frac{1}{\sin x \cos x}\right)^4$$
$$= \sec^4 x \csc^4 x$$

91. $x = \sin\theta$; then $\dfrac{x}{\sqrt{1-x^2}} = \dfrac{\sin\theta}{\sqrt{1-\sin^2\theta}} = \dfrac{\sin\theta}{\sqrt{\cos^2\theta}} = \dfrac{\sin\theta}{\cos\theta} = \tan\theta$ (since $\cos\theta \geq 0$ for $0 \leq \theta \leq \frac{\pi}{2}$).

93. $x = \sec\theta$; then
$\sqrt{x^2-1} = \sqrt{\sec^2\theta - 1} = \sqrt{\left(\tan^2\theta + 1\right) - 1} = \sqrt{\tan^2\theta} = \tan\theta$ (since $\tan\theta \geq 0$ for $0 \leq \theta < \frac{\pi}{2}$)

95. $x = 3\sin\theta$; then
$\sqrt{9-x^2} = \sqrt{9-\left(3\sin\theta\right)^2} = \sqrt{9-9\sin^2\theta} = \sqrt{9\left(1-\sin^2\theta\right)} = 3\sqrt{\cos^2\theta} = 3\cos\theta$
(since $\cos\theta \geq 0$ for $0 \leq \theta < \frac{\pi}{2}$).

97. $f\left(x\right) = \cos^2 x - \sin^2 x$, $g\left(x\right) = 1 - 2\sin^2 x$. From the graph, $f\left(x\right) = g\left(x\right)$ this appears to be an identity. *Proof:*
$f\left(x\right) = \cos^2 x - \sin^2 x = \cos^2 x + \sin^2 x - 2\sin^2 x = 1 - 2\sin^2 x = g\left(x\right)$. Since $f\left(x\right) = g\left(x\right)$ for all x , this is an identity.

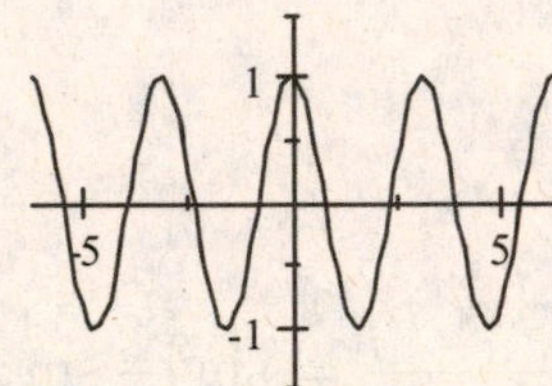

99. $f\left(x\right) = \left(\sin x + \cos x\right)^2$, $g\left(x\right) = 1$. From the graph, $f\left(x\right) = g\left(x\right)$ does not appear to be an identity. In order to show this, we can set $x = \frac{\pi}{4}$. Then we have
$f\left(\frac{\pi}{4}\right) = \left(\frac{1}{\sqrt{2}} + \frac{1}{\sqrt{2}}\right)^2 = \left(\frac{2}{\sqrt{2}}\right)^2 = \left(\sqrt{2}\right)^2 = 2 \neq 1 = g\left(\frac{\pi}{4}\right)$. Since $f\left(\frac{\pi}{4}\right) \neq g\left(\frac{\pi}{4}\right)$, this is not an identity.

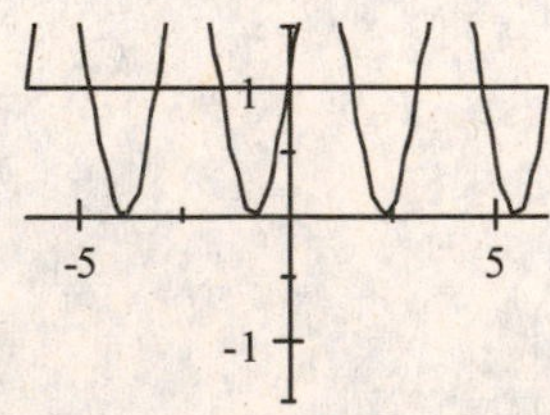

101. (a) Choose $x = \frac{\pi}{2}$. Then $\sin 2x = \sin \pi = 0$ whereas $2\sin x = 2\sin\frac{\pi}{2} = 2$.

(b) Choose $x = \frac{\pi}{4}$ and $y = \frac{\pi}{4}$. Then $\sin(x+y) = \sin\frac{\pi}{2} = 1$ whereas $\sin x + \sin y = \sin\frac{\pi}{4} + \sin\frac{\pi}{4} = \frac{1}{\sqrt{2}} + \frac{1}{\sqrt{2}} = \frac{2}{\sqrt{2}}$. Since these are not equal, the equation is not an identity.

(c) Choose $\theta = \frac{\pi}{4}$. Then $\sec^2\theta + \csc^2\theta = \left(\sqrt{2}\right)^2 + \left(\sqrt{2}\right)^2 = 4 \neq 1$.

(d) Choose $x = \frac{\pi}{4}$. Then $\dfrac{1}{\sin x + \cos x} = \dfrac{1}{\sin\frac{\pi}{4} + \cos\frac{\pi}{4}} = \dfrac{1}{\frac{1}{\sqrt{2}} + \frac{1}{\sqrt{2}}} = \frac{1}{\sqrt{2}}$ whereas $\csc x + \sec x = \csc\frac{\pi}{4} + \sec\frac{\pi}{4} = \sqrt{2} + \sqrt{2}$. Since these are not equal, the equation is not an identity.

103. No. All this proves is that $f(x) = g(x)$ for x in the range of the viewing rectangle. It does not prove that these functions are equal for all values of x. For example, let $f(x) = 1 - \dfrac{x^2}{2} + \dfrac{x^4}{24} - \dfrac{x^6}{720}$ and $g(x) = \cos x$. In the first viewing rectangle the graphs of these two functions appear identical. However, when the domain is expanded in the second viewing rectangle, you can see that these two functions are not identical.

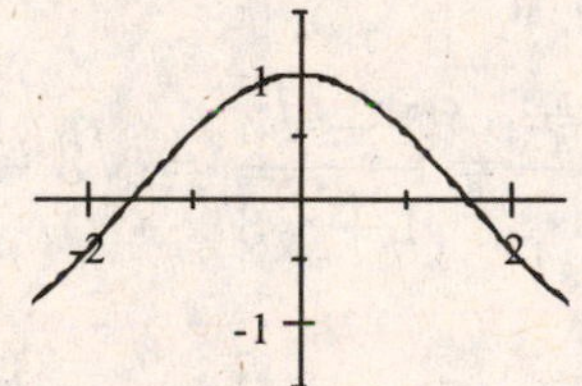

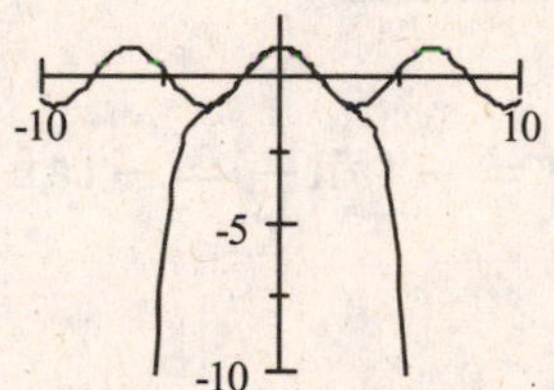

4.2 Addition and Subtraction Formulas

1. If we know the values of the sine and cosine of x and y we can find the value of $\sin\left(x+y\right)$ using the *addition* formula for sine, $\sin\left(x+y\right)=\sin x\cos y+\cos x\sin y$.

3.

$$\sin 75^\circ=\sin\left(45^\circ+30^\circ\right)=\sin 45^\circ\cos 30^\circ+\cos 45^\circ\sin 30^\circ=\tfrac{\sqrt{2}}{2}\cdot\tfrac{\sqrt{3}}{2}+\tfrac{\sqrt{2}}{2}\cdot\tfrac{1}{2}=\tfrac{\sqrt{6}+\sqrt{2}}{4}$$

5.

$$\cos 105^\circ=\cos\left(60^\circ+45^\circ\right)=\cos 60^\circ\cos 45^\circ-\sin 60^\circ\sin 45^\circ=\tfrac{1}{2}\cdot\tfrac{\sqrt{2}}{2}-\tfrac{\sqrt{3}}{2}\cdot\tfrac{\sqrt{2}}{2}=\tfrac{\sqrt{2}-\sqrt{6}}{4}$$

7.

$$\tan 15^\circ=\tan\left(45^\circ-30^\circ\right)=\frac{\tan 45^\circ-\tan 30^\circ}{1+\tan 45^\circ\tan 30^\circ}=\frac{1-\frac{\sqrt{3}}{3}}{1+1\cdot\frac{\sqrt{3}}{3}}=\frac{3-\sqrt{3}}{3+\sqrt{3}}=2-\sqrt{3}$$

9.

$$\begin{aligned}\sin\tfrac{19\pi}{12}&=-\sin\tfrac{7\pi}{12}=-\sin\left(\tfrac{\pi}{4}+\tfrac{\pi}{3}\right)=-\sin\tfrac{\pi}{4}\cos\tfrac{\pi}{3}-\cos\tfrac{\pi}{4}\sin\tfrac{\pi}{3}\\&=-\tfrac{\sqrt{2}}{2}\cdot\tfrac{1}{2}-\tfrac{\sqrt{2}}{2}\cdot\tfrac{\sqrt{3}}{2}\\&=-\tfrac{\sqrt{6}+\sqrt{2}}{4}\end{aligned}$$

11. $$\tan\left(-\tfrac{\pi}{12}\right)=-\tan\tfrac{\pi}{12}=-\tan\left(\tfrac{\pi}{3}-\tfrac{\pi}{4}\right)=-\frac{\tan\frac{\pi}{3}-\tan\frac{\pi}{4}}{1+\tan\frac{\pi}{3}\tan\frac{\pi}{4}}=\frac{1-\sqrt{3}}{1+\sqrt{3}}=\sqrt{3}-2$$

13.

$$\begin{aligned}\cos\tfrac{11\pi}{12}&=-\cos\tfrac{\pi}{12}=-\cos\left(\tfrac{\pi}{3}-\tfrac{\pi}{4}\right)=-\cos\tfrac{\pi}{3}\cos\tfrac{\pi}{4}-\sin\tfrac{\pi}{3}\sin\tfrac{\pi}{4}=-\tfrac{\sqrt{3}}{2}\cdot\tfrac{\sqrt{2}}{2}-\tfrac{1}{2}\cdot\tfrac{\sqrt{2}}{2}\\&=-\tfrac{\sqrt{6}+\sqrt{2}}{4}\end{aligned}$$

15. $\sin 18^\circ\cos 27^\circ+\cos 18^\circ\sin 27^\circ=\sin\left(18^\circ+27^\circ\right)=\sin 45^\circ=\frac{1}{\sqrt{2}}=\frac{\sqrt{2}}{2}$

17. $\cos\frac{3\pi}{7}\cos\frac{2\pi}{21}+\sin\frac{3\pi}{7}\sin\frac{2\pi}{21}=\cos\left(\frac{3\pi}{7}-\frac{2\pi}{21}\right)=\cos\frac{7\pi}{21}=\cos\frac{\pi}{3}=\frac{1}{2}$

19. $$\frac{\tan 73^\circ-\tan 13^\circ}{1+\tan 73^\circ\tan 13^\circ}=\tan\left(73^\circ-13^\circ\right)=\tan 60^\circ=\sqrt{3}$$

21.

$$\begin{aligned}\tan\left(\tfrac{\pi}{2}-u\right)&=\frac{\sin\left(\frac{\pi}{2}-u\right)}{\cos\left(\frac{\pi}{2}-u\right)}=\frac{\sin\frac{\pi}{2}\cos u-\cos\frac{\pi}{2}\sin u}{\cos\frac{\pi}{2}\cos u+\sin\frac{\pi}{2}\sin u}=\frac{1\cdot\cos u-0\cdot\sin u}{0\cdot\cos u+1\cdot\sin u}=\frac{\cos u}{\sin u}\\&=\cot u\end{aligned}$$

23.

$$\sec\left(\tfrac{\pi}{2}-u\right)=\frac{1}{\cos\left(\frac{\pi}{2}-u\right)}=\frac{1}{\cos\frac{\pi}{2}\cos u+\sin\frac{\pi}{2}\sin u}=\frac{1}{0\cdot\cos u+1\cdot\sin u}=\frac{1}{\sin u}$$
$$=\csc u$$

25. $\sin\left(x-\frac{\pi}{2}\right)=\sin x\cos\frac{\pi}{2}-\cos x\sin\frac{\pi}{2}=0\cdot\sin x-1\cdot\cos x=-\cos x$

27. $\sin\left(x-\pi\right)=\sin x\cos\pi-\cos x\sin\pi=-1\cdot\sin x-0\cdot\cos x=-\sin x$

29. $\tan\left(x-\pi\right)=\dfrac{\tan x-\tan\pi}{1+\tan x\tan\pi}=\dfrac{\tan x-0}{1+\tan x\cdot 0}=\tan x$

31. $\cos\left(x+\frac{\pi}{6}\right)+\sin\left(x-\frac{\pi}{3}\right)=\cos x\cos\frac{\pi}{6}-\sin x\sin\frac{\pi}{6}+\sin x\cos\frac{\pi}{3}-\cos x\sin\frac{\pi}{3}$

33.

$$\begin{aligned}\sin\left(x+y\right)-\sin\left(x-y\right)&=\sin x\cos y+\cos x\sin y-\left(\sin x\cos y-\cos x\sin y\right)\\&=2\cos x\sin y\end{aligned}$$

35.

$$\begin{aligned}\cot\left(x-y\right)&=\frac{1}{\tan\left(x-y\right)}=\frac{1+\tan x\tan y}{\tan x-\tan y}=\frac{1+\dfrac{1}{\cot x}\dfrac{1}{\cot y}}{\dfrac{1}{\cot x}-\dfrac{1}{\cot y}}\cdot\frac{\cot x\cot y}{\cot x\cot y}\\&=\frac{\cot x\cot y+1}{\cot y-\cot x}\end{aligned}$$

37. $\tan x-\tan y=\dfrac{\sin x}{\cos x}-\dfrac{\sin y}{\cos y}=\dfrac{\sin x\cos y-\cos x\sin y}{\cos x\cos y}=\dfrac{\sin\left(x-y\right)}{\cos x\cos y}$

39.

$$\begin{aligned}\frac{\sin\left(x+y\right)-\sin\left(x-y\right)}{\cos\left(x+y\right)+\cos\left(x-y\right)}&=\frac{\sin x\cos y+\cos x\sin y-\left(\sin x\cos y-\cos x\sin y\right)}{\cos x\cos y-\sin x\sin y+\cos x\cos y+\sin x\sin y}\\&=\frac{2\cos x\sin y}{2\cos x\cos y}\\&=\tan y\end{aligned}$$

41.

$$\begin{aligned}\sin\left(x+y+z\right)&=\sin\left(\left(x+y\right)+z\right)=\sin\left(x+y\right)\cos z+\cos\left(x+y\right)\sin z\\&=\cos z\left(\sin x\cos y+\cos x\sin y\right)+\sin z\left(\cos x\cos y-\sin x\sin y\right)\\&=\sin x\cos y\cos z+\cos x\sin y\cos z+\cos x\cos y\sin z-\sin x\sin y\sin z\end{aligned}$$

43. We want to write $\cos(\sin^{-1} x - \tan^{-1} y)$ in terms of x and y only. We let $\theta = \sin^{-1} x$ and $\phi = \tan^{-1} y$ and sketch triangles with angles θ and ϕ such that $\sin\theta = x$ and $\tan\phi = y$.

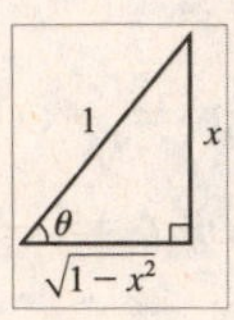

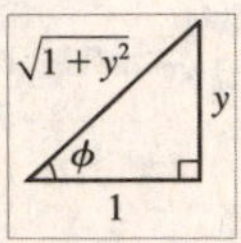

$\sin\theta = x$ $\tan\phi = y$

From the triangles, we have $\cos\theta = \sqrt{1-x^2}$, $\cos\phi = \dfrac{1}{\sqrt{1+y^2}}$, and $\sin\phi = \dfrac{y}{\sqrt{1+y^2}}$.

From the subtraction formula for cosine we have

$$\begin{aligned}\cos\left(\sin^{-1} x - \tan^{-1} y\right) &= \cos\left(\theta - \phi\right) = \cos\theta\cos\phi + \sin\theta\sin\phi \\ &= \sqrt{1-x^2}\cdot\frac{1}{\sqrt{1+y^2}} + x\cdot\frac{y}{\sqrt{1+y^2}} \\ &= \frac{\sqrt{1-x^2}+xy}{\sqrt{1+y^2}}\end{aligned}$$

45. Let $\theta = \tan^{-1} x$ and $\phi = \tan^{-1} y$.

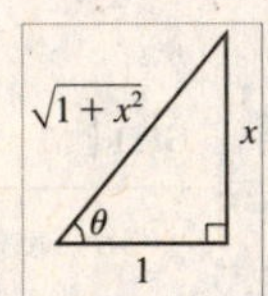

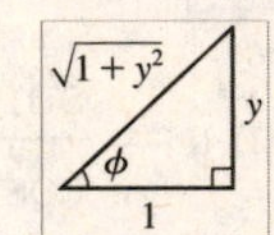

$\tan\theta = x$ $\tan\phi = y$

From the triangles, $\cos\theta = \dfrac{1}{\sqrt{1+x^2}}$, $\sin\theta = \dfrac{x}{\sqrt{1+x^2}}$, $\cos\phi = \dfrac{1}{\sqrt{1+y^2}}$, and

$\sin\phi = \dfrac{y}{\sqrt{1+y^2}}$, so using the subtraction formula for sine, we have

$$\begin{aligned}\sin\left(\tan^{-1} x - \tan^{-1} y\right) &= \sin\left(\theta - \phi\right) = \sin\theta\cos\phi - \cos\theta\sin\phi \\ &= \frac{x}{\sqrt{1+x^2}}\cdot\frac{1}{\sqrt{1+y^2}} - \frac{1}{\sqrt{1+x^2}}\cdot\frac{y}{\sqrt{1+y^2}} \\ &= \frac{x-y}{\sqrt{1+x^2}\sqrt{1+y^2}}\end{aligned}$$

47. We know that $\cos^{-1}\frac{1}{2}=\frac{\pi}{3}$ and $\tan^{-1}1=\frac{\pi}{4}$, so the addition formula for sine gives

$$\sin\left(\cos^{-1}\tfrac{1}{2}+\tan^{-1}1\right)=\sin\left(\tfrac{\pi}{3}+\tfrac{\pi}{4}\right)=\sin\tfrac{\pi}{3}\cos\tfrac{\pi}{4}+\cos\tfrac{\pi}{3}\sin\tfrac{\pi}{4}=\tfrac{\sqrt{3}}{2}\cdot\tfrac{\sqrt{2}}{2}+\tfrac{1}{2}\cdot\tfrac{\sqrt{2}}{2}$$
$$=\frac{\sqrt{6}+\sqrt{2}}{4}$$

49. We sketch triangles such that $\theta=\sin^{-1}\frac{3}{4}$ and $\phi=\cos^{-1}\frac{1}{3}$.

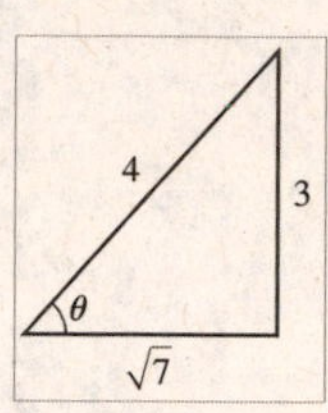

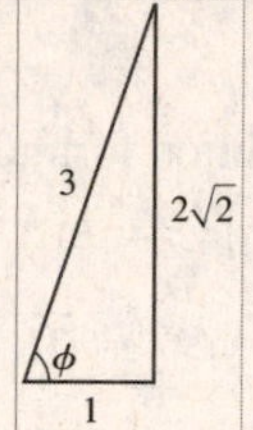

$\sin\theta=\frac{3}{4}$ $\qquad$ $\cos\phi=\frac{1}{3}$

From the triangles, we have $\tan\theta=\frac{3}{\sqrt{7}}$ and $\tan\phi=2\sqrt{2}$, so the subtraction formula for tangent gives

$$\tan\left(\sin^{-1}\tfrac{3}{4}-\cos^{-1}\tfrac{1}{3}\right)=\frac{\tan\theta-\tan\phi}{1+\tan\theta\tan\phi}=\frac{\frac{3}{\sqrt{7}}-2\sqrt{2}}{1+\frac{3}{\sqrt{7}}\cdot2\sqrt{2}}$$
$$=\frac{3-2\sqrt{14}}{\sqrt{7}+6\sqrt{2}}$$

51. As in Example 7, we sketch the angles θ and ϕ in standard position with terminal sides in the appropriate quadrants and find the remaining sides using the Pythagorean Theorem.

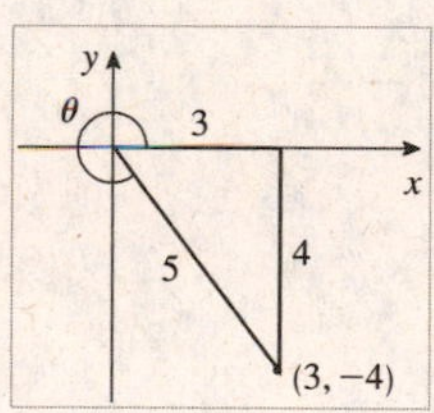

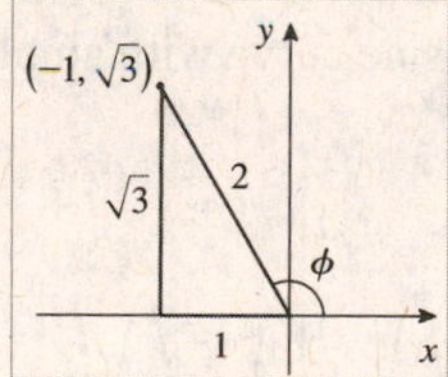

$\cos\theta=\frac{3}{5}$ $\qquad$ $\tan\phi=-\sqrt{3}$

To find $\cos(\theta-\phi)$, we use the addition formula for sine and the triangles we have sketched:

$$\cos(\theta-\phi)=\cos\theta\cos\phi+\sin\theta\sin\phi$$
$$=\tfrac{3}{5}\left(-\tfrac{1}{2}\right)+\left(-\tfrac{4}{5}\right)\tfrac{\sqrt{3}}{2}=-\frac{3+4\sqrt{3}}{10}$$

53.

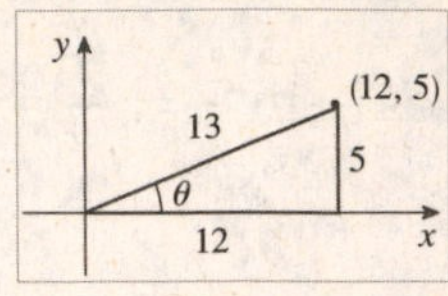

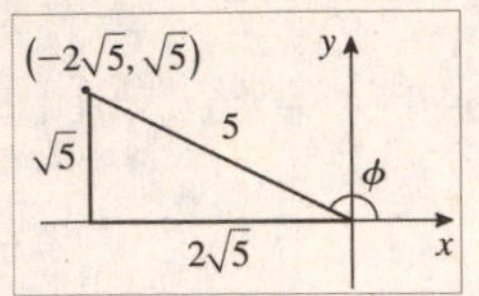

$\sin\theta = \frac{5}{13}$ $\qquad\qquad$ $\cos\phi = -\frac{2\sqrt{5}}{5}$

Using the addition formula for sine and the triangles shown, we have

$$\begin{aligned}\sin(\theta+\phi) &= \sin\theta\cos\phi + \cos\theta\sin\phi \\ &= \frac{5}{13}\left(-\frac{2\sqrt{5}}{5}\right) + \frac{12}{13}\left(\frac{\sqrt{5}}{5}\right) = \frac{2\sqrt{5}}{65}\end{aligned}$$

55. $k = \sqrt{A^2+B^2} = \sqrt{\left(-\sqrt{3}\right)^2 + 1^2} = \sqrt{4} = 2$. Thus, $\sin\phi = \frac{1}{2}$ and $\cos\phi = \frac{-\sqrt{3}}{2}$ $\Rightarrow$ $\phi = \frac{5\pi}{6}$, so $-\sqrt{3}\sin x + \cos x = k\sin(x+\phi) = 2\sin\left(x + \frac{5\pi}{6}\right)$.

57. $k = \sqrt{A^2+B^2} = \sqrt{5^2 + (-5)^2} = \sqrt{50} = 5\sqrt{2}$. Thus, $\sin\phi = -\frac{5}{5\sqrt{2}} = -\frac{1}{\sqrt{2}}$ and $\cos\phi = \frac{5}{5\sqrt{2}} = \frac{1}{\sqrt{2}}$ $\Rightarrow$ $\phi = \frac{7\pi}{4}$, so

$5(\sin 2x - \cos 2x) = k\sin(2x+\phi) = 5\sqrt{2}\sin\left(2x + \frac{7\pi}{4}\right)$.

59. (a) $g(x) = \cos 2x + \sqrt{3}\sin 2x$ $\Rightarrow$ $k = \sqrt{1^2 + \left(\sqrt{3}\right)^2} = \sqrt{4} = 2$, and ϕ satisfies $\sin\phi = \frac{1}{2}$, $\cos\phi = \frac{\sqrt{3}}{2}$ $\Rightarrow$ $\phi = \frac{\pi}{6}$. Thus, we can write

$g(x) = k\sin(2x+\phi) = 2\sin\left(2x + \frac{\pi}{6}\right) = 2\sin 2\left(x + \frac{\pi}{12}\right)$.

(b) This is a sine curve with amplitude 2 , period π , and phase shift $-\frac{\pi}{12}$.

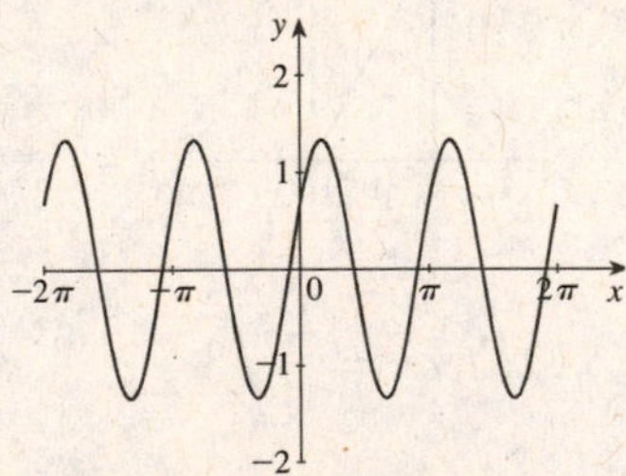

61. $g(x) = \cos x$. Now

$$\begin{aligned}\frac{g(x+h) - g(x)}{h} &= \frac{\cos(x+h) - \cos x}{h} = \frac{\cos x\cos h - \sin x\sin h - \cos x}{h} \\ &= \frac{-\cos x(1-\cos h) - \sin h(\sin x)}{h} = -\cos x\left(\frac{1-\cos h}{h}\right) - \left(\frac{\sin h}{h}\right)\sin x\end{aligned}$$

63. Let $\angle A$ and $\angle B$ be the two angles shown in the diagram.

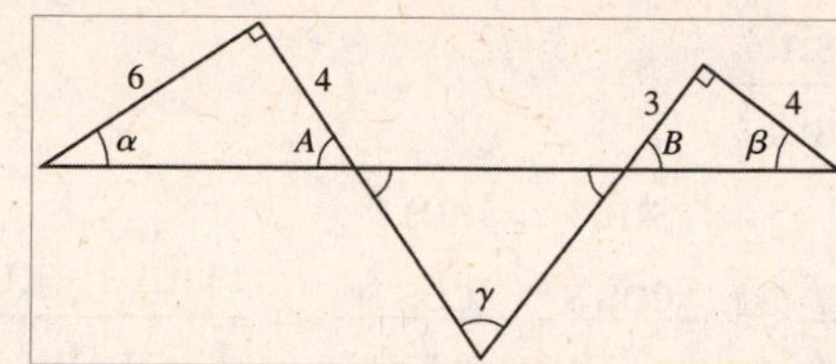

Then $180° = \gamma + A + B$, $90° = \alpha + A$, and $90° = \beta + B$. Subtracting the second and third equation from the first, we get

$180° - 90° - 90° = \gamma + A + B - (\alpha + A) - (\beta + B) \Leftrightarrow \alpha + \beta = \gamma$. Then

$$\tan\gamma = \tan(\alpha + \beta) = \frac{\tan\alpha + \tan\beta}{1 - \tan\alpha\tan\beta} = \frac{\frac{4}{6} + \frac{3}{4}}{1 - \frac{4}{6}\cdot\frac{3}{4}} = \frac{\frac{8}{12} + \frac{9}{12}}{1 - \frac{1}{2}} = 2\cdot\frac{17}{12} = \frac{17}{6}.$$

65. (a) $y = \sin^2\left(x + \frac{\pi}{4}\right) + \sin^2\left(x - \frac{\pi}{4}\right)$. From the graph we see that the value of y seems to always be equal to 1.

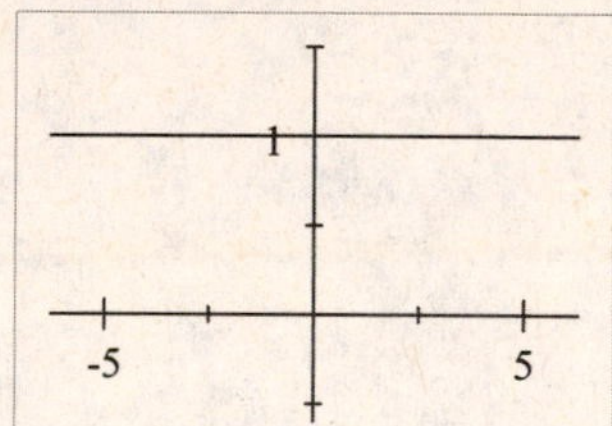

(b)

$$\begin{aligned} y &= \sin^2\left(x + \tfrac{\pi}{4}\right) + \sin^2\left(x - \tfrac{\pi}{4}\right) = \left(\sin x\cos\tfrac{\pi}{4} + \cos x\sin\tfrac{\pi}{4}\right)^2 + \left(\sin x\cos\tfrac{\pi}{4} - \cos x\sin\tfrac{\pi}{4}\right)^2 \\ &= \left[\tfrac{1}{\sqrt{2}}(\sin x + \cos x)\right]^2 + \left[\tfrac{1}{\sqrt{2}}(\sin x - \cos x)\right]^2 = \tfrac{1}{2}\left[(\sin x + \cos x)^2 + (\sin x - \cos x)^2\right] \\ &= \tfrac{1}{2}\left[\left(\sin^2 x + 2\sin x\cos x + \cos^2 x\right) + \left(\sin^2 x - 2\sin x\cos x + \cos^2 x\right)\right] \\ &= \tfrac{1}{2}\left[(1 + 2\sin x\cos x) + (1 - 2\sin x\cos x)\right] = \tfrac{1}{2}\cdot 2 = 1 \end{aligned}$$

67. Clearly $C = \frac{\pi}{4}$. Now $\tan(A + B) = \dfrac{\tan A + \tan B}{1 - \tan A\tan B} = \dfrac{\frac{1}{3} + \frac{1}{2}}{1 - \frac{1}{3}\cdot\frac{1}{2}} = 1$. Thus $A + B = \frac{\pi}{4}$, so $A + B + C = \frac{\pi}{4} + \frac{\pi}{4} = \frac{\pi}{2}$.

69. (a)

$$\begin{aligned} f(t) &= C\sin\omega t + C\sin(\omega t + \alpha) = C\sin\omega t + C(\sin\omega t\cos\alpha + \cos\omega t\sin\alpha) \\ &= C(1 + \cos\alpha)\sin\omega t + C\sin\alpha\cos\omega t = A\sin\omega t + B\cos\omega t \end{aligned}$$

where $A = C(1 + \cos\alpha)$ and $B = C\sin\alpha$.

(b) In this case, $f(t) = 10\left(1 + \cos\frac{\pi}{3}\right)\sin\omega t + 10\sin\frac{\pi}{3}\cos\omega t = 15\sin\omega t + 5\sqrt{3}\cos\omega t$.

Thus $k = \sqrt{15^2 + \left(5\sqrt{3}\right)^2} = 10\sqrt{3}$ and ϕ has $\cos\phi = \frac{15}{10\sqrt{3}} = \frac{\sqrt{3}}{2}$ and $\sin\phi = \frac{5\sqrt{3}}{10\sqrt{3}} = \frac{1}{2}$, so $\phi = \frac{\pi}{6}$. Therefore, $f(t) = 10\sqrt{3}\sin\left(\omega t + \frac{\pi}{6}\right)$.

71.

$$\tan\left(s+t\right) = \frac{\sin\left(s+t\right)}{\cos\left(s+t\right)} = \frac{\sin s\cos t+\cos s\sin t}{\cos s\cos t+\sin s\sin t}$$

$$= \frac{\sin s\cos t+\cos s\sin t}{\cos s\cos t-\sin s\sin t}\cdot\frac{\dfrac{1}{\cos s\cos t}}{\dfrac{1}{\cos s\cos t}} = \frac{\dfrac{\sin s}{\cos s}+\dfrac{\sin t}{\cos t}}{1-\dfrac{\sin s}{\cos s}\cdot\dfrac{\sin t}{\cos t}} = \frac{\tan s+\tan t}{1-\tan s\tan t}$$

4.3 Double-Angle, Half-Angle, and Product-Sum Formulas

1. If we know the values of $\sin x$ and $\cos x$, we can find the value of $\sin 2x$ using the *double-angle* formula for sine, $\sin 2x = 2\sin x\cos x$.

3. $\sin x = \frac{5}{13}$, x in quadrant I $\Rightarrow$ $\cos x = \frac{12}{13}$ and $\tan x = \frac{5}{12}$. Thus,

$\sin 2x = 2\sin x\cos x = 2\left(\frac{5}{13}\right)\left(\frac{12}{13}\right) = \frac{120}{169}$,

$\cos 2x = \cos^2 x - \sin^2 x = \left(\frac{12}{13}\right)^2 - \left(\frac{5}{13}\right)^2 = \frac{144-25}{169} = \frac{119}{169}$, and

$$\tan 2x = \frac{\sin 2x}{\cos 2x} = \frac{\frac{120}{169}}{\frac{119}{169}} = \frac{120}{169}\cdot\frac{169}{119} = \frac{120}{119}.$$

5. $\cos x = \frac{4}{5}$. Then $\sin x = -\frac{3}{5}$ ($\csc x < 0$) and $\tan x = -\frac{3}{4}$. Thus,

$\sin 2x = 2\sin x\cos x = 2\left(-\frac{3}{5}\right)\cdot\frac{4}{5} = -\frac{24}{25}$,

$\cos 2x = \cos^2 x - \sin^2 x = \left(\frac{4}{5}\right)^2 - \left(-\frac{3}{5}\right)^2 = \frac{16-9}{25} = \frac{7}{25}$, and

$$\tan 2x = \frac{\sin 2x}{\cos 2x} = \frac{-\frac{24}{25}}{\frac{7}{25}} = -\frac{24}{25}\cdot\frac{25}{7} = -\frac{24}{7}.$$

7. $\sin x = -\frac{3}{5}$. Then, $\cos x = -\frac{4}{5}$ and $\tan x = \frac{3}{4}$ (x is in quadrant III). Thus,

$\sin 2x = 2\sin x\cos x = 2\left(-\frac{3}{5}\right)\left(-\frac{4}{5}\right) = \frac{24}{25}$,

$\cos 2x = \cos^2 x - \sin^2 x = \left(-\frac{4}{5}\right)^2 - \left(-\frac{3}{5}\right)^2 = \frac{16-9}{25} = \frac{7}{25}$, and

$$\tan 2x = \frac{\sin 2x}{\cos 2x} = \frac{\frac{24}{25}}{\frac{7}{25}} = \frac{24}{25}\cdot\frac{25}{7} = \frac{24}{7}.$$

9. $\tan x = -\frac{1}{3}$ and $\cos x > 0$, so $\sin x < 0$. Thus, $\sin x = -\frac{1}{\sqrt{10}}$ and $\cos x = \frac{3}{\sqrt{10}}$. Thus,

$\sin 2x = 2\sin x\cos x = 2\left(-\frac{1}{\sqrt{10}}\right)\left(\frac{3}{\sqrt{10}}\right) = -\frac{6}{10} = -\frac{3}{5}$,

$\cos 2x = \cos^2 x - \sin^2 x = \left(\frac{3}{\sqrt{10}}\right)^2 - \left(-\frac{1}{\sqrt{10}}\right)^2 = \frac{8}{10} = \frac{4}{5}$, and

$$\tan 2x = \frac{\sin 2x}{\cos 2x} = \frac{-\frac{3}{5}}{\frac{4}{5}} = -\frac{3}{5}\cdot\frac{5}{4} = -\frac{3}{4}.$$

11.
$$\begin{aligned}\sin^4 x &= \left(\sin^2 x\right)^2 = \left(\frac{1-\cos 2x}{2}\right)^2 = \tfrac{1}{4} - \tfrac{1}{2}\cos 2x + \tfrac{1}{4}\cos^2 2x \\ &= \tfrac{1}{4} - \tfrac{1}{2}\cos 2x + \tfrac{1}{4}\cdot\frac{1+\cos 4x}{2} = \tfrac{1}{4} - \tfrac{1}{2}\cos 2x + \tfrac{1}{8} + \tfrac{1}{8}\cos 4x = \tfrac{3}{8} - \tfrac{1}{2}\cos 2x + \tfrac{1}{8}\cos 4x \\ &= \tfrac{1}{2}\left(\tfrac{3}{4} - \cos 2x + \tfrac{1}{4}\cos 4x\right)\end{aligned}$$

13. We use the result of Example 4 to get

$$\cos^2 x\sin^4 x = \left(\sin^2 x\cos^2 x\right)\sin^2 x = \left(\tfrac{1}{8} - \tfrac{1}{8}\cos 4x\right)\cdot\left(\tfrac{1}{2} - \tfrac{1}{2}\cos 2x\right) = \tfrac{1}{16}\left(1 - \cos 2x - \cos 4x + \cos 2x\cos 4x\right)$$

.

15. Since $\sin^4 x\cos^4 x = \left(\sin^2 x\cos^2 x\right)^2$ we can use the result of Example 4 to get

$$\begin{aligned}\sin^4 x\cos^4 x &= \left(\tfrac{1}{8}-\tfrac{1}{8}\cos 4x\right)^2 = \tfrac{1}{64}-\tfrac{1}{32}\cos 4x+\tfrac{1}{64}\cos^2 4x \\ &= \tfrac{1}{64}-\tfrac{1}{32}\cos 4x+\tfrac{1}{64}\cdot\tfrac{1}{2}\left(1+\cos 8x\right) = \tfrac{1}{64}-\tfrac{1}{32}\cos 4x+\tfrac{1}{128}+\tfrac{1}{128}\cos 8x \\ &= \tfrac{3}{128}-\tfrac{1}{32}\cos 4x+\tfrac{1}{128}\cos 8x = \tfrac{1}{32}\left(\tfrac{3}{4}-\cos 4x+\tfrac{1}{4}\cos 8x\right)\end{aligned}$$

17. $\sin 15^\circ = \sqrt{\tfrac{1}{2}\left(1-\cos 30^\circ\right)} = \sqrt{\tfrac{1}{2}\left(1-\tfrac{\sqrt{3}}{2}\right)} = \sqrt{\tfrac{1}{4}\left(2-\sqrt{3}\right)} = \tfrac{1}{2}\sqrt{2-\sqrt{3}}$

19. $\tan 22.5^\circ = \dfrac{1-\cos 45^\circ}{\sin 45^\circ} = \dfrac{1-\frac{\sqrt{2}}{2}}{\frac{\sqrt{2}}{2}} = \sqrt{2}-1$

21. $\cos 165^\circ = -\sqrt{\tfrac{1}{2}\left(1+\cos 330^\circ\right)} = -\sqrt{\tfrac{1}{2}\left(1+\cos 30^\circ\right)} = -\sqrt{\tfrac{1}{2}\left(1+\tfrac{\sqrt{3}}{2}\right)} = -\tfrac{1}{2}\sqrt{2+\sqrt{3}}$

23. $\tan\frac{\pi}{8} = \dfrac{1-\cos\frac{\pi}{4}}{\sin\frac{\pi}{4}} = \dfrac{1-\frac{\sqrt{2}}{2}}{\frac{\sqrt{2}}{2}} = \sqrt{2}-1$

25. $\cos\frac{\pi}{12} = \sqrt{\tfrac{1}{2}\left(1+\cos\frac{\pi}{6}\right)} = \sqrt{\tfrac{1}{2}\left(1+\tfrac{\sqrt{3}}{2}\right)} = \tfrac{1}{2}\sqrt{2+\sqrt{3}}$

27. $\sin\frac{9\pi}{8} = -\sqrt{\tfrac{1}{2}\left(1-\cos\frac{9\pi}{4}\right)} = -\sqrt{\tfrac{1}{2}\left(1-\tfrac{\sqrt{2}}{2}\right)} = -\tfrac{1}{2}\sqrt{2-\sqrt{2}}$. We have chosen the negative root because $\frac{9\pi}{8}$ is in quadrant III, so $\sin\frac{9\pi}{8} < 0$.

29. (a) $2\sin 18^\circ\cos 18^\circ = \sin 36^\circ$

(b) $2\sin 3\theta\cos 3\theta = \sin 6\theta$

31. (a) $\cos^2 34^\circ - \sin^2 34^\circ = \cos 68^\circ$

(b) $\cos^2 5\theta - \sin^2 5\theta = \cos 10\theta$

33. (a) $\dfrac{\sin 8^\circ}{1+\cos 8^\circ} = \tan\dfrac{8^\circ}{2} = \tan 4^\circ$

(b) $\dfrac{1-\cos 4\theta}{\sin 4\theta} = \tan\dfrac{4\theta}{2} = \tan 2\theta$

35. $\sin\left(x+x\right) = \sin x\cos x + \cos x\sin x = 2\sin x\cos x$

37. $\sin x = \frac{3}{5}$. Since x is in quadrant I, $\cos x = \frac{4}{5}$ and $\frac{x}{2}$ is also in quadrant I. Thus,

$\sin\frac{x}{2} = \sqrt{\tfrac{1}{2}\left(1-\cos x\right)} = \sqrt{\tfrac{1}{2}\left(1-\tfrac{4}{5}\right)} = \tfrac{1}{\sqrt{10}} = \tfrac{\sqrt{10}}{10}$,

$\cos\frac{x}{2} = \sqrt{\tfrac{1}{2}\left(1+\cos x\right)} = \sqrt{\tfrac{1}{2}\left(1+\tfrac{4}{5}\right)} = \tfrac{3}{\sqrt{10}} = \tfrac{3\sqrt{10}}{10}$, and $\tan\frac{x}{2} = \dfrac{\sin\frac{x}{2}}{\cos\frac{x}{2}} = \tfrac{1}{\sqrt{10}}\cdot\tfrac{\sqrt{10}}{3} = \tfrac{1}{3}$.

39. $\csc x = 3$. Then, $\sin x = \frac{1}{3}$ and since x is in quadrant II, $\cos x = -\dfrac{2\sqrt{2}}{3}$. Since $90^\circ \le x \le 180^\circ$, we have $45^\circ \le \frac{x}{2} \le 90^\circ$ and so $\frac{x}{2}$ is in quadrant I. Thus,

$$\sin\tfrac{x}{2} = \sqrt{\tfrac{1}{2}(1-\cos x)} = \sqrt{\tfrac{1}{2}\left(1+\tfrac{2\sqrt{2}}{3}\right)} = \sqrt{\tfrac{1}{6}\left(3+2\sqrt{2}\right)},$$

$$\cos\tfrac{x}{2} = \sqrt{\tfrac{1}{2}(1+\cos x)} = \sqrt{\tfrac{1}{2}\left(1-\tfrac{2\sqrt{2}}{3}\right)} = \sqrt{\tfrac{1}{6}\left(3-2\sqrt{2}\right)}, \text{ and}$$

$$\tan\tfrac{x}{2} = \frac{\sin\frac{x}{2}}{\cos\frac{x}{2}} = \sqrt{\tfrac{3+2\sqrt{2}}{3-2\sqrt{2}}} = 3+2\sqrt{2}.$$

41. $\sec x = \frac{3}{2}$. Then $\cos x = \frac{2}{3}$ and since x is in quadrant IV, $\sin x = -\frac{\sqrt{5}}{3}$. Since $270^\circ \le x \le 360^\circ$, we have $135^\circ \le \frac{x}{2} \le 180^\circ$ and so $\frac{x}{2}$ is in quadrant II. Thus,

$$\sin\tfrac{x}{2} = \sqrt{\tfrac{1}{2}(1-\cos x)} = \sqrt{\tfrac{1}{2}\left(1-\tfrac{2}{3}\right)} = \tfrac{1}{\sqrt{6}} = \tfrac{\sqrt{6}}{6},$$

$$\cos\tfrac{x}{2} = -\sqrt{\tfrac{1}{2}(1+\cos x)} = -\sqrt{\tfrac{1}{2}\left(1+\tfrac{2}{3}\right)} = -\tfrac{\sqrt{5}}{\sqrt{6}} = -\frac{\sqrt{30}}{6}, \text{ and}$$

$$\tan\tfrac{x}{2} = \frac{\sin\frac{x}{2}}{\cos\frac{x}{2}} = \tfrac{1}{\sqrt{6}} \cdot \tfrac{\sqrt{6}}{-\sqrt{5}} = -\tfrac{1}{\sqrt{5}} = -\tfrac{\sqrt{5}}{5}.$$

43. To write $\sin\left(2\tan^{-1} x\right)$ as an algebraic expression in x, we let $\theta = \tan^{-1} x$ and sketch a suitable triangle. We see that $\sin\theta = \dfrac{x}{\sqrt{1+x^2}}$ and $\cos\theta = \dfrac{1}{\sqrt{1+x^2}}$, so using the double-angle formula for sine, we have

$$\sin\left(2\tan^{-1} x\right) = \sin 2\theta = 2\sin\theta\cos\theta = 2 \cdot \frac{x}{\sqrt{1+x^2}} \cdot \frac{1}{\sqrt{1+x^2}} = \frac{2x}{1+x^2}.$$

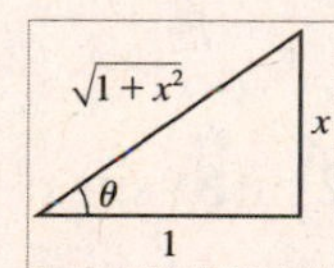

45. Using the half-angle formula for sine, we have $\sin\left(\frac{1}{2}\cos^{-1} x\right) = \pm\sqrt{\dfrac{1-\cos\left(\cos^{-1} x\right)}{2}} = \pm\sqrt{\dfrac{1-x}{2}}$. Because $\cos^{-1}$ has range $[0,\pi]$, $\frac{1}{2}\cos^{-1} x$ lies in $\left[0,\frac{\pi}{2}\right]$ and so $\sin\left(\frac{1}{2}\cos^{-1} x\right)$ is positive. Thus, $\sin\left(\frac{1}{2}\cos^{-1} x\right) = \sqrt{\dfrac{1-x}{2}}$.

47. We sketch a triangle with $\theta = \cos^{-1}\frac{7}{25}$ and find that $\sin\theta = \frac{24}{25}$.

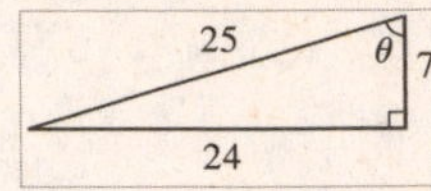

$\cos\theta = \frac{7}{25}$

Thus, using the double-angle formula for sine,

$\sin\left(2\cos^{-1}\frac{7}{25}\right) = \sin 2\theta = 2\sin\theta\cos\theta = 2\cdot\frac{24}{25}\cdot\frac{7}{25} = \frac{336}{625}$.

49. Rewriting the given expression and using a double-angle formula for cosine, we have

$$\sec\left(2\sin^{-1}\tfrac{1}{4}\right) = \frac{1}{\cos\left(2\sin^{-1}\frac{1}{4}\right)} = \frac{1}{1-2\sin^2\left(\sin^{-1}\frac{1}{4}\right)} = \frac{1}{1-2\left(\frac{1}{4}\right)^2} = \frac{8}{7}.$$

51. Using a double-angle formula for cosine, we have $\cos 2\theta = 1-2\sin^2\theta = 1-2\left(-\frac{3}{5}\right)^2 = \frac{7}{25}$.

53. To evaluate $\sin 2\theta$, we first sketch the angle θ in standard position with terminal side in quadrant II and find the remaining side using the Pythagorean Theorem.

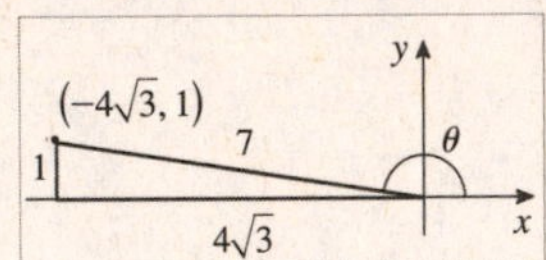

Using the double-angle formula for sine, we have $\sin 2\theta = 2\sin\theta\cos\theta = 2\left(\frac{1}{7}\right)\left(-\frac{4\sqrt{3}}{7}\right) = -\frac{8\sqrt{3}}{49}$.

55. $\sin 2x\cos 3x = \frac{1}{2}\left[\sin(2x+3x)+\sin(2x-3x)\right] = \frac{1}{2}(\sin 5x - \sin x)$

57. $\cos x\sin 4x = \frac{1}{2}\left[\sin(4x+x)+\sin(4x-x)\right] = \frac{1}{2}(\sin 5x + \sin 3x)$

59. $3\cos 4x\cos 7x = 3\cdot\frac{1}{2}\left[\cos(4x+7x)+\cos(4x-7x)\right] = \frac{3}{2}(\cos 11x + \cos 3x)$

61. $\sin 5x + \sin 3x = 2\sin\left(\dfrac{5x+3x}{2}\right)\cos\left(\dfrac{5x-3x}{2}\right) = 2\sin 4x\cos x$

63. $\cos 4x - \cos 6x = -2\sin\left(\dfrac{4x+6x}{2}\right)\sin\left(\dfrac{4x-6x}{2}\right) = -2\sin 5x\sin(-x) = 2\sin 5x\sin x$

65. $\sin 2x - \sin 7x = 2\cos\left(\dfrac{2x+7x}{2}\right)\sin\left(\dfrac{2x-7x}{2}\right) = 2\cos\dfrac{9x}{2}\sin\left(-\dfrac{5x}{2}\right) = -2\cos\dfrac{9x}{2}\sin\dfrac{5x}{2}$

67. $2\sin 52.5^\circ\sin 97.5^\circ = 2\cdot\frac{1}{2}\left[\cos(52.5^\circ-97.5^\circ)-\cos(52.5^\circ+97.5^\circ)\right] = \cos(-45^\circ)-\cos 150^\circ$

$= \cos 45^\circ - \cos 150^\circ = \frac{\sqrt{2}}{2}+\frac{\sqrt{3}}{2} = \frac{1}{2}\left(\sqrt{2}+\sqrt{3}\right)$

69. $\cos 37.5^\circ\sin 7.5^\circ = \frac{1}{2}(\sin 45^\circ - \sin 30^\circ) = \frac{1}{2}\left(\frac{\sqrt{2}}{2}-\frac{1}{2}\right) = \frac{1}{4}\left(\sqrt{2}-1\right)$

71. $\cos 255^\circ - \cos 195^\circ = -2\sin\left(\dfrac{255^\circ+195^\circ}{2}\right)\sin\left(\dfrac{255^\circ-195^\circ}{2}\right) = -2\sin 225^\circ\sin 30^\circ$

$= -2\left(-\frac{\sqrt{2}}{2}\right)\frac{1}{2} = \frac{\sqrt{2}}{2}$

73. $\cos^2 5x - \sin^2 5x = \cos(2\cdot 5x) = \cos 10x$

75. $(\sin x + \cos x)^2 = \sin^2 x + 2\sin x\cos x + \cos^2 x = 1 + 2\sin x\cos x = 1 + \sin 2x$

77. $\dfrac{\sin 4x}{\sin x} = \dfrac{2\sin 2x\cos 2x}{\sin x} = \dfrac{2(2\sin x\cos x)(\cos 2x)}{\sin x} = 4\cos x\cos 2x$

79. $\dfrac{2(\tan x - \cot x)}{\tan^2 x - \cot^2 x} = \dfrac{2(\tan x - \cot x)}{(\tan x + \cot x)(\tan x - \cot x)} = \dfrac{2}{\tan x + \cot x} = \dfrac{2}{\dfrac{\sin x}{\cos x} + \dfrac{\cos x}{\sin x}}$

$$= \frac{2}{\dfrac{\sin x}{\cos x} + \dfrac{\cos x}{\sin x}} \cdot \frac{\sin x\cos x}{\sin x\cos x} = \frac{2\sin x\cos x}{\sin^2 x + \cos^2 x} = 2\sin x\cos x = \sin 2x$$

81.

$$\tan 3x = \tan(2x + x) = \frac{\tan 2x + \tan x}{1 - \tan 2x\tan x} = \frac{\dfrac{2\tan x}{1-\tan^2 x} + \tan x}{1 - \dfrac{2\tan x}{1-\tan^2 x}\tan x} = \frac{2\tan x + \tan x(1-\tan^2 x)}{1 - \tan^2 x - 2\tan x\tan x}$$

$$= \frac{3\tan x - \tan^3 x}{1 - 3\tan^2 x}$$

83. $\cos^4 x - \sin^4 x = (\cos^2 x + \sin^2 x)(\cos^2 x - \sin^2 x) = \cos^2 x - \sin^2 x = \cos 2x$

85. $\dfrac{\sin x + \sin 5x}{\cos x + \cos 5x} = \dfrac{2\sin 3x\cos 2x}{2\cos 3x\cos 2x} = \dfrac{\sin 3x}{\cos 3x} = \tan 3x$

87. $\dfrac{\sin 10x}{\sin 9x + \sin x} = \dfrac{2\sin 5x\cos 5x}{2\sin 5x\cos 4x} = \dfrac{\cos 5x}{\cos 4x}$

89.

$$\frac{\sin x + \sin y}{\cos x + \cos y} = \frac{2\sin\left(\dfrac{x+y}{2}\right)\cos\left(\dfrac{x-y}{2}\right)}{2\cos\left(\dfrac{x+y}{2}\right)\cos\left(\dfrac{x-y}{2}\right)} = \frac{\sin\left(\dfrac{x+y}{2}\right)}{\cos\left(\dfrac{x+y}{2}\right)} = \tan\left(\frac{x+y}{2}\right)$$

91. $\sin 130^\circ - \sin 110^\circ = 2\cos\dfrac{130^\circ + 110^\circ}{2}\sin\dfrac{130^\circ - 110^\circ}{2} = 2\cos 120^\circ \sin 10^\circ = 2\left(-\tfrac{1}{2}\right)\sin 10^\circ$

$$= -\sin 10^\circ$$

93. $\sin 45^\circ + \sin 15^\circ = 2\sin\left(\dfrac{45^\circ + 15^\circ}{2}\right)\cos\left(\dfrac{45^\circ - 15^\circ}{2}\right) = 2\sin 30^\circ \cos 15^\circ = 2\cdot\tfrac{1}{2}\cdot\cos 15^\circ$

$$= \cos 15^\circ = \sin(90^\circ - 15^\circ) = \sin 75^\circ$$

95. $\dfrac{\sin x + \sin 2x + \sin 3x + \sin 4x + \sin 5x}{\cos x + \cos 2x + \cos 3x + \cos 4x + \cos 5x} = \dfrac{(\sin x + \sin 5x) + (\sin 2x + \sin 4x) + \sin 3x}{(\cos x + \cos 5x) + (\cos 2x + \cos 4x) + \cos 3x}$

$$= \frac{2\sin 3x\cos 2x + 2\sin 3x\cos x + \sin 3x}{2\cos 3x\cos 2x + 2\cos 3x\cos x + \cos 3x} = \frac{\sin 3x(2\cos 2x + 2\cos x + 1)}{\cos 3x(2\cos 2x + 2\cos x + 1)} = \tan 3x$$

97. (a) $f(x) = \dfrac{\sin 3x}{\sin x} - \dfrac{\cos 3x}{\cos x}$

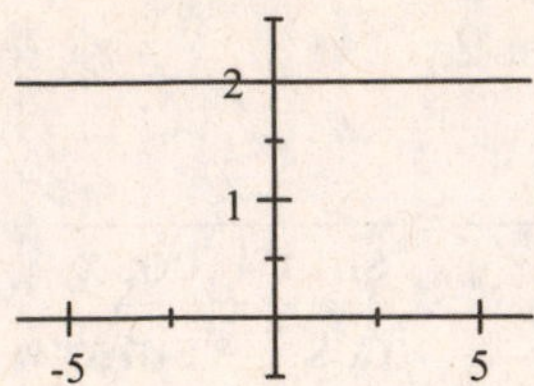

The function appears to have a constant value of 2 wherever it is defined.

(b) $f(x) = \dfrac{\sin 3x}{\sin x} - \dfrac{\cos 3x}{\cos x}$

$$= \frac{\sin 3x \cos x - \cos 3x \sin x}{\sin x \cos x} = \frac{\sin(3x - x)}{\sin x \cos x}$$

$$= \frac{\sin 2x}{\sin x \cos x} = \frac{2\sin x \cos x}{\sin x \cos x} = 2$$

for all x for which the function is defined.

99. (a) $y = \sin 6x + \sin 7x$

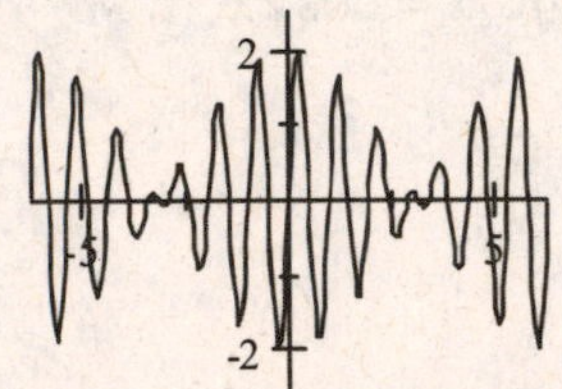

(b) By a sum-to-product formula, $y = \sin 6x + \sin 7x$

$$= 2\sin\left(\frac{6x+7x}{2}\right)\cos\left(\frac{6x-7x}{2}\right)$$

$$= 2\sin\left(\tfrac{13}{2}x\right)\cos\left(-\tfrac{1}{2}x\right)$$

$$= 2\sin\tfrac{13}{2}x\cos\tfrac{1}{2}x$$

(c) We graph $y = \sin 6x + \sin 7x$, $y = 2\cos\left(\frac{1}{2}x\right)$, and $y = -2\cos\left(\frac{1}{2}x\right)$.

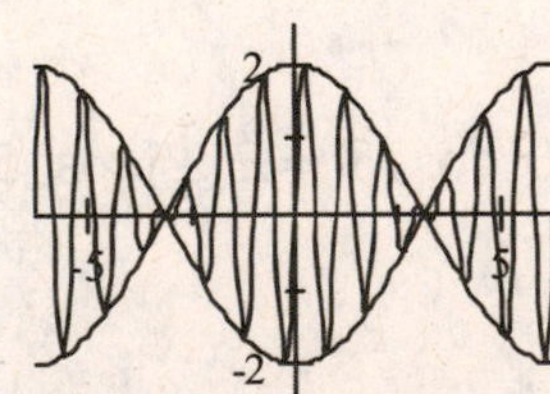

The graph of $y = f(x)$ lies between the other two graphs.

101. (a) $\cos 4x = \cos(2x+2x) = 2\cos^2 2x - 1 = 2(2\cos^2 x - 1)^2 - 1 = 8\cos^4 x - 8\cos^2 x + 1$

Thus the desired polynomial is $P(t) = 8t^4 - 8t^2 + 1$.

(b)

$$\cos 5x = \cos(4x+x) = \cos 4x \cos x - \sin 4x \sin x = \cos x(8\cos^4 x - 8\cos^2 x + 1) - 2\sin 2x \cos 2x \sin x$$
$$= 8\cos^5 x - 8\cos^3 + \cos x - 4\sin x \cos x(2\cos^2 x - 1)\sin x$$
$$= 8\cos^5 x - 8\cos^3 x + \cos x - 4\cos x(2\cos^2 x - 1)\sin^2 x$$
$$= 8\cos^5 x - 8\cos^3 x + \cos x - 4\cos x(2\cos^2 x - 1)(1 - \cos^2 x)$$
$$= 8\cos^5 x - 8\cos^3 x + \cos x + 8\cos^5 x - 12\cos^3 x + 4\cos x = 16\cos^5 x - 20\cos^3 x + 5\cos x$$

Thus, the desired polynomial is $P(t) = 16t^5 - 20t^3 + 5t$.

103. Using a product-to-sum formula,

$$RHS = 4\sin A \sin B \sin C = 4\sin A\left\{\tfrac{1}{2}\left[\cos(B-C) - \cos(B+C)\right]\right\}$$
$$= 2\sin A\cos(B-C) - 2\sin A\cos(B+C)$$

Using another product-to-sum formula, this is equal to

$$2\left\{\frac{1}{2}\left[\sin(A+B-C) + \sin(A-B+C)\right]\right\} - 2\left\{\frac{1}{2}\left[\sin(A+B+C) + \sin(A-B-C)\right]\right\}$$
$$= \sin(A+B-C) + \sin(A-B+C) - \sin(A+B+C) - \sin(A-B-C)$$

Now $A+B+C=\pi$, so $A+B-C=\pi-2C$, $A-B+C=\pi-2B$, and $A-B-C=2A-\pi$.

Thus our expression simplifies to

$$\sin(A+B-C) + \sin(A-B+C) - \sin(A+B+C) - \sin(A-B-C)$$
$$= \sin(\pi-2C) + \sin(\pi-2B) + 0 - \sin(2A-\pi) = \sin 2C + \sin 2B + \sin 2A = LHS$$

105. (a) In both logs the length of the adjacent side is $20\cos\theta$ and the length of the opposite side is $20\sin\theta$.

Thus the cross-sectional area of the beam is modeled by

$$A(\theta) = (20\cos\theta)(20\sin\theta) = 400\sin\theta\cos\theta = 200(2\sin\theta\cos\theta) = 200\sin 2\theta .$$

(b) The function $y = \sin u$ is maximized when $u = \frac{\pi}{2}$. So $2\theta = \frac{\pi}{2} \Leftrightarrow \theta = \frac{\pi}{4}$. Thus the maximum cross-sectional area is $A\left(\frac{\pi}{4}\right) = 200\sin 2\left(\frac{\pi}{4}\right) = 200$.

107. (a) $y = f_1(t) + f_2(t) = \cos 11t + \cos 13t$

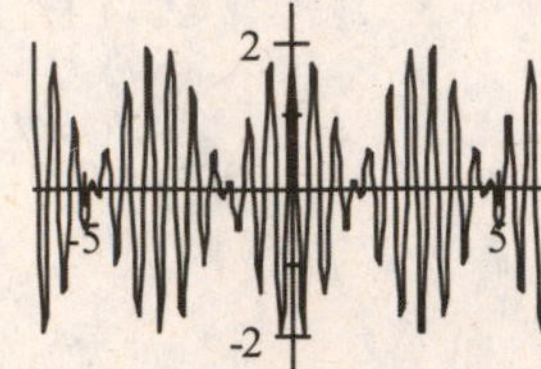

(b) Using the identity $\cos\alpha + \cos y = 2\cdot\cos\left(\frac{\alpha + y}{2}\right)\cos\left(\frac{\alpha - y}{2}\right)$, we have

$$f(t) = \cos 11t + \cos 13t = 2\cdot\cos\left(\frac{11t + 13t}{2}\right)\cos\left(\frac{11t - 13t}{2}\right)$$

$$= 2\cdot\cos 12t\cdot\cos(-t) = 2\cos 12t\cos t$$

(c) We graph $y = \cos 11t + \cos 13t$, $y = 2\cos t$, and $y = -2\cos t$.

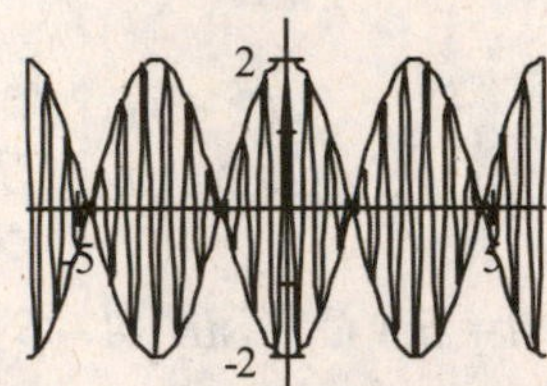

The graph of f lies between the graphs of $y = 2\cos t$ and $y = -2\cos t$. Thus, the loudness of the sound varies between $y = \pm 2\cos t$.

109.

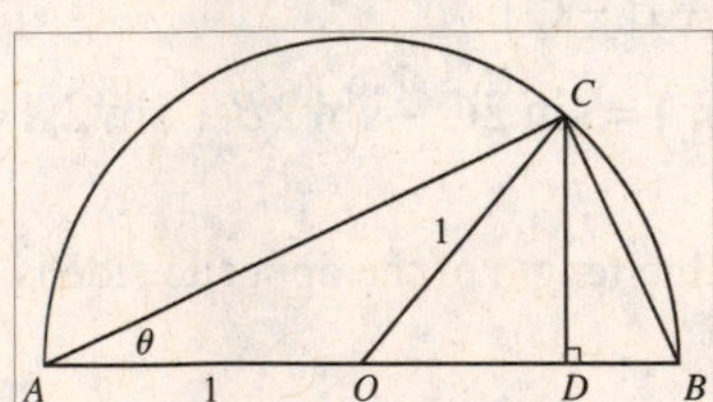

We find the area of ΔABC in two different ways. First, let AB be the base and CD be the height. Since $\angle BOC = 2\theta$ we see that $CD = \sin 2\theta$. So the area is $\frac{1}{2}(\text{base})(\text{height}) = \frac{1}{2}\cdot 2\cdot\sin 2\theta = \sin 2\theta$. On the other hand, in ΔABC we see that $\angle C$ is a right angle. So $BC = 2\sin\theta$ and $AC = 2\cos\theta$, and the area is $\frac{1}{2}(\text{base})(\text{height}) = \frac{1}{2}\cdot(2\sin\theta)(2\cos\theta) = 2\sin\theta\cos\theta$. Equating the two expressions for the area of ΔABC, we get $\sin 2\theta = 2\sin\theta\cos\theta$.

4.4 Basic Trigonometric Equations

1. Because the trigonometric functions are periodic, if a basic trigonometric equation has one solution, it has *infinitely many* solutions.

3. We can find some of the solutions of $\sin x = 0.3$ graphically by graphing $y = \sin x$ and $y = 0.3$. The solutions shown are $x \approx -9.729$, $x \approx -5.978$, $x \approx -3.446$, $x \approx 0.3047$, $x \approx 2.837$, $x \approx 6.588$, and $x \approx 9.120$.

5. Because sine has period 2π, we first find the solutions in the interval $[0, 2\pi)$.

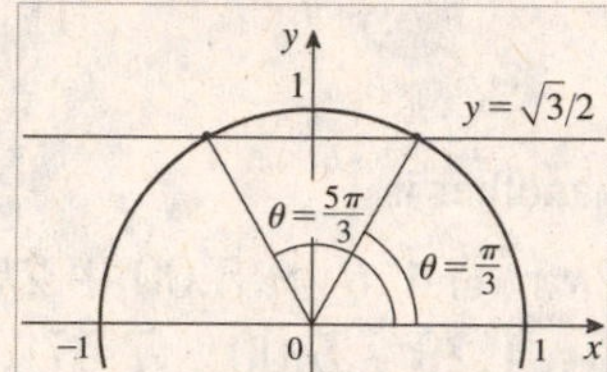

From the unit circle shown, we see that $\sin\theta = \frac{\sqrt{3}}{2}$ in quadrants I and II, so the solutions are $\theta = \frac{\pi}{3}$ and $\theta = \frac{2\pi}{3}$. We get all solutions of the equation by adding integer multiples of 2π to these solutions: $\theta = \frac{\pi}{3} + 2k\pi$ and $\theta = \frac{2\pi}{3} + 2k\pi$ for any integer k.

7. The cosine function is negative in quadrants II and III, so the solution of $\cos\theta = -1$ on the interval $[0, 2\pi)$ is $\theta = \pi$. Adding integer multiples of 2π to this solution gives all solutions: $\theta = \pi + 2k\pi = (2k+1)\pi$ for any integer k.

9. The cosine function is positive in quadrants I and IV, so the solutions of $\cos\theta = \frac{1}{4}$ on the interval $[0, 2\pi)$ are $\theta = \cos^{-1}\frac{1}{4} \approx 1.32$ and $\theta = 2\pi - \cos^{-1}\frac{1}{4} \approx 4.97$. Adding integer multiples of 2π to these solutions gives all solutions: $\theta \approx 1.32 + 2k\pi$, $4.97 + 2k\pi$ for any integer k.

11. The sine function is negative in quadrants III and IV, so the solutions of $\sin\theta = -0.45$ on the interval $[0, 2\pi)$ are $\theta = \pi + \sin^{-1}(0.45) \approx 3.61$ and $\theta = 2\pi - \sin^{-1}(0.45) \approx 5.82$. Adding integer multiples of 2π to these solutions gives all solutions, $\theta \approx 3.61 + 2k\pi$, $5.82 + 2k\pi$ for any integer k.

13. We first find one solution by taking $\tan^{-1}$ of each side of the equation: $\theta = \tan^{-1}(-\sqrt{3}) = -\frac{\pi}{3}$. By definition, this is the only solution in the interval $\left(-\frac{\pi}{2}, \frac{\pi}{2}\right)$. Since tangent has period π, we get all solutions of the equation by adding integer multiples of π: $\theta = -\frac{\pi}{3} + k\pi$ for any integer k.

15. One solution of $\tan\theta = 5$ is $\theta = \tan^{-1} 5 \approx 1.37$. Adding integer multiples of π to this solution gives all solutions: $\theta \approx 1.37 + k\pi$ for any integer k.

17. One solution of $\cos\theta = -\frac{\sqrt{3}}{2}$ is $\theta = \cos^{-1}\left(-\frac{\sqrt{3}}{2}\right) = \frac{5\pi}{6}$ and another is $\theta = 2\pi - \frac{5\pi}{6} = \frac{7\pi}{6}$. All solutions are $\theta = \frac{5\pi}{6} + 2k\pi$, $\frac{7\pi}{6} + 2k\pi$ for any integer k. Specific solutions include $\theta = \frac{5\pi}{6} - 2\pi = -\frac{7\pi}{6}$, $\theta = \frac{7\pi}{6} - 2\pi = -\frac{5\pi}{6}$, $\theta = \frac{5\pi}{6}$, $\theta = \frac{7\pi}{6}$, $\theta = \frac{5\pi}{6} + 2\pi = \frac{17\pi}{6}$, and $\theta = \frac{7\pi}{6} + 2\pi = \frac{19\pi}{6}$.

19. One solution of $\sin\theta = \frac{\sqrt{2}}{2}$ is $\theta = \sin^{-1}\frac{\sqrt{2}}{2} = \frac{\pi}{4}$ and another is $\theta = \pi - \frac{\pi}{4} = \frac{3\pi}{4}$. All solutions are $\theta = \frac{\pi}{4} + 2k\pi$ and $\theta = \frac{3\pi}{4} + 2k\pi$ for any integer k. Specific solutions include $\theta = -\frac{7\pi}{4}$, $-\frac{5\pi}{4}$, $\frac{\pi}{4}$, $\frac{3\pi}{4}$, $\frac{9\pi}{4}$, and $\frac{11\pi}{4}$.

21. One solution of $\cos\theta = 0.28$ is $\theta = \cos^{-1}0.28 \approx 1.29$ and another is $\theta = 2\pi - \cos^{-1}0.28 \approx 5.00$. All solutions are $\theta \approx 1.29 + 2k\pi$ and $\theta \approx 5.00 + 2k\pi$ for any integer k. Specific solutions include $\theta \approx -5.00$, -1.28, 1.29, 5.00, 7.57, and 11.28.

23. One solution of $\tan\theta = -10$ is $\theta = \tan^{-1}(-10) \approx -1.47$. All solutions are $\theta \approx -1.47 + k\pi$ for any integer k. Specific solutions include $\theta \approx -7.75$, -4.61, -1.47, 1.67, 4.81, and 7.95.

25. $\cos\theta + 1 = 0 \Leftrightarrow \cos\theta = -1$. In the interval $[0, 2\pi)$ the only solution is $\theta = \pi$. Thus the solutions are $\theta = (2k+1)\pi$ for any integer k.

27. $\sqrt{2}\sin\theta + 1 = 0 \Rightarrow \sqrt{2}\sin\theta = -1 \Leftrightarrow \sin\theta = -\frac{1}{\sqrt{2}}$ The solutions in the interval $[0, 2\pi)$ are $\theta = \frac{5\pi}{4}$, $\frac{7\pi}{4}$. Thus the solutions are $\theta = \frac{5\pi}{4} + 2k\pi$, $\frac{7\pi}{4} + 2k\pi$ for any integer k.

29. $5\sin\theta - 1 = 0 \Leftrightarrow \sin\theta = \frac{1}{5}$. The solutions in the interval $[0, 2\pi)$ are $\theta = \sin^{-1}\frac{1}{5} \approx 0.20$ and $\theta = \pi - \sin^{-1}\frac{1}{5} \approx 2.94$. Thus the solutions are $\theta \approx 0.20 + 2k\pi$, $2.94 + 2k\pi$ for any integer k.

31. $3\tan^2\theta - 1 = 0 \Leftrightarrow \tan^2\theta = \frac{1}{3} \Leftrightarrow \tan\theta = \pm\frac{\sqrt{3}}{3}$. The solutions in the interval $\left(-\frac{\pi}{2}, \frac{\pi}{2}\right)$ are $\theta = \pm\frac{\pi}{6}$, so all solutions are $\theta = -\frac{\pi}{6} + k\pi$, $\frac{\pi}{6} + k\pi$ for any integer k.

33. $2\cos^2\theta - 1 = 0 \Leftrightarrow \cos^2\theta = \frac{1}{2} \Leftrightarrow \cos\theta = \pm\frac{1}{\sqrt{2}} \Leftrightarrow \theta = \frac{\pi}{4}, \frac{3\pi}{4}, \frac{5\pi}{4}, \frac{7\pi}{4}$ in $[0, 2\pi)$. Thus, the solutions are $\theta = \frac{\pi}{4} + k\pi$, $\frac{3\pi}{4} + k\pi$ for any integer k.

35. $\tan^2\theta - 4 = 0 \Leftrightarrow \tan^2\theta = 4 \Leftrightarrow \tan\theta = \pm 2 \Leftrightarrow$ $\theta = \tan^{-1}(-2) \approx -1.11$ or $\theta = \tan^{-1}2 \approx 1.11$ in $\left(-\frac{\pi}{2}, \frac{\pi}{2}\right)$. Thus, the solutions are $\theta \approx -1.11 + k\pi$, $1.11 + k\pi$ for any integer k.

37. $\sec^2\theta - 2 = 0 \Leftrightarrow \sec^2\theta = 2 \Leftrightarrow \sec\theta = \pm\sqrt{2}$. In the interval $[0, 2\pi)$ the solutions are $\theta = \frac{\pi}{4}$, $\frac{3\pi}{4}$, $\frac{5\pi}{4}$, $\frac{7\pi}{4}$. Thus, the solutions are $\theta = (2k+1)\frac{\pi}{4}$ for any integer k.

39. $\left(\tan^2\theta-4\right)\left(2\cos\theta+1\right)=0 \quad\Leftrightarrow\quad \tan^2\theta=4$ or $2\cos\theta=-1$. From Exercise 35, we know that the first equation has solutions $\theta\approx-1.11+k\pi$, $1.11+k\pi$ for any integer k. $2\cos\theta=-1 \quad\Leftrightarrow\quad \cos\theta=-\frac{1}{2}$ has solutions $\cos^{-1}\left(-\frac{1}{2}\right)=\frac{2\pi}{3}$ and $\frac{4\pi}{3}$ on $\left[0,2\pi\right)$, so all solutions are $\theta\approx\frac{2\pi}{3}+2k\pi$ and $\frac{4\pi}{3}+2k\pi$ for any integer k. Thus, the original equation has solutions $\theta\approx-1.11+k\pi$, $1.11+k\pi$, $\frac{2\pi}{3}+2k\pi$, and $\frac{4\pi}{3}+2k\pi$ for any integer k.

41. $4\cos^2\theta-4\cos\theta+1=0 \quad\Leftrightarrow\quad \left(2\cos\theta-1\right)^2=0 \quad\Leftrightarrow\quad 2\cos\theta-1=0 \quad\Leftrightarrow\quad \cos\theta=\frac{1}{2} \quad\Leftrightarrow\quad \theta=\frac{\pi}{3}+2k\pi$, $\frac{5\pi}{3}+2k\pi$ for any integer k.

43. $3\sin^2\theta-7\sin\theta+2=0 \quad\Rightarrow\quad \left(3\sin\theta-1\right)\left(\sin\theta-2\right)=0 \quad\Rightarrow\quad 3\sin\theta-1=0$ or $\sin\theta-2=0$. Since $\left|\sin\theta\right|\leq 1$, $\sin\theta-2=0$ has no solution. Thus $3\sin\theta-1=0 \quad\Rightarrow\quad \sin\theta=\frac{1}{3} \quad\Rightarrow\quad \theta\approx 0.33984$ and $\theta\approx\pi-0.33984\approx 2.80176$ are the solutions in $\left[0,2\pi\right)$, and all solutions are $\theta\approx 0.33984+2k\pi$, $2.80176+2k\pi$ for any integer k.

45. $2\cos^2\theta-7\cos\theta+3=0 \quad\Leftrightarrow\quad \left(2\cos\theta-1\right)\left(\cos\theta-3\right)=0 \quad\Leftrightarrow\quad \cos\theta=\frac{1}{2}$ or $\cos\theta=3$ (which is inadmissible) $\Leftrightarrow\quad \theta=\frac{\pi}{3}$, $\frac{5\pi}{3}$. Therefore, the solutions are $\theta=\frac{\pi}{3}+2k\pi$, $\frac{5\pi}{3}+2k\pi$ for any integer k.

47. $\cos^2\theta-\cos\theta-6=0 \quad\Leftrightarrow\quad \left(\cos\theta+2\right)\left(\cos x-3\right)=0 \quad\Leftrightarrow\quad \cos x=-2$ or $\cos x=3$, neither of which has a solution. Thus, the original equation has no solution.

49. $\sin^2\theta=2\sin\theta+3 \quad\Leftrightarrow\quad \sin^2\theta-2\sin\theta-3=0 \quad\Leftrightarrow\quad \left(\sin\theta-3\right)\left(\sin\theta+1\right)=0 \quad\Leftrightarrow\quad \sin\theta-3=0$ or $\sin\theta+1=0$. Since $\left|\sin\theta\right|\leq 1$ for all θ, there is no solution for $\sin\theta-3=0$. Hence $\sin\theta+1=0 \quad\Leftrightarrow\quad \sin\theta=-1 \quad\Leftrightarrow\quad \theta=\frac{3\pi}{2}+2k\pi$ for any integer k.

51. $\cos\theta\left(2\sin\theta+1\right)=0 \quad\Leftrightarrow\quad \cos\theta=0$ or $\sin\theta=-\frac{1}{2} \quad\Leftrightarrow\quad \theta=\frac{\pi}{2}+k\pi$, $\frac{7\pi}{6}+2k\pi$, $\frac{11\pi}{6}+2k\pi$ for any integer k.

53. $\cos\theta\sin\theta-2\cos\theta=0 \quad\Leftrightarrow\quad \cos\theta\left(\sin\theta-2\right)=0 \quad\Leftrightarrow\quad \cos\theta=0$ or $\sin\theta-2=0$. Since $\left|\sin\theta\right|\leq 1$ for all θ, there is no solution for $\sin\theta-2=0$. Hence, $\cos\theta=0 \quad\Leftrightarrow\quad \theta=\frac{\pi}{2}+2k\pi$, $\frac{3\pi}{2}+2k\pi \quad\Leftrightarrow\quad \theta=\frac{\pi}{2}+k\pi$ for any integer k.

55. $3\tan\theta\sin\theta-2\tan\theta=0 \quad\Leftrightarrow\quad \tan\theta\left(3\sin\theta-2\right)=0 \quad\Leftrightarrow\quad \tan\theta=0$ or $\sin\theta=\frac{2}{3}$. $\tan\theta=0$ has solution $\theta=0$ on $\left(-\frac{\pi}{2},\frac{\pi}{2}\right)$ and $\sin\theta=\frac{2}{3}$ has solutions $\theta=\sin^{-1}\frac{2}{3}\approx 0.73$ and $\theta=\pi-\sin^{-1}\frac{2}{3}\approx 2.41$ on $\left[0,2\pi\right)$, so the original equation has solutions $\theta=k\pi$, $\theta\approx 0.73+2k\pi$, $2.41+2k\pi$ for any integer k.

57. We substitute $\theta_1 = 70^\circ$ and $\dfrac{v_1}{v_2} = 1.33$ into Snell's Law to get $\dfrac{\sin 70^\circ}{\sin\theta_2} = 1.33 \Leftrightarrow$

$\sin\theta_2 = \dfrac{\sin 70^\circ}{1.33} = 0.7065 \Rightarrow \theta_2 \approx 44.95^\circ$.

59. (a) $F = \frac{1}{2}\left(1-\cos\theta\right) = 0 \Rightarrow \cos\theta = 1 \Rightarrow \theta = 0$

(b) $F = \frac{1}{2}\left(1-\cos\theta\right) = 0.25 \Rightarrow 1-\cos\theta = 0.5 \Rightarrow \cos\theta = 0.5 \Rightarrow \theta = 60^\circ$ or 120°

(c) $F = \frac{1}{2}\left(1-\cos\theta\right) = 0.5 \Rightarrow 1-\cos\theta = 1 \Rightarrow \cos\theta = 0 \Rightarrow \theta = 90^\circ$ or 270°

(d) $F = \frac{1}{2}\left(1-\cos\theta\right) = 1 \Rightarrow 1-\cos\theta = 2 \Rightarrow \cos\theta = -1 \Rightarrow \theta = 180^\circ$

4.5 More Trigonometric Equations

1. Using a Pythagorean identity, we calculate $\sin x + \sin^2 x + \cos^2 x = 1 \Leftrightarrow \sin x + 1 = 1 \Leftrightarrow \sin x = 0$, whose solutions are $x = k\pi$ for any integer k.

3. $2\cos^2\theta + \sin\theta = 1 \Leftrightarrow 2\left(1 - \sin^2\theta\right) + \sin\theta - 1 = 0 \Leftrightarrow -2\sin^2\theta + \sin\theta + 1 = 0 \Leftrightarrow 2\sin^2\theta - \sin\theta - 1 = 0$. From Exercise 7.4.42, the solutions are $\theta = \frac{7\pi}{6} + 2k\pi$, $\frac{11\pi}{6} + 2k\pi$, $\frac{\pi}{2} + 2k\pi$ for any integer k.

5. $\tan^2\theta - 2\sec\theta = 2 \Leftrightarrow \sec^2\theta - 1 - 2\sec\theta = 2 \Leftrightarrow \sec^2\theta - 2\sec\theta - 3 = 0 \Leftrightarrow \left(\sec\theta - 3\right)\left(\sec\theta + 1\right) = 0 \Leftrightarrow \sec\theta = 3$ or $\sec\theta = -1$. If $\sec\theta = 3$, then $\cos\theta = \frac{1}{3}$, which has solutions $\theta = \cos^{-1}\frac{1}{3} \approx 1.23$ and $\theta = 2\pi - \cos^{-1}\frac{1}{3} \approx 5.05$ on $\left[0, 2\pi\right)$. If $\sec\theta = -1$, then $\cos\theta = -1$, which has solution $\theta = \pi$ on $\left[0, 2\pi\right)$. Thus, solutions are $\theta = \left(2k + 1\right)\pi$, $\theta \approx 1.23 + 2k\pi$, $5.05 + 2k\pi$ for any integer k.

7. $2\sin 2\theta - 3\sin\theta = 0 \Leftrightarrow 2\left(2\sin\theta\cos\theta\right) - 3\sin\theta = 0 \Leftrightarrow \sin\theta\left(4\cos\theta - 3\right) = 0 \Leftrightarrow \sin\theta = 0$ or $\cos\theta = \frac{3}{4}$. The first equation has solutions $\theta = 0$, π on $\left[0, 2\pi\right)$, and the second has solutions $\theta = \cos^{-1}\frac{3}{4} \approx 0.72$ and $\theta = 2\pi - \cos^{-1}\frac{3}{4} \approx 5.56$ on $\left[0, 2\pi\right)$. Thus, solutions are $\theta = k\pi$, $\theta \approx 0.72 + 2k\pi$, $5.56 + 2k\pi$ for any integer k.

9. $\cos 2\theta = 3\sin\theta - 1 \Leftrightarrow 1 - 2\sin^2\theta = 3\sin\theta - 1 \Leftrightarrow 2\sin^2\theta + 3\sin\theta - 2 = 0 \Leftrightarrow \left(\sin\theta + 2\right)\left(2\sin\theta - 1\right) = 0 \Leftrightarrow \sin\theta = -2$ or $\sin\theta = \frac{1}{2}$. The first equation has no solution and the second has solutions $\theta = \sin^{-1}\frac{1}{2} = \frac{\pi}{6}$ and $\theta = \pi - \sin^{-1}\frac{1}{2} = \frac{5\pi}{6}$ on $\left[0, 2\pi\right)$, so the original equation has solutions $\theta = \frac{\pi}{6} + 2k\pi$, $\frac{5\pi}{6} + 2k\pi$ for any integer k.

11. $2\sin^2\theta - \cos\theta = 1 \Leftrightarrow 2\left(1 - \cos^2\theta\right) - \cos\theta - 1 = 0 \Leftrightarrow -2\cos^2\theta - \cos\theta + 1 = 0 \Leftrightarrow \left(2\cos\theta - 1\right)\left(\cos\theta + 1\right) = 0 \Leftrightarrow 2\cos\theta - 1 = 0$ or $\cos\theta + 1 = 0 \Leftrightarrow \cos\theta = \frac{1}{2}$ or $\cos\theta = -1 \Leftrightarrow \theta = \frac{\pi}{3} + 2k\pi$, $\frac{5\pi}{3} + 2k\pi$, $\left(2k + 1\right)\pi$ for any integer k.

13. $\sin\theta - 1 = \cos\theta \Leftrightarrow \sin\theta + \cos\theta = 1$. Squaring both sides, we have $\sin^2\theta + \cos^2\theta + 2\sin\theta\cos\theta = 1 \Leftrightarrow \sin 2\theta = 0$, which has solutions $\theta = 0$, $\frac{\pi}{2}$, π, $\frac{3\pi}{2}$ in $\left[0, 2\pi\right)$. Checking in the original equation, we see that only $\theta = \frac{\pi}{2}$ and $\theta = \pi$ are valid. (The extraneous solutions were introduced by squaring both sides.) Thus, the solutions are $\theta = \left(2k + 1\right)\pi$, $\frac{\pi}{2} + 2k\pi$ for any integer k.

15. $\tan\theta + 1 = \sec\theta \iff \dfrac{\sin\theta}{\cos\theta} + 1 = \dfrac{1}{\cos\theta} \iff \sin\theta + \cos\theta = 1$. Squaring both sides, we have $\sin^2\theta + \cos^2\theta + 2\sin\theta\cos\theta = 1 \iff \sin 2\theta = 0$, which has solutions $\theta = 0$, $\frac{\pi}{2}$, π , $\frac{3\pi}{2}$ on $[0, 2\pi)$. Checking in the original equation, we see that only $\theta = 0$ is valid. Thus, the solutions are $\theta = 2k\pi$ for any integer k .

17. (a) $2\cos 3\theta = 1 \iff \cos 3\theta = \frac{1}{2} \Rightarrow 3\theta = \frac{\pi}{3}$, $\frac{5\pi}{3}$ for 3θ in $[0, 2\pi)$. Thus, solutions are $\frac{\pi}{9} + \frac{2}{3}k\pi$, $\frac{5\pi}{9} + \frac{2}{3}k\pi$ for any integer k .

(b) We take $k = 0$, 1 , 2 in the expressions in part (a) to obtain the solutions $\theta = \frac{\pi}{9}$, $\frac{5\pi}{9}$, $\frac{7\pi}{9}$, $\frac{11\pi}{9}$, $\frac{13\pi}{9}$, $\frac{17\pi}{9}$ in $[0, 2\pi)$.

19. (a) $2\cos 2\theta + 1 = 0 \iff \cos 2\theta = -\frac{1}{2} \iff 2\theta = \frac{2\pi}{3} + 2k\pi$, $\frac{4\pi}{3} + 2k\pi \iff \theta = \frac{\pi}{3} + k\pi$, $\frac{2\pi}{3} + k\pi$ for any integer k .

(b) The solutions in $[0, 2\pi)$ are $\frac{\pi}{3}$, $\frac{2\pi}{3}$, $\frac{4\pi}{3}$, $\frac{5\pi}{3}$.

21. (a) $\sqrt{3}\tan 3\theta + 1 = 0 \iff \tan 3\theta = -\frac{1}{\sqrt{3}} \iff 3\theta = \frac{5\pi}{6} + k\pi \iff \theta = \frac{5\pi}{18} + \frac{1}{3}k\pi$ for any integer k .

(b) The solutions in $[0, 2\pi)$ are $\frac{5\pi}{18}$, $\frac{11\pi}{18}$, $\frac{17\pi}{18}$, $\frac{23\pi}{18}$, $\frac{29\pi}{18}$, $\frac{35\pi}{18}$.

23. (a) $\cos\frac{\theta}{2} - 1 = 0 \iff \cos\frac{\theta}{2} = 1 \iff \frac{\theta}{2} = 2k\pi \iff \theta = 4k\pi$ for any integer k .

(b) The only solution in $[0, 2\pi)$ is $\theta = 0$.

25. (a) $2\sin\frac{\theta}{3} + \sqrt{3} = 0 \iff 2\sin\frac{\theta}{3} = -\sqrt{3} \iff \sin\frac{\theta}{3} = -\frac{\sqrt{3}}{2} \iff \frac{\theta}{3} = \frac{4\pi}{3} + 2k\pi$, $\frac{5\pi}{3} + 2k\pi \iff \theta = 4\pi + 6k\pi$, $5\pi + 6k\pi$ for any integer k .

(b) There is no solution in $[0, 2\pi)$.

27. (a) $\sin 2\theta = 3\cos 2\theta \iff \tan 2\theta = 3 \iff \theta = \frac{1}{2}\tan^{-1} 3 \approx 0.62$ on $\left(-\frac{\pi}{4}, \frac{\pi}{4}\right)$. Thus, solutions are $\theta \approx 0.62 + \frac{1}{2}k\pi$ for any integer k .

(b) The solutions in $[0, 2\pi)$ are $\theta \approx 0.62$, 2.19 , 3.76 , 5.33 .

29. (a) $\sec\theta - \tan\theta = \cos\theta \iff \cos\theta(\sec\theta - \tan\theta) = \cos\theta(\cos\theta) \iff$
$1 - \sin\theta = \cos^2\theta \iff 1 - \sin\theta = 1 - \sin^2\theta \iff \sin\theta = \sin^2\theta \iff$
$\sin^2\theta - \sin\theta = 0 \iff \sin\theta(\sin\theta - 1) = 0 \iff \sin\theta = 0$ or $\sin\theta = 1 \iff$
$\theta = 0, \pi$ or $\theta = \frac{\pi}{2}$ in $[0, 2\pi)$. However, since the equation is undefined when $\theta = \frac{\pi}{2}$, the solutions are $\theta = k\pi$ for any integer k .

(b) The solutions in $[0, 2\pi)$ are $\theta = 0$, π .

31. (a) $3\tan^3\theta - 3\tan^2\theta - \tan\theta + 1 = 0 \iff (\tan\theta - 1)(3\tan^2\theta - 1) = 0 \iff$ $\tan\theta = 1$ or $3\tan^2\theta = 1 \iff \tan\theta = 1$ or $\tan\theta = \pm\frac{1}{\sqrt{3}} \iff \theta = \frac{\pi}{6} + k\pi$, $\frac{\pi}{4} + k\pi$, $\frac{5\pi}{6} + k\pi$ for any integer k.

(b) The solutions in $[0, 2\pi)$ are $\theta = \frac{\pi}{6}, \frac{\pi}{4}, \frac{5\pi}{6}, \frac{7\pi}{6}, \frac{5\pi}{4}, \frac{11\pi}{6}$.

33. (a) $2\sin\theta\tan\theta - \tan\theta = 1 - 2\sin\theta \iff 2\sin\theta\tan\theta - \tan\theta + 2\sin\theta - 1 = 0$ $\iff (2\sin\theta - 1)(\tan\theta + 1) = 0 \iff 2\sin\theta - 1 = 0$ or $\tan\theta + 1 = 0 \iff$ $\sin\theta = \frac{1}{2}$ or $\tan\theta = -1 \iff \theta = \frac{\pi}{6} + 2k\pi$, $\frac{5\pi}{6} + 2k\pi$, $\frac{3\pi}{4} + k\pi$ for any integer k.

(b) The solutions in $[0, 2\pi)$ are $\frac{\pi}{6}, \frac{3\pi}{4}, \frac{5\pi}{6}, \frac{7\pi}{4}$.

35. (a)

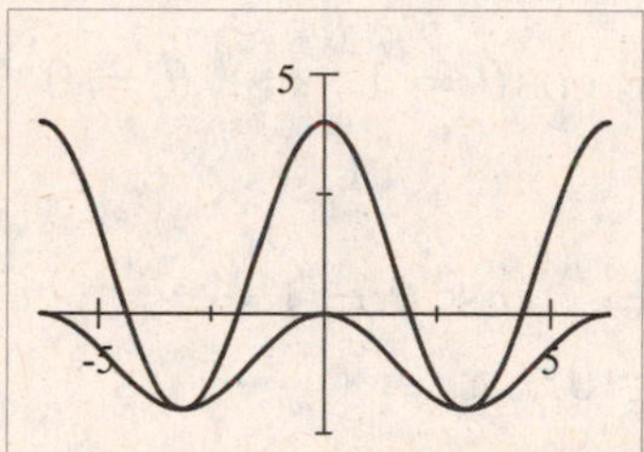

The points of intersection are approximately $(\pm 3.14, -2)$.

(b) $f(x) = 3\cos x + 1$; $g(x) = \cos x - 1$. $f(x) = g(x)$ when $3\cos x + 1 = \cos x - 1 \iff 2\cos x = -2 \iff \cos x = -1 \iff$ $x = \pi + 2k\pi = (2k+1)\pi$. The points of intersection are $((2k+1)\pi, -2)$ for any integer k.

37. (a)

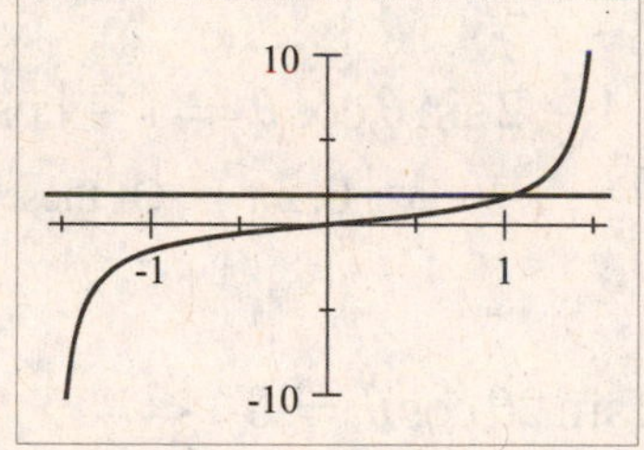

The point of intersection is approximately $(1.04, 1.73)$.

(b) $f(x) = \tan x$; $g(x) = \sqrt{3}$. $f(x) = g(x)$ when $\tan x = \sqrt{3} \iff$ $x = \frac{\pi}{3} + k\pi$. The intersection points are $\left(\frac{\pi}{3} + k\pi, \sqrt{3}\right)$ for any integer k.

39. $\cos\theta\cos 3\theta - \sin\theta\sin 3\theta = 0 \Leftrightarrow \cos(\theta + 3\theta) = 0 \Leftrightarrow \cos 4\theta = 0 \Leftrightarrow$ $4\theta = \frac{\pi}{2}, \frac{3\pi}{2}, \frac{5\pi}{2}, \frac{7\pi}{2}, \frac{9\pi}{2}, \frac{11\pi}{2}, \frac{13\pi}{2}, \frac{15\pi}{2}$ in $[0, 8\pi) \Leftrightarrow \theta = \frac{\pi}{8}, \frac{3\pi}{8}, \frac{5\pi}{8}, \frac{7\pi}{8}$ $, \frac{9\pi}{8}, \frac{11\pi}{8}, \frac{13\pi}{8}, \frac{15\pi}{8}$ in $[0, 2\pi)$.

41. $\sin 2\theta\cos\theta - \cos 2\theta\sin\theta = \frac{\sqrt{3}}{2} \Leftrightarrow \sin(2\theta - \theta) = \frac{\sqrt{3}}{2} \Leftrightarrow \sin\theta = \frac{\sqrt{3}}{2} \Leftrightarrow$ $\theta = \frac{\pi}{3}, \frac{2\pi}{3}$ in $[0, 2\pi)$.

43. $\sin 2\theta + \cos\theta = 0 \Leftrightarrow 2\sin\theta\cos\theta + \cos\theta = 0 \Leftrightarrow \cos\theta(2\sin\theta + 1) = 0 \Leftrightarrow$ $\cos\theta = 0$ or $\sin\theta = -\frac{1}{2} \Leftrightarrow \theta = \frac{\pi}{2}, \frac{7\pi}{6}, \frac{3\pi}{2}, \frac{11\pi}{6}$ in $[0, 2\pi)$.

45. $\cos 2\theta + \cos\theta = 2 \Leftrightarrow 2\cos^2\theta - 1 + \cos\theta - 2 = 0 \Leftrightarrow$ $2\cos^2\theta + \cos\theta - 3 = 0 \Leftrightarrow (2\cos\theta + 3)(\cos\theta - 1) = 0 \Leftrightarrow 2\cos\theta + 3 = 0$ or $\cos\theta - 1 = 0 \Leftrightarrow \cos\theta = -\frac{3}{2}$ (which is impossible) or $\cos\theta = 1 \Leftrightarrow \theta = 0$ in $[0, 2\pi)$.

47. $\cos 2\theta - \cos^2\theta = 0 \Leftrightarrow 2\cos^2\theta - 1 - \cos^2\theta = 0 \Leftrightarrow \cos^2\theta = 1 \Leftrightarrow$ $\theta = k\pi$ for any integer k. On $[0, 2\pi)$, the solutions are $\theta = 0$, π.

49. $\cos 2\theta - \cos 4\theta = 0 \Leftrightarrow \cos 2\theta - (2\cos^2 2\theta - 1) = 0 \Leftrightarrow$ $(\cos 2\theta - 1)(2\cos 2\theta + 1) = 0$. The first factor has zeros at $\theta = 0$, π and the second has zeros at $\theta = \frac{\pi}{3}, \frac{2\pi}{3}, \frac{4\pi}{3}, \frac{5\pi}{3}$. Thus, solutions of the original equation are are $\theta = 0, \frac{\pi}{3}$, $\frac{2\pi}{3}, \pi, \frac{4\pi}{3}, \frac{5\pi}{3}$ in $[0, 2\pi)$.

51. $\cos\theta - \sin\theta = \sqrt{2}\sin\frac{\theta}{2} \Leftrightarrow \cos\theta - \sin\theta = \sqrt{2}\left(\pm\sqrt{\frac{1 - \cos\theta}{2}}\right)$. Squaring both sides, we have $\cos^2\theta + \sin^2\theta - 2\sin\theta\cos\theta = 1 - \cos\theta \Leftrightarrow 1 - 2\sin\theta\cos\theta = 1 - \cos\theta$ $\Leftrightarrow$ either $2\sin\theta = 1$ or $\cos\theta = 0 \Leftrightarrow \theta = \frac{\pi}{6}, \frac{\pi}{2}, \frac{5\pi}{6}, \frac{3\pi}{2}$ in $[0, 2\pi)$. Of these, only $\frac{\pi}{6}$ and $\frac{3\pi}{2}$ satisfy the original equation.

53. $\sin\theta + \sin 3\theta = 0 \Leftrightarrow 2\sin 2\theta\cos(-\theta) = 0 \Leftrightarrow 2\sin 2\theta\cos\theta = 0 \Leftrightarrow$ $\sin 2\theta = 0$ or $\cos\theta = 0 \Leftrightarrow 2\theta = k\pi$ or $\theta = k\frac{\pi}{2} \Leftrightarrow \theta = \frac{1}{2}k\pi$ for any integer k

55. $\cos 4\theta + \cos 2\theta = \cos\theta \Leftrightarrow 2\cos 3\theta\cos\theta = \cos\theta \Leftrightarrow \cos\theta(2\cos 3\theta - 1) = 0$ $\Leftrightarrow \cos\theta = 0$ or $\cos 3\theta = \frac{1}{2} \Leftrightarrow \theta = \frac{\pi}{2}$ or $3\theta = \frac{\pi}{3} + 2k\pi, \frac{5\pi}{3} + 2k\pi$, $\frac{7\pi}{3} + 2k\pi, \frac{11\pi}{3} + 2k\pi, \frac{13\pi}{3} + 2k\pi, \frac{17\pi}{3} + 2k\pi \Leftrightarrow \theta = \frac{\pi}{2} + k\pi, \frac{\pi}{9} + \frac{2}{3}k\pi$, $\frac{5\pi}{9} + \frac{2}{3}k\pi$ for any integer k.

57. $\sin 2x = x$

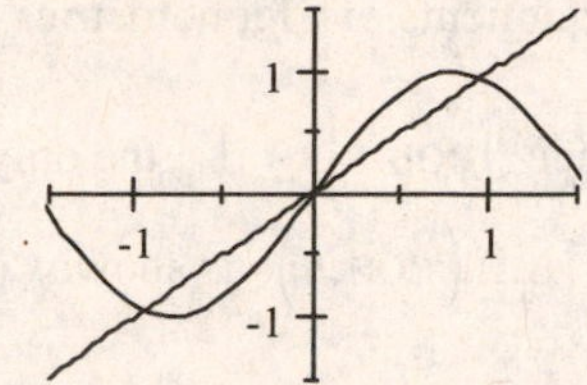

The three solutions are $x = 0$ and $x \approx \pm 0.95$.

59. $2^{\sin x} = x$

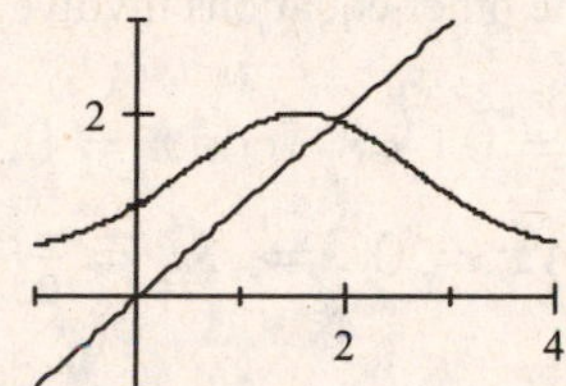

The only solution is $x \approx 1.92$.

61. $\dfrac{\cos x}{1 + x^2} = x^2$

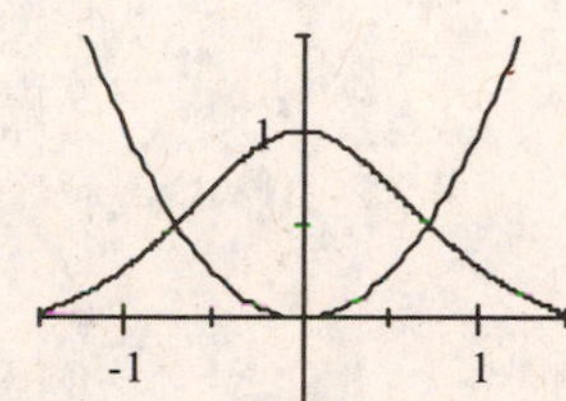

The two solutions are $x \approx \pm 0.71$.

63. We substitute $v_0 = 2200$ and $R(\theta) = 5000$ and solve for θ . So

$$5000 = \frac{(2200)^2 \sin 2\theta}{32} \Leftrightarrow 5000 = 151250 \sin 2\theta \Leftrightarrow \sin 2\theta = 0.03308 \Rightarrow$$

$2\theta = 1.89442°$ or $2\theta = 180° - 1.89442° = 178.10558°$. If $2\theta = 1.89442°$, then $\theta = 0.94721°$, and if $2\theta = 178.10558°$, then $\theta = 89.05279°$.

65. (a) $10 = 12 + 2.83 \sin\left(\frac{2\pi}{3}(t - 80)\right) \Leftrightarrow 2.83 \sin\left(\frac{2\pi}{3}(t - 80)\right) = -2 \Leftrightarrow$ $\sin\left(\frac{2\pi}{3}(t - 80)\right) = -0.70671$. Now $\sin\theta = -0.70671$ and $\theta = -0.78484$. If $\frac{2\pi}{3}(t - 80) = -0.78484 \Leftrightarrow t - 80 = 45.6 \Leftrightarrow t = 34.4$. Now in the interval $[0, 2\pi)$, we have $\theta = \pi + 0.78484 \approx 3.92644$ and $\theta = 2\pi - 0.78484 \approx 5.49834$. If $\frac{2\pi}{3}(t - 80) = 3.92644 \Leftrightarrow t - 80 = 228.1 \Leftrightarrow t = 308.1$. And if $\frac{2\pi}{3}(t - 80) = 5.49834 \Leftrightarrow t - 80 = 319.4 \Leftrightarrow$ $t = 399.4$ $(399.4 - 365 = 34.4)$. So according to this model, there should be 10 hours of sunshine on the 34 th day (February 3) and on the 308 th day (November 4).

(b) Since $L(t) = 12 + 2.83 \sin\left(\frac{2\pi}{3}(t - 80)\right) \geq 10$ for $t \in [34, 308]$, the number of days with more than 10 hours of daylight is $308 - 34 + 1 = 275$ days.

67. $\sin(\cos x)$ is a function of a function, that is, a composition of trigonometric functions (see Section 2.6). Most of the other equations involve sums, products, differences, or quotients of trigonometric functions.

$\sin(\cos x) = 0 \Leftrightarrow \cos x = 0$ or $\cos x = \pi$. However, since $|\cos x| \leq 1$, the only solution is $\cos x = 0 \Rightarrow x = \frac{\pi}{2} + k\pi$. The graph of $f(x) = \sin(\cos x)$ is shown.

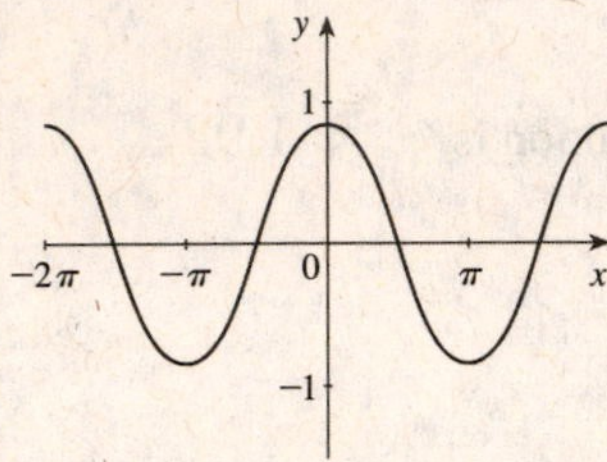

4 Review

1. $\sin\theta(\cot\theta+\tan\theta)=\sin\theta\left(\dfrac{\cos\theta}{\sin\theta}+\dfrac{\sin\theta}{\cos\theta}\right)=\cos\theta+\dfrac{\sin^2\theta}{\cos\theta}=\dfrac{\cos^2\theta+\sin^2\theta}{\cos\theta}=\dfrac{1}{\cos\theta}=\sec\theta$

3. $\cos^2 x\csc x-\csc x=\left(1-\sin^2 x\right)\csc x-\csc x=\csc x-\sin^2 x\csc x-\csc x=-\sin^2 x\cdot\dfrac{1}{\sin x}=-\sin x$

5. $\dfrac{\cos^2 x-\tan^2 x}{\sin^2 x}=\dfrac{\cos^2 x}{\sin^2 x}-\dfrac{\tan^2 x}{\sin^2 x}=\cot^2 x-\dfrac{1}{\cos^2 x}=\cot^2 x-\sec^2 x$

7. $\dfrac{\cos^2 x}{1-\sin x}=\dfrac{\cos x}{\dfrac{1}{\cos x}(1-\sin x)}=\dfrac{\cos x}{\dfrac{1}{\cos x}-\dfrac{\sin x}{\cos x}}=\dfrac{\cos x}{\sec x-\tan x}$

9. $\sin^2 x\cot^2 x+\cos^2 x\tan^2 x=\sin^2 x\cdot\dfrac{\cos^2 x}{\sin^2 x}+\cos^2 x\cdot\dfrac{\sin^2 x}{\cos^2 x}=\cos^2 x+\sin^2 x=1$

11. $\dfrac{\sin 2x}{1+\cos 2x}=\dfrac{2\sin x\cos x}{1+2\cos^2 x-1}=\dfrac{2\sin x\cos x}{2\cos^2 x}=\dfrac{2\sin x}{2\cos x}=\tan x$

13. $\tan\dfrac{x}{2}=\dfrac{1-\cos x}{\sin x}=\dfrac{1}{\sin x}-\dfrac{\cos x}{\sin x}=\csc x-\cot x$

15. $$\begin{aligned}\sin(x+y)\sin(x-y)&=\tfrac{1}{2}\left[\cos((x+y)-(x-y))-\cos((x+y)+(x-y))\right]=\tfrac{1}{2}(\cos 2y-\cos 2x)\\&=\tfrac{1}{2}\left[1-2\sin^2 y-\left(1-2\sin^2 x\right)\right]=\tfrac{1}{2}\left(2\sin^2 x-2\sin^2 y\right)=\sin^2 x-\sin^2 y\end{aligned}$$

17. $1+\tan x\tan\dfrac{x}{2}=1+\dfrac{\sin x}{\cos x}\cdot\dfrac{1-\cos x}{\sin x}=1+\dfrac{1-\cos x}{\cos x}=1+\dfrac{1}{\cos x}-1=\dfrac{1}{\cos x}=\sec x$

19. $$\begin{aligned}\left(\cos\frac{x}{2}-\sin\frac{x}{2}\right)^2&=\cos^2\frac{x}{2}-2\sin\frac{x}{2}\cos\frac{x}{2}+\sin^2\frac{x}{2}=\sin^2\frac{x}{2}+\cos^2\frac{x}{2}-2\sin\frac{x}{2}\cos\frac{x}{2}=1-\sin\left(2\cdot\frac{x}{2}\right)\\&=1-\sin x\end{aligned}$$

21. $\dfrac{\sin 2x}{\sin x}-\dfrac{\cos 2x}{\cos x}=\dfrac{2\sin x\cos x}{\sin x}-\dfrac{2\cos^2 x-1}{\cos x}=2\cos x-2\cos x+\dfrac{1}{\cos x}=\sec x$

23. $\tan\left(x+\frac{\pi}{4}\right)=\dfrac{\tan x+\tan\frac{\pi}{4}}{1-\tan x\tan\frac{\pi}{4}}=\dfrac{1+\tan x}{1-\tan x}$

25. (a) $f(x) = 1 - \left(\cos\frac{x}{2} - \sin\frac{x}{2}\right)^2$, $g(x) = \sin x$

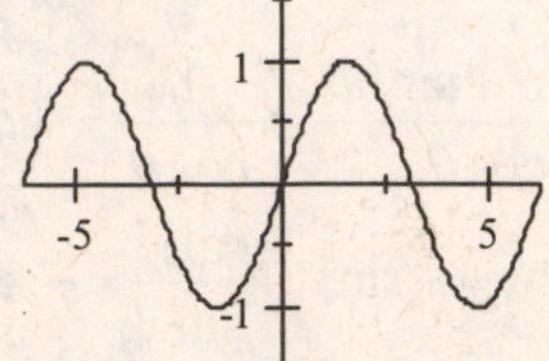

(b) The graphs suggest that $f(x) = g(x)$ is an identity. To prove this, expand $f(x)$ and simplify, using the double-angle formula for sine:

$$\begin{aligned} f(x) &= 1 - \left(\cos\tfrac{x}{2} - \sin\tfrac{x}{2}\right)^2 \\ &= 1 - \left(\cos^2\tfrac{x}{2} - 2\cos\tfrac{x}{2}\sin\tfrac{x}{2} + \sin^2\tfrac{x}{2}\right) \\ &= 1 + 2\cos\tfrac{x}{2}\sin\tfrac{x}{2} - \left(\cos^2\tfrac{x}{2} + \sin^2\tfrac{x}{2}\right) \\ &= 1 + \sin x - (1) = \sin x = g(x) \end{aligned}$$

27. (a) $f(x) = \tan x \tan\frac{x}{2}$, $g(x) = \dfrac{1}{\cos x}$

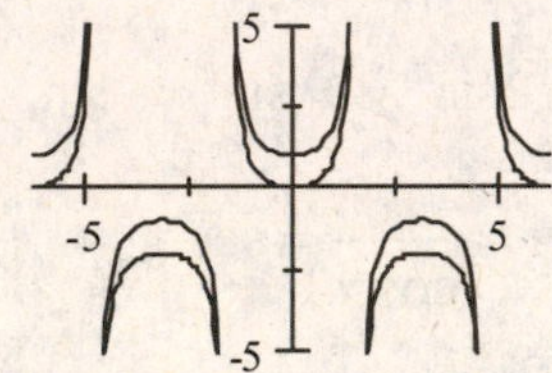

(b) The graphs suggest that $f(x) \neq g(x)$ in general. For example, choose $x = \frac{\pi}{3}$ and evaluate:

$f\left(\frac{\pi}{3}\right) = \tan\frac{\pi}{3} \quad \tan\frac{\pi}{6} = \sqrt{3} \cdot \frac{1}{\sqrt{3}} = 1$, whereas $g\left(\frac{\pi}{3}\right) = \dfrac{1}{\frac{1}{2}} = 2$, so $f(x) \neq g(x)$.

29. (a) $f(x) = 2\sin^2 3x + \cos 6x$

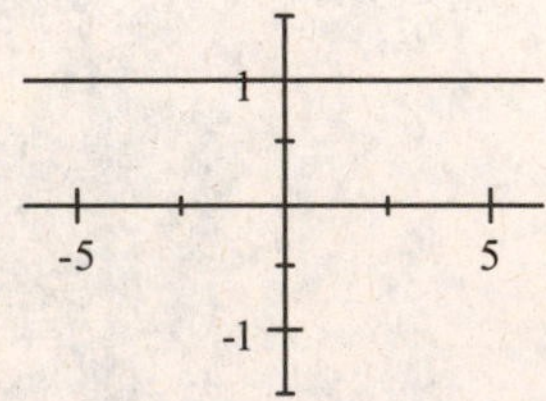

(b) The graph suggests that $f(x) = 1$ for all x . To prove this, we use the double angle formula to note that $\cos 6x = \cos(2(3x)) = 1 - 2 \quad \sin^2 3x$, so $f(x) = 2\sin^2 3x + \left(1 - 2\sin^2 3x\right) = 1$.

31. $4\sin\theta - 3 = 0 \iff 4\sin\theta = 3 \iff \sin\theta = \frac{3}{4} \iff \theta = \sin^{-1}\frac{3}{4} \approx 0.8481$ or $\theta = \pi - \sin^{-1}\frac{3}{4} \approx 2.2935$.

33. $\cos x \sin x - \sin x = 0 \iff \sin x(\cos x - 1) = 0 \iff \sin x = 0$ or $\cos x = 1 \iff x = 0$, π or $x = 0$. Therefore, the solutions are $x = 0$ and π.

35. $2\sin^2 x - 5\sin x + 2 = 0 \iff (2\sin x - 1)(\sin x - 2) = 0 \iff \sin x = \frac{1}{2}$ or $\sin x = 2$ (which is inadmissible) $\iff x = \frac{\pi}{6}$, $\frac{5\pi}{6}$. Thus, the solutions in $[0, 2\pi)$ are $x = \frac{\pi}{6}$ and $\frac{5\pi}{6}$.

37. $2\cos^2 x - 7\cos x + 3 = 0 \iff (2\cos x - 1)(\cos x - 3) = 0 \iff \cos x = \frac{1}{2}$ or $\cos x = 3$ (which is inadmissible) $\iff x = \frac{\pi}{3}$, $\frac{5\pi}{3}$. Therefore, the solutions in $[0, 2\pi)$ are $x = \frac{\pi}{3}$, $\frac{5\pi}{3}$.

39. Note that $x = \pi$ is not a solution because the denominator is zero. $\dfrac{1 - \cos x}{1 + \cos x} = 3 \iff 1 - \cos x = 3 + 3\cos x \iff -4\cos x = 2 \iff \cos x = -\frac{1}{2} \iff x = \frac{2\pi}{3}$, $\frac{4\pi}{3}$ in $[0, 2\pi)$.

41. Factor by grouping: $\tan^3 x + \tan^2 x - 3\tan x - 3 = 0 \iff (\tan x + 1)(\tan^2 x - 3) = 0 \iff \tan x = -1$ or $\tan x = \pm\sqrt{3} \iff x = \frac{3\pi}{4}$, $\frac{7\pi}{4}$ or $x = \frac{\pi}{3}$, $\frac{2\pi}{3}$, $\frac{4\pi}{3}$, $\frac{5\pi}{3}$. Therefore, the solutions in $[0, 2\pi)$ are $x = \frac{\pi}{3}$, $\frac{2\pi}{3}$, $\frac{3\pi}{4}$, $\frac{4\pi}{3}$, $\frac{5\pi}{3}$, $\frac{7\pi}{4}$.

43. $\tan\frac{1}{2}x + 2\sin 2x = \csc x \iff \dfrac{1 - \cos x}{\sin x} + 4\sin x \cos x = \dfrac{1}{\sin x} \iff 1 - \cos x + 4\sin^2 x \cos x = 1 \iff 4\sin^2 x \cos x - \cos x = 0 \iff \cos x(4\sin^2 x - 1) = 0 \iff \cos x = 0$ or $\sin x = \pm\frac{1}{2} \iff x = \frac{\pi}{2}$, $\frac{3\pi}{2}$ or $x = \frac{\pi}{6}$, $\frac{5\pi}{6}$, $\frac{7\pi}{6}$, $\frac{11\pi}{6}$. Thus, the solutions in $[0, 2\pi)$ are $x = \frac{\pi}{6}$, $\frac{\pi}{2}$, $\frac{5\pi}{6}$, $\frac{7\pi}{6}$, $\frac{3\pi}{2}$, $\frac{11\pi}{6}$.

45. $\tan x + \sec x = \sqrt{3} \iff \dfrac{\sin x}{\cos x} + \dfrac{1}{\cos x} = \sqrt{3} \iff \sin x + 1 = \sqrt{3}\cos x \iff \sqrt{3}\cos x - \sin x = 1 \iff \frac{\sqrt{3}}{2}\cos x - \frac{1}{2}\sin x = \frac{1}{2} \iff \cos\frac{\pi}{6}\cos x - \sin\frac{\pi}{6}\sin x = \frac{1}{2} \iff \cos\left(x + \frac{\pi}{6}\right) = \frac{1}{2} \iff x + \frac{\pi}{6} = \frac{\pi}{3}$, $\frac{5\pi}{3} \iff x = \frac{\pi}{6}$, $\frac{3\pi}{2}$. However, $x = \frac{3\pi}{2}$ is inadmissible because $\sec\frac{3\pi}{2}$ is undefined. Thus, the only solution in $[0, 2\pi)$ is $x = \frac{\pi}{6}$.

47. We graph $f(x) = \cos x$ and $g(x) = x^2 - 1$ in the viewing rectangle $[0, 6.5]$ by $[-2, 2]$. The two functions intersect at only one point, $x \approx 1.18$.

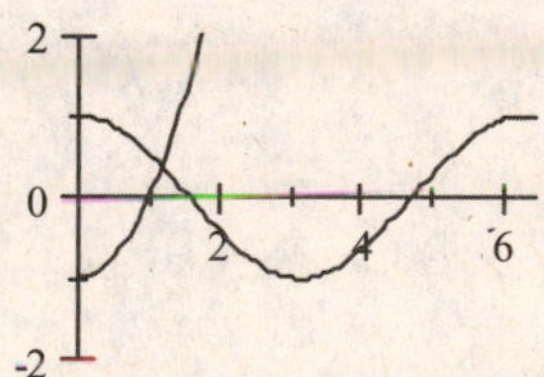

49. (a) $2000=\dfrac{(400)^2\sin^2\theta}{64} \iff \sin^2\theta=0.8 \iff \sin\theta\approx 0.8944 \iff \theta\approx 63.4^\circ$

(b) $\dfrac{(400)^2\sin^2\theta}{64}=2500\sin^2\theta\le 2500$. Therefore it is impossible for the projectile to reach a height of 3000 ft.

(c) The function $M(\theta)=2500\sin^2\theta$ is maximized when $\sin^2\theta=1$, so $\theta=90^\circ$. The projectile will travel the highest when it is shot straight up.

51. Since 15° is in quadrant I, $\cos 15^\circ=\sqrt{\dfrac{1+\cos 30^\circ}{2}}=\sqrt{\dfrac{2+\sqrt{3}}{4}}=\frac{1}{2}\sqrt{2+\sqrt{3}}$.

53. $\tan\frac{\pi}{8}=\dfrac{1-\cos\frac{\pi}{4}}{\sin\frac{\pi}{4}}=\dfrac{1-\frac{1}{\sqrt{2}}}{\frac{1}{\sqrt{2}}}=\left(1-\frac{1}{\sqrt{2}}\right)\sqrt{2}=\sqrt{2}-1$

55. $\sin 5^\circ\cos 40^\circ+\cos 5^\circ\sin 40^\circ=\sin\left(5^\circ+40^\circ\right)=\sin 45^\circ=\frac{1}{\sqrt{2}}=\frac{\sqrt{2}}{2}$

57. $\cos^2\frac{\pi}{8}-\sin^2\frac{\pi}{8}=\cos\left(2\left(\frac{\pi}{8}\right)\right)=\cos\frac{\pi}{4}=\frac{1}{\sqrt{2}}=\frac{\sqrt{2}}{2}$

59. We use a product-to-sum formula:

$\cos 37.5^\circ\cos 7.5^\circ=\frac{1}{2}\left(\cos 45^\circ+\cos 30^\circ\right)=\frac{1}{2}\left(\frac{\sqrt{2}}{2}+\frac{\sqrt{3}}{2}\right)=\frac{1}{4}\left(\sqrt{2}+\sqrt{3}\right)$.

In Exercises 61--66, x and y are in quadrant I, so we know that $\sec x=\frac{3}{2} \Rightarrow \cos x=\frac{2}{3}$ **, so** $\sin x=\frac{\sqrt{5}}{3}$ **and** $\tan x=\frac{\sqrt{5}}{2}$ **. Also,** $\csc y=3 \Rightarrow \sin y=\frac{1}{3}$ **, and so** $\cos y=\frac{2\sqrt{2}}{3}$ **, and** $\tan y=\frac{1}{2\sqrt{2}}=\frac{\sqrt{2}}{4}$.

61. $\sin(x+y)=\sin x\cos y+\cos x\sin y=\frac{\sqrt{5}}{3}\cdot\frac{2\sqrt{2}}{3}+\frac{2}{3}\cdot\frac{1}{3}=\frac{2}{9}\left(1+\sqrt{10}\right)$.

63.

$$\tan(x+y)=\frac{\tan x+\tan y}{1-\tan x\tan y}=\frac{\frac{\sqrt{5}}{2}+\frac{\sqrt{2}}{4}}{1-\left(\frac{\sqrt{5}}{2}\right)\left(\frac{\sqrt{2}}{4}\right)}=\frac{\frac{\sqrt{5}}{2}+\frac{\sqrt{2}}{4}}{1-\left(\frac{\sqrt{5}}{2}\right)\left(\frac{\sqrt{2}}{4}\right)}\cdot\frac{8}{8}=\frac{2\left(2\sqrt{5}+\sqrt{2}\right)}{8-\sqrt{10}}\cdot\frac{8+\sqrt{10}}{8+\sqrt{10}}=\frac{2}{3}\left(\sqrt{2}+\sqrt{5}\right)$$

65. $\cos\dfrac{y}{2}=\sqrt{\dfrac{1+\cos y}{2}}=\sqrt{\dfrac{1+\left(\frac{2\sqrt{2}}{3}\right)}{2}}=\sqrt{\dfrac{3+2\sqrt{2}}{6}}$ (since cosine is positive in quadrant I)

67. We sketch a triangle such that $\theta = \cos^{-1} \frac{3}{7}$.

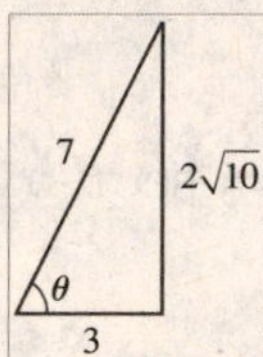

$\cos\theta = \frac{3}{7}$

We see that $\tan\theta = \frac{2\sqrt{10}}{3}$, and the double-angle formula for tangent gives

$$\tan 2\theta = \frac{2\tan\theta}{1-\tan^2\theta} = \frac{2\cdot\frac{2\sqrt{10}}{3}}{1-\left(\frac{2\sqrt{10}}{3}\right)^2} = \frac{\frac{4\sqrt{10}}{3}}{1-\frac{40}{9}} = -\frac{12\sqrt{10}}{31} .$$

69. The double-angle formula for tangent gives $\tan\left(2\tan^{-1}x\right) = \dfrac{2\tan\left(\tan^{-1}x\right)}{1-\tan^2\left(\tan^{-1}x\right)} = \dfrac{2x}{1-x^2}$.

71. (a) $\tan\theta = \dfrac{10}{x} \Leftrightarrow \theta = \tan^{-1}\left(\dfrac{10}{x}\right)$

(b) $\theta = \tan^{-1}\left(\dfrac{10}{x}\right)$, for $x > 0$. Since the road sign can first be seen when $\theta = 2^\circ$, we have

$2^\circ = \tan^{-1}\left(\dfrac{10}{x}\right) \Leftrightarrow x = \dfrac{10}{\tan 2^\circ} \approx 286.4$ ft. Thus, the sign can first be seen at a height of

286.4 ft.

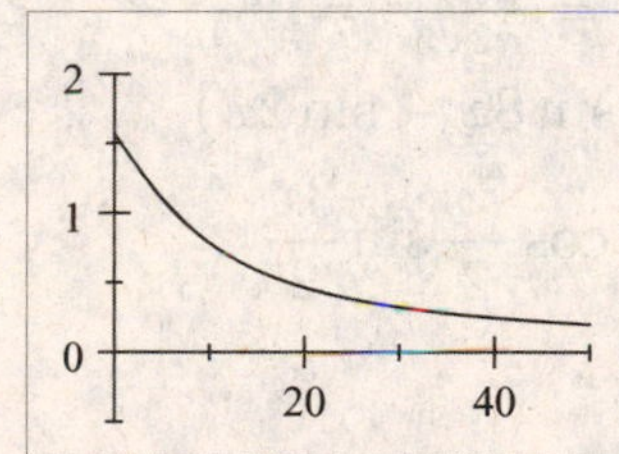

4 Test

1. (a) $\tan\theta\sin\theta+\cos\theta=\dfrac{\sin\theta}{\cos\theta}\sin\theta+\cos\theta=\dfrac{\sin^2\theta}{\cos\theta}+\dfrac{\cos^2\theta}{\cos\theta}=\dfrac{1}{\cos\theta}=\sec\theta$

(b)

$$\frac{\tan x}{1-\cos x}=\frac{\tan x}{1-\cos x}\cdot\frac{1+\cos x}{1+\cos x}=\frac{\tan x\left(1+\cos x\right)}{1-\cos^2 x}=\frac{\frac{\sin x}{\cos x}\left(1+\cos x\right)}{\sin^2 x}$$
$$=\frac{1}{\sin x}\cdot\frac{1+\cos x}{\cos x}=\csc x\left(1+\sec x\right)$$

(c) $\dfrac{2\tan x}{1+\tan^2 x}=\dfrac{2\tan x}{\sec^2 x}=\dfrac{2\sin x}{\cos x}\cdot\cos^2 x=2\sin x\cos x=\sin 2x$

2. $\dfrac{x}{\sqrt{4-x^2}}=\dfrac{2\sin\theta}{\sqrt{4-(2\sin\theta)^2}}=\dfrac{2\sin\theta}{\sqrt{4-4\sin^2\theta}}=\dfrac{2\sin\theta}{2\sqrt{1-\sin^2\theta}}=\dfrac{\sin\theta}{\sqrt{\cos^2\theta}}=\dfrac{\sin\theta}{|\cos\theta|}=\dfrac{\sin\theta}{\cos\theta}=\tan\theta$

(because $\cos\theta>0$ for $-\frac{\pi}{2}<\theta<\frac{\pi}{2}$)

3. (a) $\sin 8^\circ\cos 22^\circ+\cos 8^\circ\sin 22^\circ=\sin(8^\circ+22^\circ)=\sin 30^\circ=\frac{1}{2}$

(b) $\sin 75^\circ=\sin\left(45^\circ+30^\circ\right)=\sin 45^\circ\cos 30^\circ+\cos 45^\circ\sin 30^\circ=\frac{\sqrt{2}}{2}\cdot\frac{\sqrt{3}}{2}+\frac{\sqrt{2}}{2}\cdot\frac{1}{2}$
$=\frac{1}{4}\left(\sqrt{6}+\sqrt{2}\right)$

(c) $\sin\frac{\pi}{12}=\sin\left(\frac{\pi}{2}\right)=\sqrt{\dfrac{1-\cos\frac{\pi}{6}}{2}}=\sqrt{\dfrac{1-\frac{\sqrt{3}}{2}}{2}}=\sqrt{\dfrac{2-\sqrt{3}}{4}}=\frac{1}{2}\sqrt{2-\sqrt{3}}$

4. From the figures, we have
$\cos(\alpha+\beta)=\cos\alpha\cos\beta-\sin\alpha\sin\beta=\frac{2}{\sqrt{5}}\cdot\frac{\sqrt{5}}{3}-\frac{1}{\sqrt{5}}\cdot\frac{2}{3}=\frac{2\sqrt{5}-2}{3\sqrt{5}}=\frac{10-2\sqrt{5}}{15}$.

5. (a) $\sin 3x\cos 5x=\frac{1}{2}[\sin(3x+5x)+\sin(3x-5x)]=\frac{1}{2}(\sin 8x-\sin 2x)$

(b) $\sin 2x-\sin 5x=2\cos\left(\dfrac{2x+5x}{2}\right)\sin\left(\dfrac{2x-5x}{2}\right)=-2\cos\dfrac{7x}{2}\sin\dfrac{3x}{2}$

6. $\sin\theta=-\frac{4}{5}$. Since θ is in quadrant III, $\cos\theta=-\frac{3}{5}$. Then

$$\tan\frac{\theta}{2}=\frac{1-\cos\theta}{\sin\theta}=\frac{1-(-\frac{3}{5})}{-\frac{4}{5}}=-\frac{1+\frac{3}{5}}{\frac{4}{5}}=-\frac{5+3}{4}=-2\ .$$

7. (a) $3\sin\theta-1=0\ \Leftrightarrow\ 3\sin\theta=1\ \Leftrightarrow\ \sin\theta=\frac{1}{3}\ \Leftrightarrow\ \theta=\sin^{-1}\frac{1}{3}\approx 0.34$ or
$\theta=\pi-\sin^{-1}\frac{1}{3}\approx 2.80$ on $[0,2\pi)$.

(b) $(2\cos\theta-1)(\sin\theta-1)=0\ \Leftrightarrow\ \cos\theta=\frac{1}{2}$ or $\sin\theta=1$. The first equation has solutions $\theta=\frac{\pi}{3}\approx 1.05$ and $\theta=\frac{5\pi}{3}\approx 5.24$ on $[0,2\pi)$, while the second has the solution $\theta=\frac{\pi}{2}\approx 1.57$.

(c) $2\cos^2\theta+5\cos\theta+2=0\ \Leftrightarrow\ (2\cos\theta+1)(\cos\theta+2)=0\ \Leftrightarrow\ \cos\theta=-\frac{1}{2}$ or $\cos\theta=-2$ (which is impossible). So in the interval $[0,2\pi)$, the solutions are $\theta=\frac{2\pi}{3}\approx 2.09$, $\frac{4\pi}{3}\approx 4.19$.

(d) $\sin 2\theta-\cos\theta=0\ \Leftrightarrow\ 2\sin\theta\cos\theta-\cos\theta=0\ \Leftrightarrow\ \cos\theta(2\sin\theta-1)=0$
$\Leftrightarrow\ \cos\theta=0$ or $\sin\theta=\frac{1}{2}\ \Leftrightarrow\ \theta=\frac{\pi}{2}$, $\frac{3\pi}{2}$ or $\theta=\frac{\pi}{6}$, $\frac{5\pi}{6}$. Therefore, the solutions

in $[0,2\pi)$ are $\theta = \frac{\pi}{6} \approx 0.52$, $\frac{\pi}{2} \approx 1.57$, $\frac{5\pi}{6} \approx 2.62$, $\frac{3\pi}{2} \approx 4.71$.

8. $5\cos 2\theta = 2 \quad\Leftrightarrow\quad \cos 2\theta = \frac{2}{5} \quad\Leftrightarrow\quad 2\theta = \cos^{-1} 0.4 \approx 1.159279$. The solutions in $[0,4\pi)$ are $2\theta \approx 1.159279$, $2\pi - 1.159279$, $2\pi + 1.159279$, $4\pi - 1.159279 \quad\Leftrightarrow$ $2\theta \approx 1.159279$, 5.123906 , 7.442465 , $11.407091 \quad\Leftrightarrow\quad \theta \approx 0.57964$, 2.56195 , 3.72123 , 5.70355 in $[0,2\pi)$.

9. Let $u = \tan^{-1} \frac{9}{40}$ so $\tan u = \frac{9}{40}$.

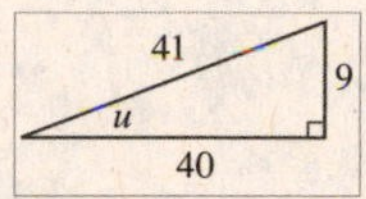

From the triangle, $\cos u = \frac{40}{41}$, so using a double-angle formula for cosine,

$\cos 2u = 2\cos^2 u - 1 = 2\left(\frac{40}{41}\right)^2 - 1 = \frac{1519}{1681}$.

10. We sketch triangles such that $\theta = \cos^{-1} x$ and $\phi = \tan^{-1} y$.

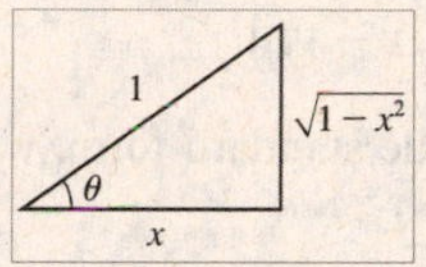

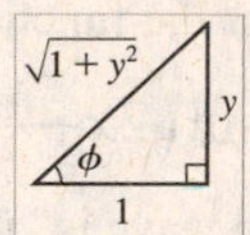

$\cos\theta = x$ $\qquad$ $\tan\phi = y$

From the triangles, we have $\sin\theta = \sqrt{1-x^2}$, $\sin\phi = \dfrac{y}{\sqrt{1+y^2}}$, and $\cos\phi = \dfrac{1}{\sqrt{1+y^2}}$, so the addition formula for sine gives

$$\begin{aligned}\sin(\theta - \phi) &= \sin\theta\cos\phi - \cos\theta\sin\phi \\ &= \sqrt{1-x^2}\cdot\frac{1}{\sqrt{1+y^2}} - x\cdot\frac{y}{\sqrt{1+y^2}} = \frac{\sqrt{1-x^2} - xy}{\sqrt{1+y^2}}\end{aligned}$$

1. (a) Substituting $x=0$, we get $y(0,t)=5\sin\left(2\cdot 0-\frac{\pi}{2}t\right)=5\sin\left(-\frac{\pi}{2}t\right)=-5\sin\frac{\pi}{2}t$.

(b)

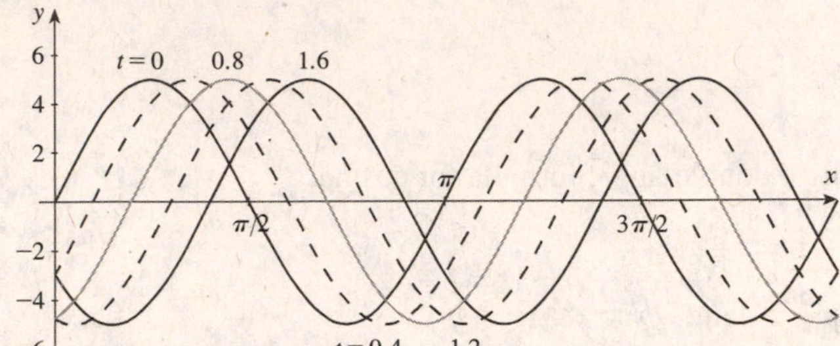

(c) We express the function in the standard form $y(x,t)=A\sin k(x-vt)$:

$y(x,t)=5\sin\left(2x-\frac{\pi}{2}t\right)=5\sin 2\left(x-\frac{\pi}{4}t\right)$. Comparing this to the standard form, we see that the velocity of the wave is $v=\frac{\pi}{4}$.

3. From the graph, we see that the amplitude is $A=2.7$ and the period is 9.2, so $k=\frac{2\pi}{9.2}\approx 0.68$. Since $v=6$, we have $kv=\frac{2\pi}{9.2}\cdot 6\approx 4.10$, so the equation we seek is $y(x,t)=2.7\sin(0.68x-4.10t)$.

5. From the graphs, we see that the amplitude is $A=0.6$. The nodes occur at $x=0$, 1, 2, 3. Since $\sin\alpha x=0$ when $\alpha x=k\pi$ (k any integer), we have $\alpha=\pi$. Then since the frequency is $\beta/2\pi$, we get $20=\beta/2\pi \quad\Leftrightarrow\quad \beta=40\pi$. Thus, an equation for this model is $f(x,t)=0.6\sin\pi x\cos 40\pi t$.

7. (a) The first standing wave has $\alpha=1$, the second has $\alpha=2$, the third has $\alpha=3$, and the fourth has $\alpha=4$.

(b) α is equal to the number of nodes minus 1. The first string has two nodes and $\alpha=1$; the second string has three nodes and $\alpha=2$, and so forth.

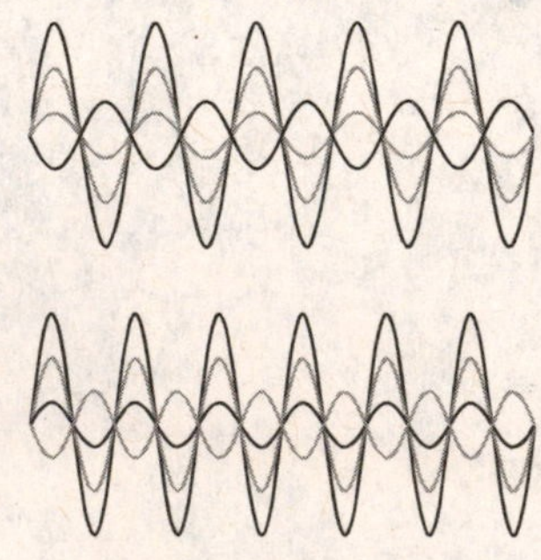

(c) Since the frequency is $\beta/2\pi$, we have $440=\beta/2\pi \quad\Leftrightarrow\quad \beta=880\pi$.

(d) The first standing wave has equation $y=\sin x\cos 880\pi t$, the second has equation $y=\sin 2x\cos 880\pi t$, the third has equation $y=\sin 3x\cos 880\pi t$, and the fourth has equation $y=\sin 4x\cos 880\pi t$.

5 Polar Coordinates and Parametric Equations

5.1 Polar Coordinates

1. We can describe the location of a point in the plane using different *coordinate* systems. The point P shown in the figure has rectangular coordinates $(1,1)$ and polar coordinates $\left(\sqrt{2},\frac{\pi}{4}\right)$.

3.

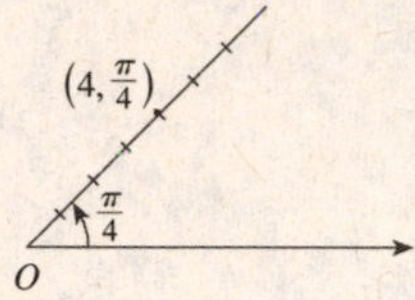

5.

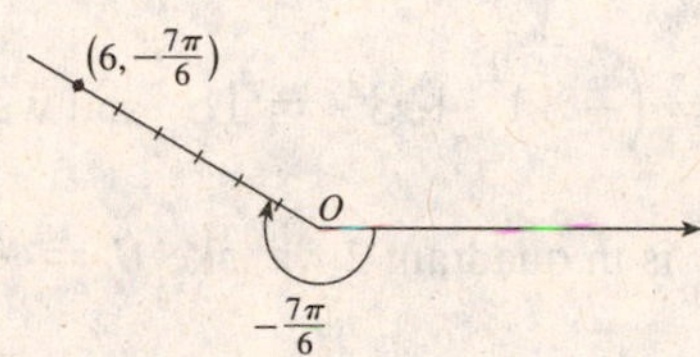

7.

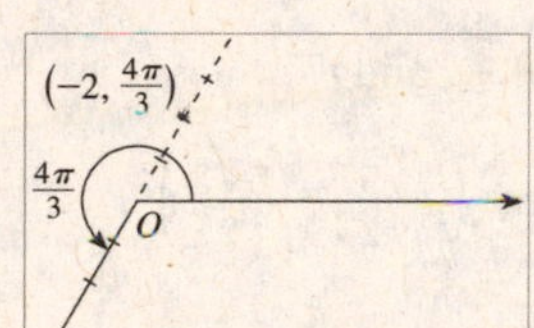

Answers to Exercises 9--14 will vary.

9. $\left(3,\frac{\pi}{2}\right)$ has polar coordinates $\left(3,\frac{5\pi}{2}\right)$ or $\left(-3,\frac{3\pi}{2}\right)$

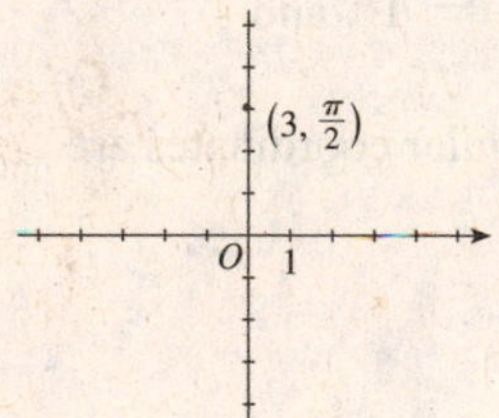

11. $\left(-1,\frac{7\pi}{6}\right)$ has polar coordinates $\left(1,\frac{\pi}{6}\right)$ or $\left(-1,-\frac{5\pi}{6}\right)$.

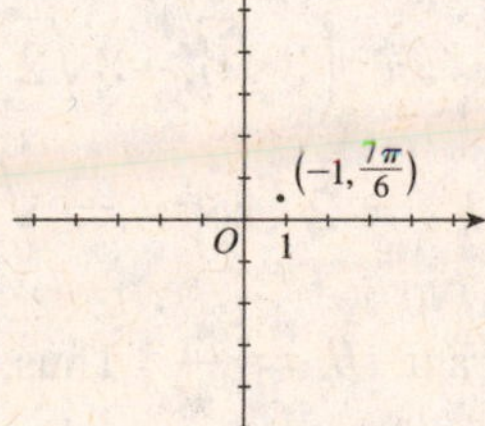

13. $(-5,0)$ has polar coordinates $(5,\pi)$ or $(-5,2\pi)$.

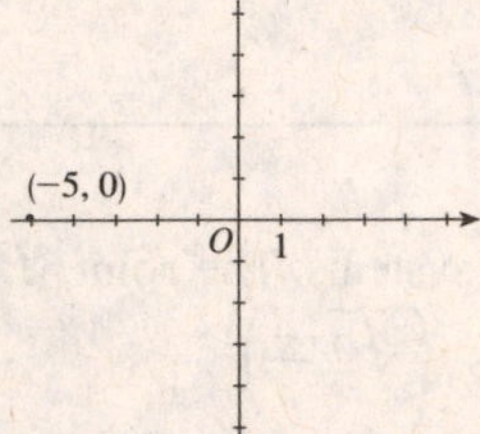

15. Q has coordinates $\left(4,\frac{3\pi}{4}\right)$.

17. Q has coordinates $\left(-4,-\frac{\pi}{4}\right)=\left(4,\frac{3\pi}{4}\right)$.

19. P has coordinates $\left(4,-\frac{23\pi}{4}\right)=\left(4,\frac{\pi}{4}\right)$.

21. P has coordinates $\left(-4,\frac{101\pi}{4}\right)=\left(-4,\frac{5\pi}{4}\right)=\left(4,\frac{\pi}{4}\right)$.

23. $P=(-3,3)$ in rectangular coordinates, so $r^2=x^2+y^2=(-3)^2+3^2=18$ and we can take $r=3\sqrt{2}$. $\tan\theta=\dfrac{y}{x}=\dfrac{3}{-3}=-1$, so since P is in quadrant 2 we take $\theta=\frac{3\pi}{4}$. Thus, polar coordinates for P are $\left(3\sqrt{2},\frac{3\pi}{4}\right)$.

25. Here $r=5$ and $\theta=-\frac{2\pi}{3}$, so $x=r\cos\theta=5\cos\left(-\frac{2\pi}{3}\right)=-\frac{5}{2}$ and $y=r\sin\theta=5\sin\left(-\frac{2\pi}{3}\right)=-\frac{5\sqrt{3}}{2}$. R has rectangular coordinates $\left(-\frac{5}{2},-\frac{5\sqrt{3}}{2}\right)$.

27. $(r,\theta)=\left(4,\frac{\pi}{6}\right)$. So $x=r\cos\theta=4\cos\frac{\pi}{6}=4\cdot\frac{\sqrt{3}}{2}=2\sqrt{3}$ and $y=r\sin\theta=4\sin\frac{\pi}{6}=4\cdot\frac{1}{2}=2$. Thus, the rectangular coordinates are $\left(2\sqrt{3},2\right)$.

29. $(r,\theta)=\left(\sqrt{2},-\frac{\pi}{4}\right)$. So $x=r\cos\theta=\sqrt{2}\cos\left(-\frac{\pi}{4}\right)=\sqrt{2}\cdot\frac{1}{\sqrt{2}}=1$, and $y=r\sin\theta=\sqrt{2}\sin\left(-\frac{\pi}{4}\right)=\sqrt{2}\left(-\frac{1}{\sqrt{2}}\right)=-1$. Thus, the rectangular coordinates are $(1,-1)$.

31. $(r,\theta)=(5,5\pi)$. So $x=r\cos\ \theta=5\cos 5\pi=-5$, and $y=r\sin\theta=5\sin 5\pi=0$. Thus, the rectangular coordinates are $(-5,0)$.

33. $(r,\theta)=\left(6\sqrt{2},\frac{11\pi}{6}\right)$. So $x=r\cos\theta=6\sqrt{2}\cos\frac{11\pi}{6}=3\sqrt{6}$ and $y=r\sin\theta=6\sqrt{2}\sin\frac{11\pi}{6}=-3\sqrt{2}$. Thus, the rectangular coordinates are $\left(3\sqrt{6},-3\sqrt{2}\right)$.

35. $(x,y)=(-1,1)$. Since $r^2=x^2+y^2$, we have $r^2=(-1)^2+1^2=2$, so $r=\sqrt{2}$. Now $\tan\theta=\dfrac{y}{x}=\dfrac{1}{-1}=-1$, so, since the point is in the second quadrant, $\theta=\frac{3\pi}{4}$. Thus, polar coordinates are $\left(\sqrt{2},\frac{3\pi}{4}\right)$.

37. $(x,y)=(\sqrt{8},\sqrt{8})$. Since $r^2=x^2+y^2$, we have $r^2=(\sqrt{8})^2+(\sqrt{8})^2=16$, so $r=4$. Now $\tan\theta=\dfrac{y}{x}=\frac{\sqrt{8}}{\sqrt{8}}=1$, so, since the point is in the first quadrant, $\theta=\frac{\pi}{4}$. Thus, polar coordinates are $\left(4,\frac{\pi}{4}\right)$.

39. $(x,y)=(3,4)$. Since $r^2=x^2+y^2$, we have $r^2=3^2+4^2=25$, so $r=5$. Now $\tan\theta=\dfrac{y}{x}=\frac{4}{3}$, so, since the point is in the first quadrant, $\theta=\tan^{-1}\frac{4}{3}$. Thus, polar coordinates are $\left(5,\tan^{-1}\frac{4}{3}\right)$.

41. $(x,y)=(-6,0)$. $r^2=(-6^2)=36$, so $r=6$. Now $\tan\theta=\dfrac{y}{x}=0$, so since the point is on the negative x -axis, $\theta=\pi$. Thus, polar coordinates are $(6,\pi)$.

43. $x=y \quad\Leftrightarrow\quad r\cos\theta=r\sin\theta \quad\Leftrightarrow\quad \tan\theta=1$, and so $\theta=\frac{\pi}{4}$.

45. $y=x^2$. We substitute and then solve for r : $r\sin\theta=(r\cos\theta)^2=r^2\cos^2\theta \quad\Leftrightarrow$ $\sin\theta=r\cos^2\theta \quad\Leftrightarrow\quad r=\dfrac{\sin\theta}{\cos^2\theta}=\tan\theta\sec\theta$.

47. $x=4$. We substitute and then solve for r : $r\cos\theta=4 \quad\Leftrightarrow\quad r=\dfrac{4}{\cos\theta}=4\sec\theta$.

49. $r=7$. But $r^2=x^2+y^2$, so $x^2+y^2=r^2=49$. Hence, the equivalent equation in rectangular coordinates is $x^2+y^2=49$.

51. $\theta=-\frac{\pi}{2} \quad\Rightarrow\quad \cos\theta=0$, so an equivalent equation in rectangular coordinates is $x=0$.

53. $r\cos\theta=6$. But $x=r\cos\theta$, and so $x=6$ is an equivalent rectangular equation.

55. $r=4\sin\theta \quad\Leftrightarrow\quad r^2=4r\sin\theta$. Thus, $x^2+y^2=4y$ is an equivalent rectangular equation. Completing the square, it can be written as $x^2+(y-2)^2=4$.

57. $r=1+\cos\theta$. If we multiply both sides of this equation by r we get $r^2=r+r\cos\theta$. Thus $r^2-r\cos\theta=r$, and squaring both sides gives $(r^2-r\cos\theta)^2=r^2$, or $(x^2+y^2-x)^2=x^2+y^2$ in rectangular coordinates.

59. $r=1+2\sin\theta$. If we multiply both sides of this equation by r we get $r^2=r+2r\sin\theta$. Thus $r^2-2r\sin\theta=r$, and squaring both sides gives $(r^2-2r\sin\theta)^2=r^2$, or $(x^2+y^2-2y)^2=x^2+y^2$ in rectangular coordinates.

61. $r=\dfrac{1}{\sin\theta-\cos\theta} \quad\Rightarrow\quad r(\sin\theta-\cos\theta)=1 \quad\Leftrightarrow\quad r\sin\theta-r\cos\theta=1$, and since $r\cos\theta=x$ and $r\sin\theta=y$, we get $y-x=1$.

63. $r = \dfrac{4}{1+2\sin\theta} \Leftrightarrow r\left(1+2\sin\theta\right) = 4 \Leftrightarrow r + 2r\sin\theta = 4$. Thus

$r = 4 - 2r\sin\theta$. Squaring both sides, we get $r^2 = \left(4 - 2r\sin\theta\right)^2$. Substituting,

$x^2 + y^2 = \left(4 - 2y\right)^2 \Leftrightarrow x^2 + y^2 = 16 - 16y + 4y^2 \Leftrightarrow$

$x^2 - 3y^2 + 16y - 16 = 0$.

65. $r^2 = \tan\theta$. Substituting $r^2 = x^2 + y^2$ and $\tan\theta = \dfrac{y}{x}$, we get $x^2 + y^2 = \dfrac{y}{x}$.

67. $\sec\theta = 2 \Leftrightarrow \cos\theta = \frac{1}{2} \Leftrightarrow \theta = \pm\frac{\pi}{3} \Leftrightarrow \tan\theta = \pm\sqrt{3} \Leftrightarrow \dfrac{y}{x} = \pm\sqrt{3}$

$\Leftrightarrow y = \pm\sqrt{3}x$.

69. (a) In rectangular coordinates, the points $\left(r_1, \theta_1\right)$ and $\left(r_2, \theta_2\right)$ are $\left(x_1, y_1\right) = \left(r_1\cos\theta_1, r_1\sin\theta_1\right)$ and $\left(x_2, y_2\right) = \left(r_2\cos\theta_2, r_2\sin\theta_2\right)$. Then, the distance between the points is

$$\begin{aligned} D &= \sqrt{\left(x_1 - x_2\right)^2 + \left(y_1 - y_2\right)^2} = \sqrt{\left(r_1\cos\theta_1 - r_2\cos\theta_2\right)^2 + \left(r_1\sin\theta_1 - r_2\sin\theta_2\right)^2} \\ &= \sqrt{r_1^2\left(\cos^2\theta_1 + \sin^2\theta_1\right) + r_2^2\left(\cos^2\theta_2 + \sin^2\theta_2\right) - 2r_1r_2\left(\cos\theta_1\cos\theta_2 + \sin\theta_1\sin\theta_2\right)} \\ &= \sqrt{r_1^2 + r_2^2 - 2r_1r_2\cos\left(\theta_2 - \theta_1\right)} \end{aligned}$$

(b) The distance between the points $\left(3, \frac{3\pi}{4}\right)$ and $\left(-1, \frac{7\pi}{6}\right)$ is

$$D = \sqrt{3^2 + \left(-1\right)^2 - 2\left(3\right)\left(-1\right)\cos\left(\tfrac{7\pi}{6} - \tfrac{3\pi}{4}\right)} = \sqrt{9 + 1 + 6\cos\tfrac{5\pi}{12}} \approx 3.40$$

5.2 Graphs of Polar Coordinates

1. To plot points in polar coordinates we use a grid consisting of *circles* centered at the pole and *rays* emanating from the pole.

3. VI

5. II

7. I

9. Polar axis: $2-\sin(-\theta)=2+\sin\theta\neq r$, so the graph is not symmetric about the polar axis.

Pole:

$2-\sin(\theta+\pi)=2-(\sin\pi\cos\theta+\cos\pi\sin\theta)=2-(-\sin\theta)=2+\sin\theta\neq r$,

so the graph is not symmetric about the pole.

Line $\theta=\frac{\pi}{2}$: $2-\sin(\pi-\theta)=2-(\sin\pi\cos\theta-\cos\pi\sin\theta)=2-\sin\theta=r$, so

the graph is symmetric about $\theta=\frac{\pi}{2}$.

11. Polar axis: $3\sec(-\theta)=3\sec\theta=r$, so the graph is symmetric about the polar axis.

Pole:

$$3\sec(\theta+\pi)=\frac{3}{\cos(\theta+\pi)}=\frac{1}{\cos\pi\cos\theta-\sin\pi\sin\theta}=\frac{3}{-\cos\theta}=-3\sec\theta\neq r\ ,$$

so the graph is not symmetric about the pole.

Line $\theta=\frac{\pi}{2}$:

$$3\sec(\pi-\theta)=\frac{3}{\cos(\pi-\theta)}=\frac{1}{\cos\pi\cos\theta+\sin\pi\sin\theta}=\frac{3}{-\cos\theta}=-3\sec\theta\neq r\ ,$$

so the graph is not symmetric about $\theta=\frac{\pi}{2}$.

13. Polar axis: $\dfrac{4}{3-2\sin(-\theta)}=\dfrac{4}{3+2\sin\theta}\neq r$, so the graph is not symmetric about the polar axis.

Pole:

$$\frac{4}{3-2\sin(\theta+\pi)}=\frac{4}{3-2(\sin\pi\cos\theta+\cos\pi\sin\theta)}=\frac{4}{3-2(-\sin\theta)}=\frac{4}{3+2\sin\theta}\neq r$$

so the graph is not symmetric about the pole.

Line $\theta=\frac{\pi}{2}$:

$$\frac{4}{3-2\sin(\pi-\theta)}=\frac{4}{3-2(\sin\pi\cos\theta-\cos\pi\sin\theta)}=\frac{4}{3-2\sin\theta}=r$$

, so the graph is symmetric about $\theta=\frac{\pi}{2}$.

15. Polar axis: $4\cos 2(-\theta) = 4\cos 2\theta = r^2$, so the graph is symmetric about the polar axis.

Pole: $(-r)^2 = r^2$, so the graph is symmetric about the pole.

Line $\theta = \frac{\pi}{2}$: $4\cos 2(\pi - \theta) = 4\cos(2\pi - 2\theta) = 4\cos(-2\theta) = 4\cos 2\theta = r^2$, so the graph is symmetric about $\theta = \frac{\pi}{2}$.

17. $r = 2 \Rightarrow r^2 = 4 \Rightarrow x^2 + y^2 = 4$ is an equation of a circle with radius 2 centered at the origin.

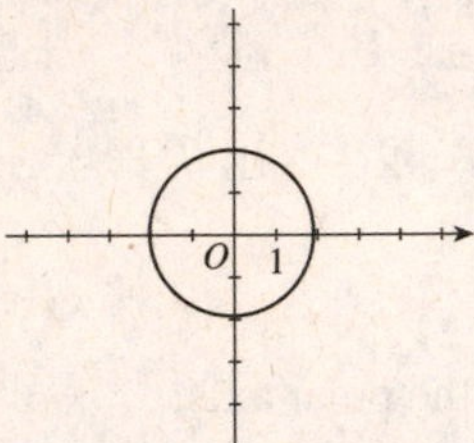

19. $\theta = -\frac{\pi}{2} \Rightarrow \cos\theta = 0 \Rightarrow x = 0$ is an equation of a vertical line.

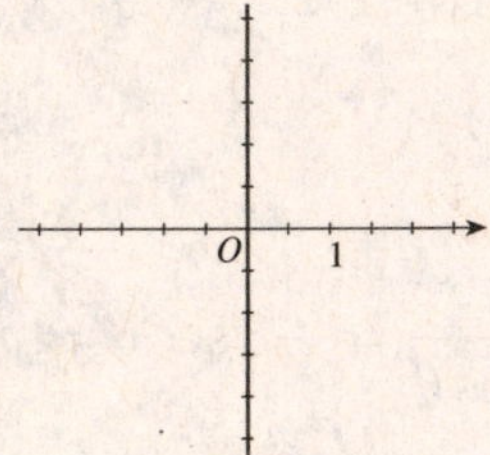

21. $r = 6\sin\theta \Rightarrow r^2 = 6r\sin\theta \Rightarrow x^2 + y^2 = 6y \Rightarrow x^2 + (y-3)^2 = 9$, a circle of radius 3 centered at $(0,3)$.

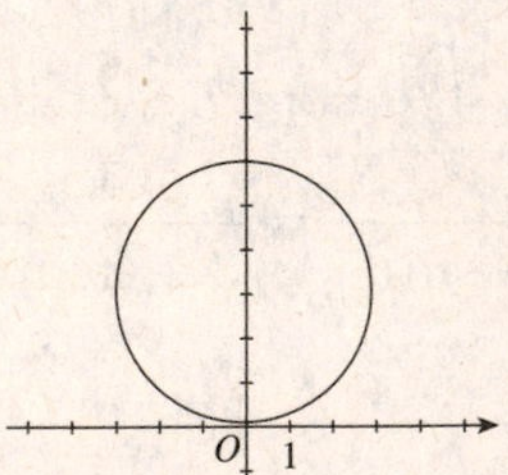

23. $r = -2\cos\theta$. Circle.

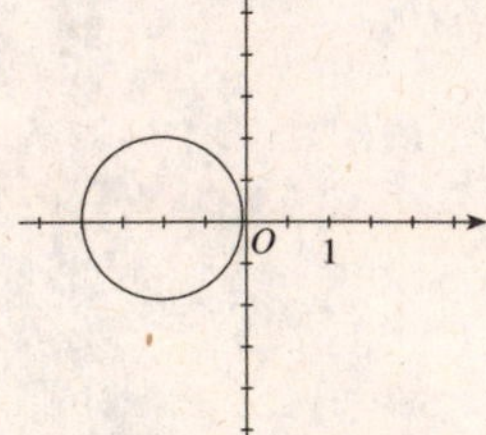

25. $r = 2 - 2\cos\theta$. Cardioid.

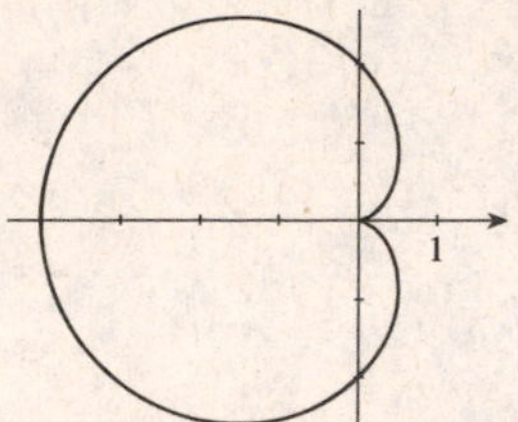

27. $r = -3\left(1 + \sin\theta\right)$. Cardioid.

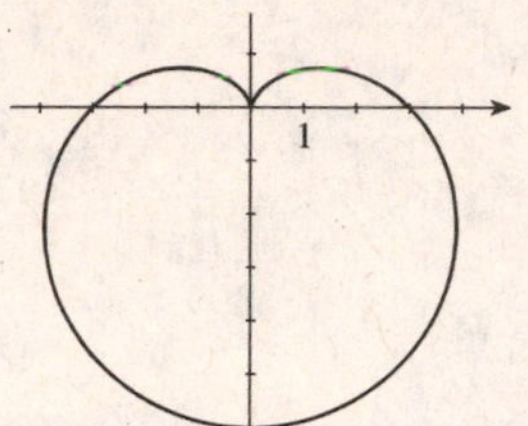

29. $r = \sin 2\theta$

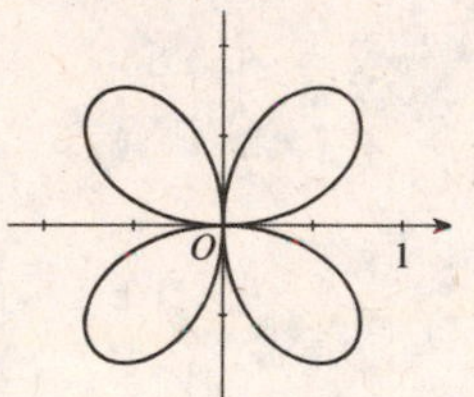

31. $r = -\cos 5\theta$

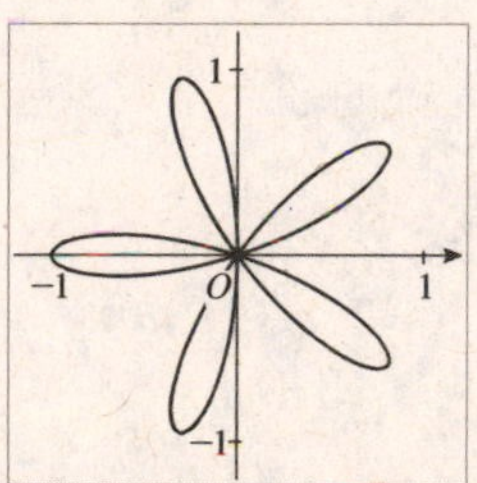

33. $r = \sqrt{3} - 2\sin\theta$

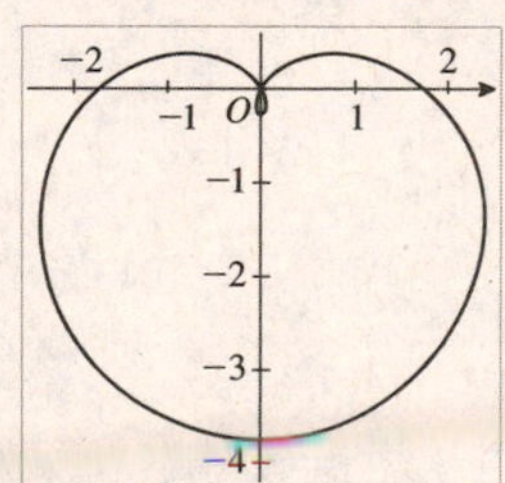

35. $r = \sqrt{3} + \cos\theta$

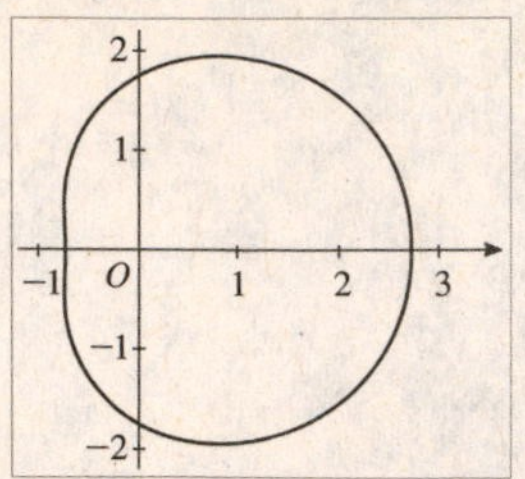

37. $r^2 = \cos 2\theta$

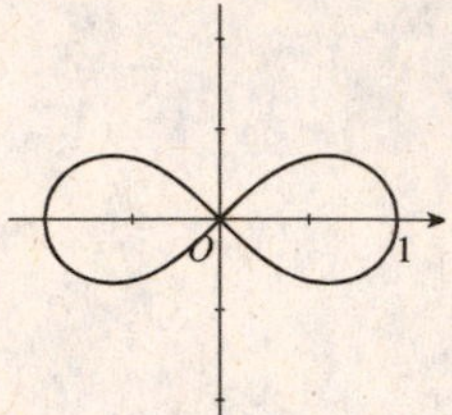

39. $r = \theta\ ,\ \theta \geq 0$

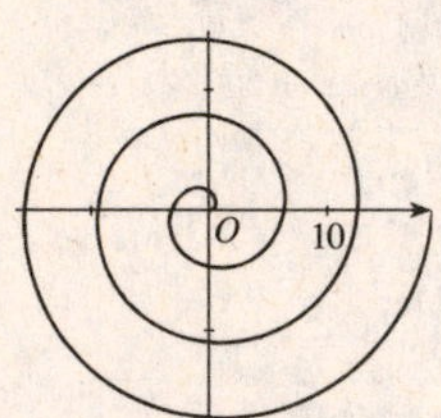

41. $r = 2 + \sec\theta$

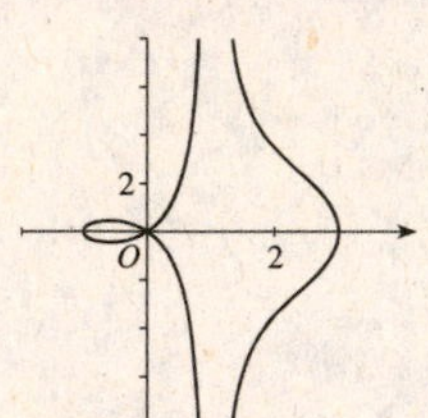

43. $r = \cos\left(\dfrac{\theta}{2}\right),\ \theta \in \left[0, 4\pi\right]$

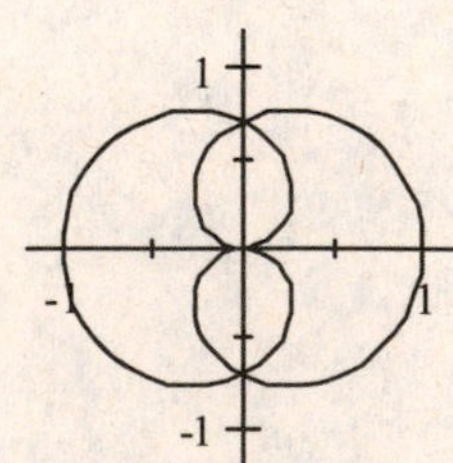

45. $r = 1 + 2\sin\left(\dfrac{\theta}{2}\right),\ \theta \in \left[0, 4\pi\right]$

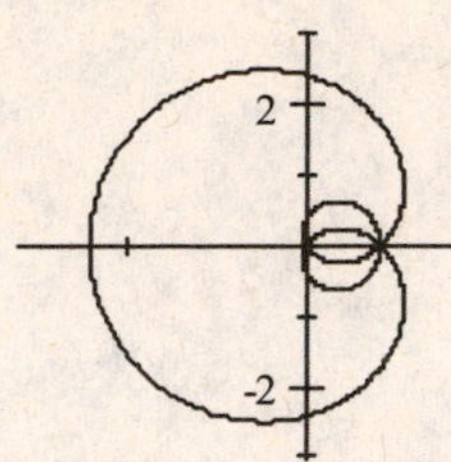

47. $r = 1 + \sin n\theta$. The number of loops is n .

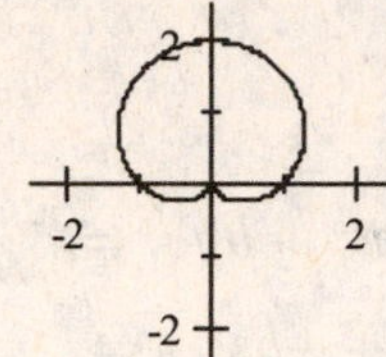

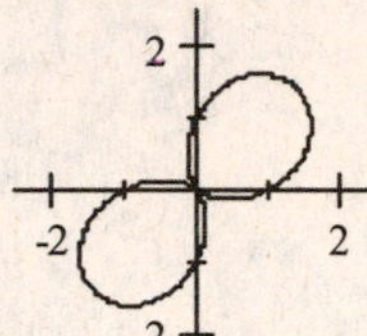

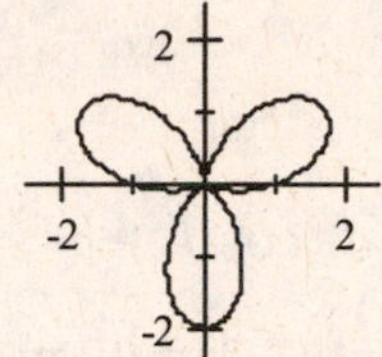

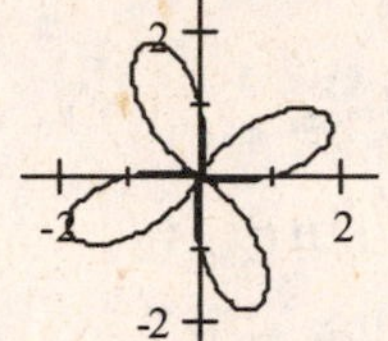

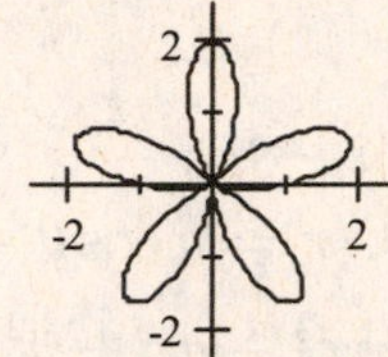

49. The graph of $r = \sin\left(\dfrac{\theta}{2}\right)$ is IV, since the graph must contain the points $(0,0), \left(\dfrac{1}{\sqrt{2}}, \frac{\pi}{2}\right), (1, \pi)$, and so on.

51. The graph of $r = \theta \sin \theta$ is III, since for $\theta = \frac{\pi}{2}, \frac{5\pi}{2}, \frac{7\pi}{2}, \ldots$ the values of r are also $\frac{\pi}{2}, \frac{5\pi}{2}, \frac{7\pi}{2}, \ldots$. Thus the graph must cross the vertical axis at an infinite number of points.

53. $\left(x^2 + y^2\right)^3 = 4x^2y^2 \iff \left(r^2\right)^3 = 4\left(r\cos\theta\right)^2\left(r\sin\theta\right)^2 \iff$

$r^6 = 4r^4\cos^2\theta\sin^2\theta \iff r^2 = 4\cos^2\theta\sin^2\theta \iff r = 2\cos\theta\sin\theta = \sin 2\theta$.
The equation is $r = \sin 2\theta$, and the graph is a rose.

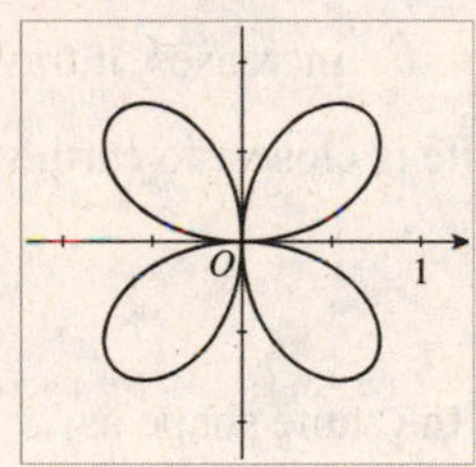

55. $\left(x^2+y^2\right)^2 = x^2 - y^2 \Leftrightarrow \left(r^2\right)^2 = \left(r\cos\theta\right)^2 - \left(r\sin\theta\right)^2 \Leftrightarrow$

$r^4 = r^2\cos^2\theta - r^2\sin^2\theta \Leftrightarrow r^4 = r^2\left(\cos^2\theta - \sin^2\theta\right) \Leftrightarrow$

$r^2 = \cos^2\theta - \sin^2\theta = \cos 2\theta$. The graph is $r^2 = \cos 2\theta$, a leminiscate.

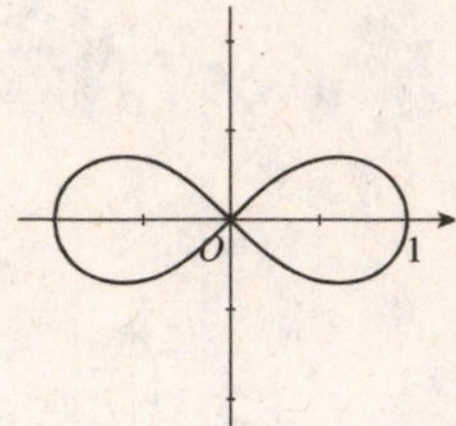

57. $r = a\cos\theta + b\sin\theta \Leftrightarrow r^2 = ar\cos\theta + br\sin\theta \Leftrightarrow x^2+y^2 = ax+by \Leftrightarrow$

$x^2 - ax + y^2 - by = 0 \Leftrightarrow x^2 - ax + \frac{1}{4}a^2 + y^2 - by + \frac{1}{4}b^2 = \frac{1}{4}a^2 + \frac{1}{4}b^2 \Leftrightarrow$

$\left(x - \frac{1}{2}a\right)^2 + \left(y - \frac{1}{2}b\right)^2 = \frac{1}{4}\left(a^2+b^2\right)$. Thus, in rectangular coordinates the center is $\left(\frac{1}{2}a, \frac{1}{2}b\right)$ and the radius is $\frac{1}{2}\sqrt{a^2+b^2}$.

59. (a)

At $\theta = 0$, the satellite is at the rightmost point in its orbit, $\left(5625, 0\right)$. As θ increases, it travels counterclockwise. Note that it is moving fastest when $\theta = \pi$.**(b)** The satellite is closest to earth when $\theta = \pi$. Its height above the earth's surface at this point is

$22500 / \left(4 - \cos\pi\right) - 3960 = 4500 - 3960 = 540$ mi.

61. The graphs of $r = 1 + \sin\left(\theta - \frac{\pi}{6}\right)$ and $r = 1 + \sin\left(\theta - \frac{\pi}{3}\right)$ have the same shape as $r = 1 + \sin\theta$, rotated through angles of $\frac{\pi}{6}$ and $\frac{\pi}{3}$, respectively. Similarly, the graph of $r = f\left(\theta - \alpha\right)$ is the graph of $r = f\left(\theta\right)$ rotated by the angle α .

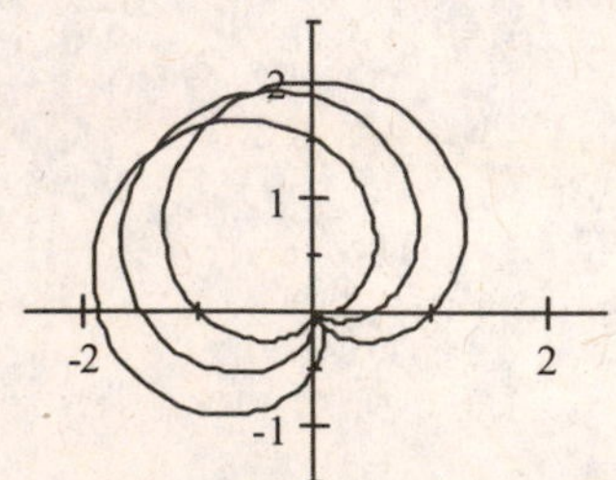

63. $y = 2 \quad \Leftrightarrow \quad r\sin\theta = 2 \quad \Leftrightarrow \quad r = 2\csc\theta$. The rectangular coordinate system gives the simpler equation here. It is easier to study lines in rectangular coordinates.

5.3 Polar Form of Complex Numbers; De Moivre's Theorem

1. A complex number $z = a + bi$ has two parts: a is the *real* part and b is the *imaginary* part. To graph $a + bi$ we graph the ordered pair (a, b) in the complex plane.

3. (a) The complex number $z = -1 + i$ in polar form is $z = \sqrt{2}\left(\cos\frac{3\pi}{4} + i\sin\frac{3\pi}{4}\right)$. The complex number $z = 2\left(\cos\frac{\pi}{6} + i\sin\frac{\pi}{6}\right)$ in rectangular form is $z = \sqrt{3} + i$.

(b) The complex number z can be expressed in rectangular form as $1 + i$ or in polar form as $\sqrt{2}\left(\cos\frac{\pi}{4} + i\sin\frac{\pi}{4}\right)$.

5. $|4i| = \sqrt{0^2 + 4^2} = 4$

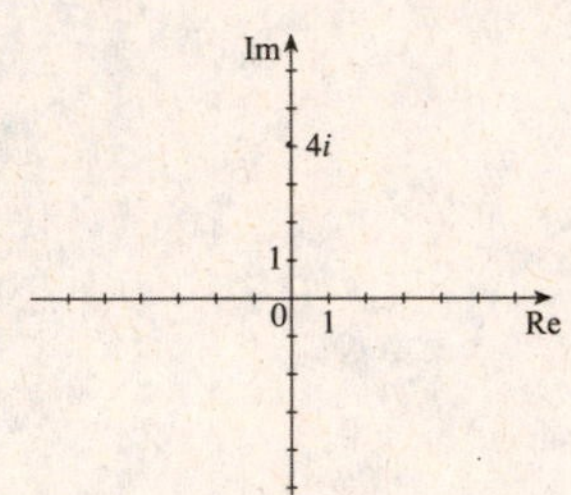

7. $|-2| = \sqrt{4 + 0} = 2$

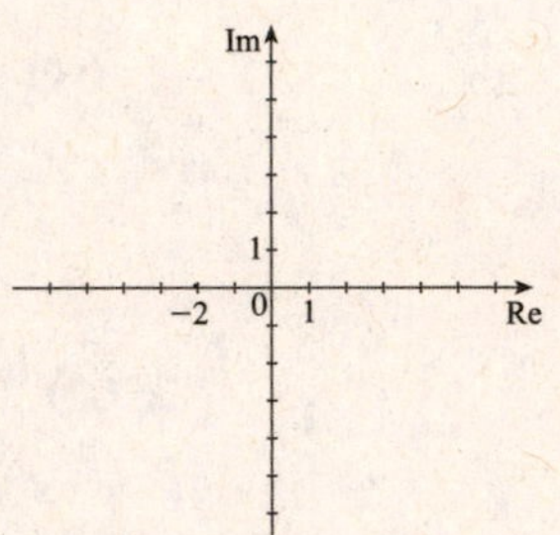

9. $|5 + 2i| = \sqrt{5^2 + 2^2} = \sqrt{29}$

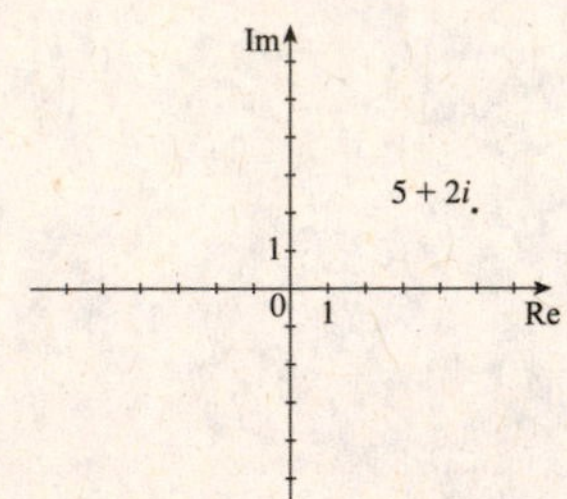

11. $\left|\sqrt{3} + i\right| = \sqrt{3 + 1} = 2$

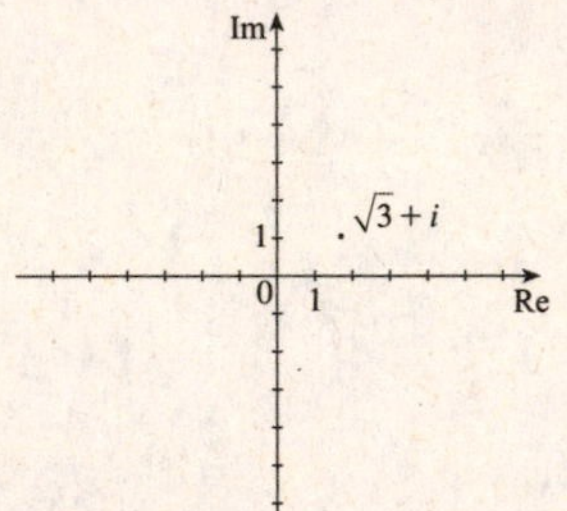

13. $\left|\dfrac{3 + 4i}{5}\right| = \sqrt{\dfrac{9}{25} + \dfrac{16}{25}} = 1$

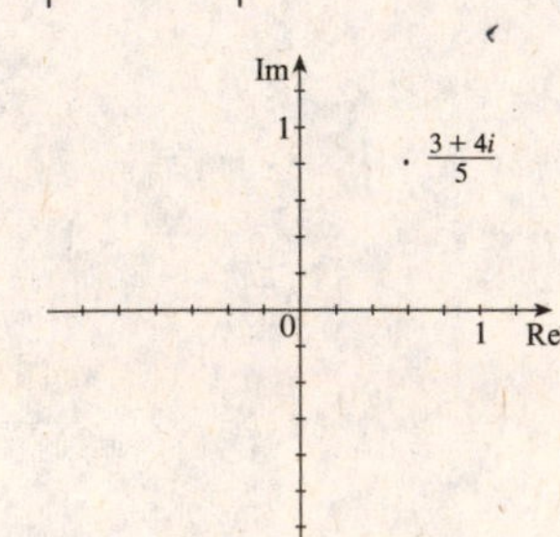

15. $z = 1 + i$, $2z = 2 + 2i$, $-z = -1 - i$, $\frac{1}{2}z = \frac{1}{2} + \frac{1}{2}i$

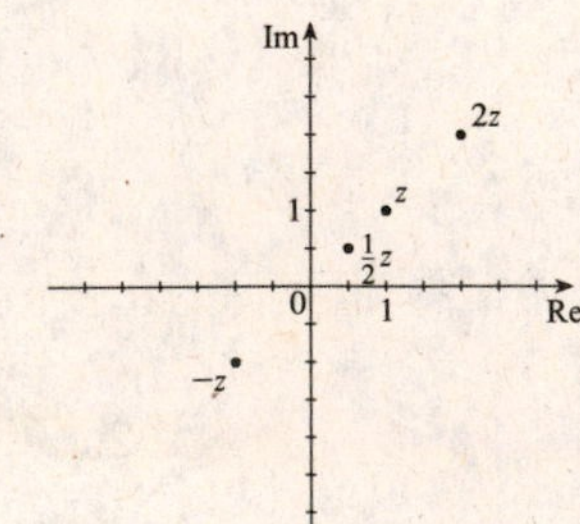

17. $z = 8 + 2i$, $\bar{z} = 8 - 2i$

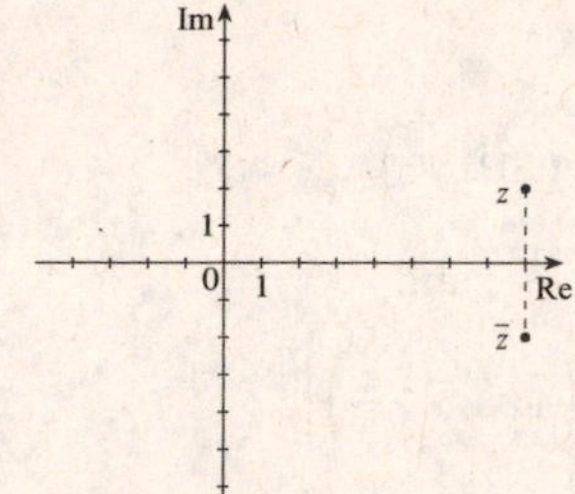

19. $z_1 = 2 - i$, $z_2 = 2 + i$,
$z_1 + z_2 = 2 - i + 2 + i = 4$,
$z_1 z_2 = (2 - i)(2 + i) = 4 - i^2 = 5$

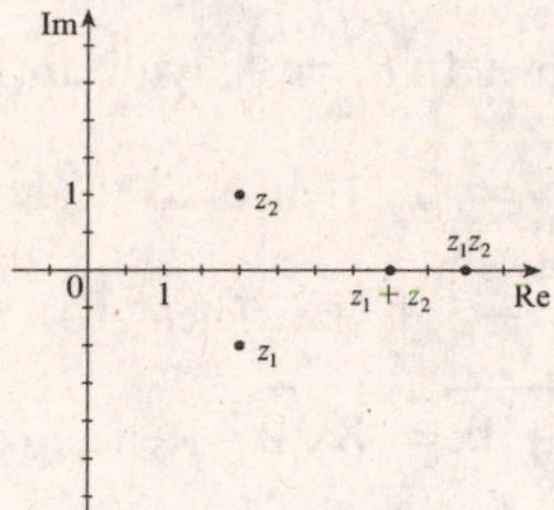

21. $\{z = a + bi \mid a \le 0, b \ge 0\}$

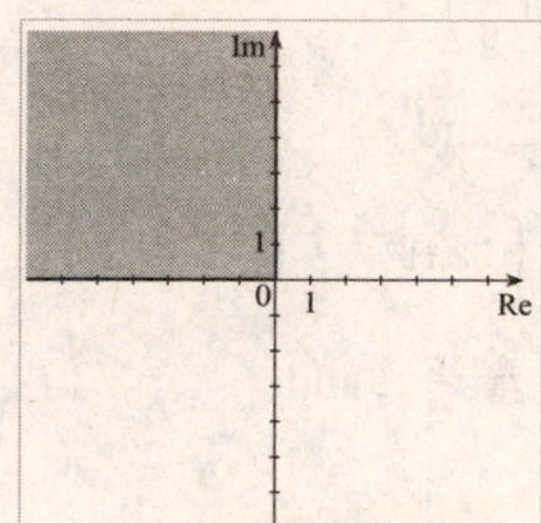

23. $\{z \mid |z| = 3\}$

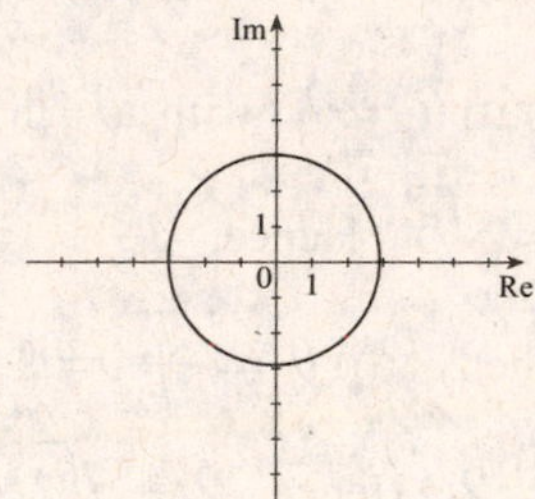

25. $\{z \mid |z| < 2\}$

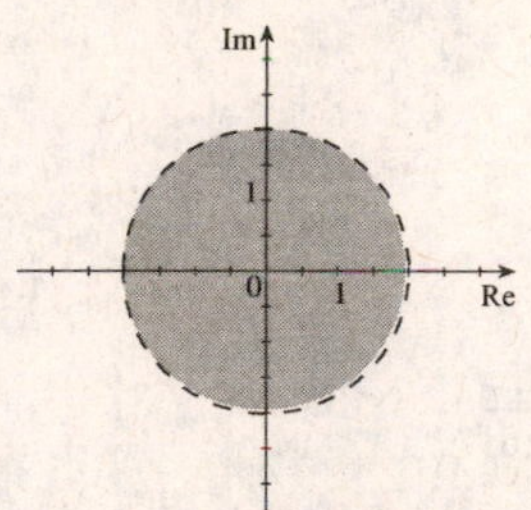

27. $\{z = a + bi \mid a + b < 2\}$

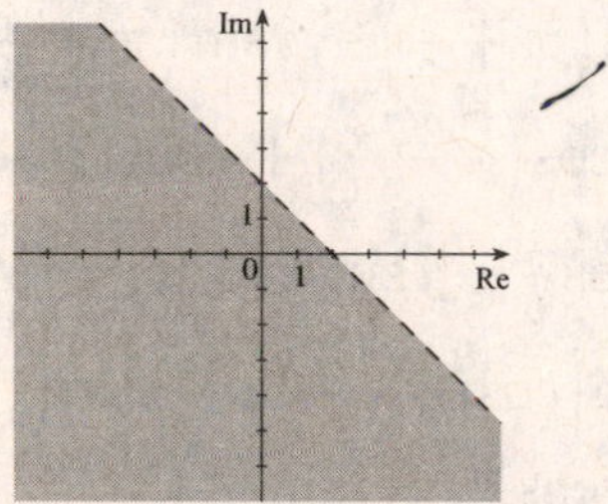

29. $1 + i$. Then $\tan\theta = \frac{1}{1} = 1$ with θ in quadrant I $\Rightarrow$ $\theta = \frac{\pi}{4}$, and $r = \sqrt{1^2 + 1^2} = \sqrt{2}$. Hence, $1 + i = \sqrt{2}\left(\cos\frac{\pi}{4} + i\sin\frac{\pi}{4}\right)$.

31. $\sqrt{2} - \sqrt{2}i$. Then $\tan\theta = \frac{-\sqrt{2}}{\sqrt{2}} = -1$ with θ in quadrant IV $\Rightarrow$ $\theta = \frac{7\pi}{4}$, and $r = \sqrt{2 + 2} = 2$. Hence, $\sqrt{2} - \sqrt{2}i = 2\left(\cos\frac{7\pi}{4} + i\sin\frac{7\pi}{4}\right)$.

33. $2\sqrt{3} - 2i$. Then $\tan\theta = \frac{-2}{2\sqrt{3}} = -\frac{1}{\sqrt{3}}$ with θ in quadrant IV $\Rightarrow$ $\theta = \frac{11\pi}{6}$, and $r = \sqrt{12 + 4} = 4$. Hence, $2\sqrt{3} - 2i = 4\left(\cos\frac{11\pi}{6} + i\sin\frac{11\pi}{6}\right)$.

35. $-3i$. Then $\theta = \frac{3\pi}{2}$, and $r = \sqrt{0 + 9} = 3$. Hence, $3i = 3\left(\cos\frac{3\pi}{2} + i\sin\frac{3\pi}{2}\right)$.

37. $5 + 5i$. Then $\tan\theta = \frac{5}{5} = 1$ with θ in quadrant I $\Rightarrow$ $\theta = \frac{\pi}{4}$, and $r = \sqrt{25 + 25} = 5\sqrt{2}$. Hence, $5 + 5i = 5\sqrt{2}\left(\cos\frac{\pi}{4} + i\sin\frac{\pi}{4}\right)$.

39. $4\sqrt{3}-4i$. Then $\tan\theta=\frac{-4}{4\sqrt{3}}=-\frac{1}{\sqrt{3}}$ with θ in quadrant IV $\Rightarrow$ $\theta=\frac{11\pi}{6}$, and $r=\sqrt{48+16}=8$. Hence, $4\sqrt{3}-4i=8\left(\cos\frac{11\pi}{6}+i\sin\frac{11\pi}{6}\right)$.

41. -20 . Then $\theta=\pi$, and $r=20$. Hence, $-20=20\left(\cos\pi+i\sin\pi\right)$.

43. $3+4i$. Then $\tan\theta=\frac{4}{3}$ with θ in quadrant I $\Rightarrow$ $\theta=\tan^{-1}\frac{4}{3}$, and $r=\sqrt{9+16}=5$. Hence, $3+4i=5\left[\cos\left(\tan^{-1}\frac{4}{3}\right)+i\sin\left(\tan^{-1}\frac{4}{3}\right)\right]$.

45. $3i\left(1+i\right)=-3+3i$. Then $\tan\theta=\frac{3}{-3}=-1$ with θ in quadrant II $\Rightarrow$ $\theta=\frac{3\pi}{4}$, and $r=\sqrt{9+9}=3\sqrt{2}$. Hence, $3i\left(1+i\right)=3\sqrt{2}\left(\cos\frac{3\pi}{4}+i\sin\frac{3\pi}{4}\right)$.

47. $4\left(\sqrt{3}+i\right)=4\sqrt{3}+4i$. Then $\tan\theta=\frac{4}{4\sqrt{3}}=\frac{1}{\sqrt{3}}$ with θ in quadrant I $\Rightarrow$ $\theta=\frac{\pi}{6}$, and $r=\sqrt{48+16}=8$. Hence, $4\left(\sqrt{3}+i\right)=8\left(\cos\frac{\pi}{6}+i\sin\frac{\pi}{6}\right)$.

49. $2+i$. Then $\tan\theta=\frac{1}{2}$ with θ in quadrant I $\Rightarrow$ $\theta=\tan^{-1}\frac{1}{2}$, and $r=\sqrt{4+1}=\sqrt{5}$. Hence, $2+i=\sqrt{5}\left[\cos\left(\tan^{-1}\frac{1}{2}\right)+i\sin\left(\tan^{-1}\frac{1}{2}\right)\right]$.

51. $\sqrt{2}+\sqrt{2}i$. Then $\tan\theta=\frac{\sqrt{2}}{\sqrt{2}}=1$ with θ in quadrant I $\Rightarrow$ $\theta=\frac{\pi}{4}$, and $r=\sqrt{2+2}=2$. Hence, $2+\sqrt{2}i=2\left(\cos\frac{\pi}{4}+i\sin\frac{\pi}{4}\right)$.

53. $z_1=\cos\pi+i\sin\pi$, $z_2=\cos\frac{\pi}{3}+i\sin\frac{\pi}{3}$,

$z_1z_2=\cos\left(\pi+\frac{\pi}{3}\right)+i\sin\left(\pi+\frac{\pi}{3}\right)=\cos\frac{4\pi}{3}+i\sin\frac{4\pi}{3}$,

$z_1/z_2=\cos\left(\pi-\frac{\pi}{3}\right)+i\sin\left(\pi-\frac{\pi}{3}\right)=\cos\frac{2\pi}{3}+i\sin\frac{2\pi}{3}$

55. $z_1=3\left(\cos\frac{\pi}{6}+i\sin\frac{\pi}{6}\right)$, $z_2=5\left(\cos\frac{4\pi}{3}+i\sin\frac{4\pi}{3}\right)$,

$$\begin{aligned}z_1z_2&=3\cdot5\left[\cos\left(\frac{\pi}{6}+\frac{4\pi}{3}\right)+i\sin\left(\frac{\pi}{6}+\frac{4\pi}{3}\right)\right]=15\left(\cos\frac{9\pi}{6}+i\sin\frac{9\pi}{6}\right)\\&=15\left(\cos\frac{3\pi}{2}+i\sin\frac{3\pi}{2}\right)\end{aligned},$$

$$\begin{aligned}z_1/z_2&=\frac{3}{5}\left[\cos\left(\frac{\pi}{6}-\frac{4\pi}{3}\right)+i\sin\left(\frac{\pi}{6}-\frac{4\pi}{3}\right)\right]=\frac{3}{5}\left[\cos\left(-\frac{7\pi}{6}\right)+i\sin\left(-\frac{7\pi}{6}\right)\right]\\&=\frac{3}{5}\left[\cos\left(\frac{7\pi}{6}\right)-i\sin\left(\frac{7\pi}{6}\right)\right]\end{aligned}$$

57. $z_1=4\left(\cos120^\circ+i\sin120^\circ\right)$, $z_2=2\left(\cos30^\circ+i\sin30^\circ\right)$,

$z_1z_2=4\cdot2\left[\cos\left(120^\circ+30^\circ\right)+i\sin\left(120^\circ+30^\circ\right)\right]=8\left(\cos150^\circ+i\sin150^\circ\right)$,

$z_1/z_2=\frac{4}{2}\left[\cos\left(120^\circ-30^\circ\right)+i\sin\left(120^\circ-30^\circ\right)\right]=2\left(\cos90^\circ+i\sin90^\circ\right)$

59. $z_1=4\left(\cos200^\circ+i\sin200^\circ\right)$, $z_2=25\left(\cos150^\circ+i\sin150^\circ\right)$,

$z_1z_2=4\cdot25\left[\cos\left(200^\circ+150^\circ\right)+i\sin\left(200^\circ+150^\circ\right)\right]=100\left(\cos350^\circ+i\sin350^\circ\right)$,

$z_1/z_2=\frac{4}{25}\left[\cos\left(200^\circ-150^\circ\right)+i\sin\left(200^\circ-150^\circ\right)\right]=\frac{4}{25}\left(\cos50^\circ+i\sin50^\circ\right)$

61. $z_1 = \sqrt{3} + i$, so $\tan\theta_1 = \frac{1}{\sqrt{3}}$ with θ_1 in quadrant I $\Rightarrow$ $\theta_1 = \frac{\pi}{6}$, and $r_1 = \sqrt{3+1} = 2$.

$z_2 = 1 + \sqrt{3}i$, so $\tan\theta_2 = \sqrt{3}$ with θ_2 in quadrant I $\Rightarrow$ $\theta_2 = \frac{\pi}{3}$, and $r_1 = \sqrt{1+3} = 2$.

Hence, $z_1 = 2\left(\cos\frac{\pi}{6} + i\sin\frac{\pi}{6}\right)$ and $z_2 = 2\left(\cos\frac{\pi}{3} + i\sin\frac{\pi}{3}\right)$.

Thus, $z_1z_2 = 2\cdot 2\left[\cos\left(\frac{\pi}{6}+\frac{\pi}{3}\right) + i\sin\left(\frac{\pi}{6}+\frac{\pi}{3}\right)\right] = 4\left(\cos\frac{\pi}{2} + i\sin\frac{\pi}{2}\right)$,

$z_1 / z_2 = \frac{2}{2}\left[\cos\left(\frac{\pi}{6}-\frac{\pi}{3}\right) + i\sin\left(\frac{\pi}{6}-\frac{\pi}{3}\right)\right] = \cos\left(-\frac{\pi}{6}\right) + i\sin\left(-\frac{\pi}{6}\right) = \cos\frac{\pi}{6} - i\sin\frac{\pi}{6}$,

and $1 / z_1 = \frac{1}{2}\left[\cos\left(-\frac{\pi}{6}\right) + i\sin\left(-\frac{\pi}{6}\right)\right] = \frac{1}{2}\left(\cos\frac{\pi}{6} - i\sin\frac{\pi}{6}\right)$.

63. $z_1 = 2\sqrt{3} - 2i$, so $\tan\theta_1 = \frac{-2}{2\sqrt{3}} = -\frac{1}{\sqrt{3}}$ with θ_1 in quadrant IV $\Rightarrow$ $\theta_1 = \frac{11\pi}{6}$, and $r_1 = \sqrt{12+4} = 4$.

$z_2 = -1 + i$, so $\tan\theta_2 = -1$ with θ_2 in quadrant II $\Rightarrow$ $\theta_2 = \frac{3\pi}{4}$, and $r_2 = \sqrt{1+1} = \sqrt{2}$.

Hence, $z_1 = 4\left(\cos\frac{11\pi}{6} + i\sin\frac{11\pi}{6}\right)$ and $z_2 = \sqrt{2}\left(\cos\frac{3\pi}{4} + i\sin\frac{3\pi}{4}\right)$.

Thus, $z_1z_2 = 4\cdot\sqrt{2}\left[\cos\left(\frac{11\pi}{6}+\frac{3\pi}{4}\right) + i\sin\left(\frac{11\pi}{6}+\frac{3\pi}{4}\right)\right] = 4\sqrt{2}\left(\cos\frac{7\pi}{12} + i\sin\frac{7\pi}{12}\right)$,

$z_1 / z_2 = \frac{4}{\sqrt{2}}\left[\cos\left(\frac{11\pi}{6}-\frac{3\pi}{4}\right) + i\sin\left(\frac{11\pi}{6}-\frac{3\pi}{4}\right)\right] = 2\sqrt{2}\left(\cos\frac{13\pi}{12} + i\sin\frac{13\pi}{12}\right)$, and

$1 / z_1 = \frac{1}{4}\left(\cos\left(-\frac{11\pi}{6}\right) + i\sin\left(-\frac{11\pi}{6}\right)\right) = \frac{1}{4}\left(\cos\frac{11\pi}{6} - i\sin\frac{11\pi}{6}\right)$.

65. $z_1 = 5 + 5i$, so $\tan\theta_1 = \frac{5}{5} = 1$ with θ_1 in quadrant I $\Rightarrow$ $\theta_1 = \frac{\pi}{4}$, and $r_1 = \sqrt{25+25} = 5\sqrt{2}$.

$z_2 = 4$, so $\theta_2 = 0$, and $r_2 = 4$.

Hence, $z_1 = 5\sqrt{2}\left(\cos\frac{\pi}{4} + i\sin\frac{\pi}{4}\right)$ and $z_2 = 4\left(\cos 0 + i\sin 0\right)$.

Thus, $z_1z_2 = 5\sqrt{2}\cdot 4\left[\cos\left(\frac{\pi}{4}+0\right) + i\sin\left(\frac{\pi}{4}+0\right)\right] = 20\sqrt{2}\left(\cos\frac{\pi}{4} + i\sin\frac{\pi}{4}\right)$,

$z_1 / z_2 = \frac{5\sqrt{2}}{4}\left(\cos\frac{\pi}{4} + i\sin\frac{\pi}{4}\right)$, and

$1 / z_1 = \frac{1}{5\sqrt{2}}\left(\cos\left(-\frac{\pi}{4}\right) + i\sin\left(-\frac{\pi}{4}\right)\right) = \frac{\sqrt{2}}{10}\left(\cos\frac{\pi}{4} - i\sin\frac{\pi}{4}\right)$.

67. $z_1 = -20$, so $\theta_1 = \pi$, and $r_1 = 20$.

$z_2 = \sqrt{3} + i$, so $\tan\theta_2 = \frac{1}{\sqrt{3}}$ with θ_2 in quadrant I $\Rightarrow$ $\theta_2 = \frac{\pi}{6}$, and

$r_2 = \sqrt{3+1} = 2$.

Hence, $z_1 = 20(\cos\pi + i\sin\pi)$ and $z_2 = 2\left(\cos\frac{\pi}{6} + i\sin\frac{\pi}{6}\right)$.

Thus, $z_1 z_2 = 20\cdot 2\left[\cos\left(\pi + \frac{\pi}{6}\right) + i\sin\left(\pi + \frac{\pi}{6}\right)\right] = 40\left(\cos\frac{7\pi}{6} + i\sin\frac{7\pi}{6}\right)$,

$z_1 / z_2 = \frac{20}{2}\left[\cos\left(\pi - \frac{\pi}{6}\right) + i\sin\left(\pi - \frac{\pi}{6}\right)\right] = 10\left(\cos\frac{5\pi}{6} + i\sin\frac{5\pi}{6}\right)$, and

$1 / z_1 = \frac{1}{20}\left[\cos(-\pi) + i\sin(-\pi)\right] = \frac{1}{20}(\cos\pi - i\sin\pi)$.

69. From Exercise 29, $1+i = \sqrt{2}\left(\cos\frac{\pi}{4} + i\sin\frac{\pi}{4}\right)$. Thus,

$$(1+i)^{20} = \left(\sqrt{2}\right)^{20}\left[\cos 20\left(\tfrac{\pi}{4}\right) + i\sin 20\left(\tfrac{\pi}{4}\right)\right] = \left(2^{1/2}\right)^{20}(\cos 5\pi + i\sin 5\pi) = 2^{10}(-1+0i) = -1024$$

71. $r = \sqrt{12+4} = 4$ and $\tan\theta = \frac{2}{2\sqrt{3}} = \frac{1}{\sqrt{3}}$ $\Rightarrow$ $\theta = \frac{\pi}{6}$. Thus,

$2\sqrt{3} + 2i = 4\left(\cos\frac{\pi}{6} + i\sin\frac{\pi}{6}\right)$. So

$\left(2\sqrt{3} + 2i\right)^5 = 4^5\left(\cos\frac{5\pi}{6} + i\sin\frac{5\pi}{6}\right) = 1024\left(-\frac{\sqrt{3}}{2} + \frac{1}{2}i\right) = 512\left(-\sqrt{3} + i\right)$.

73. $r = \sqrt{\frac{1}{2} + \frac{1}{2}} = 1$ and $\tan\theta = 1$ $\Rightarrow$ $\theta = \frac{\pi}{4}$. Thus $\frac{\sqrt{2}}{2} + \frac{\sqrt{2}}{2}i = \cos\frac{\pi}{4} + i\sin\frac{\pi}{4}$.

Therefore, $\left(\frac{\sqrt{2}}{2} + \frac{\sqrt{2}}{2}i\right)^{12} = \cos 12\left(\frac{\pi}{4}\right) + i\sin 12\left(\frac{\pi}{4}\right) = \cos 3\pi + i\sin 3\pi = -1$.

75. $r = \sqrt{4+4} = 4\sqrt{2}$ and $\tan\theta = -1$ with θ in quadrant IV $\Rightarrow$ $\theta = \frac{7\pi}{4}$. Thus

$2 - 2i = 2\sqrt{2}\left(\cos\frac{7\pi}{4} + i\sin\frac{7\pi}{4}\right)$, so

$(2-2i)^8 = \left(2\sqrt{2}\right)^8(\cos 14\pi + i\sin 14\pi) = 4096(1 - 0i) = 4096$.

77. $r = \sqrt{1+1} = \sqrt{2}$ and $\tan\theta = 1$ with θ in quadrant III $\Rightarrow$ $\theta = \frac{5\pi}{4}$. Thus

$-1 - i = \sqrt{2}\left(\cos\frac{5\pi}{4} + i\sin\frac{5\pi}{4}\right)$, so

$$\begin{aligned}(-1-i)^7 &= \left(\sqrt{2}\right)^7\left(\cos\tfrac{35\pi}{4} + i\sin\tfrac{35\pi}{4}\right) = 8\sqrt{2}\left(\cos\tfrac{3\pi}{4} + i\sin\tfrac{3\pi}{4}\right) = 8\sqrt{2}\left(\tfrac{1}{\sqrt{2}} - i\tfrac{1}{\sqrt{2}}\right)\\ &= 8(-1+i)\end{aligned}$$

79. $r = \sqrt{12+4} = 4$ and $\tan\theta = \frac{2}{2\sqrt{3}} = \frac{1}{\sqrt{3}}$ $\Rightarrow$ $\theta = \frac{\pi}{6}$. Thus

$2\sqrt{3} + 2i = 4\left(\cos\frac{\pi}{6} + i\sin\frac{\pi}{6}\right)$, so

$\left(2\sqrt{3} + 2i\right)^{-5} = \left(\frac{1}{4}\right)^5\left(\cos\frac{-5\pi}{6} + i\sin\frac{-5\pi}{6}\right) = \frac{1}{1024}\left(-\frac{\sqrt{3}}{2} - \frac{1}{2}i\right) = \frac{1}{2048}\left(-\sqrt{3} - i\right)$

81. $r = \sqrt{48+16} = 8$ and $\tan\theta = \frac{4}{4\sqrt{3}} = \frac{1}{\sqrt{3}} \Rightarrow \theta = \frac{\pi}{6}$. Thus

$4\sqrt{3} + 4i = 8\left(\cos\frac{\pi}{6} + i\sin\frac{\pi}{6}\right)$. So,

$\left(4\sqrt{3}+4i\right)^{1/2} = \sqrt{8}\left[\cos\left(\dfrac{\pi/6+2k\pi}{2}\right) + i\sin\left(\dfrac{\pi/6+2k\pi}{2}\right)\right]$ for $k = 0, 1$. Thus

the two roots are $w_0 = 2\sqrt{2}\left(\cos\frac{\pi}{12} + i\sin\frac{\pi}{12}\right)$ and $w_1 = 2\sqrt{2}\left(\cos\frac{13\pi}{12} + i\sin\frac{13\pi}{12}\right)$.

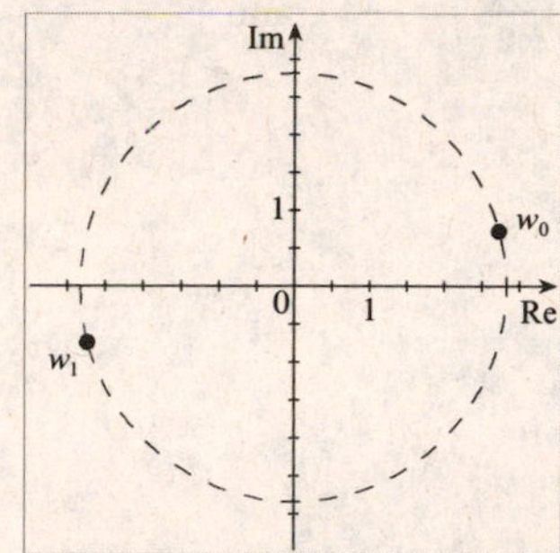

83. $-81i = 81\left(\cos\frac{3\pi}{2} + i\sin\frac{3\pi}{2}\right)$. Thus,

$\left(-81i\right)^{1/4} = 81^{1/4}\left[\cos\left(\dfrac{3\pi/2+2k\pi}{4}\right) + i\sin\left(\dfrac{3\pi/2+2k\pi}{4}\right)\right]$ for $k = 0, 1, 2,$

3. The four roots are $w_0 = 3\left(\cos\frac{3\pi}{8} + i\sin\frac{3\pi}{8}\right)$, $w_1 = 3\left(\cos\frac{7\pi}{8} + i\sin\frac{7\pi}{8}\right)$,

$w_2 = 3\left(\cos\frac{11\pi}{8} + i\sin\frac{11\pi}{8}\right)$, and $w_3 = 3\left(\cos\frac{15\pi}{8} + i\sin\frac{15\pi}{8}\right)$.

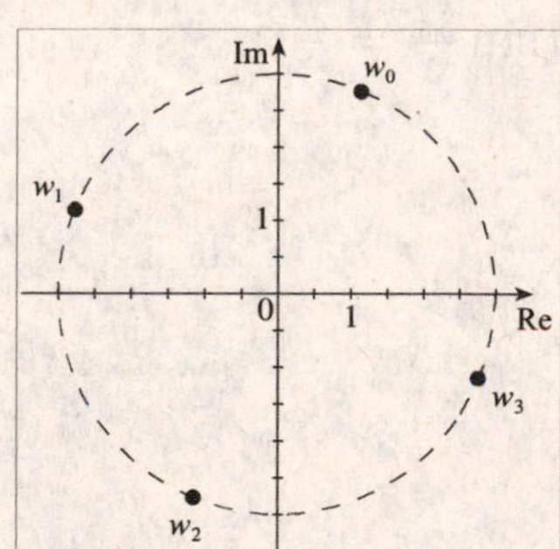

85. $1 = \cos 0 + i \sin 0$. Thus, $1^{1/8} = \cos \dfrac{2k\pi}{8} + i \sin \dfrac{2k\pi}{8}$, for $k = 0$, 1 , 2 , 3 , 4 , 5 , 6 , 7 . So the eight roots are $w_0 = \cos 0 + i \sin 0 = 1$,
$w_1 = \cos \frac{\pi}{4} + i \sin \frac{\pi}{4} = \frac{\sqrt{2}}{2} + i \frac{\sqrt{2}}{2}$, $w_2 = \cos \frac{\pi}{2} + i \sin \frac{\pi}{2} = i$,
$w_3 = \cos \frac{3\pi}{4} + i \sin \frac{3\pi}{4} = -\frac{\sqrt{2}}{2} + i \frac{\sqrt{2}}{2}$, $w_4 = \cos \pi + i \sin \pi = -1$,
$w_5 = \cos \frac{5\pi}{4} + i \sin \frac{5\pi}{4} = -\frac{\sqrt{2}}{2} - i \frac{\sqrt{2}}{2}$, $w_6 = \cos \frac{3\pi}{2} + i \sin \frac{3\pi}{2} = -i$, and
$w_7 = \cos \frac{7\pi}{4} + i \sin \frac{7\pi}{4} = \frac{\sqrt{2}}{2} - i \frac{\sqrt{2}}{2}$.

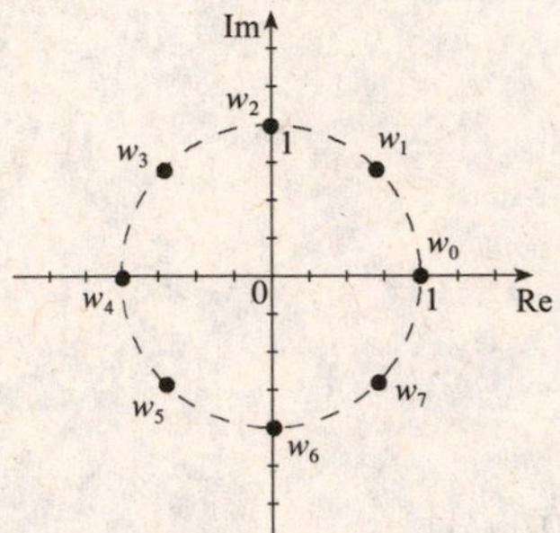

87. $i = \cos \frac{\pi}{2} + i \sin \frac{\pi}{2}$, so $i^{1/3} = \cos \left(\dfrac{\pi / 2 + 2k\pi}{3} \right) + i \sin \left(\dfrac{\pi / 2 + 2k\pi}{3} \right)$ for $k = 0$, 1 , 2 . Thus the three roots are $w_0 = \cos \frac{\pi}{6} + i \sin \frac{\pi}{6} = \frac{\sqrt{3}}{2} + \frac{1}{2} i$,
$w_1 = \cos \frac{5\pi}{6} + i \sin \frac{5\pi}{6} = -\frac{\sqrt{3}}{2} + \frac{1}{2} i$, and $w_2 = \cos \frac{3\pi}{2} + i \sin \frac{3\pi}{2} = -i$.

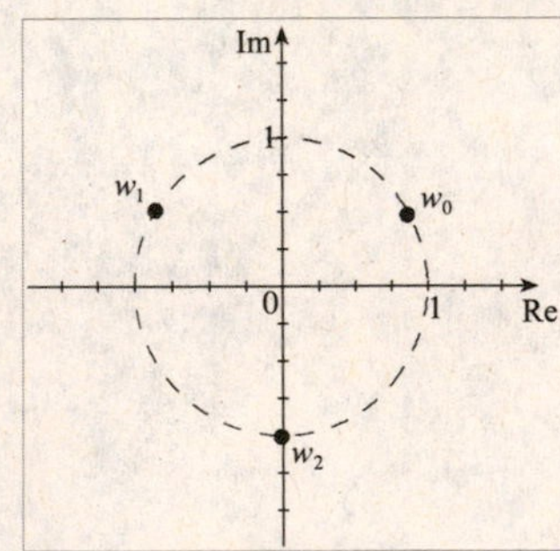

89. $-1 = \cos\pi + i\sin\pi$. Then $\left(-1\right)^{1/4} = \cos\left(\dfrac{\pi + 2k\pi}{4}\right) + i\sin\left(\dfrac{\pi + 2k\pi}{4}\right)$ for $k = 0$, 1 , 2 , 3 . So the four roots are $w_0 = \cos\frac{\pi}{4} + i\sin\frac{\pi}{4} = \frac{\sqrt{2}}{2} + i\frac{\sqrt{2}}{2}$,
$w_1 = \cos\frac{3\pi}{4} + i\sin\frac{3\pi}{4} = -\frac{\sqrt{2}}{2} + i\frac{\sqrt{2}}{2}$, $w_2 = \cos\frac{5\pi}{4} + i\sin\frac{5\pi}{4} = -\frac{\sqrt{2}}{2} - i\frac{\sqrt{2}}{2}$, and
$w_3 = \cos\frac{7\pi}{4} + i\sin\frac{7\pi}{4} = \frac{\sqrt{2}}{2} - i\frac{\sqrt{2}}{2}$.

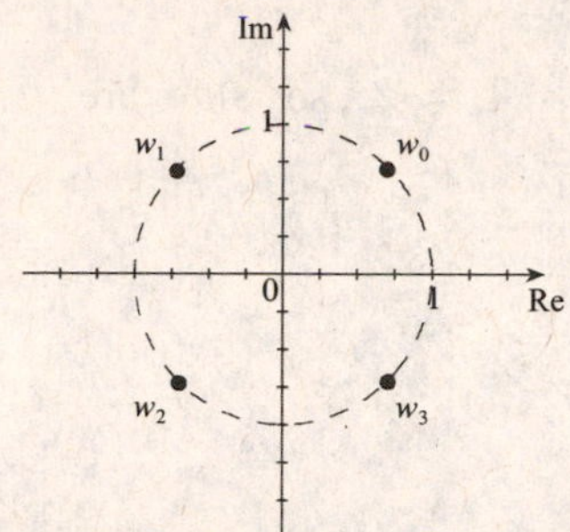

91. $z^4 + 1 = 0 \;\Leftrightarrow\; z = \left(-1\right)^{1/4} = \frac{\sqrt{2}}{2}\left(\pm 1 \pm i\right)$ (from Exercise 85)

93. $z^3 - 4\sqrt{3} - 4i = 0 \;\Leftrightarrow\; z = \left(4\sqrt{3} + 4i\right)^{1/3}$. Since
$4\sqrt{3} + 4i = 8\left(\cos\frac{\pi}{6} + i\sin\frac{\pi}{6}\right)$,
$\left(4\sqrt{3} + 4i\right)^{1/3} = 8^{1/3}\left[\cos\left(\dfrac{\pi/6 + 2k\pi}{3}\right) + i\sin\left(\dfrac{\pi/6 + 2k\pi}{3}\right)\right]$, for $k = 0$, 1 , 2
. Thus the three roots are $z = 2\left(\cos\frac{\pi}{18} + i\sin\frac{\pi}{18}\right)$, $z = 2\left(\cos\frac{13\pi}{18} + i\sin\frac{13\pi}{8}\right)$, and
$z = 2\left(\cos\frac{25\pi}{18} + i\sin\frac{25\pi}{18}\right)$.

95. $z^3 + 1 = -i \;\Rightarrow\; z = \left(-1 - i\right)^{1/3}$. Since $-1 - i = \sqrt{2}\left(\cos\frac{5\pi}{4} + i\sin\frac{5\pi}{4}\right)$,
$z = \left(-1 - i\right)^{1/3} = 2^{1/6}\left[\cos\left(\dfrac{5\pi/4 + 2k\pi}{3}\right) + i\sin\left(\dfrac{5\pi/4 + 2k\pi}{3}\right)\right]$ for $k = 0$, 1 , 2 . Thus the three solutions to this equation are $z = 2^{1/6}\left(\cos\frac{5\pi}{12} + i\sin\frac{5\pi}{12}\right)$,
$2^{1/6}\left(\cos\frac{13\pi}{12} + i\sin\frac{13\pi}{12}\right)$, and $2^{1/6}\left(\cos\frac{21\pi}{12} + i\sin\frac{21\pi}{12}\right)$.

97. (a) $w = \cos\frac{2\pi}{n} + i\sin\frac{2\pi}{n}$ for a positive integer n. Then, $w^k = \cos\frac{2k\pi}{n} + i\sin\frac{2k\pi}{n}$.

Now $w^0 = \cos 0 + i\sin 0 = 1$ and for $k \neq 0, \left(w^k\right)^n = \cos 2k\pi + i\sin 2k\pi = 1$. So the n th roots of 1 are $\cos\frac{2k\pi}{n} + i\sin\frac{2k\pi}{n} = w^k$ for $k = 0,1,2,\ldots,n-1$. In other words, the n th roots of 1 are $w^0, w^1, w^2, w^3, \ldots, w^{n-1}$ or $1, w, w^2, w^3, \ldots, w^{n-1}$.

(b) For $k = 0,1,\ldots,n-1$, we have $\left(sw^k\right)^n = s^n\left(w^k\right)^n = z \cdot 1 = z$, so sw^k are n th roots of z for $k = 0,1,\ldots,n-1$.

99. The cube roots of 1 are $w^0 = 1$, $w^1 = \cos\frac{2\pi}{3} + i\sin\frac{2\pi}{3}$, and $w^2 = \cos\frac{4\pi}{3} + i\sin\frac{4\pi}{3}$, so their product is

$w^0 \cdot w^1 \cdot w^2 = (1)\left(\cos\frac{2\pi}{3} + i\sin\frac{2\pi}{3}\right)\left(\cos\frac{4\pi}{3} + i\sin\frac{4\pi}{3}\right) = \cos 2\pi + i\sin 2\pi = 1$.

The fourth roots of 1 are $w^0 = 1$, $w^1 = i$, $w^2 = -1$, and $w^3 = -i$, so their product is $w^0 \cdot w^1 \cdot w^2 \cdot w^3 = (1)\cdot(i)\cdot(-1)\cdot(-i) = i^2 = -1$.

The fifth roots of 1 are $w^0 = 1$, $w^1 = \cos\frac{2\pi}{5} + i\sin\frac{2\pi}{5}$, $w^2 = \cos\frac{4\pi}{5} + i\sin\frac{4\pi}{5}$, $w^3 = \cos\frac{6\pi}{5} + i\sin\frac{6\pi}{5}$, and $w^4 = \cos\frac{8\pi}{5} + i\sin\frac{8\pi}{5}$, so their product is

$$1\left(\cos\frac{2\pi}{5} + i\sin\frac{2\pi}{5}\right)\left(\cos\frac{4\pi}{5} + i\sin\frac{4\pi}{5}\right)\left(\cos\frac{6\pi}{5} + i\sin\frac{6\pi}{5}\right)\left(\cos\frac{8\pi}{5} + i\sin\frac{8\pi}{5}\right) = \cos 4\pi + i\sin 4\pi = 1$$

The sixth roots of 1 are $w^0 = 1$, $w^1 = \cos\frac{\pi}{3} + i\sin\frac{\pi}{3}$, $w^2 = \cos\frac{2\pi}{3} + i\sin\frac{2\pi}{3} = -\frac{1}{2} + \frac{\sqrt{3}}{2}i$, $w^3 = -1$, $w^4 = \cos\frac{4\pi}{3} + i\sin\frac{4\pi}{3} = -\frac{1}{2} - \frac{\sqrt{3}}{2}i$, and $w^5 = \cos\frac{5\pi}{3} + i\sin\frac{5\pi}{3} = \frac{1}{2} - \frac{\sqrt{3}}{2}i$, so their product is

$$1\left(\cos\frac{\pi}{3} + i\sin\frac{\pi}{3}\right)\left(\cos\frac{2\pi}{3} + i\sin\frac{2\pi}{3}\right)(-1)\left(\cos\frac{4\pi}{3} + i\sin\frac{4\pi}{3}\right)\left(\cos\frac{5\pi}{3} + i\sin\frac{5\pi}{3}\right) = \cos 5\pi + i\sin 5\pi = -1$$

The eight roots of 1 are $w^0 = 1$, $w^1 = \cos\frac{\pi}{4} + i\sin\frac{\pi}{4}$, $w^2 = i$, $w^3 = \cos\frac{3\pi}{4} + i\sin\frac{3\pi}{4}$, $w^4 = -1$, $w^5 = \cos\frac{5\pi}{4} + i\sin\frac{5\pi}{4}$, $w^6 = -i$, $w^7 = \cos\frac{7\pi}{4} + i\sin\frac{7\pi}{4}$, so their product is

$$1\left(\cos\frac{\pi}{4} + i\sin\frac{\pi}{4}\right)i\left(\cos\frac{3\pi}{4} + i\sin\frac{3\pi}{4}\right)(-1)\left(\cos\frac{5\pi}{4} + i\sin\frac{5\pi}{4}\right)(-i)\left(\cos\frac{7\pi}{4} + i\sin\frac{7\pi}{4}\right) = i^2\cdot\left(\cos 2\pi + i\sin 2\pi\right) = -1$$

The product of the n th roots of 1 is -1 if n is even and 1 if n is odd.

The proof requires the fact that the sum of the first m integers is $\dfrac{m(m+1)}{2}$.

Let $w = \cos\dfrac{2\pi}{n} + i\sin\dfrac{2\pi}{n}$. Then $w^k = \cos\dfrac{2k\pi}{n} + i\sin\dfrac{2k\pi}{n}$ for $k = 0,1,2,\ldots,n-1$

The argument of the product of the n roots of unity can be found by adding the arguments of each w^k. So the argument of the product is

$$\theta = 0 + \frac{2(1)\pi}{n} + \frac{2(2)\pi}{n} + \frac{2(3)\pi}{n} + \cdots + \frac{2(n-2)\pi}{n} + \frac{2(n-1)\pi}{n} = \frac{2\pi}{n}\left[0+1+2+3+\cdots+(n-2)+(n-1)\right]$$

. Since this is the sum of the first $n-1$ integers, this sum is $\dfrac{2\pi}{n}\cdot\dfrac{(n-1)n}{2} = (n-1)\pi$. Thus the product of the n roots of unity is $\cos((n-1)\pi) + i\sin((n-1)\pi) = -1$ if n is even and 1 if n is odd.

5.4 Plane Curves and Parametric Equations

1. (a) The parametric equations $x = f(t)$ and $y = g(t)$ give the coordinates of a point $(x, y) = (f(t), g(t))$ for appropriate values of t. The variable t is called a *parameter*.

(b) When $t = 0$ the object is at $(0, 0^2) = (0, 0)$ and when $t = 1$ the object is at $(1, 1^2) = (1, 1)$.

(c) If we eliminate the parameter in part (b) we get the equation $y = x^2$. We see from this equation that the path of the moving object is a *parabola*.

3. (a) $x = 2t$, $y = t + 6$

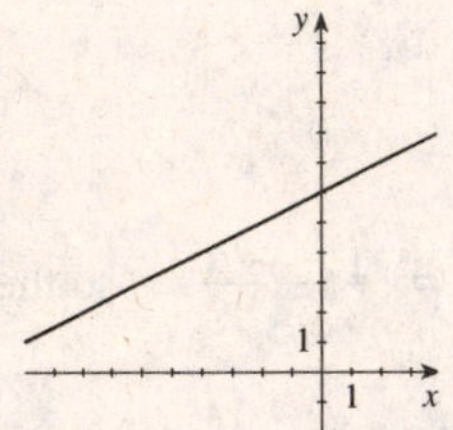

(b) Since $x = 2t$, $t = \dfrac{x}{2}$ and so $y = \dfrac{x}{2} + 6 \Leftrightarrow x - 2y + 12 = 0$.

5. (a) $x = t^2$, $y = t - 2$, $2 \le t \le 4$

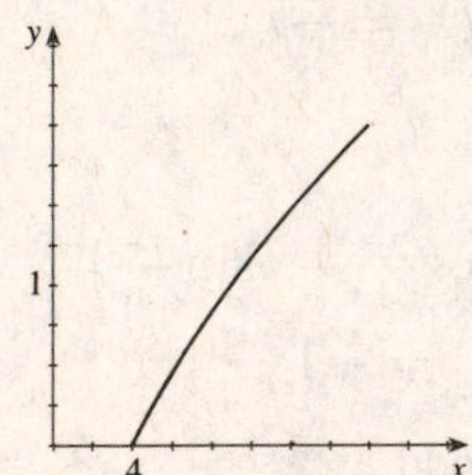

(b) Since $y = t - 2 \Leftrightarrow t = y + 2$, we have $x = t^2 \Leftrightarrow x = (y + 2)^2$, and since $2 \le t \le 4$, we have $4 \le x \le 16$.

7. (a) $x = \sqrt{t}$, $y = 1 - t \Rightarrow t \ge 0$

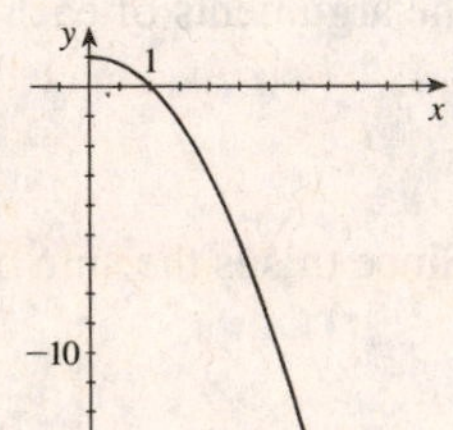

(b) Since $x = \sqrt{t}$, we have $x^2 = t$, and so $y = 1 - x^2$ with $x \ge 0$.

9. (a) $x = \dfrac{1}{t}$, $y = t + 1$

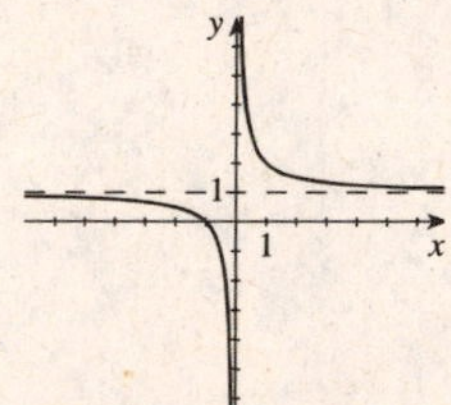

(b) Since $x = \dfrac{1}{t}$ we have $t = \dfrac{1}{x}$ and so $y = \dfrac{1}{x} + 1$.

11. (a) $x = 4t^2$, $y = 8t^3$

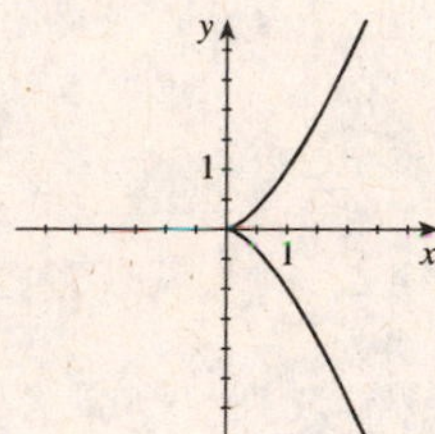

(b) Since $y = 8t^3 \Leftrightarrow y^2 = 64t^6 = \left(4t^2\right)^3 = x^3$, we have $y^2 = x^3$.

13. (a) $x = 2\sin t$, $y = 2\cos t$, $0 \le t \le \pi$

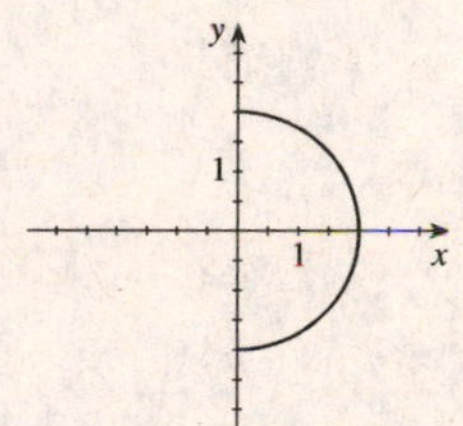

(b) $x^2 = \left(2\sin t\right)^2 = 4\sin^2 t$ and $y^2 = 4\cos^2 t$. Hence,

$x^2 + y^2 = 4\sin^2 t + 4\cos^2 t = 4 \Leftrightarrow x^2 + y^2 = 4$, where $x \ge 0$.

15. (a) $x = \sin^2 t$, $y = \sin^4 t$

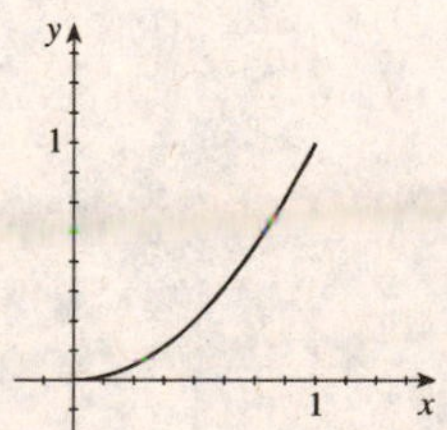

(b) Since $x = \sin^2 t$ we have $x^2 = \sin^4 t$ and so $y = x^2$. But since $0 \le \sin^2 t \le 1$ we only get the part of this parabola for which $0 \le x \le 1$.

17. (a) $x = \cos t$, $y = \cos 2t$

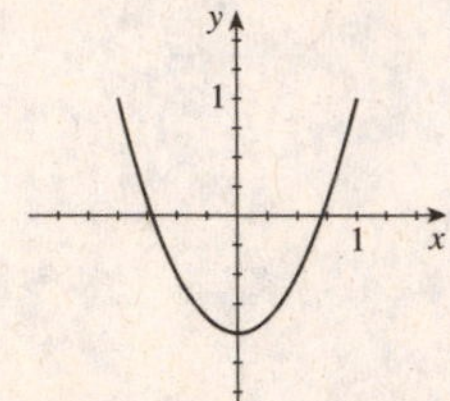

(b) Since $x = \cos t$ we have $x^2 = \cos^2 t$, so $2x^2 - 1 = 2\cos^2 t - 1 = \cos 2t = y$. Hence, the rectangular equation is $y = 2x^2 - 1$, $-1 \le x \le 1$.

19. (a) $x = \sec t$, $y = \tan t$, $0 \le t < \frac{\pi}{2}$ $\Rightarrow x \ge 1$ and $y \ge 0$.

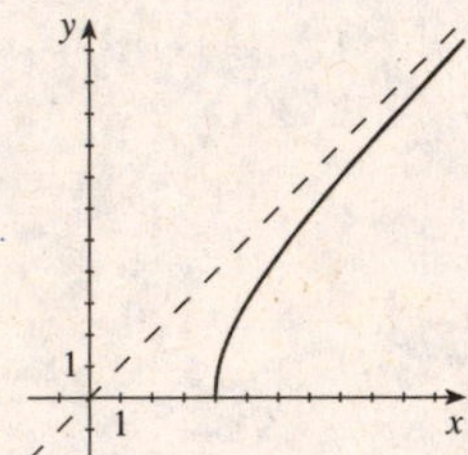

(b) $x^2 = \sec^2 t$, $y^2 = \tan^2 t$, and $y^2 + 1 = \tan^2 t + 1 = \sec^2 t = x^2$. Therefore, $y^2 + 1 = x^2 \Leftrightarrow x^2 - y^2 = 1$, $x \ge 1$, $y \ge 0$.

21. (a) $x = e^t$, $y = e^{-t}$ $\Rightarrow x > 0, y > 0$.

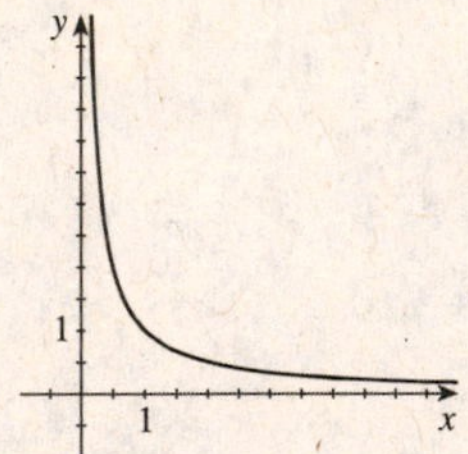

(b) $xy = e^t \cdot e^{-t} = e^0 = 1$. Hence, the equation is $xy = 1$, with $x > 0$, $y > 0$.

23. (a) $x = \cos^2 t$, $y = \sin^2 t$

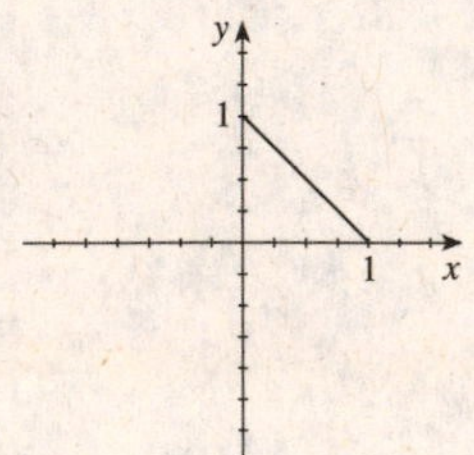

(b) $x + y = \cos^2 t + \sin^2 t = 1$. Hence, the equation is $x + y = 1$ with $0 \le x$, $y \le 1$

25. $x = 3\cos t$, $y = 3\sin t$. The radius of the circle is 3 , the position at time 0 is $\left(x(0), y(0)\right) = \left(3\cos 0, 3\sin 0\right) = \left(3, 0\right)$ and the orientation is counterclockwise (because x is decreasing and y is increasing initially). $\left(x, y\right) = \left(3, 0\right)$ again when $t = 2\pi$, so it takes 2π units of time to complete one revolution.

27. $x = \sin 2t$, $y = \cos 2t$. The radius of the circle is 1 , the position at time 0 is $\left(x(0), y(0)\right) = \left(\sin 0, \cos 0\right) = \left(0, 1\right)$ and the orientation is clockwise (because x is increasing and y is decreasing initially). $\left(x, y\right) = \left(0, 1\right)$ again when $t = \pi$, so it takes π units of time to complete one revolution.

29. Since the line passes through the point $\left(4, -1\right)$ and has slope $\frac{1}{2}$, parametric equations for the line are $x = 4 + t$, $y = -1 + \frac{1}{2}t$.

31. Since the line passes through the points $\left(6, 7\right)$ and $\left(7, 8\right)$, its slope is $\dfrac{8-7}{7-6} = 1$. Thus, parametric equations for the line are $x = 6 + t$, $y = 7 + t$.

33. Since $\cos^2 t + \sin^2 t = 1$, we have $a^2\cos^2 t + a^2\sin^2 t = a^2$. If we let $x = a\cos t$ and $y = a\sin t$, then $x^2 + y^2 = a^2$. Hence, parametric equations for the circle are $x = a\cos t$, $y = a\sin t$.

35. $x = a\tan\theta \quad\Leftrightarrow\quad \tan\theta = \dfrac{x}{a} \quad\Rightarrow \tan^2\theta = \dfrac{x^2}{a^2}$. Also, $y = b\sec\theta \quad\Leftrightarrow\quad \sec\theta = \dfrac{y}{b}$ $\Rightarrow \sec^2\theta = \dfrac{y^2}{b^2}$. Since $\tan^2\theta = \sec^2\theta - 1$, we have $\dfrac{x^2}{a^2} = \dfrac{y^2}{b^2} - 1 \quad\Leftrightarrow$ $\dfrac{y^2}{b^2} - \dfrac{x^2}{a^2} = 1$, which is the equation of a hyperbola.

36. Substituting the given values for x and y into the equation we derived in Exercise 29, we get $\dfrac{\left(b\sqrt{t+1}\right)^2}{b^2} - \dfrac{\left(a\sqrt{t}\right)^2}{a^2} = \left(t+1\right) - t = 1$. Thus the points on this curve satisfy the equation, which is that of a hyperbola. However, this hyperbola is only the part of $\dfrac{y^2}{b^2} - \dfrac{x^2}{a^2} = 1$ for which $x \geq 0$ and $y \geq 0$.

37. $x = t\cos t$, $y = t\sin t$, $t \geq 0$

t	x	y
0	0	0
$\frac{\pi}{4}$	$\frac{\pi\sqrt{2}}{8}$	$\frac{\pi\sqrt{2}}{8}$
$\frac{\pi}{2}$	0	$\frac{\pi}{2}$
$\frac{3\pi}{4}$	$-\frac{3\pi\sqrt{2}}{8}$	$\frac{3\pi\sqrt{2}}{8}$
π	$-\pi$	0

t	x	y
$\frac{5\pi}{4}$	$-\frac{5\pi\sqrt{2}}{8}$	$-\frac{5\pi\sqrt{2}}{8}$
$\frac{3\pi}{2}$	0	$-\frac{3\pi}{2}$
$\frac{7\pi}{4}$	$\frac{7\pi\sqrt{2}}{8}$	$-\frac{7\pi\sqrt{2}}{8}$
2π	2π	0

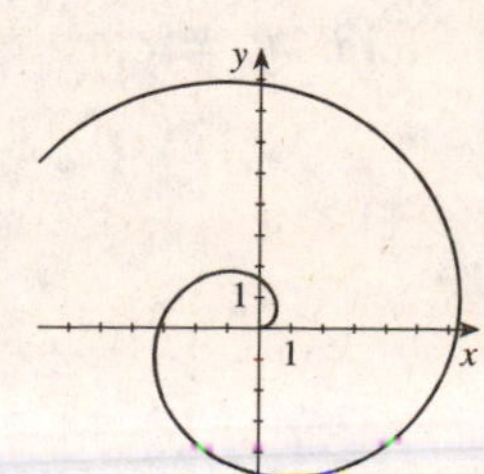

39. $x = \dfrac{3t}{1+t^3}$, $y = \dfrac{3t^2}{1+t^3}$, $t \neq -1$

t	x	y
−0.9	−9.96	8.97
−0.75	−3.89	2.92
−0.5	−1.71	0.86
0	0	0
0.5	1.33	0.67
1	1.5	1.5
1.5	1.03	1.54

t	x	y
2	0.67	1.33
2.5	0.45	1.13
3	0.32	0.96
4	0.18	0.74
5	0.12	0.60
6	0.08	0.50

t	x	y
−1.1	9.97	−10.97
−1.25	3.93	−4.92
−1.5	1.89	−2.84
−2	0.86	−1.71
−2.5	0.51	−1.28
−3	0.35	−1.04
−3.5	0.25	−0.87

t	x	y
−4	0.19	−0.76
−4.5	0.15	−0.67
−5	0.12	−0.60
−6	0.08	−0.50
−7	0.06	−0.43
−8	0.05	−0.38

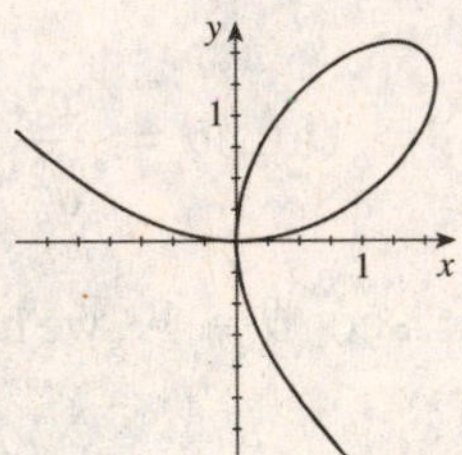

As $t \to -1^-$ we have $x \to \infty$ and $y \to -\infty$. As $t \to -1^+$ we have $x \to -\infty$ and $y \to \infty$. As $t \to \infty$ we have $x \to 0^+$ and $y \to 0^+$. As $t \to -\infty$ we have $x \to 0^+$ and $y \to 0^-$.

41. $x = \left(v_0 \cos\alpha\right)t$, $y = \left(v_0 \sin\alpha\right)t - 16t^2$. From the equation for x , $t = \dfrac{x}{v_0 \cos\alpha}$.

Substituting into the equation for y gives

$y = \left(v_0 \sin\alpha\right)\dfrac{x}{v_0 \cos\alpha} - 16\left(\dfrac{x}{v_0 \cos\alpha}\right)^2 = x\tan\alpha - \dfrac{16x^2}{v_0^2 \cos^2\alpha}$. Thus the equation is of the form $y = c_1 x - c_2 x^2$, where c_1 and c_2 are constants, so its graph is a parabola.

43. $x = \sin t$, $y = 2\cos 3t$

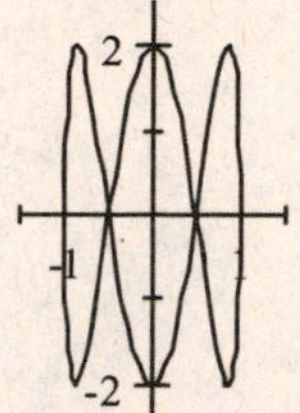

45. $x = 3\sin 5t$, $y = 5\cos 3t$

47. $x = \sin\left(\cos t\right)$, $y = \cos t^{3/2}$, $0 \le t \le 2\pi$

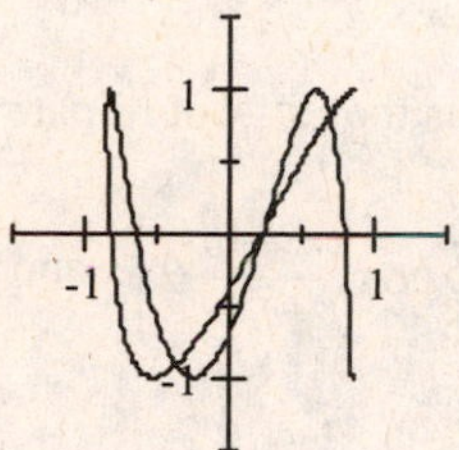

49. (a) $r = 2^{\theta/12}$, $0 \le \theta \le 4\pi$ $\Rightarrow$ $x = 2^{t/12}\cos t$, $y = 2^{t/12}\sin t$

(b)

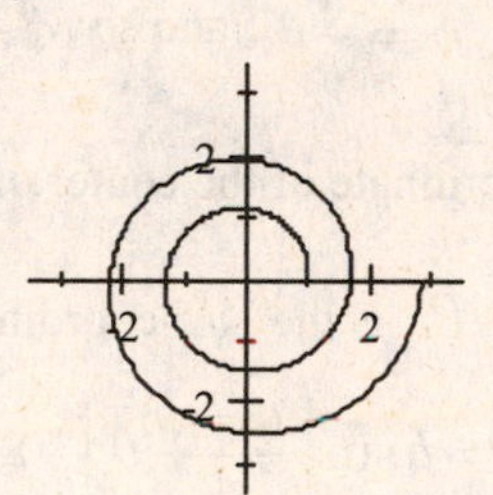

51. (a) $r = \dfrac{4}{2 - \cos\theta}$ $\Leftrightarrow$ $x = \dfrac{4\cos t}{2 - \cos t}$, $y = \dfrac{4\sin t}{2 - \cos t}$

(b)

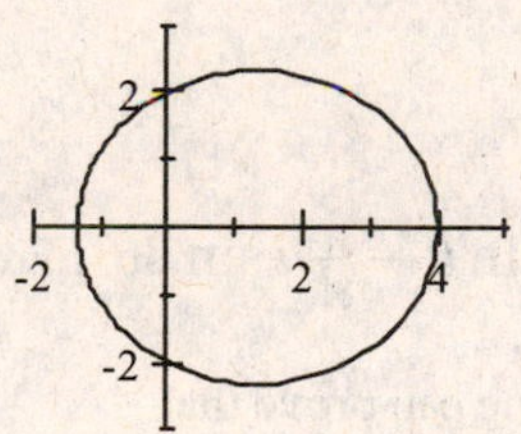

53. $x = t^3 - 2t$, $y = t^2 - t$ is Graph III, since $y = t^2 - t = \left(t^2 - t + \frac{1}{4}\right) - \frac{1}{4} = \left(t - \frac{1}{2}\right)^2 - \frac{1}{4}$, and so $y \ge -\frac{1}{4}$ on this curve, while x is unbounded.

55. $x = t + \sin 2t$, $y = t + \sin 3t$ is Graph II, since the values of x and y oscillate about their values on the line $x = t$, $y = t$ $\Leftrightarrow$ $y = x$.

57. (a) If we modify Figure 8 so that $|PC| = b$, then by the same reasoning as in Example 6, we see that $x = |OT| - |PQ| = a\theta - b\sin\theta$ and $y = |TC| - |CQ| = a - b\cos\theta$. We graph the case where $a = 3$ and $b = 2$.

(b)

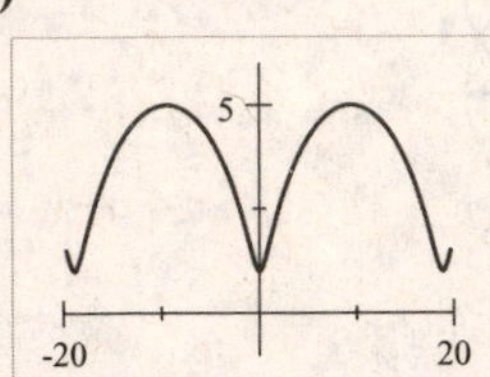

59. (a) We first note that the center of circle C (the small circle) has coordinates $([a-b]\cos\theta, [a-b]\sin\theta)$. Now the arc PQ has the same length as the arc $P'Q$, so

$b\phi = a\theta \Leftrightarrow \phi = \frac{a}{b}\theta$, and so $\phi - \theta = \frac{a}{b}\theta - \theta = \frac{a-b}{b}\theta$. Thus the x -coordinate of P is the x -coordinate of the center of circle C plus $b\cos(\phi - \theta) = b\cos\left(\frac{a-b}{b}\theta\right)$, and the y -coordinate of P is the y -coordinate of the center of circle C minus $b \cdot \sin(\phi - \theta) = b\sin\left(\frac{a-b}{b}\theta\right)$. So $x = (a-b)\cos\theta + b\cos\left(\frac{a-b}{b}\theta\right)$ and $y = (a-b)\sin\theta - b\sin\left(\frac{a-b}{b}\theta\right)$.

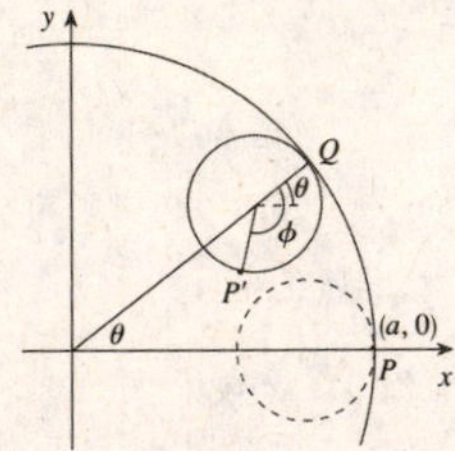

(b) If $a = 4b$, $b = \frac{a}{4}$, and $x = \frac{3}{4}a\cos\theta + \frac{1}{4}a\cos 3\theta$, $y = \frac{3}{4}a\sin\theta - \frac{1}{4}a\sin 3\theta$. From Example 2 in Section 7.3, $\cos 3\theta = 4\cos^3\theta - 3\cos\theta$. Similarly, one can prove that $\sin 3\theta = 3\sin\theta - 4\sin^3\theta$. Substituting, we get

$x = \frac{3}{4}a\cos\theta + \frac{1}{4}a(4\cos^3\theta - 3\cos\theta) = a\cos^3\theta$

$y = \frac{3}{4}a\sin\theta - \frac{1}{4}a(3\sin\theta - 4\sin^3\theta) = a\sin^3\theta$. Thus,

$x^{2/3} + y^{2/3} = a^{2/3}\cos^2\theta + a^{2/3}\sin^2\theta = a^{2/3}$, so $x^{2/3} + y^{2/3} = a^{2/3}$.

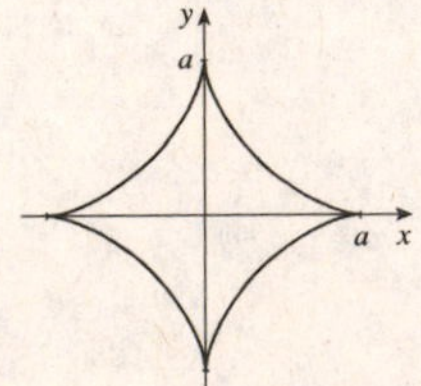

61. A polar equation for the circle is $r = 2a\sin\theta$. Thus the coordinates of Q are $x = r\cos\theta = 2a\sin\theta\cos\theta$ and $y = r\sin\theta = 2a\sin^2\theta$. The coordinates of R are $x = 2a\cot\theta$ and $y = 2a$. Since P is the midpoint of QR , we use the midpoint formula to get $x = a\left(\sin\theta\cos\theta + \cot\theta\right)$ and $y = a\left(1 + \sin^2\theta\right)$.

63. (a) A has coordinates $\left(a\cos\theta, a\sin\theta\right)$. Since OA is perpendicular to AB , ΔOAB is a right triangle and B has coordinates $\left(a\sec\theta, 0\right)$. It follows that P has coordinates $\left(a\sec\theta, b\sin\theta\right)$. Thus, the parametric equations are $x = a\sec\theta$, $y = b\sin\theta$.

(b) The right half of the curve is graphed with $a = 3$ and $b = 2$.

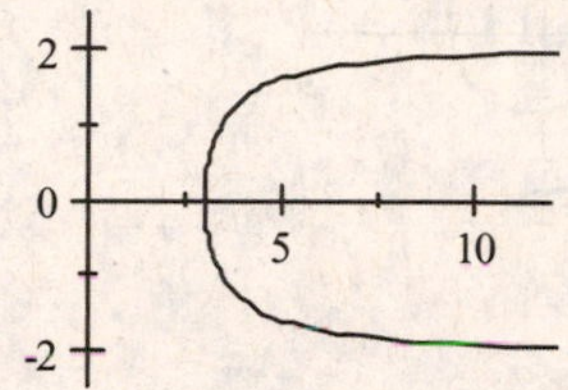

65. We use the equation for y from Example 6 and solve for θ . Thus for $0 \le \theta \le \pi$,

$y = a\left(1 - \cos\theta\right) \Leftrightarrow \dfrac{a-y}{a} = \cos\theta \Leftrightarrow \theta = \cos^{-1}\left(\dfrac{a-y}{a}\right)$. Substituting into the equation for x , we get $x = a\left[\cos^{-1}\left(\dfrac{a-y}{a}\right) - \sin\left(\cos^{-1}\left(\dfrac{a-y}{a}\right)\right)\right]$. However,

$\sin\left(\cos^{-1}\left(\dfrac{a-y}{a}\right)\right) = \sqrt{1 - \left(\dfrac{a-y}{a}\right)^2} = \dfrac{\sqrt{2ay - y^2}}{a}$. Thus,

$x = a\left[\cos^{-1}\left(\dfrac{a-y}{a}\right) - \dfrac{\sqrt{2ay - y^2}}{a}\right]$, and we have $\dfrac{\sqrt{2ay - y^2} + x}{a} = \cos^{-1}\left(\dfrac{a-y}{a}\right)$

$\Rightarrow\ 1 - \dfrac{y}{a} = \cos\left(\dfrac{\sqrt{2ay - y^2} + x}{a}\right) \Rightarrow y = a\left[1 - \cos\left(\dfrac{\sqrt{2ay - y^2} + x}{a}\right)\right]$.

67. (a) In the figure, since OQ and QT are perpendicular and OT and TD are perpendicular, the angles formed by their intersections are equal, that is, $\theta = \angle DTQ$. Now the coordinates of T are $\left(\cos\theta, \sin\theta\right)$. Since $\left|TD\right|$ is the length of the string that has been unwound from the circle, it must also have arc length θ , so $\left|TD\right| = \theta$. Thus the x -displacement from T to D is $\theta \cdot \sin\theta$ while the y -displacement from T to D is $\theta \cdot \cos\theta$. So the coordinates of D are $x = \cos\theta + \theta\sin\theta$ and $y = \sin\theta - \theta\cos\theta$.

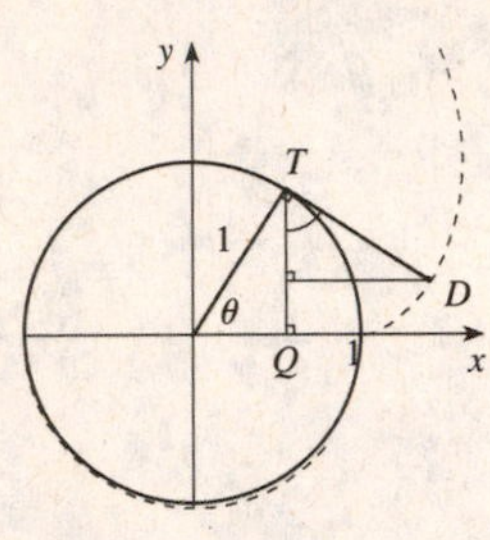

(b)

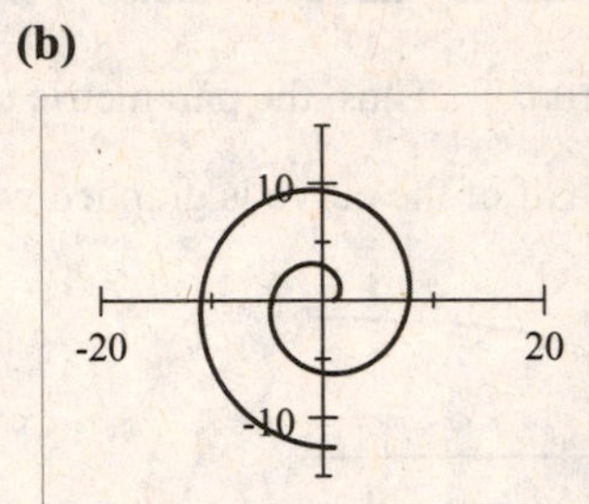

69. $C : x = t$, $y = t^2$; $D : x = \sqrt{t}$, $y = t$, $t \geq 0 E : x = \sin t$, $y = 1 - \cos^2 t F : x = e^t$, $y = e^{2t}$

(a) For C , $x = t$, $y = t^2 \Rightarrow y = x^2$.

For D , $x = \sqrt{t}$, $y = t \Rightarrow y = x^2$.

For E , $x = \sin t \Rightarrow x^2 = \sin^2 t = 1 - \cos^2 t = y$ and so $y = x^2$.

For F , $x = e^t \Rightarrow x^2 = e^{2t} = y$ and so $y = x^2$. Therefore, the points on all four curves satisfy the same rectangular equation.

(b) Curve C is the entire parabola $y = x^2$. Curve D is the right half of the parabola because $t \geq 0$ and so $x \geq 0$. Curve E is the portion of the parabola for $-1 \leq x \leq 1$. Curve F is the portion of the parabola where $x > 0$, since $e^t > 0$ for all t .

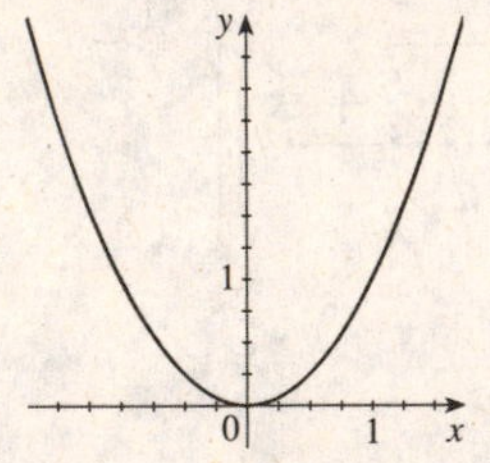

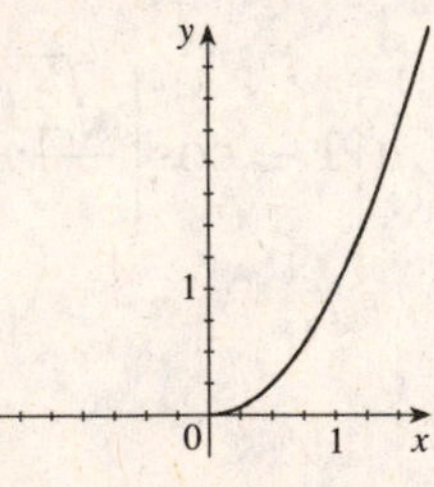

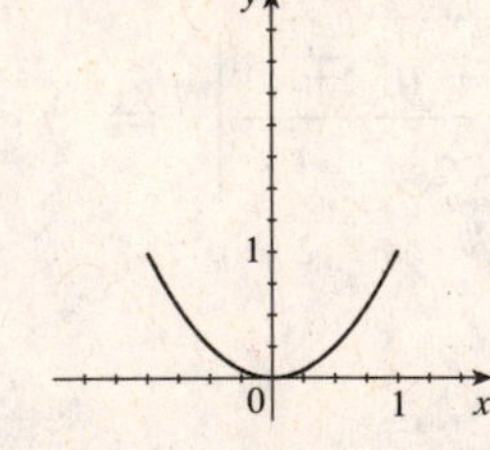

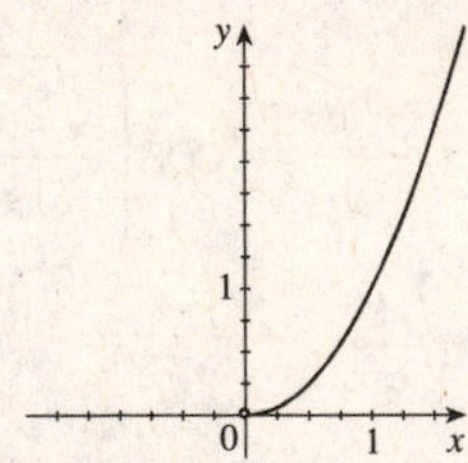

5 Review

1. (a)

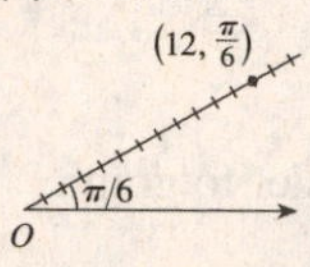

(b) $x = 12\cos\frac{\pi}{6} = 12 \cdot \frac{\sqrt{3}}{2} = 6\sqrt{3}$, $y = 12\sin\frac{\pi}{6} = 12 \cdot \frac{1}{2} = 6$. Thus, the rectangular coordinates of P are $\left(6\sqrt{3}, 6\right)$.

3. (a)

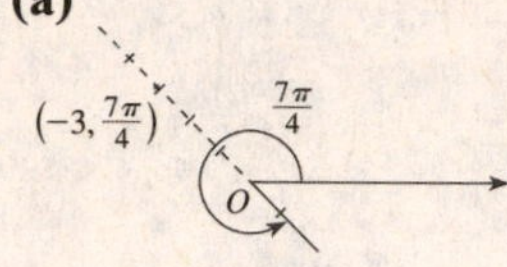

(b) $x = -3\cos\frac{7\pi}{4} = -3\left(\frac{\sqrt{2}}{2}\right) = -\frac{3\sqrt{2}}{2}$, $y = -3\sin\frac{7\pi}{4} = -3\left(-\frac{\sqrt{2}}{2}\right) = \frac{3\sqrt{2}}{2}$. Thus, the rectangular coordinates of P are $\left(-\frac{3\sqrt{2}}{2}, \frac{3\sqrt{2}}{2}\right)$.

5. (a)

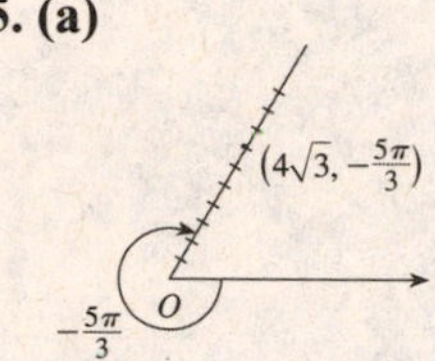

(b)
$x = 4\sqrt{3}\cos\left(-\frac{5\pi}{3}\right) = 4\sqrt{3}\left(\frac{1}{2}\right) = 2\sqrt{3}$, $y = 4\sqrt{3}\sin\left(-\frac{5\pi}{3}\right) = 4\sqrt{3}\left(\frac{\sqrt{3}}{2}\right) = 6$. Thus, the rectangular coordinates of P are $\left(2\sqrt{3}, 6\right)$.

7. (a)

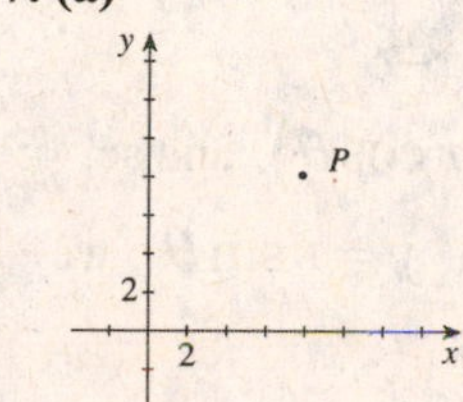

(b) $r = \sqrt{8^2 + 8^2} = \sqrt{128} = 8\sqrt{2}$ and $\overline{\theta} = \tan^{-1}\frac{8}{8}$. Since P is in quadrant I, $\theta = \frac{\pi}{4}$. Polar coordinates for P are $\left(8\sqrt{2}, \frac{\pi}{4}\right)$.

(c) $\left(-8\sqrt{2}, \frac{5\pi}{4}\right)$

9. (a)

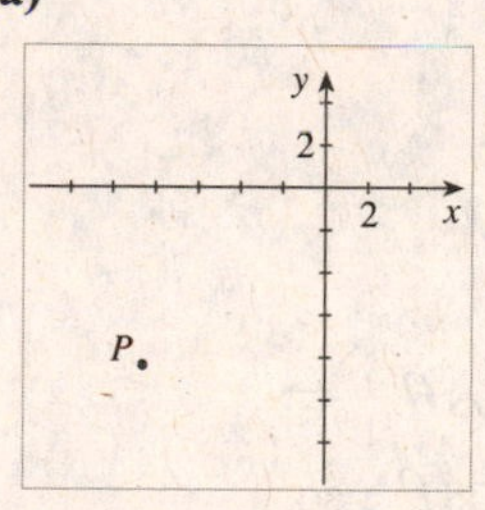

(b) $r = \sqrt{\left(-6\sqrt{2}\right)^2 + \left(-6\sqrt{2}\right)^2}$ and
$= \sqrt{144} = 12$
$\overline{\theta} = \tan^{-1}\frac{-6\sqrt{2}}{-6\sqrt{2}} = \frac{\pi}{4}$. Since P is in quadrant III, $\theta = \frac{5\pi}{4}$. Polar coordinates for P are $\left(12, \frac{5\pi}{4}\right)$.

(c) $\left(-12, \frac{\pi}{4}\right)$

11. (a)

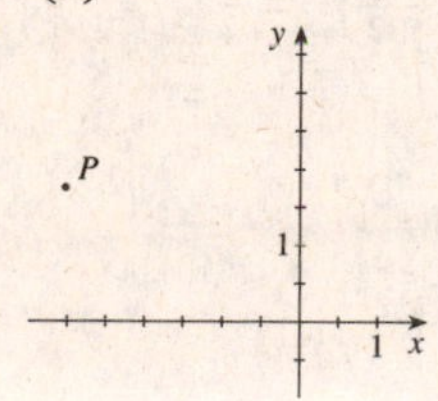

(b) $r = \sqrt{(-3)^2 + \left(\sqrt{3}\right)^2} = \sqrt{12} = 2\sqrt{3}$ and $\overline{\theta} = \tan^{-1}\frac{\sqrt{3}}{-3}$. Since P is in quadrant II, $\theta = \frac{5\pi}{6}$. Polar coordinates for P are $\left(2\sqrt{3}, \frac{5\pi}{6}\right)$.

(c) $\left(-2\sqrt{3}, -\frac{\pi}{6}\right)$

13. (a) $x+y=4 \Leftrightarrow$

$r\cos\theta + r\sin\theta = 4 \Leftrightarrow$

$r(\cos\theta+\sin\theta)=4 \Leftrightarrow$

$r=\dfrac{4}{\cos\theta+\sin\theta}$

(b) The rectangular equation is easier to graph.

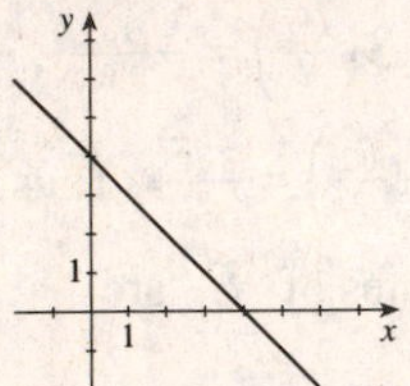

15. (a) $x^2+y^2=4x+4y \Leftrightarrow$

$r^2=4r\cos\theta+4r\sin\theta \Leftrightarrow$

$r^2=r(4\cos\theta+4\sin\theta) \Leftrightarrow$

$r=4\cos\theta+4\sin\theta$

(b) The polar equation is easier to graph.

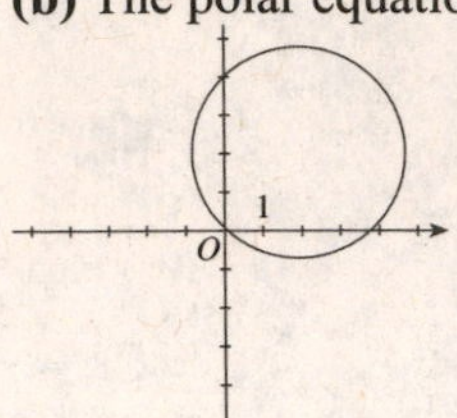

17. (a)

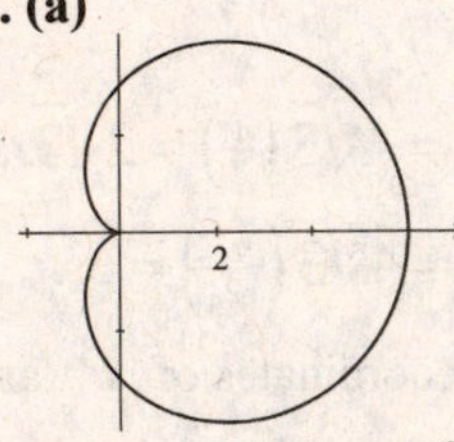

(b) $r=3+3\cos\theta \Leftrightarrow$

$r^2=3r+3r\cos\theta$, which gives

$x^2+y^2=3\sqrt{x^2+y^2}+3x \Leftrightarrow$

$x^2-3x+y^2=3\sqrt{x^2+y^2}$. Squaring both sides gives

$\left(x^2-3x+y^2\right)^2=9\left(x^2+y^2\right)$.

19. (a)

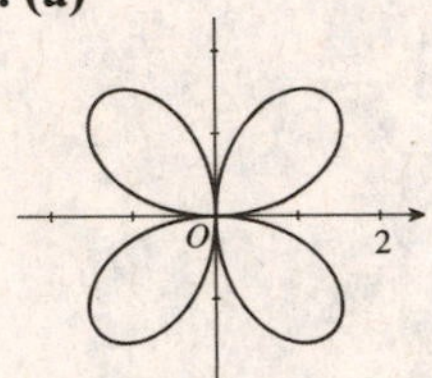

(b) $r=2\sin 2\theta \Leftrightarrow$

$r=2\cdot 2\sin\theta\cos\theta \Leftrightarrow$

$r^3=4r^2\sin\theta\cos\theta \Leftrightarrow$

$\left(r^2\right)^{3/2}=4(r\sin\theta)(r\cos\theta)$ and so, since $x=r\cos\theta$ and $y=r\sin\theta$, we get $\left(x^2+y^2\right)^3=16x^2y^2$.

21. (a)

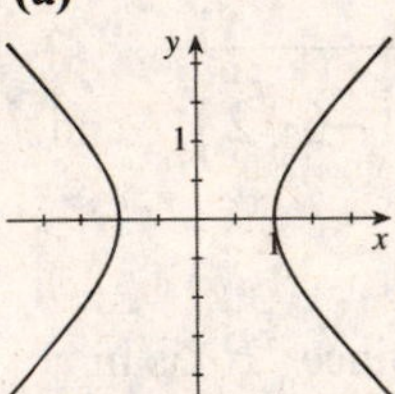

(b)

$r^2=\sec 2\theta=\dfrac{1}{\cos 2\theta}=\dfrac{1}{\cos^2\theta-\sin^2\theta}$

$\Leftrightarrow\ r^2\left(\cos^2\theta-\sin^2\theta\right)=1 \Leftrightarrow$

$r^2\cos^2\theta-r^2\sin^2\theta=1 \Leftrightarrow$

$(r\cos\theta)^2-(r\sin\theta)^2=1 \Leftrightarrow$

$x^2-y^2=1$.

23. (a)

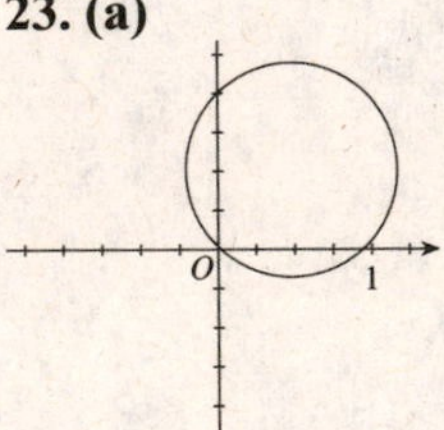

(b) $r=\sin\theta+\cos\theta \Leftrightarrow$

$r^2=r\sin\theta+r\cos\theta$, so

$x^2+y^2=y+x \Leftrightarrow \left(x^2-x+\frac{1}{4}\right)+\left(y^2-y+\frac{1}{4}\right)=\frac{1}{2}$

$\Leftrightarrow\ \left(x-\frac{1}{2}\right)^2+\left(y-\frac{1}{2}\right)^2=\frac{1}{2}$.

25. $r=\cos(\theta/3)$, $\theta\in[0,3\pi]$.

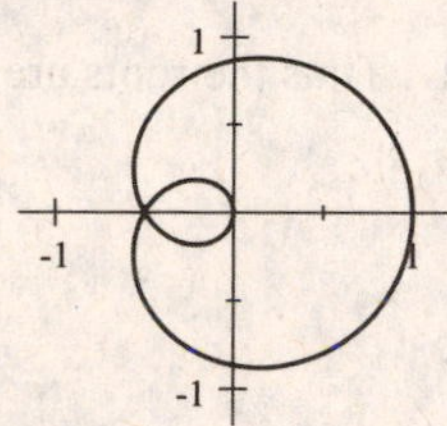

27. $r=1+4\cos(\theta/3)$, $\theta\in[0,6\pi]$.

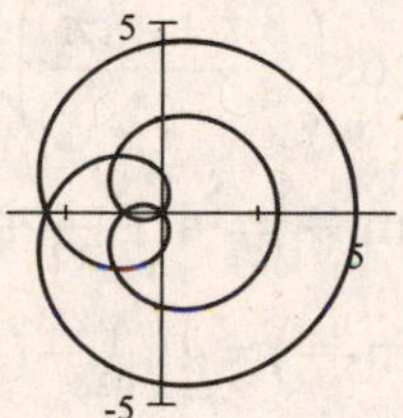

29. (a)

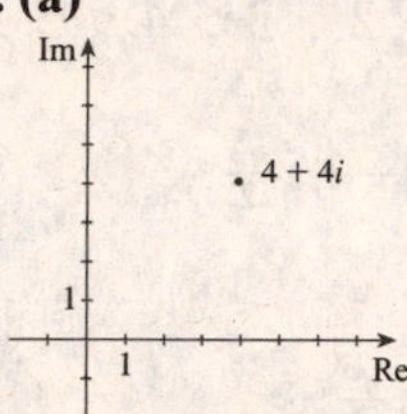

(b) $4+4i$ has $r=\sqrt{16+16}=4\sqrt{2}$, and $\theta=\tan^{-1}\frac{4}{4}=\frac{\pi}{4}$ (in quadrant I).

(c) $4+4i=4\sqrt{2}\left(\cos\frac{\pi}{4}+i\sin\frac{\pi}{4}\right)$

31. (a)

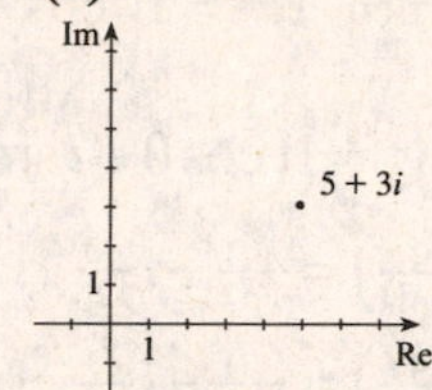

(b) $5+3i$. Then $r=\sqrt{25+9}=\sqrt{34}$, and $\theta=\tan^{-1}\frac{3}{5}$.

(c)

$5+3i=\sqrt{34}\left[\cos\left(\tan^{-1}\frac{3}{5}\right)+i\sin\left(\tan^{-1}\frac{3}{5}\right)\right]$

33. (a)

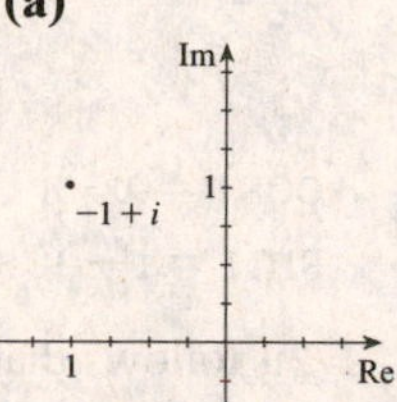

(b) $-1+i$ has $r=\sqrt{1+1}=\sqrt{2}$ and $\tan\theta=\dfrac{1}{-1}$ with θ in quadrant II $\Leftrightarrow$ $\theta=\frac{3\pi}{4}$.

(c) $-1+i=\sqrt{2}\left(\cos\frac{3\pi}{4}+i\sin\frac{3\pi}{4}\right)$

35. $1-\sqrt{3}i$ has $r=\sqrt{1+3}=2$ and $\tan\theta=\frac{-\sqrt{3}}{1}=-\sqrt{3}$ with θ in quadrant III $\Leftrightarrow$ $\theta=\frac{5\pi}{3}$. Therefore, $1-\sqrt{3}i=2\left(\cos\frac{5\pi}{3}+i\sin\frac{5\pi}{3}\right)$, and so

$$\left(1-\sqrt{3}i\right)^4=2^4\left(\cos\tfrac{20\pi}{3}+i\sin\tfrac{20\pi}{3}\right)=16\left(\cos\tfrac{2\pi}{3}+i\sin\tfrac{2\pi}{3}\right)=16\left(-\tfrac{1}{2}+i\tfrac{\sqrt{3}}{2}\right)=8\left(-1+i\sqrt{3}\right).$$

37. $\sqrt{3}+i$ has $r=\sqrt{3+1}=2$ and $\tan\theta=\frac{1}{\sqrt{3}}$ with θ in quadrant I $\Leftrightarrow$ $\theta=\frac{\pi}{6}$. Therefore, $\sqrt{3}+i=2\left(\cos\frac{\pi}{6}+i\sin\frac{\pi}{6}\right)$, and so

$$\begin{aligned}\left(\sqrt{3}+i\right)^{-4}&=2^{-4}\left(\cos\tfrac{-4\pi}{6}+i\sin\tfrac{-4\pi}{6}\right)=\tfrac{1}{16}\left(\cos\tfrac{2\pi}{3}-i\sin\tfrac{2\pi}{3}\right)=\tfrac{1}{16}\left(-\tfrac{1}{2}-i\tfrac{\sqrt{3}}{2}\right)\\&=\tfrac{1}{32}\left(-1-i\sqrt{3}\right)=-\tfrac{1}{32}\left(1+i\sqrt{3}\right)\end{aligned}$$

39. $-16i$ has $r = 16$ and $\theta = \frac{3\pi}{2}$. Thus, $-16i = 16\left(\cos\frac{3\pi}{2} + i\sin\frac{3\pi}{2}\right)$ and so

$(-16i)^{1/2} = 16^{1/2}\left[\cos\left(\dfrac{3\pi + 4k\pi}{2}\right) + i\sin\left(\dfrac{3\pi + 4k\pi}{2}\right)\right]$ for $k = 0, 1$. Thus the roots are

$w_0 = 4\left(\cos\frac{3\pi}{4} + i\sin\frac{3\pi}{4}\right) = 4\left(-\frac{1}{\sqrt{2}} + i\frac{1}{\sqrt{2}}\right) = 2\sqrt{2}(-1+i)$ and

$w_1 = 4\left(\cos\frac{7\pi}{4} + i\sin\frac{7\pi}{4}\right) = 4\left(\frac{1}{\sqrt{2}} - i\frac{1}{\sqrt{2}}\right) = 2\sqrt{2}(1-i)$.

41. $1 = \cos 0 + i\sin 0$. Then $1^{1/6} = 1\left(\cos\dfrac{2k\pi}{6} + i\sin\dfrac{2k\pi}{6}\right)$ for $k = 0, 1, 2, 3, 4, 5$.

Thus the six roots are $w_0 = 1(\cos 0 + i\sin 0) = 1$, $w_1 = 1\left(\cos\frac{\pi}{3} + i\sin\frac{\pi}{3}\right) = \frac{1}{2} + i\frac{\sqrt{3}}{2}$,

$w_2 = 1\left(\cos\frac{2\pi}{3} + i\sin\frac{2\pi}{3}\right) = -\frac{1}{2} + i\frac{\sqrt{3}}{2}$, $w_3 = 1(\cos\pi + i\sin\pi) = -1$,

$w_4 = 1\left(\cos\frac{4\pi}{3} + i\sin\frac{4\pi}{3}\right) = -\frac{1}{2} - i\frac{\sqrt{3}}{2}$, and $w_5 = 1\left(\cos\frac{5\pi}{3} + i\sin\frac{5\pi}{3}\right) = \frac{1}{2} - i\frac{\sqrt{3}}{2}$.

43. (a)

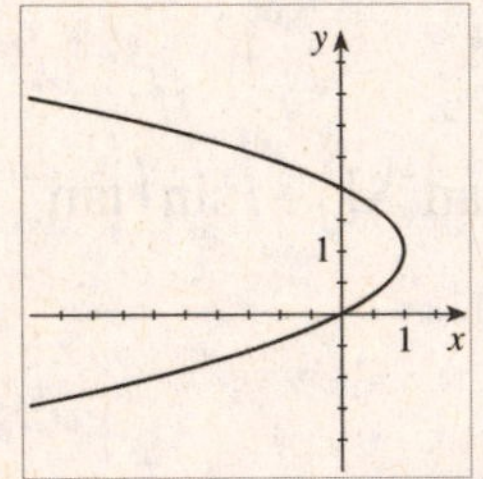

(b) $x = 1 - t^2$, $y = 1 + t$ $\Leftrightarrow$ $t = y - 1$.

Substituting for t gives $x = 1 - (y-1)^2$ $\Leftrightarrow$ $(y-1)^2 = 1 - x$ in rectangular coordinates.

45. (a)

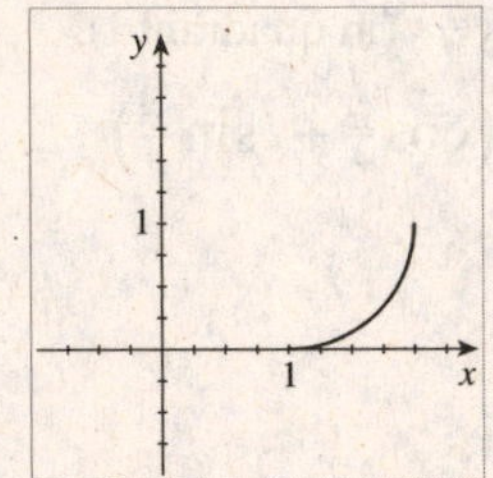

(b) $x = 1 + \cos t$ $\Leftrightarrow$ $\cos t = x - 1$, and $y = 1 - \sin t$ $\Leftrightarrow$ $\sin t = 1 - y$. Since $\cos^2 t + \sin^2 t = 1$, it follows that $(x-1)^2 + (1-y)^2 = 1$ $\Leftrightarrow$ $(x-1)^2 + (y-1)^2 = 1$. Since t is restricted by $0 \le t \le \frac{\pi}{2}$, $1 + \cos 0 \le x \le 1 + \cos\frac{\pi}{2} \Leftrightarrow 1 \le x \le 2$, and similarly, $0 \le y \le 1$. (This is the lower right quarter of the circle.)

47. $x = \cos 2t$, $y = \sin 3t$

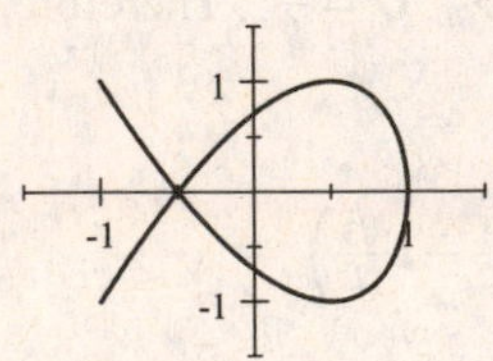

49. The coordinates of Q are $x = \cos\theta$ and $y = \sin\theta$. The coordinates of R are $x = 1$ and $y = \tan\theta$. Hence, the midpoint P is $\left(\dfrac{1+\cos\theta}{2}, \dfrac{\sin\theta + \tan\theta}{2}\right)$, so parametric equations for the curve are $x = \dfrac{1+\cos\theta}{2}$ and $y = \dfrac{\sin\theta + \tan\theta}{2}$.

5 Test

1. (a) $x = 8\cos\frac{5\pi}{4} = 8\left(-\frac{\sqrt{2}}{2}\right) = -4\sqrt{2}$, $y = 8\sin\frac{5\pi}{4} = 8\left(-\frac{\sqrt{2}}{2}\right) = -4\sqrt{2}$. So the point has rectangular coordinates $\left(-4\sqrt{2}, -4\sqrt{2}\right)$.

(b) $P = \left(-6, 2\sqrt{3}\right)$ in rectangular coordinates. So $\tan\theta = \frac{2\sqrt{3}}{-6}$ and the reference angle is $\bar{\theta} = \frac{\pi}{6}$. Since P is in quadrant II, we have $\theta = \frac{5\pi}{6}$. Next, $r^2 = (-6)^2 + \left(2\sqrt{3}\right)^2 = 36 + 12 = 48$, so $r = 4\sqrt{3}$. Thus, polar coordinates for the point are $\left(4\sqrt{3}, \frac{5\pi}{6}\right)$ or $\left(-4\sqrt{3}, \frac{11\pi}{6}\right)$.

2. (a)

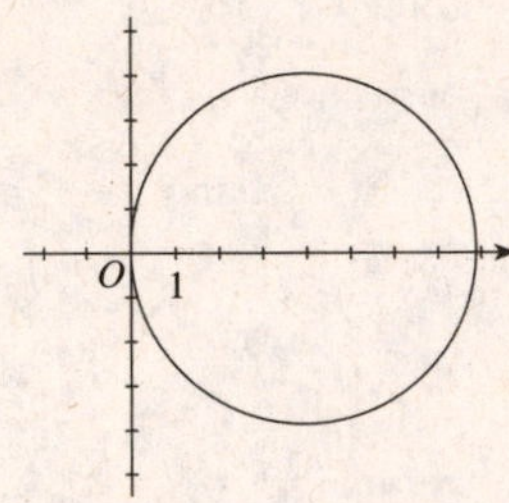

(b) $r = 8\cos\theta \Leftrightarrow r^2 = 8r\cos\theta \Leftrightarrow x^2 + y^2 = 8x \Leftrightarrow x^2 - 8x + y^2$

$$= 0 \Leftrightarrow x^2 - 8x + 16 + y^2 = 16 \Leftrightarrow (x-4)^2 + y^2 = 16$$

3.

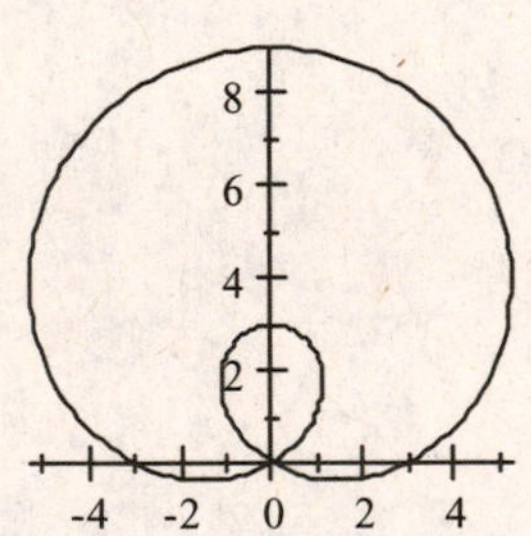

The curve is a limaçon.

4. (a)

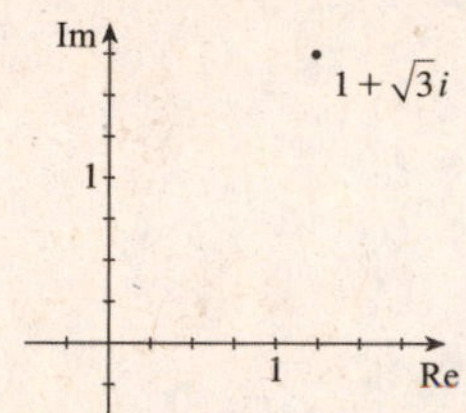

(b) $1+\sqrt{3}i$ has $r=\sqrt{1+3}=2$ and $\theta=\tan^{-1}\left(\sqrt{3}\right)=\frac{\pi}{3}$. So, in trigonometric form, $1+\sqrt{3}i=2\left(\cos\frac{\pi}{3}+i\sin\frac{\pi}{3}\right)$.

(c) $z=1+\sqrt{3}i=2\left(\cos\frac{\pi}{3}+i\sin\frac{\pi}{3}\right)\Rightarrow$

$$\begin{aligned} z^9 &= 2^9\left(\cos\tfrac{9\pi}{3}+i\sin\tfrac{9\pi}{3}\right)=512\left(\cos 3\pi+i\sin 3\pi\right)\\ &=512\left(-1+i(0)\right)=-512 \end{aligned}$$

5. $z_1=4\left(\cos\frac{7\pi}{12}+i\sin\frac{7\pi}{12}\right)$ and $z_2=2\left(\cos\frac{5\pi}{12}+i\sin\frac{5\pi}{12}\right)$.

Then $z_1z_2=4\cdot 2\left[\cos\left(\dfrac{7\pi+5\pi}{12}\right)+i\sin\left(\dfrac{7\pi+5\pi}{12}\right)\right]=8\left(\cos\pi+i\sin\pi\right)=-8$ and

$$\begin{aligned} z_1/z_2 &= \tfrac{4}{2}\left[\cos\left(\frac{7\pi-5\pi}{12}\right)+i\sin\left(\frac{7\pi-5\pi}{12}\right)\right]=2\left(\cos\tfrac{\pi}{6}+i\sin\tfrac{\pi}{6}\right)\\ &=2\left(\tfrac{\sqrt{3}}{2}+\tfrac{1}{2}i\right)=\sqrt{3}+i \end{aligned}$$

6. $27i$ has $r=27$ and $\theta=\frac{\pi}{2}$, so $27i=27\left(\cos\frac{\pi}{2}+i\sin\frac{\pi}{2}\right)$. Thus,

$$\begin{aligned} (27i)^{1/3} &= \sqrt[3]{27}\left[\cos\left(\frac{\frac{\pi}{2}+2k\pi}{3}\right)+i\sin\left(\frac{\frac{\pi}{2}+2k\pi}{3}\right)\right]\\ &=3\left[\cos\left(\frac{\pi+4k\pi}{6}\right)+i\sin\left(\frac{\pi+4k\pi}{6}\right)\right] \end{aligned}$$

for $k=0,1,2$. Thus, the three roots are

$w_0=3\left(\cos\frac{\pi}{6}+i\sin\frac{\pi}{6}\right)=3\left(\frac{\sqrt{3}}{2}+\frac{1}{2}i\right)=\frac{3}{2}\left(\sqrt{3}+i\right)$,

$w_1=3\left(\cos\frac{5\pi}{6}+i\sin\frac{5\pi}{6}\right)=3\left(-\frac{\sqrt{3}}{2}+\frac{1}{2}i\right)=\frac{3}{2}\left(-\sqrt{3}+i\right)$, and

$w_2=3\left(\cos\frac{9\pi}{6}+i\sin\frac{9\pi}{6}\right)=-3i$.

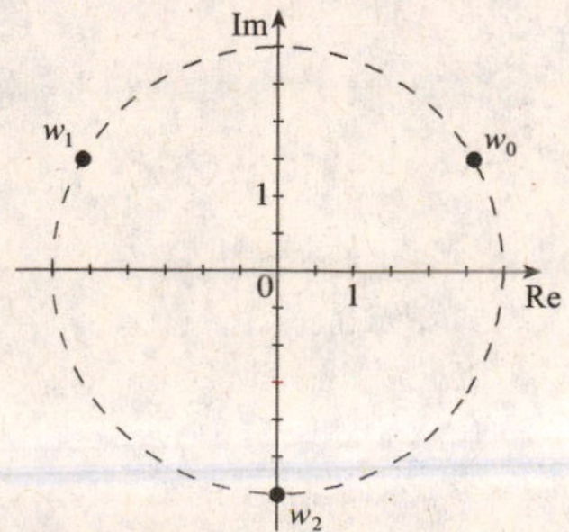

7. (a) $x = 3\sin t + 3$, $y = 2\cos t$, $0 \le t \le \pi$. From the work of part (b), we see that this is the half-ellipse shown.

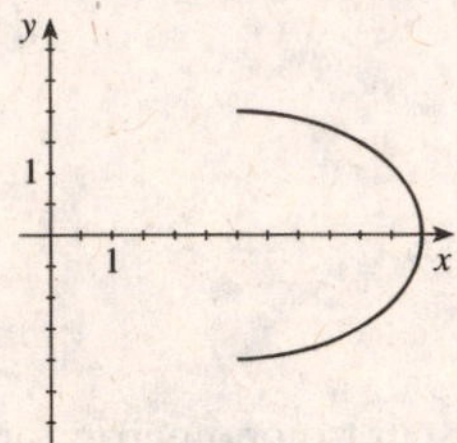

(b) $x = 3\sin t + 3 \iff x - 3 = 3\sin t \iff \dfrac{x-3}{3} = \sin t$. Squaring both sides gives $\dfrac{(x-3)^2}{9} = \sin^2 t$. Similarly, $y = 2\cos t \iff \dfrac{y}{2} = \cos t$, and squaring both sides gives $\dfrac{y^2}{4} = \cos^2 t$. Since $\sin^2 t + \cos^2 t = 1$, it follows that $\dfrac{(x-3)^2}{9} + \dfrac{y^2}{4} = 1$. Since $0 \le t \le \pi$, $\sin t \ge 0$, so $3\sin t \ge 0 \Rightarrow 3\sin t + 3 \ge 3$, and so $x \ge 3$. Thus the curve consists of only the right half of the ellipse.

8. We start at the point $(3,5)$. Because the line has slope 2 , for every 1 unit we move to the right, we must move up 2 units. Therefore, parametric equations are $x = 3 + t$, $y = 5 + 2t$.

1. From $x = (v_0 \cos\theta)t$, we get $t = \dfrac{x}{v_0 \cos\theta}$. Substituting this value for t into the equation for y, we get $y = (v_0 \sin\theta)t - \frac{1}{2}gt^2 \Leftrightarrow y = (v_0 \sin\theta)\left(\dfrac{x}{v_0 \cos\theta}\right) - \frac{1}{2}g\left(\dfrac{x}{v_0 \cos\theta}\right)^2 \Leftrightarrow y = (\tan\theta)x - \dfrac{g}{2v_0^2 \cos^2\theta}x^2$

This shows that y is a quadratic function of x, so its graph is a parabola as long as $\theta \neq 90°$. When $\theta = 90°$, the path of the projectile is a straight line up (then down).

3. (a) We use the equation $t = \frac{2v_0 \sin\theta}{g}$. Substituting $g \approx 32$ ft / s ^{2}, $\theta = 5°$, and $v_0 = 1000$, we get $t = \frac{2 \cdot 1000 \cdot \sin 5°}{32} \approx 5.447$ seconds.

(b) Substituting the given values into $y = (v_0 \sin\theta)t - \frac{1}{2}gt^2$, we get $y = 87.2t - 16t^2$. The maximum value of y is attained at the vertex of the parabola; thus

$y = 87.2t - 16t^2 = -16(t^2 - 5.45t) \Leftrightarrow y = -16\left[t^2 - 2(2.725)t + 7.425625\right] + 118.7$

Thus the greatest height is 118.7 ft.

(c) The rocket hits the ground after 5.447 s, so substituting this into the expression for the horizontal distance gives $x = (1000 \cos 5°)5.447 = 5426$ ft.

(d)

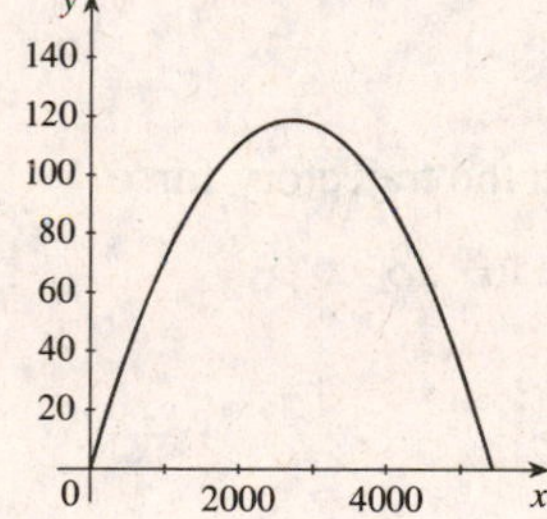

5. We use the equation of the parabola from Exercise 1 and find its vertex:

$$y = \left(\tan\theta\right)x - \frac{g}{2v_0^2\cos^2\theta}x^2 \quad \Leftrightarrow$$

$$y = -\frac{g}{2v_0^2\cos^2\theta}\left[x^2 - \frac{2v_0^2\sin\theta\cos\theta x}{g}\right] \Leftrightarrow y$$

$$= -\frac{g}{2v_0^2\cos^2\theta}\left[x^2 - \frac{2v_0^2\sin\theta\cos\theta x}{g} + \left(\frac{v_0^2\sin\theta\cos\theta}{g}\right)^2\right] + \frac{g}{2v_0^2\cos^2\theta}\cdot\left(\frac{v_0^2\sin\theta\cos\theta}{g}\right)^2 \Leftrightarrow y$$

$$= -\frac{g}{2v_0^2\cos^2\theta}\left[x - \frac{v_0^2\sin\theta\cos\theta}{g}\right]^2 + \frac{v_0^2\sin^2\theta}{2g}$$

Thus the vertex is at $\left(\dfrac{v_0^2\sin\theta\cos\theta}{g}, \dfrac{v_0^2\sin^2\theta}{2g}\right)$, so the maximum height is $\dfrac{v_0^2\sin^2\theta}{2g}$.

7. In Exercise 6 we derived the equations $x = \left(v_0\cos\theta - w\right)t$, $y = \left(v_0\sin\theta\right)t - \frac{1}{2}gt^2$. We plot the graphs for the given values of v_0 , w , and θ in the figure to the right. The projectile will be blown backwards if the horizontal component of its velocity is less than the speed of the wind, that is, $32\cos\theta < 24 \Leftrightarrow \cos\theta < \frac{3}{4} \Rightarrow \theta > 41.4^\circ$.

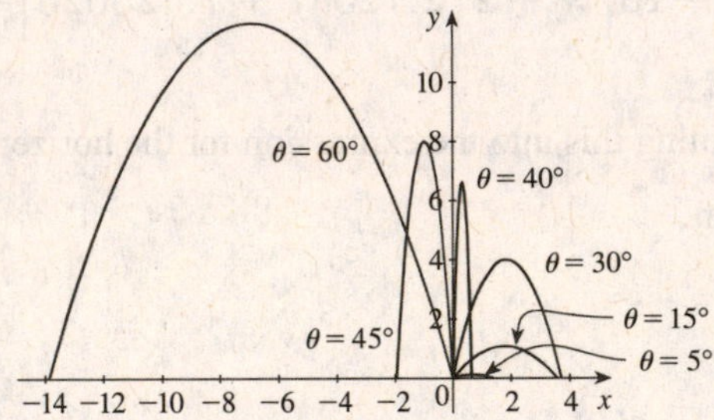

The optimal firing angle appears to be between 15° and 30° . We graph the trajectory for $\theta = 20^\circ$, $\theta = 23^\circ$, and $\theta = 25^\circ$. The solution appears to be close to 23° .

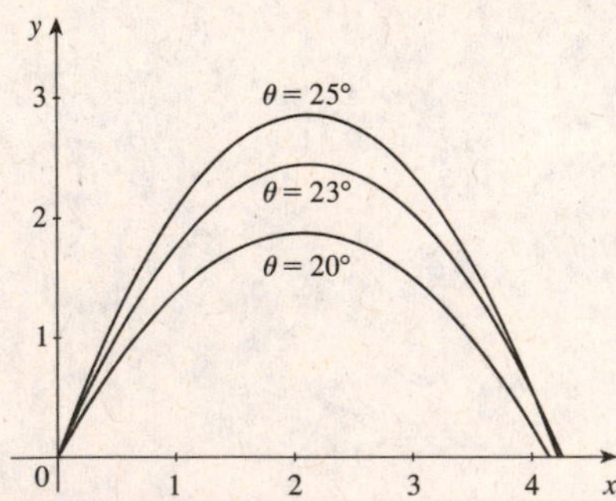

6 Vectors in Two and Three Dimensions

6.1 Vectors in Two Dimensions

1. (a) The vector $\mathbf{u}$ has initial point A and terminal point B.

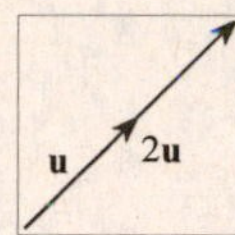

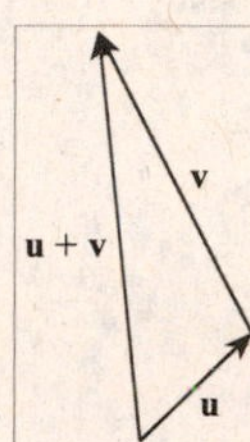

(b) The vector $\mathbf{u}$ has initial point $(2,1)$ and terminal point $(4,3)$. In component form we write $\mathbf{u} = \langle 2,2 \rangle$ and $\mathbf{v} = \langle -3,6 \rangle$. Then $2\mathbf{u} = \langle 4,4 \rangle$ and $\mathbf{u} + \mathbf{v} = \langle -1,8 \rangle$.

3. $2\mathbf{u} = 2\langle -2,3 \rangle = \langle -4,6 \rangle$

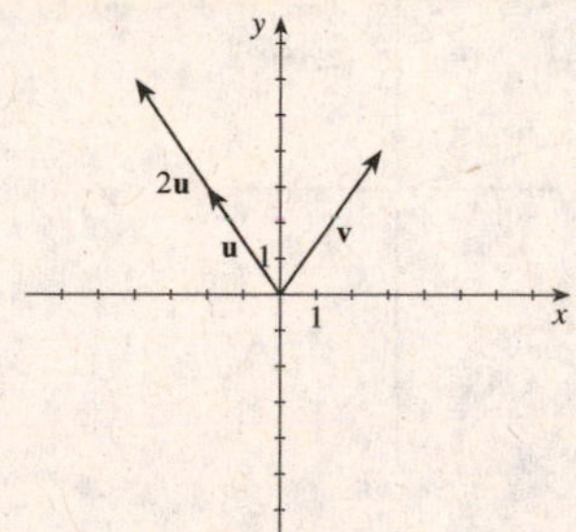

5. $\mathbf{u} + \mathbf{v} = \langle -2,3 \rangle + \langle 3,4 \rangle$
$= \langle -2+3, 3+4 \rangle = \langle 1,7 \rangle$

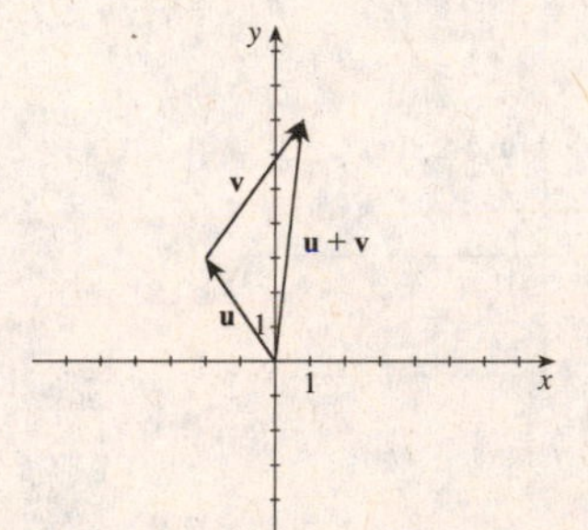

7.
$$\begin{aligned}\mathbf{v} - 2\mathbf{u} &= \langle 3,4 \rangle - 2\langle -2,3 \rangle \\ &= \langle 3 - 2(-2), 4 - 2(3) \rangle \\ &= \langle 7,-2 \rangle\end{aligned}$$

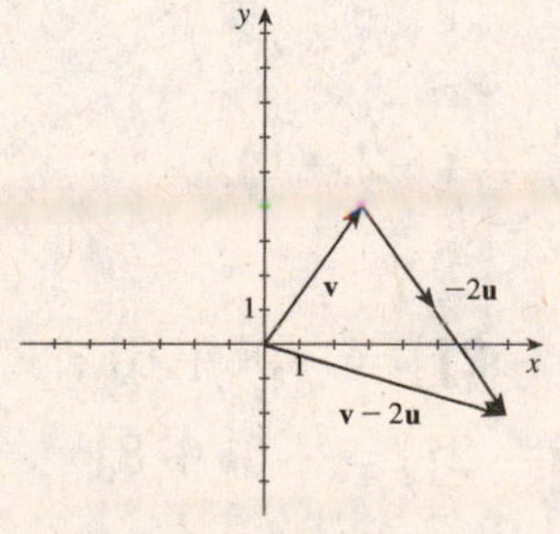

In Solutions 9--18, $\mathbf{v}$ represents the vector with initial point P and terminal point Q.

9. $P(2,1)$, $Q(5,4)$. $\mathbf{v} = \langle 5-2, 4-1 \rangle = \langle 3,3 \rangle$

11. $P(1,2)$, $Q(4,1)$. $\mathbf{v} = \langle 4-1, 1-2 \rangle = \langle 3,-1 \rangle$

13. $P(3,2)$, $Q(8,9)$. $\mathbf{v} = \langle 8-3, 9-2 \rangle = \langle 5,7 \rangle$

15. $P(5,3)$, $Q(1,0)$. $\mathbf{v} = \langle 1-5, 0-3 \rangle = \langle -4,-3 \rangle$

17. $P(-1,-1)$, $Q(-1,1)$. $\mathbf{v} = \langle -1-(-1), 1-(-1) \rangle = \langle 0,2 \rangle$

19.

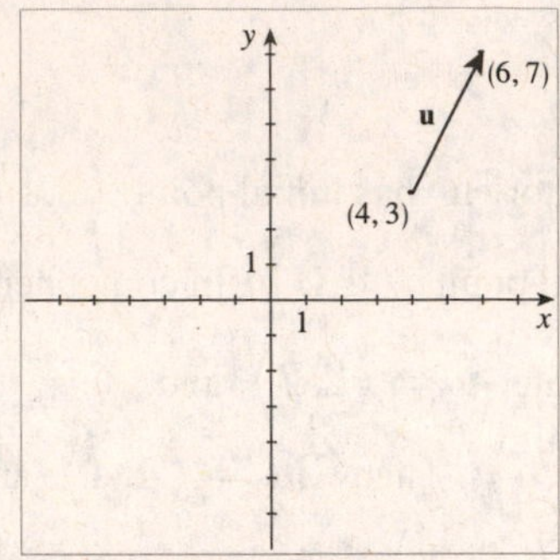

The terminal point is $(4+2, 3+4) = (6,7)$.

21.

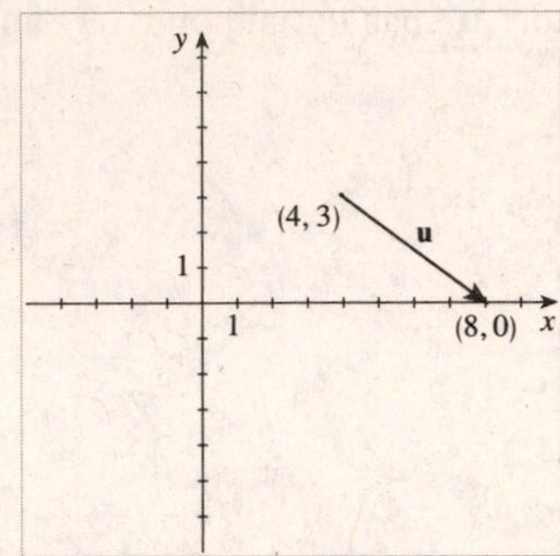

The terminal point is $(4+4, 3-3) = (8,0)$.

23.

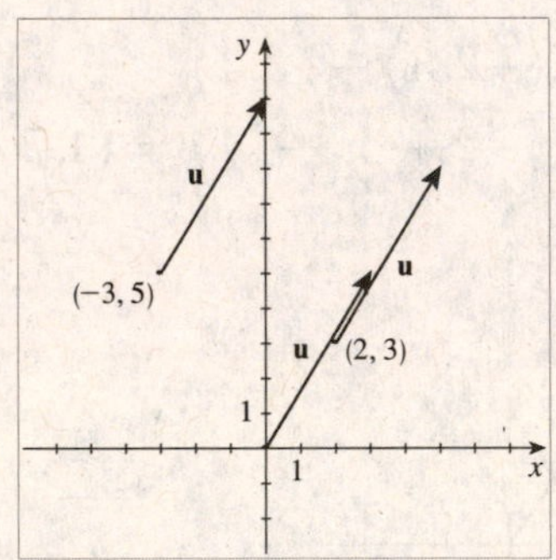

25.

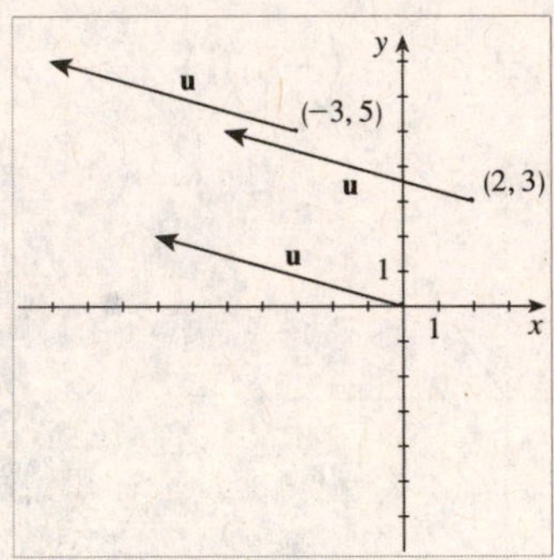

27. $\mathbf{u} = \langle 1,4 \rangle = \mathbf{i} + 4\mathbf{j}$

29. $\mathbf{u} = \langle 3,0 \rangle = 3\mathbf{i}$

31. $\mathbf{u} = \langle 2,7 \rangle$, $\mathbf{v} = \langle 3,1 \rangle$. $2\mathbf{u} = 2 \cdot \langle 2,7 \rangle = \langle 4,14 \rangle$;
$-3\mathbf{v} = -3 \cdot \langle 3,1 \rangle = \langle -9,-3 \rangle$; $\mathbf{u} + \mathbf{v} = \langle 2,7 \rangle + \langle 3,1 \rangle = \langle 5,8 \rangle$;
$3\mathbf{u} - 4\mathbf{v} = \langle 6,21 \rangle - \langle 12,4 \rangle = \langle -6,17 \rangle$

33. $\mathbf{u} = \langle 0,-1 \rangle$, $\mathbf{v} = \langle -2,0 \rangle$. $2\mathbf{u} = 2 \cdot \langle 0,-1 \rangle = \langle 0,-2 \rangle$;
$-3\mathbf{v} = -3 \cdot \langle -2,0 \rangle = \langle 6,0 \rangle$; $\mathbf{u} + \mathbf{v} = \langle 0,-1 \rangle + \langle -2,0 \rangle = \langle -2,-1 \rangle$;
$3\mathbf{u} - 4\mathbf{v} = \langle 0,-3 \rangle - \langle -8,0 \rangle = \langle 8,-3 \rangle$

35. $\mathbf{u} = 2\mathbf{i}$, $\mathbf{v} = 3\mathbf{i} - 2\mathbf{j}$. $2\mathbf{u} = 2 \cdot 2\mathbf{i} = 4\mathbf{i}$; $-3\mathbf{v} = -3(3\mathbf{i} - 2\mathbf{j}) = -9\mathbf{i} + 6\mathbf{j}$;
$\mathbf{u} + \mathbf{v} = 2\mathbf{i} + 3\mathbf{i} - 2\mathbf{j} = 5\mathbf{i} - 2\mathbf{j}$; $3\mathbf{u} - 4\mathbf{v} = 3 \cdot 2\mathbf{i} - 4(3\mathbf{i} - 2\mathbf{j}) = -6\mathbf{i} + 8\mathbf{j}$

37. $\mathbf{u} = 2\mathbf{i} + \mathbf{j}, \mathbf{v} = 3\mathbf{i} - 2\mathbf{j}$. Then $|\mathbf{u}| = \sqrt{2^2 + 1^2} = \sqrt{5}$; $|\mathbf{v}| = \sqrt{3^2 + 2^2} = \sqrt{13}$;

$2\mathbf{u} = 4\mathbf{i} + 2\mathbf{j}$; $|2\mathbf{u}| = \sqrt{4^2 + 2^2} = 2\sqrt{5}$; $\frac{1}{2}\mathbf{v} = \frac{3}{2}\mathbf{i} - \mathbf{j}$;

$\left|\frac{1}{2}\mathbf{v}\right| = \sqrt{\left(\frac{3}{2}\right)^2 + 1^2} = \frac{1}{2}\sqrt{13}$; $\mathbf{u} + \mathbf{v} = 5\mathbf{i} - \mathbf{j}$; $|\mathbf{u} + \mathbf{v}| = \sqrt{5^2 + 1^2} = \sqrt{26}$;

$\mathbf{u} - \mathbf{v} = 2\mathbf{i} + \mathbf{j} - 3\mathbf{i} + 2\mathbf{j} = -\mathbf{i} + 3\mathbf{j}$; $|\mathbf{u} - \mathbf{v}| = \sqrt{1^2 + 3^2} = \sqrt{10}$;

$|\mathbf{u}| - |\mathbf{v}| = \sqrt{5} - \sqrt{13}$

39. $\mathbf{u} = \langle 10, -1 \rangle$, $\mathbf{v} = \langle -2, -2 \rangle$. Then $|\mathbf{u}| = \sqrt{10^2 + 1^2} = \sqrt{101}$;

$|\mathbf{v}| = \sqrt{(-2)^2 + (-2)^2} = 2\sqrt{2}$; $2\mathbf{u} = \langle 20, -2 \rangle$;

$|2\mathbf{u}| = \sqrt{20^2 + 2^2} = \sqrt{404} = 2\sqrt{101}$; $\frac{1}{2}\mathbf{v} = \langle -1, -1 \rangle$;

$\left|\frac{1}{2}\mathbf{v}\right| = \sqrt{(-1)^2 + (-1)^2} = \sqrt{2}$; $\mathbf{u} + \mathbf{v} = \langle 8, -3 \rangle$; $|\mathbf{u} + \mathbf{v}| = \sqrt{8^2 + 3^2} = \sqrt{73}$;

$\mathbf{u} - \mathbf{v} = \langle 12, 1 \rangle$; $|\mathbf{u} - \mathbf{v}| = \sqrt{12^2 + 1^2} = \sqrt{145}$; $|\mathbf{u}| - |\mathbf{v}| = \sqrt{101} - 2\sqrt{2}$

In Solutions 41--46, x represents the horizontal component and y the vertical component.

41. $|\mathbf{v}| = 40$, direction $\theta = 30°$. $x = 40\cos 30° = 20\sqrt{3}$ and $y = 40\sin 30° = 20$.

Thus, $\mathbf{v} = x\mathbf{i} + y\mathbf{j} = 20\sqrt{3}\mathbf{i} + 20\mathbf{j}$.

43. $|\mathbf{v}| = 1$, direction $\theta = 225°$. $x = \cos 225° = -\frac{1}{\sqrt{2}}$ and $y = \sin 225° = -\frac{1}{\sqrt{2}}$. Thus,

$\mathbf{v} = x\mathbf{i} + y\mathbf{j} = -\frac{1}{\sqrt{2}}\mathbf{i} - \frac{1}{\sqrt{2}}\mathbf{j} = -\frac{\sqrt{2}}{2}\mathbf{i} - \frac{\sqrt{2}}{2}\mathbf{j}$.

45. $|\mathbf{v}| = 4$, direction $\theta = 10°$. $x = 4\cos 10° \approx 3.94$ and $y = 4\sin 10° \approx 0.69$. Thus,

$\mathbf{v} = x\mathbf{i} + y\mathbf{j} = (4\cos 10°)\mathbf{i} + (4\sin 10°)\mathbf{j} \approx 3.94\mathbf{i} + 0.69\mathbf{j}$.

47. $\mathbf{v} = \langle 3, 4 \rangle$. The magnitude is $|\mathbf{v}| = \sqrt{3^2 + 4^2} = 5$. The direction is θ , where

$\tan\theta = \frac{4}{3} \Leftrightarrow \theta = \tan^{-1}\left(\frac{4}{3}\right) \approx 53.13°$.

49. $\mathbf{v} = \langle -12, 5 \rangle$. The magnitude is $|\mathbf{v}| = \sqrt{(-12)^2 + 5^2} = \sqrt{169} = 13$. The direction is

θ , where $\tan\theta = -\frac{5}{12}$ with θ in quadrant II $\Leftrightarrow$ $\theta = \pi + \tan^{-1}\left(-\frac{5}{12}\right) \approx 157.38°$.

51. $\mathbf{v} = \mathbf{i} + \sqrt{3}\mathbf{j}$. The magnitude is $|\mathbf{v}| = \sqrt{1^2 + (\sqrt{3})^2} = 2$. The direction is θ , where

$\tan\theta = \sqrt{3}$ with θ in quadrant I $\Leftrightarrow$ $\theta = \tan^{-1}\sqrt{3} = 60°$.

53. $|\mathbf{v}| = 30$, direction $\theta = 30°$. $x = 30\cos 30° = 30 \cdot \dfrac{\sqrt{3}}{2} \approx 25.98$,

$y = 30\sin 30° = 15$. So the horizontal component of force is $15\sqrt{3}$ lb and the vertical component is -15 lb.

55. The flow of the river can be represented by the vector $\mathbf{v} = -3\mathbf{j}$ and the swimmer can be represented by the vector $\mathbf{u} = 2\mathbf{i}$. Therefore the true velocity is $\mathbf{u} + \mathbf{v} = 2\mathbf{i} - 3\mathbf{j}$.

57.

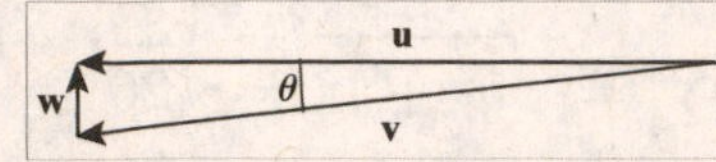

The speed of the airplane is 300 mi/h, so its velocity relative to the air is $\mathbf{v} = (-300\cos\theta)\mathbf{i} + (-300\sin\theta)\mathbf{j}$. The wind has velocity $\mathbf{w} = 30\mathbf{j}$, so the true course of the airplane is given by $\mathbf{u} = \mathbf{v} + \mathbf{w} = (-300\cos\theta)\mathbf{i} + (-300\sin\theta + 30)\mathbf{j}$. We want the y -component of the airplane's velocity to be 0 , so we solve $-300\sin\theta + 30 = 0 \;\Leftrightarrow$ $\sin\theta = \frac{1}{10} \;\Leftrightarrow\; \theta \approx 5.74°$. Therefore, the airplane should head in the direction $185.74°$ (or S $84.26°$ W).

59. (a) The velocity of the wind is $40\mathbf{j}$.

(b) The velocity of the jet relative to the air is $425\mathbf{i}$.

(c) The true velocity of the jet is $\mathbf{v} = 425\mathbf{i} + 40\mathbf{j} = \langle 425, 40\rangle$.

(d) The true speed of the jet is $|\mathbf{v}| = \sqrt{425^2 + 40^2} \approx 427$ mi/h, and the true direction is $\theta = \tan^{-1}\left(\frac{40}{425}\right) \approx 5.4° \;\Rightarrow\; \theta$ is N $84.6°$ E.

61. If the direction of the plane is N $30°$ W, the airplane's velocity is $\mathbf{u} = \langle \mathbf{u}_x, \mathbf{u}_y\rangle$ where $\mathbf{u}_x = -765\cos 60° = -382.5$, and $\mathbf{u}_y = 765\sin 60° \approx 662.51$. If the direction of the wind is N $30°$ E, the wind velocity is $\mathbf{w} = \langle w_x, w_y\rangle$ where $w_x = 55\cos 60° = 27.5$, and $w_y = 55\sin 60° \approx 47.63$. Thus, the actual flight path is $\mathbf{v} = \mathbf{u} + \mathbf{w} = \langle -382.5 + 27.5, 662.51 + 47.63\rangle = \langle -355, 710.14\rangle$, and so the true speed is $|\mathbf{v}| = \sqrt{355^2 + 710.14^2} \approx 794$ mi / h, and the true direction is $\theta = \tan^{-1}\left(-\frac{710.14}{355}\right) \approx 116.6°$ so θ is N $26.6°$ W.

63. (a) The velocity of the river is represented by the vector $\mathbf{r} = \langle 10, 0\rangle$.

(b) Since the boater direction is $60°$ from the shore at 20 mi / h, the velocity of the boat is represented by the vector $\mathbf{b} = \langle 20\cos 60°, 20\sin 60°\rangle \approx \langle 10, 17.32\rangle$.

(c) $\mathbf{w} = \mathbf{r} + \mathbf{b} = \langle 10 + 10, 0 + 17.32\rangle = \langle 20, 17.32\rangle$

(d) The true speed of the boat is $|\mathbf{w}| = \sqrt{20^2 + 17.32^2} \approx 26.5$ mi / h, and the true direction is $\theta = \tan^{-1}\left(\frac{17.32}{20}\right) \approx 40.9° \approx$ N $49.1°$ E.

65. (a) Let $\mathbf{b} = \left\langle b_x, b_y \right\rangle$ represent the velocity of the boat relative to the water. Then $\mathbf{b} = \left\langle 24\cos 18°, 24\sin 18° \right\rangle$.

(b) Let $\mathbf{w} = \left\langle w_x, w_y \right\rangle$ represent the velocity of the water. Then $\mathbf{w} = \left\langle 0, w \right\rangle$ where w is the speed of the water. So the true velocity of the boat is $\mathbf{b} + \mathbf{w} = \left\langle 24\cos 18°, 24\sin 18° - w \right\rangle$. For the direction to be due east, we must have $24\sin 18° - w = 0 \Leftrightarrow w = 7.42$ mi / h. Therefore, the true speed of the water is 7.4 mi/h. Since $\mathbf{b} + \mathbf{w} = \left\langle 24\cos 18°, 0 \right\rangle$, the true speed of the boat is $\left|\mathbf{b} + \mathbf{w}\right| = 24\cos 18° \approx 22.8$ mi / h.

67. $\mathbf{F}_1 = \left\langle 2, 5 \right\rangle$ and $\mathbf{F}_2 = \left\langle 3, -8 \right\rangle$.

(a) $\mathbf{F}_1 + \mathbf{F}_2 = \left\langle 2 + 3, 5 - 8 \right\rangle = \left\langle 5, -3 \right\rangle$

(b) The additional force required is $\mathbf{F}_3 = \left\langle 0, 0 \right\rangle - \left\langle 5, -3 \right\rangle = \left\langle -5, 3 \right\rangle$.

69. $\mathbf{F}_1 = 4\mathbf{i} - \mathbf{j}$, $\mathbf{F}_2 = 3\mathbf{i} - 7\mathbf{j}$, $\mathbf{F}_3 = -8\mathbf{i} + 3\mathbf{j}$, and $\mathbf{F}_4 = \mathbf{i} + \mathbf{j}$.

(a) $\mathbf{F}_1 + \mathbf{F}_2 + \mathbf{F}_3 + \mathbf{F}_4 = (4 + 3 - 8 + 1)\mathbf{i} + (-1 - 7 + 3 + 1)\mathbf{j} = 0\mathbf{i} - 4\mathbf{j}$

(b) The additional force required is $\mathbf{F}_5 = 0\mathbf{i} + 0\mathbf{j} - (0\mathbf{i} - 4\mathbf{j}) = 4\mathbf{j}$.

71. $\mathbf{F}_1 = \left\langle 10\cos 60°, 10\sin 60° \right\rangle = \left\langle 5, 5\sqrt{3} \right\rangle$,

$\mathbf{F}_2 = \left\langle -8\cos 30°, 8\sin 30° \right\rangle = \left\langle -4\sqrt{3}, 4 \right\rangle$, and

$\mathbf{F}_3 = \left\langle -6\cos 20°, -6\sin 20° \right\rangle \approx \left\langle -5.638, -2.052 \right\rangle$.

(a) $\mathbf{F}_1 + \mathbf{F}_2 + \mathbf{F}_3 = \left\langle 5 - 4\sqrt{3} - 5.638, 5\sqrt{3} + 4 - 2.052 \right\rangle \approx \left\langle -7.57, 10.61 \right\rangle$.

(b) The additional force required is $\mathbf{F}_4 = \left\langle 0, 0 \right\rangle - \left\langle -7.57, 10.61 \right\rangle = \left\langle 7.57, -10.61 \right\rangle$.

73. From the figure we see that $\mathbf{T}_1 = -|\mathbf{T}_1|\cos 50°\mathbf{i} + |\mathbf{T}_1|\sin 50°\mathbf{j}$ and $\mathbf{T}_2 = |\mathbf{T}_2|\cos 30°\mathbf{i} + |\mathbf{T}_2|\sin 30°\mathbf{j}$. Since $\mathbf{T}_1 + \mathbf{T}_2 = 100\mathbf{j}$ we get $-|\mathbf{T}_1|\cos 50° + |\mathbf{T}_2|\cos 30° = 0$ and $|\mathbf{T}_1|\sin 50° + |\mathbf{T}_2|\sin 30° = 100$. From the first equation, $|\mathbf{T}_2| = |\mathbf{T}_1|\dfrac{\cos 50°}{\cos 30°}$, and substituting into the second equation gives

$$|\mathbf{T}_1|\sin 50° + |\mathbf{T}_1|\frac{\cos 50° \sin 30°}{\cos 30°} = 100 \Leftrightarrow$$

$$|\mathbf{T}_1|\left(\sin 50° \cos 30° + \cos 50° \sin 30°\right) = 100\cos 30° \Leftrightarrow$$

$$|\mathbf{T}_1|\sin\left(50° + 30°\right) = 100\cos 30° \Leftrightarrow |\mathbf{T}_1| = 100\frac{\cos 30°}{\sin 80°} \approx 87.9385 .$$

Similarly, solving for $|\mathbf{T}_1|$ in the first equation gives $|\mathbf{T}_1| = |\mathbf{T}_2|\dfrac{\cos 30°}{\cos 50°}$ and substituting gives

$$|\mathbf{T}_2|\frac{\cos 30° \sin 50°}{\cos 50°} + |\mathbf{T}_2|\sin 30° = 100 \Leftrightarrow$$

$$|\mathbf{T}_2|\left(\cos 30° \sin 50° + \cos 50° \sin 30°\right) = 100\cos 50° \Leftrightarrow$$

$$|\mathbf{T}_2| = \frac{100\cos 50°}{\sin 80°} \approx 65.2704 .$$ Thus,

$$\mathbf{T}_1 \approx \left(-87.9416\cos 50°\right)\mathbf{i} + \left(87.9416\sin 50°\right)\mathbf{j} \approx -56.5\mathbf{i} + 67.4\mathbf{j}$$ and

$$\mathbf{T}_2 \approx \left(65.2704\cos 30°\right)\mathbf{i} + \left(65.2704\sin 30°\right)\mathbf{j} \approx 56.5\mathbf{i} + 32.6\mathbf{j} .$$

75. When we add two (or more vectors), the resultant vector can be found by first placing the initial point of the second vector at the terminal point of the first vector. The resultant vector can then found by using the new terminal point of the second vector and the initial point of the first vector. When the n vectors are placed head to tail in the plane so that they form a polygon, the initial point and the terminal point are the same. Thus the sum of these n vectors is the zero vector.

6.2 The Dot Product

1. The dot product of $\mathbf{a} = \langle a_1, a_2 \rangle$ and $\mathbf{b} = \langle b_1, b_2 \rangle$ is defined by $\mathbf{a} \cdot \mathbf{b} = a_1 a_2 + b_1 b_2$. The dot product of two vectors is a *real number, or scalar*, not a vector.

3. (a) The component of $\mathbf{a}$ along $\mathbf{b}$ is the scalar $|\mathbf{a}| \cos\theta$ and can be expressed in terms of the dot product as $\dfrac{\mathbf{a} \cdot \mathbf{b}}{|\mathbf{b}|}$. **(b)** The projection of $\mathbf{a}$ onto $\mathbf{b}$ is the vector $\operatorname{proj}_{\mathbf{b}} \mathbf{a} = \left(\dfrac{\mathbf{a} \cdot \mathbf{b}}{|\mathbf{b}|^2} \right) \mathbf{b}$.

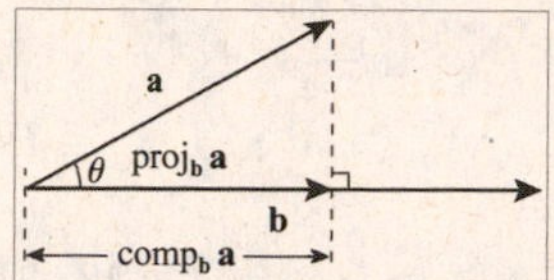

5. (a) $\mathbf{u} \cdot \mathbf{v} = \langle 2, 0 \rangle \cdot \langle 1, 1 \rangle = 2 + 0 = 2$

(b) $\cos\theta = \dfrac{\mathbf{u} \cdot \mathbf{v}}{|\mathbf{u}||\mathbf{v}|} = \frac{2}{2 \cdot \sqrt{2}} = \frac{1}{\sqrt{2}} \Rightarrow \theta = 45°$

7. (a) $\mathbf{u} \cdot \mathbf{v} = \langle 2, 7 \rangle \cdot \langle 3, 1 \rangle = 6 + 7 = 13$

(b) $\cos\theta = \dfrac{\mathbf{u} \cdot \mathbf{v}}{|\mathbf{u}||\mathbf{v}|} = \frac{13}{\sqrt{53} \cdot \sqrt{10}} \Rightarrow \theta \approx 56°$

9. (a) $\mathbf{u} \cdot \mathbf{v} = \langle 3, -2 \rangle \cdot \langle 1, 2 \rangle = 3 + (-4) = -1$

(b) $\cos\theta = \dfrac{\mathbf{u} \cdot \mathbf{v}}{|\mathbf{u}||\mathbf{v}|} = \frac{-1}{\sqrt{13} \cdot \sqrt{5}} \Rightarrow \theta \approx 97°$

11. (a) $\mathbf{u} \cdot \mathbf{v} = \langle 0, -5 \rangle \cdot \langle -1, -\sqrt{3} \rangle = 0 + 5\sqrt{3} = 5\sqrt{3}$ **(b)** $\cos\theta = \dfrac{\mathbf{u} \cdot \mathbf{v}}{|\mathbf{u}||\mathbf{v}|} = \frac{5\sqrt{3}}{5 \cdot 2} = \frac{\sqrt{3}}{2}$

$\Rightarrow \theta = 30°$

13. (a) $\mathbf{u} \cdot \mathbf{v} = (\mathbf{i} + 3\mathbf{j}) \cdot (4\mathbf{i} - \mathbf{j}) = 4 - 3 = 1$

(b) $\cos\theta = \dfrac{\mathbf{u} \cdot \mathbf{v}}{|\mathbf{u}||\mathbf{v}|} = \frac{1}{\sqrt{10} \cdot \sqrt{17}} \Rightarrow \theta \approx 85.6°$

15. $\mathbf{u} \cdot \mathbf{v} = -12 + 12 = 0 \Rightarrow$ vectors are orthogonal

17. $\mathbf{u} \cdot \mathbf{v} = -8 + 12 = 4 \neq 0 \Rightarrow$ vectors are not orthogonal

19. $\mathbf{u} \cdot \mathbf{v} = -24 + 24 = 0 \Rightarrow$ vectors are orthogonal

21. $$\begin{aligned} \mathbf{u} \cdot \mathbf{v} + \mathbf{u} \cdot \mathbf{w} &= \langle 2, 1 \rangle \cdot \langle 1, -3 \rangle + \langle 2, 1 \rangle \cdot \langle 3, 4 \rangle \\ &= 2 - 3 + 6 + 4 = 9 \end{aligned}$$

23. $$\begin{aligned} (\mathbf{u} + \mathbf{v}) \cdot (\mathbf{u} - \mathbf{v}) &= [\langle 2, 1 \rangle + \langle 1, -3 \rangle] \cdot [\langle 2, 1 \rangle - \langle 1, -3 \rangle] \\ &= \langle 3, -2 \rangle \cdot \langle 1, 4 \rangle = 3 - 8 = -5 \end{aligned}$$

25. $x = \dfrac{\mathbf{u}\cdot\mathbf{v}}{|\mathbf{v}|} = \dfrac{12-24}{5} = -\dfrac{12}{5}$

27. $x = \dfrac{\mathbf{u}\cdot\mathbf{v}}{|\mathbf{v}|} = \dfrac{0-24}{1} = -24$

29. (a) $\mathbf{u}_1 = \text{proj}_{\mathbf{v}}\,\mathbf{u} = \left(\dfrac{\mathbf{u}\cdot\mathbf{v}}{|\mathbf{v}|^2}\right)\mathbf{v} = \left(\dfrac{\langle -2,4\rangle\cdot\langle 1,1\rangle}{1^2+1^2}\right)\langle 1,1\rangle = \langle 1,1\rangle$.

(b) $\mathbf{u}_2 = \mathbf{u}-\mathbf{u}_1 = \langle -2,4\rangle - \langle 1,1\rangle = \langle -3,3\rangle$

31. (a) $\mathbf{u}_1 = \text{proj}_{\mathbf{v}}\,\mathbf{u} = \left(\dfrac{\mathbf{u}\cdot\mathbf{v}}{|\mathbf{v}|^2}\right)\mathbf{v} = \left(\dfrac{\langle 1,2\rangle\cdot\langle 1,-3\rangle}{1^2+(-3)^2}\right)\langle 1,-3\rangle = -\tfrac{1}{2}\langle 1,-3\rangle = \left\langle -\tfrac{1}{2},\tfrac{3}{2}\right\rangle$

(b) $\mathbf{u}_2 = \mathbf{u}-\mathbf{u}_1 = \langle 1,2\rangle - \left\langle -\tfrac{1}{2},\tfrac{3}{2}\right\rangle = \left\langle \tfrac{3}{2},\tfrac{1}{2}\right\rangle$

33. (a) $\mathbf{u}_1 = \text{proj}_{\mathbf{v}}\,\mathbf{u} = \left(\dfrac{\mathbf{u}\cdot\mathbf{v}}{|\mathbf{v}|^2}\right)\mathbf{v} = \left(\dfrac{\langle 2,9\rangle\cdot\langle -3,4\rangle}{(-3)^3+4^2}\right)\langle -3,4\rangle = \tfrac{6}{5}\langle -3,4\rangle = \left\langle -\tfrac{18}{5},\tfrac{24}{5}\right\rangle$

(b) $\mathbf{u}_2 = \mathbf{u}-\mathbf{u}_1 = \langle 2,9\rangle - \left\langle -\tfrac{18}{5},\tfrac{24}{5}\right\rangle = \left\langle \tfrac{28}{5},\tfrac{21}{5}\right\rangle$

35. $W = \mathbf{F}\cdot\mathbf{d} = \langle 4,-5\rangle\cdot\langle 3,8\rangle = -28$

37. $W = \mathbf{F}\cdot\mathbf{d} = \langle 10,3\rangle\cdot\langle 4,-5\rangle = 25$

39. Let $\mathbf{u} = \langle \mathbf{u}_1,\mathbf{u}_2\rangle$ and $\mathbf{v} = \langle \mathbf{v}_1,\mathbf{v}_2\rangle$. Then

$$\mathbf{u}\cdot\mathbf{v} = \langle u_1,u_2\rangle\cdot\langle v_1,v_2\rangle = u_1v_1+u_2v_2 = v_1u_1+v_2u_2 = \langle v_1,v_2\rangle\cdot\langle u_1,u_2\rangle = \mathbf{v}\cdot\mathbf{u}$$

41. Let $\mathbf{u} = \langle u_1,u_2\rangle$, $\mathbf{v} = \langle v_1,v_2\rangle$, and $\mathbf{w} = \langle w_1,w_2\rangle$. Then

$$\begin{aligned}(\mathbf{u}+\mathbf{v})\cdot\mathbf{w} &= (\langle u_1,u_2\rangle+\langle v_1,v_2\rangle)\cdot\langle w_1,w_2\rangle = \langle u_1+v_1,u_2+v_2\rangle\cdot\langle w_1,w_2\rangle \\ &= u_1w_1+v_1w_1+u_2w_2+v_2w_2 = u_1w_1+u_2w_2+v_1w_1+v_2w_2 \\ &= \langle u_1,u_2\rangle\cdot\langle w_1,w_2\rangle+\langle v_1,v_2\rangle\cdot\langle w_1,w_2\rangle = \mathbf{u}\cdot\mathbf{w}+\mathbf{v}\cdot\mathbf{w}\end{aligned}$$

43. We use the definition that $\text{proj}_{\mathbf{v}}\,\mathbf{u} = \left(\dfrac{\mathbf{u}\cdot\mathbf{v}}{|\mathbf{v}|^2}\right)\mathbf{v}$. Then

$$\begin{aligned}\text{proj}_{\mathbf{v}}\,\mathbf{u}\cdot(\mathbf{u}-\text{proj}_{\mathbf{v}}\,\mathbf{u}) &= \left(\frac{\mathbf{u}\cdot\mathbf{v}}{|\mathbf{v}|^2}\right)\mathbf{v}\cdot\left[\mathbf{u}-\left(\frac{\mathbf{u}\cdot\mathbf{v}}{|\mathbf{v}|^2}\right)\mathbf{v}\right] = \left(\frac{\mathbf{u}\cdot\mathbf{v}}{|\mathbf{v}|^2}\right)(\mathbf{v}\cdot\mathbf{u})-\left(\frac{\mathbf{u}\cdot\mathbf{v}}{|\mathbf{v}|^2}\right)\mathbf{v}\cdot\left(\frac{\mathbf{u}\cdot\mathbf{v}}{|\mathbf{v}|^2}\right)\mathbf{v} \\ &= \frac{(\mathbf{u}\cdot\mathbf{v})^2}{|\mathbf{v}|^2}-\frac{(\mathbf{u}\cdot\mathbf{v})^2}{|\mathbf{v}|^4}|\mathbf{v}|^2 = \frac{(\mathbf{u}\cdot\mathbf{v})^2}{|\mathbf{v}|^2}-\frac{(\mathbf{u}\cdot\mathbf{v})^2}{|\mathbf{v}|^2} = 0\end{aligned}$$

Thus $\mathbf{u}$ and $\mathbf{u}-\text{proj}_{\mathbf{v}}\,\mathbf{u}$ are orthogonal.

45. $W = \mathbf{F}\cdot\mathbf{d} = \langle 4,-7\rangle\cdot\langle 4,0\rangle = 16$ ft-lb

47. The distance vector is $\mathbf{D} = \langle 200, 0 \rangle$ and the force vector is $\mathbf{F} = \langle 50\cos 30°, 50\sin 30° \rangle$. Hence, the work done is

$$W = \mathbf{F} \cdot \mathbf{D} = \langle 200, 0 \rangle \cdot \langle 50\cos 30°, 50\sin 30° \rangle = 200 \cdot 50\cos 30° \approx 8660 \text{ ft-lb.}$$

49. Since the weight of the car is 2755 lb, the force exerted perpendicular to the earth is 2755 lb. Resolving this into a force $\mathbf{u}$ perpendicular to the driveway gives $|\mathbf{u}| = 2766\cos 65° \approx 1164$. lb. Thus, a force of about 1164 lb is required.

51. Since the force required parallel to the plane is 80 lb and the weight of the package is 200 lb, it follows that $80 = 200\sin\theta$, where θ is the angle of inclination of the plane. Then $\theta = \sin^{-1}\left(\frac{80}{200}\right) \approx 23.58°$, and so the angle of inclination is approximately $23.6°$.

53. (a) $2(0) + 4(2) = 8$, so $Q(0,2)$ lies on L. $2(2) + 4(1) = 4 + 4 = 8$, so $R(2,1)$ lies on L.

(b) $\mathbf{u} = \overrightarrow{QP} = \langle 0,2 \rangle - \langle 3,4 \rangle = \langle -3,-2 \rangle$. $\mathbf{v} = \overrightarrow{QR} = \langle 0,2 \rangle - \langle 2,1 \rangle = \langle -2,1 \rangle$.

$$\begin{aligned} \mathbf{w} &= \text{proj}_{\mathbf{v}}\,\mathbf{u} = \left(\frac{\mathbf{u} \cdot \mathbf{v}}{|\mathbf{v}|^2}\right)\mathbf{v} = \frac{\langle -3,-2 \rangle \cdot \langle 2,1 \rangle}{(-2)^2 + 1^2}\langle -2,1 \rangle \\ &= -\tfrac{8}{5}\langle -2,1 \rangle = \left\langle \tfrac{16}{5}, -\tfrac{8}{5} \right\rangle \end{aligned}$$

(c) From the graph, we can see that $\mathbf{u} - \mathbf{w}$ is orthogonal to $\mathbf{v}$ (and thus to L). Thus, the distance from P to L is $|\mathbf{u} - \mathbf{w}|$.

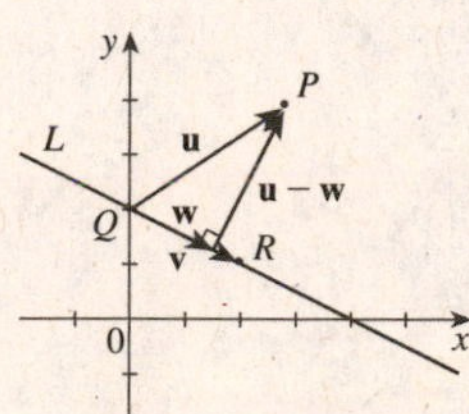

6.3 Three-Dimensional Coordinate Geometry

1. In a three-dimensional coordinate system the three mutually perpendicular axes are called the x -axis, the y -axis, and the z -axis. The point P has coordinates $(5,2,3)$. The equation of the plane passing through P and parallel to the xz -plane is $y=2$.

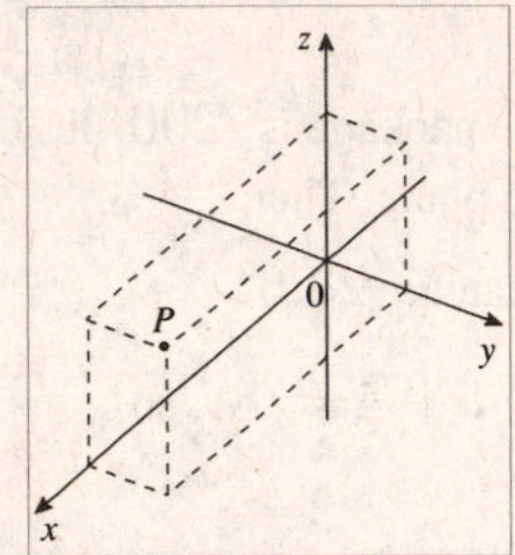

3. (a)

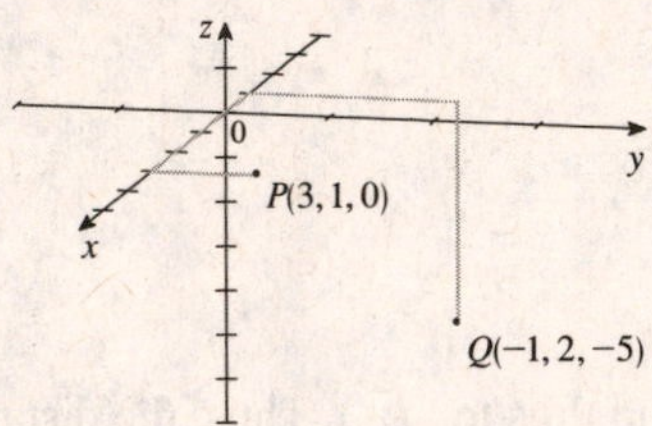

(b) $d(P,Q)=\sqrt{(-1-3)^2+(2-1)^2+(-5-0)^2}=\sqrt{42}$

5. (a)

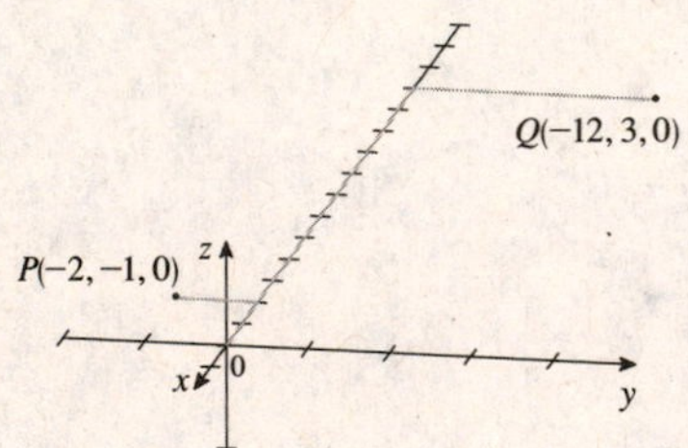

(b) $d(P,Q)=\sqrt{(-12+2)^2+(3+1)^2+(0-0)^2}=2\sqrt{29}$

7. $x=4$ is a plane parallel to the yz -plane.

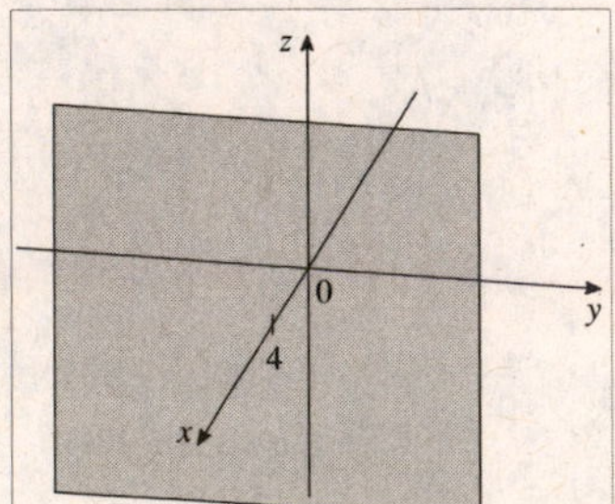

9. $z = 8$ is a plane parallel to the xy -plane.

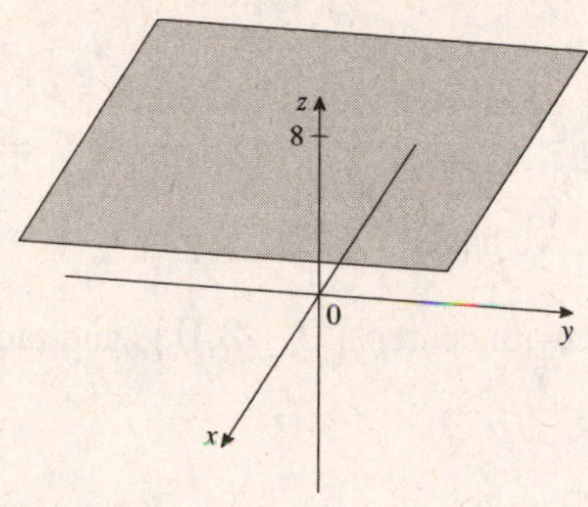

11. A sphere with radius $r = 5$ and center $C(2,-5,3)$ has equation

$(x-2)^2 + [y-(-5)]^2 + (z-3)^2 = 5^2$, or $(x-2)^2 + (y+5)^2 + (z-3)^2 = 25$

.

13. A sphere with radius $r = \sqrt{6}$ and center $C(3,-1,0)$ has equation

$(x-3)^2 + (y+1)^2 + z^2 = 6$.

15. We complete the squares in x , y , and z : $x^2 + y^2 + z^2 - 10x + 2y + 8z = 9 \Leftrightarrow$

$(x^2 - 10x + 25) + (y^2 + 2y + 1) + (z^2 + 8z + 16) = 9 + 25 + 1 + 16 \Leftrightarrow$

$(x-5)^2 + (y+1)^2 + (z+4)^2 = 51$. This is an equation of a sphere with center $(5,-1,-4)$ and radius $\sqrt{51}$.

17. We complete the squares in x , y , and z : $x^2 + y^2 + z^2 = 12x + 2y \Leftrightarrow$

$(x^2 - 12x + 36) + (y^2 - 2y + 1) + z^2 = 36 + 1 \Leftrightarrow$

$(x-6)^2 + (y-1)^2 + z^2 = 37$. This is an equation of a sphere with center $(6,1,0)$ and radius $\sqrt{37}$.

19. (a) To find the trace in the yz -plane, we set $x = 0$:

$(0+1)^2 + (y-2)^2 + (z+10)^2 = 100 \Leftrightarrow (y-2)^2 + (z+10)^2 = 99$. This represents a circle with center $(0,2,-10)$ and radius $3\sqrt{11}$.

(b) We set $x = 4$ and find $(4+1)^2 + (y-2)^2 + (z+10)^2 = 100 \Leftrightarrow$

$(y-2)^2 + (z+10)^2 = 75$. This represents a circle with center $(4,2,-10)$ and radius $5\sqrt{3}$.

21. With the origin at its center, an equation of the tank is $x^2 + y^2 + z^2 = 25$. The metal circle is the trace in the plane $z = -4$, so its equation is $x^2 + y^2 + (-4)^2 = 25$ or $x^2 + y^2 = 9$. Therefore, its radius is 3 .

24. Using the Distance Formula (squaring both sides first), we have

$$x^2 + (y-3)^2 + z^2 = 2(x^2 + y^2 + z^2) \quad \Leftrightarrow$$

$$x^2 + (y-3)^2 + z^2 = 2x^2 + 2y^2 + 2z^2 \quad \Leftrightarrow \quad x^2 + 2y^2 - (y-3)^2 + z^2 = 0 \quad \Leftrightarrow$$

$x^2 + y^2 + 6y - 9 + z^2 = 0$. Completing the square in y , we have

$x^2 + (y+3)^2 + z^2 = 18$. This is an equation of the sphere with center $(0,-3,0)$ and radius $\sqrt{18} = 3\sqrt{2}$.

6.4 Vectors in Three Dimensions

1. A vector in three dimensions can be written in terms of the *unit* vectors $\mathbf{i}$, $\mathbf{j}$, and $\mathbf{k}$ as $a = a_1\mathbf{i} + a_2\mathbf{j} + a_3\mathbf{k}$. The magnitude of the vector $\mathbf{a}$ is $|a| = \sqrt{a_1^2 + a_2^2 + a_3^2}$. So $\langle 4,-2,4\rangle = 4\mathbf{i} + (-2)\mathbf{j} + 4\mathbf{k}$ and $7\mathbf{j} - 24\mathbf{k} = \langle 0,7,-24\rangle$.

3. The vector with initial point $P(1,-1,0)$ and terminal point $Q(0,-2,5)$ is $\mathbf{v} = \langle 0-1,-2-(-1),5-0\rangle = \langle -1,-1,5\rangle$.

5. The vector with initial point $P(6,-1,0)$ and terminal point $Q(0,-3,0)$ is $\mathbf{v} = \langle 0-6,-3-(-1),0-0\rangle = \langle -6,-2,0\rangle$.

7. If the vector $\mathbf{v} = \langle 3,4,-2\rangle$ has initial point $P(2,0,1)$, its terminal point is $(2+3,0+4,1-2) = (5,4,-1)$.

9. If the vector $\mathbf{v} = \langle -2,0,2\rangle$ has initial point $P(3,0,-3)$, its terminal point is $(3-2,0+0,-3+2) = (1,0,-1)$.

11. $|\langle -2,1,2\rangle| = \sqrt{(-2)^2 + 1^2 + 2^2} = 3$

13. $|\langle 3,5,-4\rangle| = \sqrt{3^2 + 5^2 + (-4)^2} = 5\sqrt{2}$

15. If $\mathbf{u} = \langle 2,-7,3\rangle$ and $\mathbf{v} = \langle 0,4,-1\rangle$, then
$\mathbf{u} + \mathbf{v} = \langle 2+0,-7+4,3-1\rangle = \langle 2,-3,2\rangle$,
$\mathbf{u} - \mathbf{v} = \langle 2-0,-7-4,3-(-1)\rangle = \langle 2,-11,4\rangle$, and
$3\mathbf{u} - \frac{1}{2}\mathbf{v} = \langle 3(2)-\frac{1}{2}(0),3(-7)-\frac{1}{2}(4),3(3)-\frac{1}{2}(-1)\rangle = \langle 6,-23,\frac{19}{2}\rangle$.

17. If $\mathbf{u} = \mathbf{i} + \mathbf{j}$ and $\mathbf{v} = -\mathbf{j} - 2\mathbf{k}$, then $\mathbf{u} + \mathbf{v} = \mathbf{i} + \mathbf{j} - \mathbf{j} - 2\mathbf{k} = \mathbf{i} - 2\mathbf{k}$,
$\mathbf{u} - \mathbf{v} = \mathbf{i} + \mathbf{j} - (-\mathbf{j} - 2\mathbf{k}) = \mathbf{i} + 2\mathbf{j} + 2\mathbf{k}$, and
$3\mathbf{u} - \frac{1}{2}\mathbf{v} = 3(\mathbf{i} + \mathbf{j}) - \frac{1}{2}(-\mathbf{j} - 2\mathbf{k}) = 3\mathbf{i} + \frac{7}{2}\mathbf{j} + \mathbf{k}$.

19. $\langle 12,0,2\rangle = 12\mathbf{i} + 2\mathbf{k}$

21. $\langle 3,-3,0\rangle = 3\mathbf{i} - 3\mathbf{j}$

23. (a) $-2\mathbf{u} + 3\mathbf{v} = -2\langle 0,-2,1\rangle + 3\langle 1,-1,0\rangle = \langle 3,1,-2\rangle$

(b) $-2\mathbf{u} + 3\mathbf{v} = 3\mathbf{i} + \mathbf{j} - 2\mathbf{k}$

25. $\mathbf{u}\cdot\mathbf{v} = \langle 2,5,0\rangle\cdot\langle \frac{1}{2},-1,10\rangle = 2(\frac{1}{2}) + 5(-1) + 0(10) = -4$

27. $\mathbf{u}\cdot\mathbf{v} = (6\mathbf{i} - 4\mathbf{j} - 2\mathbf{k})\cdot(\frac{5}{6}\mathbf{i} + \frac{3}{2}\mathbf{j} - \mathbf{k}) = 6(\frac{5}{6}) - 4(\frac{3}{2}) - 2(-1) = 1$

29. $\langle 4,-2,-4\rangle\cdot\langle 1,-2,2\rangle = 4(1) - 2(-2) - 4(2) = 0$, so the vectors are perpendicular.

31. $\langle 0.3,1.2,-0.9\rangle\cdot\langle 10,-5,10\rangle = 0.3(10) + 1.2(-5) - 0.9(10) = -12$, so the vectors are not perpendicular.

33. $\cos\theta = \dfrac{\mathbf{u}\cdot\mathbf{v}}{|\mathbf{u}||\mathbf{v}|} = \dfrac{\langle 2,-2,-1\rangle\cdot\langle 1,2,2\rangle}{|\langle 2,-2,-1\rangle||\langle 1,2,2\rangle|} = \dfrac{2(1)-2(2)-1(2)}{\sqrt{2^2+(-2)^2+(-1)^2}\sqrt{1^2+2^2+2^2}} = -\dfrac{4}{9}$,

so $\theta = \cos^{-1}\left(-\frac{4}{9}\right) \approx 116.4^\circ$.

35. $\cos\theta = \dfrac{\mathbf{u}\cdot\mathbf{v}}{|\mathbf{u}||\mathbf{v}|} = \dfrac{(\mathbf{j}+\mathbf{k})\cdot(\mathbf{i}+2\mathbf{j}-3\mathbf{k})}{|\mathbf{j}+\mathbf{k}||\mathbf{i}+2\mathbf{j}-3\mathbf{k}|} = \dfrac{1(2)+1(-3)}{\sqrt{1^2+1^2}\sqrt{1^2+2^2+(-3)^2}} = -\dfrac{\sqrt{28}}{28}$,

so $\theta = \cos^{-1}\left(-\frac{\sqrt{28}}{28}\right) \approx 100.9^\circ$.

37. The length of the vector $3\mathbf{i}+4\mathbf{j}+5\mathbf{k}$ is $\sqrt{3^2+4^2+5^2} = 5\sqrt{2}$, so by definition, its direction angles satisfy $\cos\alpha = \frac{3}{5\sqrt{2}}$, $\cos\beta = \frac{4}{5\sqrt{2}}$, and $\cos\gamma = \frac{1}{\sqrt{2}}$. Thus,

$\alpha = \cos^{-1}\frac{3}{5\sqrt{2}} \approx 65^\circ$, $\beta = \cos^{-1}\frac{2\sqrt{2}}{5} \approx 56^\circ$, and $\gamma = \cos^{-1}\frac{1}{\sqrt{2}} = 45^\circ$.

39. $|\langle 2,3,-6\rangle| = \sqrt{2^2+3^2+(-6)^2} = 7$, so $\cos\alpha = \frac{2}{7}$, $\cos\beta = \frac{3}{7}$, and $\cos\gamma = \frac{-6}{7}$

$\Leftrightarrow$ $\alpha = \cos^{-1}\frac{2}{7} \approx 73^\circ$, $\beta = \cos^{-1}\frac{3}{7} \approx 65^\circ$, and $\gamma = \cos^{-1}\left(-\frac{6}{7}\right) \approx 149^\circ$.

41. We are given that $\alpha = \frac{\pi}{3}$, $\gamma = \frac{2\pi}{3}$, and β is acute. Using the property of direction cosines

$\cos^2\alpha + \cos^2\beta + \cos^2\gamma = 1$, we have $\left(\cos\frac{\pi}{3}\right)^2 + \cos^2\beta + \left(\cos\frac{2\pi}{3}\right)^2 = 1$ $\Leftrightarrow$

$\left(\frac{1}{2}\right)^2 + \cos^2\beta + \left(-\frac{1}{2}\right)^2 = 1$ $\Leftrightarrow$ $\cos^2\beta = \frac{1}{2}$ $\Leftrightarrow$ $\cos\beta = \pm\frac{1}{\sqrt{2}}$. Because β is acute, we have $\beta = \cos^{-1}\frac{1}{\sqrt{2}} = 45^\circ$.

43. We are given that $\alpha = 60^\circ$, $\beta = 50^\circ$, and γ is obtuse, so

$\cos^2 60^\circ + \cos^2 50^\circ + \cos^2\gamma = 1$ $\Leftrightarrow$ $\cos^2\gamma = 1 - \cos^2 60^\circ - \cos^2 50^\circ \approx 0.337$.

Because γ is obtuse, $\gamma \approx \cos^{-1}\left(-\sqrt{0.337}\right) \approx 125^\circ$.

45. Here $\cos^2\alpha + \cos^2\beta = \cos^2 20^\circ + \cos^2 45^\circ \approx 1.38 > 1$, so there is no angle γ satisfying the property of direction cosines $\cos^2\alpha + \cos^2\beta + \cos^2\gamma = 1$.

47. (a) The second and third forces are $\mathbf{F}_2 = 24\mathbf{j}$ and $\mathbf{F}_3 = -25\mathbf{k}$. Therefore,

$\mathbf{F}_1 + \mathbf{F}_2 + \mathbf{F}_3 + \mathbf{F}_4 = 0$ $\Leftrightarrow$ $7\mathbf{i} + 24\mathbf{j} - 25\mathbf{k} + \mathbf{F}_4 = 0$ $\Leftrightarrow$

$\mathbf{F}_4 = -7\mathbf{i} - 24\mathbf{j} + 25\mathbf{k}$.

(b) $|\mathbf{F}_4| = \sqrt{(-7)^2 + (-24)^2 + (-25)^2} = 25\sqrt{2}$

49. (a) We solve $\mathbf{v} = a\mathbf{u} \Leftrightarrow \langle -6,4,-8\rangle = a\langle 3,-2,4\rangle \Leftrightarrow a = -2$. Therefore, the vectors are parallel and $\mathbf{v} = -2\mathbf{u}$.

(b) $\mathbf{v} = a\mathbf{u} \Leftrightarrow \langle 12,8,-16\rangle = a\langle -9,-6,12\rangle \Leftrightarrow a = -\frac{4}{3}$, so the vectors are parallel and $\mathbf{v} = -\frac{4}{3}\mathbf{u}$.

(c) $\mathbf{v} = a\mathbf{u} \Leftrightarrow 2\mathbf{i} + 2\mathbf{j} - 2\mathbf{k} = a(\mathbf{i} + \mathbf{j} + \mathbf{k})$ has no solution, so the vectors are not parallel.

51. (a) $(\mathbf{r} - \mathbf{a})\cdot(\mathbf{r} - \mathbf{b}) = 0 \Leftrightarrow$

$\langle x-2, y-2, z-2\rangle\cdot\langle x-(-2), y-(-2), z-0\rangle = 0 \Leftrightarrow$

$(x-2)(x+2) + (y-2)(y+2) + (z-2)z = 0 \Leftrightarrow$

$x^2 - 4 + y^2 - 4 + z^2 - 2z = 0 \Leftrightarrow x^2 + y^2 + (z-1)^2 = 4+4+1 = 9$.

(b) The sphere with equation $x^2 + y^2 + (z-1)^2 = 9$ has center $(0,0,1)$ and radius 3 .

(c) The diagram shows the plane determined by $\mathbf{a}$, $\mathbf{b}$, and $\mathbf{r}$, along with the trace of the sphere in that plane. We see that the equation $(\mathbf{r} - \mathbf{a})\cdot(\mathbf{r} - \mathbf{b}) = 0$ states the fact that lines from the ends of a diameter of a circle to any point on its surface meet at right angles.

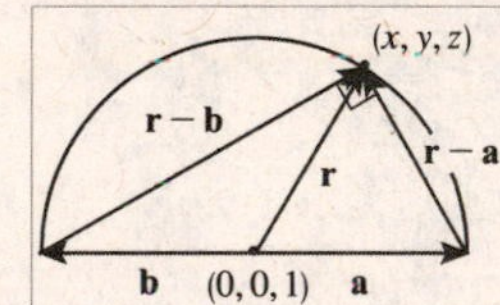

(d) Let $\mathbf{a} = \langle 0,1,3\rangle$ and $\mathbf{b} = \langle 2,-1,4\rangle$. Then $(\mathbf{r} - \mathbf{a})\cdot(\mathbf{r} - \mathbf{b}) = 0 \Leftrightarrow$

$\langle x, y-1, z-3\rangle\cdot\langle x-2, y+1, z-4\rangle = 0 \Leftrightarrow$

$x(x-2) + (y-1)(y+1) + (z-3)(z-4) = 0 \Leftrightarrow$

$x^2 - 2x + y^2 - 1 + z^2 - 7x + 12 = 0 \Leftrightarrow$

$(x-1)^2 + y^2 + \left(z - \frac{7}{2}\right)^2 = 1 - 12 - 1 + \frac{49}{4} = \frac{1}{4}$, an equation of a circle with center $\left(1,0,\frac{7}{2}\right)$ and radius $\frac{1}{2}$.

6.5 The Cross Product

1. The cross product of the vectors $\mathbf{a} = \langle a_1, a_2, a_3 \rangle$ and $\mathbf{b} = \langle b_1, b_2, b_3 \rangle$ is the vector

$$\mathbf{a} \times \mathbf{b} = \begin{vmatrix} \mathbf{i} & \mathbf{j} & \mathbf{k} \\ a_1 & a_2 & a_3 \\ b_1 & b_2 & b_3 \end{vmatrix} = (a_2 b_3 - a_3 b_2)\mathbf{i} + (a_3 b_1 - a_1 b_3)\mathbf{j} + (a_1 b_2 - a_2 b_1)\mathbf{k}$$

. So the cross product of $\mathbf{a} = \langle 1,0,1 \rangle$ and $\mathbf{b} = \langle 2,3,0 \rangle$ is

$$\mathbf{a} \times \mathbf{b} = \begin{vmatrix} \mathbf{i} & \mathbf{j} & \mathbf{k} \\ 1 & 0 & 1 \\ 2 & 3 & 0 \end{vmatrix} = -3\mathbf{i} + 2\mathbf{j} + 3\mathbf{k}.$$

3. $\mathbf{a} \times \mathbf{b} = \begin{vmatrix} \mathbf{i} & \mathbf{j} & \mathbf{k} \\ 1 & 0 & -3 \\ 2 & 3 & 0 \end{vmatrix} = 9\mathbf{i} - 6\mathbf{j} + 3\mathbf{k}$

5. $\mathbf{a} \times \mathbf{b} = \begin{vmatrix} \mathbf{i} & \mathbf{j} & \mathbf{k} \\ 6 & -2 & 8 \\ -9 & 3 & -12 \end{vmatrix} = \mathbf{0}$

7. $\mathbf{a} \times \mathbf{b} = \begin{vmatrix} \mathbf{i} & \mathbf{j} & \mathbf{k} \\ 1 & 1 & 1 \\ 3 & 0 & -4 \end{vmatrix} = -4\mathbf{i} + 7\mathbf{j} - 3\mathbf{k}$

9. (a) $\mathbf{a} \times \mathbf{b} = \begin{vmatrix} \mathbf{i} & \mathbf{j} & \mathbf{k} \\ 1 & 1 & -1 \\ -1 & 1 & -1 \end{vmatrix} = \langle 0,2,2 \rangle$ is perpendicular to both $\mathbf{a}$ and $\mathbf{b}$.

(b) $\dfrac{\mathbf{a} \times \mathbf{b}}{|\mathbf{a} \times \mathbf{b}|} = \dfrac{\langle 0,2,2 \rangle}{\sqrt{2^2 + 2^2}} = \left\langle 0, \frac{\sqrt{2}}{2}, \frac{\sqrt{2}}{2} \right\rangle$ is a unit vector perpendicular to both $\mathbf{a}$ and $\mathbf{b}$.

11. (a) $\mathbf{a} \times \mathbf{b} = \begin{vmatrix} \mathbf{i} & \mathbf{j} & \mathbf{k} \\ \frac{1}{2} & -1 & \frac{2}{3} \\ 6 & -12 & -6 \end{vmatrix} = \langle 14,7,0 \rangle$ is perpendicular to both $\mathbf{a}$ and $\mathbf{b}$.

(b) $\dfrac{\mathbf{a} \times \mathbf{b}}{|\mathbf{a} \times \mathbf{b}|} = \dfrac{\langle 14,7,0 \rangle}{\sqrt{14^2 + 7^2}} = \left\langle \frac{2\sqrt{5}}{5}, \frac{\sqrt{5}}{5}, 0 \right\rangle$ is a unit vector perpendicular to both $\mathbf{a}$ and $\mathbf{b}$.

13. $|\mathbf{a} \times \mathbf{b}| = |\mathbf{a}||\mathbf{b}|\sin\theta = 6\left(\frac{1}{2}\right)\sin 60^\circ = \frac{3\sqrt{3}}{2}$

15. $|\mathbf{a} \times \mathbf{b}| = |\mathbf{a}||\mathbf{b}|\sin\theta = 10(10)\sin 90^\circ = 100$

17. $\overrightarrow{PQ} \times \overrightarrow{PR} = \langle 1,1,-1 \rangle \times \langle -2,0,0 \rangle = \langle 0,2,2 \rangle$ is perpendicular to the plane passing through P, Q, and R.

19. $\overrightarrow{PQ} \times \overrightarrow{PR} = \langle 1,1,5 \rangle \times \langle -1,-1,5 \rangle = \langle 10,-10,0 \rangle$ is perpendicular to the plane passing through P, Q, and R.

21. The area of the parallelogram determined by $\mathbf{u} = \langle 3,2,1 \rangle$ and $\mathbf{v} = \langle 1,2,3 \rangle$ is $|\mathbf{u} \times \mathbf{v}| = |\langle 4,-8,4 \rangle| = 4\sqrt{6}$.

23. The area of the parallelogram determined by $\mathbf{u} = 2\mathbf{i} - \mathbf{j} + 4\mathbf{k}$ and $\mathbf{v} = \frac{1}{2}\mathbf{i} + 2\mathbf{j} - \frac{3}{2}\mathbf{k}$ is $|\mathbf{u} \times \mathbf{v}| = \left|\left\langle -\frac{13}{2}, 5, \frac{9}{2} \right\rangle\right| = \frac{5\sqrt{14}}{2}$.

25. The area of triangle PQR is one-half the area of the parallelogram determined by $\overrightarrow{PQ}$ and $\overrightarrow{PR}$, that is, $\frac{1}{2}\left|\overrightarrow{PQ} \times \overrightarrow{PR}\right| = \frac{1}{2}|\langle -1,1,-1 \rangle \times \langle 1,3,3 \rangle| = \frac{1}{2}\sqrt{6^2 + 2^2 + (-4)^2} = \sqrt{14}$.

27. The area of triangle PQR is one-half the area of the parallelogram determined by $\overrightarrow{PQ}$ and $\overrightarrow{PR}$, that is,

$$\frac{1}{2}\left|\overrightarrow{PQ} \times \overrightarrow{PR}\right| = \frac{1}{2}|\langle -6,-6,0 \rangle \times \langle -6,0,-6 \rangle| = \frac{1}{2}\sqrt{36^2 + (-36)^2 + (-36)^2} = 18\sqrt{3}$$

.

29. **(a)** $\mathbf{a} \cdot (\mathbf{b} \times \mathbf{c}) = \langle 1,2,3 \rangle \cdot (\langle -3,2,1 \rangle \times \langle 0,8,10 \rangle) = \langle 1,2,3 \rangle \cdot \langle 12,30,-24 \rangle = 0$

(b) Because their scalar triple product is 0, the vectors are coplanar.

31. **(a)** $\mathbf{a} \cdot (\mathbf{b} \times \mathbf{c}) = \langle 2,3,-2 \rangle \cdot (\langle -1,4,0 \rangle \times \langle 3,-1,3 \rangle) = \langle 2,3,-2 \rangle \cdot \langle 12,3,-11 \rangle = 55$

(b) Because their scalar triple product is nonzero, the vectors are not coplanar. The volume of the parallelepiped that they determine is $|\mathbf{a} \cdot (\mathbf{b} \times \mathbf{c})| = 55$.

33. **(a)** $\mathbf{a} \cdot (\mathbf{b} \times \mathbf{c}) = \langle 1,-1,1 \rangle \cdot (\langle 0,-1,1 \rangle \times \langle 1,1,1 \rangle) = \langle 1,-1,1 \rangle \cdot \langle -2,1,1 \rangle = -2$

(b) Because their scalar triple product is nonzero, the vectors are not coplanar. The volume of the parallelepiped that they determine is $|\mathbf{a} \cdot (\mathbf{b} \times \mathbf{c})| = 2$.

35. **(a)** We have $|\mathbf{a}| = 120$ cm, $|\mathbf{b}| = 150$ cm, $|\mathbf{c}| = 300$ cm, the angle between $\mathbf{b}$ and $\mathbf{c}$ is $90° - 30° = 60°$, and the angle between $\mathbf{a}$ and $\mathbf{b} \times \mathbf{c}$ is $0°$ (because $\mathbf{a}$ is perpendicular to both $\mathbf{b}$ and $\mathbf{c}$). Therefore,

$\mathbf{a} \cdot (\mathbf{b} \times \mathbf{c}) = |\mathbf{a}||\mathbf{b} \times \mathbf{c}|\cos 0° = 120(150 \cdot 300 \cdot \sin 60°) = 2,700,000\sqrt{3} \approx 4,676,537$.

(b) The capacity in liters is approximately $\dfrac{4{,}676{,}537 \text{ cm}^3}{1000 \text{ cm}^3 / \text{L}} \approx 4677$ liters.

37. **(a)** $\mathbf{u} \cdot (\mathbf{v} \times \mathbf{w}) = \langle 0,1,1 \rangle \cdot (\langle 1,0,1 \rangle \times \langle 1,1,0 \rangle) = \langle 0,1,1 \rangle \cdot \langle -1,1,1 \rangle = 2$,

$\mathbf{u} \cdot (\mathbf{w} \times \mathbf{v}) = \langle 0,1,1 \rangle \cdot (\langle 1,1,0 \rangle \times \langle 1,0,1 \rangle) = \langle 0,1,1 \rangle \cdot \langle 1,-1,-1 \rangle = -2$,

$\mathbf{v} \cdot (\mathbf{u} \times \mathbf{w}) = \langle 1,0,1 \rangle \cdot (\langle 0,1,1 \rangle \times \langle 1,1,0 \rangle) = \langle 1,0,1 \rangle \cdot \langle -1,1,-1 \rangle = -2$,

$\mathbf{v} \cdot (\mathbf{w} \times \mathbf{u}) = \langle 1,0,1 \rangle \cdot (\langle 1,1,0 \rangle \times \langle 0,1,1 \rangle) = \langle 1,0,1 \rangle \cdot \langle 1,-1,1 \rangle = 2$,

$\mathbf{w}\cdot(\mathbf{u}\times\mathbf{v}) = \langle 1,1,0\rangle\cdot(\langle 0,1,1\rangle\times\langle 1,0,1\rangle) = \langle 1,1,0\rangle\cdot\langle 1,1,-1\rangle = 2$, and

$\mathbf{w}\cdot(\mathbf{v}\times\mathbf{u}) = \langle 1,1,0\rangle\cdot(\langle 1,0,1\rangle\times\langle 0,1,1\rangle) = \langle 1,1,0\rangle\cdot\langle -1,-1,1\rangle = -2$.

(b) It appears that

$\mathbf{u}\cdot(\mathbf{v}\times\mathbf{w}) = \mathbf{v}\cdot(\mathbf{w}\times\mathbf{u}) = \mathbf{w}\cdot(\mathbf{u}\times\mathbf{v}) = -\mathbf{u}\cdot(\mathbf{w}\times\mathbf{v}) = -\mathbf{v}\cdot(\mathbf{u}\times\mathbf{w}) = -\mathbf{w}\cdot(\mathbf{v}\times\mathbf{u})$

.

(c) We know that the absolute values of the six scalar triple products must be equal because they all represent the volume of the parallelepiped determined by $\mathbf{u}$, $\mathbf{v}$, and $\mathbf{w}$. The fact that $\mathbf{a}\times\mathbf{b} = -(\mathbf{b}\times\mathbf{a})$ completes the proof.

6.6 Equations of Lines and Planes

1. A line in space is described algebraically by using *parametric* equations. The line that passes through the point $P(x_0, y_0, z_0)$ and is parallel to the vector $\mathbf{v} = \langle a, b, c \rangle$ is described by the equations $x = x_0 + at$, $y = y_0 + bt$, $z = z_0 + ct$.

3. The line passing through $P(1, 0, -2)$ parallel to $\mathbf{v} = \langle 3, 2, -3 \rangle$ has parametric equations $x = 1 + 3t$, $y = 2t$, $z = -2 - 3t$.

5. $x = 3$, $y = 2 - 4t$, $z = 1 + 2t$

7. $x = 1 + 2t$, $y = 0$, $z = -2 - 5t$

9. We first find a vector determined by $P(1, -3, 2)$ and $Q(2, 1, -1)$:
$\mathbf{v} = \langle 2 - 1, 1 - (-3), -1 - 2 \rangle = \langle 1, 4, -3 \rangle$. Now we use $\mathbf{v}$ and the point $(1, -3, 2)$ to find parametric equations: $x = 1 + t$, $y = -3 + 4t$, $z = 2 - 3t$ where t is any real number.

11. A vector determined by $P(1, 1, 0)$ and $Q(0, 2, 2)$ is $\langle -1, 1, 2 \rangle$, so parametric equations are $x = 1 - t$, $y = 1 + t$, $z = 2t$.

13. A vector determined by $P(3, 7, -5)$ and $Q(7, 3, -5)$ is $\langle 4, -4, 0 \rangle$, so parametric equations are $x = 3 + 4t$, $y = 7 - 4t$, $z = -5$.

15. An equation of the plane with normal vector $\mathbf{n} = \langle 1, 1, -1 \rangle$ that passes through $P(0, 2, -3)$ is $1(x - 0) + 1(y - 2) + (-1)[z - (-3)] = 0$ or $x + y - z = 5$.

(b) Setting $y = z = 0$, we find $x = 5$, so the x -intercept is 5 . Similarly, the y -intercept is 5 and the z -intercept is -5 .

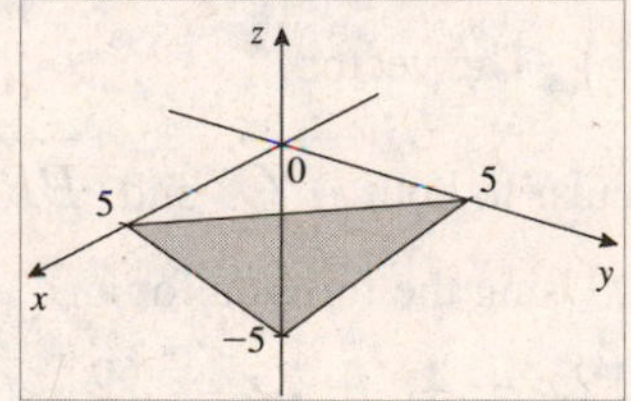

17. (a) $3(x - 2) - \frac{1}{2}(z - 8) = 0 \iff 6x - z = 4$

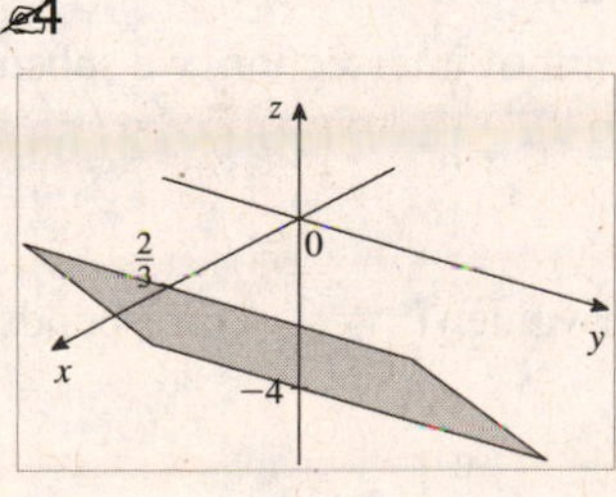

(b) x -intercept $\frac{2}{3}$, no y -intercept, z -intercept

19. (a) $3x-(y-2)+2(z+3)=0 \Leftrightarrow 3x-y+2z=-8$

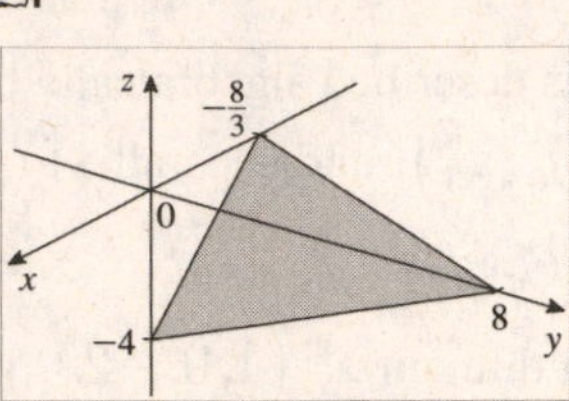

(b) x -intercept $-\frac{8}{3}$, y -intercept 8 , z -intercept

21. The vector $\overrightarrow{PQ}\times\overrightarrow{PR}=\langle -1,-1,-2\rangle\times\langle 1,2,-1\rangle=\langle 5,-3,-1\rangle$ is perpendicular to both $\overrightarrow{PQ}$ and $\overrightarrow{PR}$ and is therefore perpendicular to the plane through P , Q , and R . Using the formula for an equation of a plane with the point P , we have $5(x-6)-3(y+2)-(z-1)=0 \Leftrightarrow 5x-3y-z=35$.

23. $\overrightarrow{PQ}\times\overrightarrow{PR}=\langle 1,\frac{1}{3},2\rangle\times\langle -1,-\frac{1}{3},6\rangle=\langle \frac{8}{3},-8,0\rangle$, so an equation is $\frac{8}{3}(x-3)-8(y+\frac{1}{3})=0 \Leftrightarrow x-3y=2$.

25. $\overrightarrow{PQ}\times\overrightarrow{PR}=\langle -3,1,-1\rangle\times\langle -6,-1,-1\rangle=\langle -2,3,9\rangle$, so an equation is $-2(x-6)+3(y-1)+9(z-1)=0 \Leftrightarrow 2x-3y-9z=0$

27. The line passes through $(0,0,4)$ and $(2,5,0)$. A vector determined by these two points is $\mathbf{v}=\langle 2-0,5-0,0-4\rangle=\langle 2,5,-4\rangle$. Now we use $\mathbf{v}$ and the point $(0,0,4)$ to find parametric equations: $x=2t$, $y=5t$, $z=4-4t$, where t is any real number.

29. The line passes through $(2,-1,5)$ and is parallel to $\mathbf{j}$, so equations are $x=2$, $y=-1+t$, $z=5$, where t is any real number.

31. The plane passes through $P(1,0,0)$, $Q(0,3,0)$, and $R(0,0,4)$. The vector $\overrightarrow{PQ}\times\overrightarrow{PR}=\langle -1,3,0\rangle\times\langle -1,0,4\rangle=\langle 12,4,3\rangle$ is perpendicular to both $\overrightarrow{PQ}$ and $\overrightarrow{PR}$ and is therefore perpendicular to the plane through P , Q , and R . Using the formula for an equation of a plane, we have $12(x-1)+4y+3z=0 \Leftrightarrow 12x+4y+3z=12$.

33. This plane has the same normal vector as $x-2y+4z=6$, and because it contains the origin, its equation is $x-2y+4z=0$.

35. (a) To find the point of intersection, we substitute the parametric equations of the line into the equation of the plane: $5(2+t)-2(3t)-2(5-t)=1 \Leftrightarrow 10+5t-6t-10+2t=1 \Leftrightarrow t=1$.

(b) The parameter value $t=1$ corresponds to the point $(3,3,4)$.

37. (a) Setting $t = 0$ in the equation for Line 1 gives the point $P\left(1, 0, -6\right)$. Setting $t = 1$ gives $Q\left(0, 3, -1\right)$. If we set $t = 1$ in the equation for Line 2, we have the point $P'\left(1, 0, -6\right) = P$, and if we set $t = \frac{1}{2}$, we get $Q'\left(0, 3, -1\right) = Q$.

(b) Setting $t = 0$ in the equation for Line 1 gives the point $\left(0, 3, -5\right)$. But if a point on Line 4 has x-coordinate 0, it must have $x = 8 - 2t = 0 \Leftrightarrow t = 4$. But this parameter value gives the point $\left(0, 3, 2\right)$ on Line 2, so the two lines are not the same.

6 Review

1. $\mathbf{u} = \langle -2,3 \rangle$, $\mathbf{v} = \langle 8,1 \rangle$. $|\mathbf{u}| = \sqrt{(-2)^2 + 3^2} = \sqrt{13}$, $\mathbf{u}+\mathbf{v} = \langle -2+8, 3+1 \rangle = \langle 6,4 \rangle$, $\mathbf{u}-\mathbf{v} = \langle -2-8, 3-1 \rangle = \langle -10,2 \rangle$, $2\mathbf{u} = \langle 2(-2), 2(3) \rangle = \langle -4,6 \rangle$, and $3\mathbf{u}-2\mathbf{v} = \langle 3(-2)-2(8), 3(3)-2(1) \rangle = \langle -22,7 \rangle$.

3. $\mathbf{u} = 2\mathbf{i}+\mathbf{j}$, $\mathbf{v} = \mathbf{i}-2\mathbf{j}$. $|\mathbf{u}| = \sqrt{2^2+1^2} = \sqrt{5}$, $\mathbf{u}+\mathbf{v} = (2+1)\mathbf{i}+(1-2)\mathbf{j} = 3\mathbf{i}-\mathbf{j}$, $\mathbf{u}-\mathbf{v} = (2-1)\mathbf{i}+(1+2)\mathbf{j} = \mathbf{i}+3\mathbf{j}$, $2\mathbf{u} = 4\mathbf{i}+2\mathbf{j}$, and $3\mathbf{u}-2\mathbf{v} = 3(2\mathbf{i}+\mathbf{j})-2(\mathbf{i}-2\mathbf{j}) = 4\mathbf{i}+7\mathbf{j}$.

5. The vector with initial point $P(0,3)$ and terminal point $Q(3,-1)$ is $\langle 3-0, -1-3 \rangle = \langle 3,-4 \rangle$.

7. $\mathbf{u} = \langle -2, 2\sqrt{3} \rangle$ has length $\sqrt{(-2)^2 + (2\sqrt{3})^2} = 4$. Its direction is given by $\tan\theta = \dfrac{2\sqrt{3}}{-2} = -\sqrt{3}$ with θ in quadrant II, so $\theta = \pi + \tan^{-1}(-\sqrt{3}) = 120^\circ$.

9. $\mathbf{u} = \langle |\mathbf{u}|\cos\theta, |\mathbf{u}|\sin\theta \rangle = \langle 20\cos 60^\circ, 20\sin 60^\circ \rangle = \langle 10, 10\sqrt{3} \rangle$.

11. **(a)** The force exerted by the first tugboat can be expressed in component form as $\mathbf{u} = \langle 2.0\times10^4 \cos 40^\circ, 2.0\times10^4 \sin 40^\circ \rangle \approx \langle 15321, 12856 \rangle$, and that of the second tugboat is $\langle 3.4\times10^4 \cos(-15^\circ), 3.4\times10^4 \sin(-15^\circ) \rangle \approx \langle 32841, -8800 \rangle$. Therefore, the resultant force is $\mathbf{w} = \mathbf{u}+\mathbf{v} \approx \langle 15321+32841, 12856-8800 \rangle = \langle 48162, 4056 \rangle$.

(b) The magnitude of the resultant force is $\sqrt{48162^2 + 4056^2} \approx 48,332$ lb. Its direction is given by $\tan\theta = \dfrac{4056}{48{,}162} \approx 0.084$, so $\theta \approx \tan^{-1} 0.084 \approx 4.8^\circ$ or N 85.2° E.

13. $\mathbf{u} = \langle 4,-3 \rangle$, $\mathbf{v} = \langle 9,-8 \rangle$. $|\mathbf{u}| = \sqrt{4^2 + (-3)^2} = 5$, $\mathbf{u}\cdot\mathbf{u} = 4^2 + (-3)^2 = 25$, and $\mathbf{u}\cdot\mathbf{v} = 4(9)+(-3)(-8) = 60$.

15. $\mathbf{u} = -2\mathbf{i}+2\mathbf{j}$, $\mathbf{v} = \mathbf{i}+\mathbf{j}$. $|\mathbf{u}| = \sqrt{(-2)^2 + 2^2} = 2\sqrt{2}$, $\mathbf{u}\cdot\mathbf{u} = (-2)^2 + 2^2 = 8$, and $\mathbf{u}\cdot\mathbf{v} = -2(1)+2(1) = 0$.

17. $\mathbf{u}\cdot\mathbf{v} = \langle -4,2 \rangle \cdot \langle 3,6 \rangle = -4(3)+2(6) = 0$, so the vectors are perpendicular.

19. $\mathbf{u}\cdot\mathbf{v} = (2\mathbf{i}+\mathbf{j})\cdot(\mathbf{i}+3\mathbf{j}) = 2(1)+1(3) = 5$, so the vectors are not perpendicular. The angle between them is given by $\cos\theta = \dfrac{\mathbf{u}\cdot\mathbf{v}}{|\mathbf{u}||\mathbf{v}|} = \dfrac{5}{\sqrt{2^2+1^2}\sqrt{1^2+3^2}} = \dfrac{\sqrt{2}}{2}$, so $\theta = \cos^{-1}\frac{\sqrt{2}}{2} = 45^\circ$.

21. (a) $\mathbf{u} = \langle 3,1 \rangle$, $\mathbf{v} = \langle 6,-1 \rangle$. The component of $\mathbf{u}$ along $\mathbf{v}$ is

$$\frac{\mathbf{u}\cdot\mathbf{v}}{|\mathbf{v}|} = \frac{3(6)+1(-1)}{\sqrt{6^2+1^2}} = \frac{17\sqrt{37}}{37} .$$

(b) $\operatorname{proj}_{\mathbf{v}} \mathbf{u} = \left(\dfrac{\mathbf{u}\cdot\mathbf{v}}{|\mathbf{v}|^2}\right)\mathbf{v} = \frac{17}{37}\langle 6,-1\rangle = \left\langle \frac{102}{37}, -\frac{17}{37} \right\rangle$

(c) $\mathbf{u}_1 = \operatorname{proj}_{\mathbf{v}} \mathbf{u} = \left\langle \frac{102}{37}, -\frac{17}{37} \right\rangle$ and $\mathbf{u}_2 = \mathbf{u} - \operatorname{proj}_{\mathbf{v}} \mathbf{u} = \langle 3,1 \rangle - \left\langle \frac{102}{37}, -\frac{17}{37} \right\rangle = \left\langle \frac{9}{37}, \frac{54}{37} \right\rangle$.

23. (a) $\mathbf{u} = \mathbf{i} + 2\mathbf{j}$, $\mathbf{v} = 4\mathbf{i} - 9\mathbf{j}$. The component of $\mathbf{u}$ along $\mathbf{v}$ is

$$\frac{\mathbf{u}\cdot\mathbf{v}}{|\mathbf{v}|} = \frac{1(4)+2(-9)}{\sqrt{4^2+(-9)^2}} = -\frac{14\sqrt{97}}{97} .$$

(b) $\operatorname{proj}_{\mathbf{v}} \mathbf{u} = \left(\dfrac{\mathbf{u}\cdot\mathbf{v}}{|\mathbf{v}|^2}\right)\mathbf{v} = \frac{-14}{97}(4\mathbf{i} - 9\mathbf{j}) = -\frac{56}{97}\mathbf{i} + \frac{126}{97}\mathbf{j}$

(c) $\mathbf{u}_1 = \operatorname{proj}_{\mathbf{v}} \mathbf{u} = -\frac{56}{97}\mathbf{i} + \frac{126}{97}\mathbf{j}$ and $\mathbf{u}_2 = \mathbf{u} - \operatorname{proj}_{\mathbf{v}} \mathbf{u} = \mathbf{i} + 2\mathbf{j} - \left(-\frac{56}{97}\mathbf{i} + \frac{126}{97}\mathbf{j}\right) = \frac{153}{97}\mathbf{i} + \frac{68}{97}\mathbf{j}$.

25.

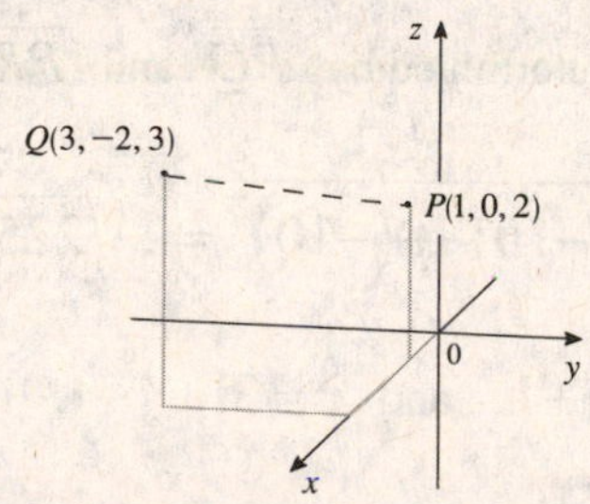

The distance between P and Q is $\sqrt{(3-1)^2 + (-2-0)^2 + (3-2)^2} = 3$.

27. The sphere with radius $r = 6$ and center $C(0,0,0)$ has equation

$(x-0)^2 + (y-0)^2 + (z-0)^2 = 6^2 \iff x^2 + y^2 + z^2 = 36$.

29. We complete the squares to find $x^2 + y^2 + z^2 - 2x - 6y + 4z = 2 \iff$

$(x^2 - 2x + 1) + (y^2 - 6y + 9) + (z^2 + 4z + 4) = 2 + 1 + 9 + 4 \iff$

$(x-1)^2 + (y-3)^2 + (z+2)^2 = 16$, an equation of the sphere with center $(1,3,-2)$ and radius 4 .

31. $\mathbf{u} = \langle 4,-2,4 \rangle$ and $\mathbf{v} = \langle 2,3,-1 \rangle$, so $|\mathbf{u}| = \sqrt{4^2 + (-2)^2 + (4)^2} = 6$,

$\mathbf{u} + \mathbf{v} = \langle 4+2, -2+3, 4+(-1) \rangle = \langle 6,1,3 \rangle$, $\mathbf{u} - \mathbf{v} = \langle 4-2, -2-3, 4-(-1) \rangle = \langle 2,-5,5 \rangle$,

and $\frac{3}{4}\mathbf{u} - 2\mathbf{v} = \frac{3}{4}\langle 4,-2,4 \rangle - 2\langle 2,3,-1 \rangle = \left\langle -1, -\frac{15}{2}, 5 \right\rangle$.

33. (a) $\mathbf{u}\cdot\mathbf{v}=\langle 3,-2,4\rangle\cdot\langle 3,1,-2\rangle=3(3)-2(1)+4(-2)=-1$

(b) $\mathbf{u}\cdot\mathbf{v}\neq 0$, so the vectors are not perpendicular. The angle between them is given by

$$\cos\theta=\frac{\mathbf{u}\cdot\mathbf{v}}{|\mathbf{u}||\mathbf{v}|}=\frac{-1}{\sqrt{3^2+(-2)^2+4^2}\sqrt{3^2+1^2+(-2)^2}}=-\frac{\sqrt{406}}{406}\text{, so}$$

$\theta=\cos^{-1}\left(-\frac{\sqrt{406}}{406}\right)\approx 92.8^\circ$.

35. (a) $\mathbf{u}\cdot\mathbf{v}=(2\mathbf{i}-\mathbf{j}+4\mathbf{k})\cdot(3\mathbf{i}+2\mathbf{j}-\mathbf{k})=2(3)-1(2)+4(-1)=0$

(b) $\mathbf{u}\cdot\mathbf{v}=0$, so the vectors are perpendicular.

37. (a) $\mathbf{u}\times\mathbf{v}=\langle 1,1,3\rangle\times\langle 5,0,-2\rangle=(-2-0)\mathbf{i}-(-2-15)\mathbf{j}+(0-5)\mathbf{k}=\langle -2,17,-5\rangle$

(b) A unit vector perpendicular to $\mathbf{u}$ and $\mathbf{v}$ is

$$\frac{\mathbf{u}\times\mathbf{v}}{|\mathbf{u}\times\mathbf{v}|}=\frac{\langle -2,17,-5\rangle}{\sqrt{(-2)^2+17^2+(-5)^2}}=\left\langle -\tfrac{\sqrt{318}}{159},\tfrac{17\sqrt{318}}{318},-\tfrac{5\sqrt{318}}{318}\right\rangle .$$

39. (a) $\mathbf{u}\times\mathbf{v}=(\mathbf{i}-\mathbf{j})\times(2\mathbf{j}-\mathbf{k})=(1-0)\mathbf{i}-(-1-0)\mathbf{j}+(2-0)\mathbf{k}=\mathbf{i}+\mathbf{j}+2\mathbf{k}$

(b) A unit vector perpendicular to $\mathbf{u}$ and $\mathbf{v}$ is $\dfrac{\mathbf{u}\times\mathbf{v}}{|\mathbf{u}\times\mathbf{v}|}=\dfrac{\mathbf{i}+\mathbf{j}+2\mathbf{k}}{\sqrt{1^2+1^2+2^2}}=\frac{\sqrt{6}}{6}\mathbf{i}+\frac{\sqrt{6}}{6}\mathbf{j}+\frac{\sqrt{6}}{3}\mathbf{k}$.

41. The area of triangle PQR is one-half the area of the parallelogram determined by $\overrightarrow{PQ}$ and $\overrightarrow{PR}$, that is,

$$\tfrac{1}{2}\left|\overrightarrow{PQ}\times\overrightarrow{PR}\right|=\tfrac{1}{2}\left|(1-6)\mathbf{i}-(2+8)\mathbf{j}+(-6-4)\mathbf{k}\right|=\tfrac{1}{2}\sqrt{(-5)^2+(-10)^2+(-10)^2}=\tfrac{1}{2}\sqrt{225}=\tfrac{15}{2}$$

43. The volume of the parallelepiped determined by $\mathbf{a}=2\mathbf{i}-\mathbf{j}$, $\mathbf{b}=2\mathbf{j}+\mathbf{k}$, and $\mathbf{c}=3\mathbf{i}+\mathbf{j}-\mathbf{k}$ is the absolute value of their scalar triple product:

$V=|\mathbf{a}\cdot(\mathbf{b}\times\mathbf{c})|=|(2\mathbf{i}-\mathbf{j})\cdot(-3\mathbf{i}+3\mathbf{j}-6\mathbf{k})|=|-6-3|=9$.

45. The line that passes through $P(2,0,-6)$ and is parallel to $\mathbf{v}=\langle 3,1,0\rangle$ has parametric equations $x=2+3t$, $y=t$, $z=-6$.

47. A vector determined by $P(6,-2,-3)$ and $Q(4,1,-2)$ is $\langle -2,3,1\rangle$, so parametric equations are $x=6-2t$, $y=-2+3t$, $z=-3+t$.

49. Using the formula for an equation of a plane, the plane with normal vector $\mathbf{n}=\langle 2,3,-5\rangle$ passing through $P(2,1,1)$ has equation $2(x-2)+3(y-1)-5(z-1)=0 \Leftrightarrow 2x+3y-5z=2$.

51. The plane passes through $P(1,1,1)$, $Q(3,-4,2)$, and $R(6,-1,0)$.

$\overrightarrow{PQ}\times\overrightarrow{PR}=\langle 2,-5,1\rangle\times\langle 5,-2,-1\rangle=\langle 7,7,6\rangle$, so an equation is

$7(x-1)+7(y-1)+6(z-1)=0 \Leftrightarrow 7x+7y+6z=20$.

53. The line passes through the points $(2,0,0)$ and $(0,0,-4)$. A vector determined by these points is $\langle -2,0,-4\rangle$ or $\langle 1,0,2\rangle$, so parametric equations are $x=2+t$, $y=0$, $z=2t$.

6 Test

1. (a)

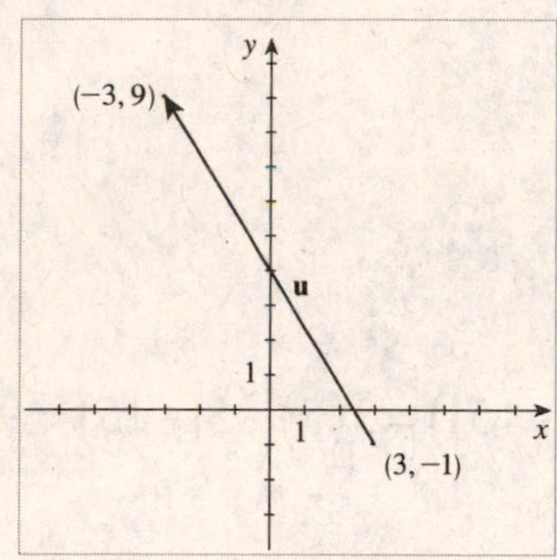

(b) $\mathbf{u}=(-3-3)\mathbf{i}+[9-(-1)]\mathbf{j}=-6\mathbf{i}+10\mathbf{j}$

(c) $|\mathbf{u}|=\sqrt{(-6)^2+10^2}=2\sqrt{34}$

2. (a) $\mathbf{u}-3\mathbf{v}=\langle 1,3\rangle-3\langle -6,2\rangle=\langle 1-3(-6),3-3(2)\rangle=\langle 19,-3\rangle$

(b) $|\mathbf{u}+\mathbf{v}|=|\langle 1,3\rangle+\langle -6,2\rangle|=|\langle -5,5\rangle|=\sqrt{(-5)^2+5^2}=5\sqrt{2}$

(c) $\mathbf{u}\cdot\mathbf{v}=\langle 1,3\rangle\cdot\langle -6,2\rangle=1(-6)+3(2)=0$

(d) Because $\mathbf{u}\cdot\mathbf{v}=0$, $\mathbf{u}$ and $\mathbf{v}$ are perpendicular.

3. (a)

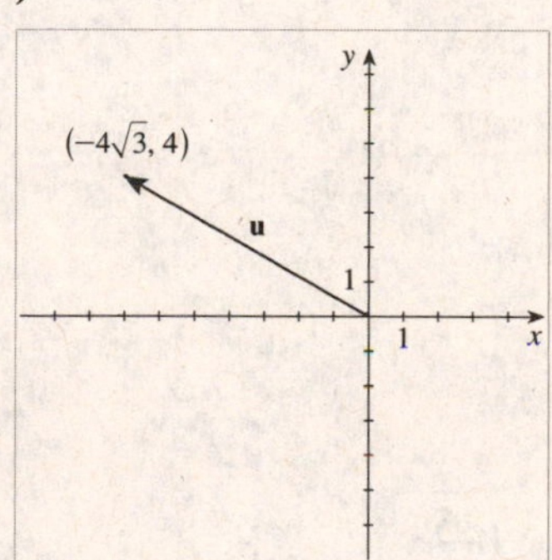

(b) The length of $\mathbf{u}$ is $|\mathbf{u}|=\sqrt{\left(-4\sqrt{3}\right)^2+4^2}=8$. Its direction is given by $\tan\theta=\frac{4}{-4\sqrt{3}}=-\frac{\sqrt{3}}{3}$ with θ in quadrant II, so $\theta=180^\circ-\tan^{-1}\left(\frac{\sqrt{3}}{3}\right)=150^\circ$.

4. (a) The river's current can be represented by the vector $\mathbf{u}=\langle 8,0\rangle$ and the motorboat's velocity relative to the water by the vector $\mathbf{v}=\langle 12\cos 60^\circ,12\sin 60^\circ\rangle=\langle 6,6\sqrt{3}\rangle$. Thus, the true velocity is $\mathbf{w}=\mathbf{u}+\mathbf{v}=\langle 14,6\sqrt{3}\rangle$.

(b) The true speed is $|\mathbf{w}|=\sqrt{14^2+\left(6\sqrt{3}\right)^2}\approx 17.4$ mi/h. The direction is given by

$\tan\theta=\dfrac{6\sqrt{13}}{14}\approx 0.742$, so $\theta\approx\tan^{-1}0.742\approx 36.6^\circ$ or N 53.4° E.

5. (a) $\cos\theta = \dfrac{\mathbf{u}\cdot\mathbf{v}}{|\mathbf{u}||\mathbf{v}|} = \dfrac{3(5)+2(-1)}{\sqrt{3^2+2^2}\sqrt{5^2+(-1)^2}} = \dfrac{\sqrt{338}}{26}$, so $\theta = \cos^{-1}\frac{\sqrt{338}}{26} \approx 45.0^\circ$.

(b) The component of $\mathbf{u}$ along $\mathbf{v}$ is $\dfrac{\mathbf{u}\cdot\mathbf{v}}{|\mathbf{v}|} = \dfrac{13}{\sqrt{26}} = \dfrac{\sqrt{26}}{2}$.

(c) $\text{proj}_{\mathbf{v}}\,\mathbf{u} = \left(\dfrac{\mathbf{u}\cdot\mathbf{v}}{|\mathbf{v}|^2}\right)\mathbf{v} = \frac{13}{26}(5\mathbf{i}-\mathbf{j}) = \frac{5}{2}\mathbf{i} - \frac{1}{2}\mathbf{j}$

6. The work is

$W = \mathbf{F}\cdot\mathbf{d} = (3\mathbf{i}-5\mathbf{j})\cdot[(7-2)\mathbf{i}+(-13-2)\mathbf{j}] = (3\mathbf{i}-5\mathbf{j})\cdot(5\mathbf{i}-15\mathbf{j}) = 3(5)-5(-15) = 90$.

7. (a) The distance between $P(4,3,-1)$ and $Q(6,-1,3)$ is

$d = \sqrt{(6-4)^2+(-1-3)^2+[3-(-1)]^2} = 6$.

(b) An equation is $(x-4)^2+(y-3)^2+[z-(-1)]^2 = 6^2 \Leftrightarrow$

$(x-4)^2+(y-3)^2+(z+1)^2 = 36$.

(c) $\mathbf{u} = \langle 6-4, -1-3, 3-(-1)\rangle = \langle 2,-4,4\rangle = 2\mathbf{i}-4\mathbf{j}+4\mathbf{k}$

8. $\mathbf{a} = \mathbf{i}+\mathbf{j}-2\mathbf{k}$, $\mathbf{b} = 3\mathbf{i}-2\mathbf{j}+\mathbf{k}$, and $\mathbf{c} = \mathbf{j}-5\mathbf{k}$.

(a) $2\mathbf{a}+3\mathbf{b} = [2(1)+3(3)]\mathbf{i}+[2(1)+3(-2)]\mathbf{j}+[2(-2)+3(1)]\mathbf{k} = 11\mathbf{i}-4\mathbf{j}-\mathbf{k}$

(b) $|\mathbf{a}| = \sqrt{1^2+1^2+(-2)^2} = \sqrt{6}$

(c) $\mathbf{a}\cdot\mathbf{b} = 1(3)+1(-2)-2(1) = -1$

(d) $\mathbf{a}\times\mathbf{b} = \begin{vmatrix} \mathbf{i} & \mathbf{j} & \mathbf{k} \\ 1 & 1 & -2 \\ 3 & -2 & 1 \end{vmatrix} = -3\mathbf{i}-7\mathbf{j}-5\mathbf{k}$

(e) $|\mathbf{b}\times\mathbf{c}| = \left\|\begin{matrix} \mathbf{i} & \mathbf{j} & \mathbf{k} \\ 3 & -2 & 1 \\ 0 & 1 & -5 \end{matrix}\right\| = |9\mathbf{i}+15\mathbf{j}+3\mathbf{k}| = \sqrt{9^2+15^2+3^2} = 3\sqrt{35}$

(f) $\mathbf{a}\cdot(\mathbf{b}\times\mathbf{c}) = (\mathbf{i}+\mathbf{j}-2\mathbf{k})\cdot(9\mathbf{i}+15\mathbf{j}+3\mathbf{k}) = 1(9)+1(15)-2(3) = 18$

(g) $\cos\theta = \dfrac{\mathbf{a}\cdot\mathbf{b}}{|\mathbf{a}||\mathbf{b}|} = \dfrac{-1}{\sqrt{6}\sqrt{3^2+(-2)^2+1^2}} = -\dfrac{\sqrt{84}}{84}$, so $\theta = \cos^{-1}\left(-\frac{\sqrt{84}}{84}\right) \approx 96.3^\circ$.

9. A vector perpendicular to both $\mathbf{u} = \mathbf{j}+2\mathbf{k}$ and $\mathbf{v} = \mathbf{i}-2\mathbf{j}+3\mathbf{k}$ is

$\mathbf{u}\times\mathbf{v} = \langle 0,1,2\rangle\times\langle 1,-2,3\rangle = \langle 7,2,-1\rangle$, so two unit vectors perpendicular to $\mathbf{u}$ and $\mathbf{v}$ are

$\dfrac{\mathbf{u}\times\mathbf{v}}{|\mathbf{u}\times\mathbf{v}|} = \dfrac{\langle 7,2,-1\rangle}{\sqrt{7^2+2^2+(-1)^2}} = \left\langle \frac{7\sqrt{6}}{18}, \frac{\sqrt{6}}{9}, -\frac{\sqrt{6}}{18}\right\rangle$ and $\left\langle -\frac{7\sqrt{6}}{18}, -\frac{\sqrt{6}}{9}, \frac{\sqrt{6}}{18}\right\rangle$.

10. (a) A vector perpendicular to the plane that contains the points $P(1,0,0)$, $Q(2,0,-1)$, and $R(1,4,3)$ is $\overrightarrow{PQ}\times\overrightarrow{PR}=\langle 1,0,-1\rangle\times\langle 0,4,3\rangle=\langle 4,-3,4\rangle$.

(b) An equation of the plane is $4(x-1)-3y+4z=0 \quad\Leftrightarrow\quad 4x-3y+4z=4$.

(c) The area of triangle PQR is half the area of the parallelogram determined by $\overrightarrow{PQ}$ and $\overrightarrow{PR}$, that is, $\frac{1}{2}\left|\overrightarrow{PQ}\times\overrightarrow{PR}\right|=\frac{1}{2}\sqrt{4^2+(-3)^2+4^2}=\frac{\sqrt{41}}{2}$.

11. A vector determined by the two points is $\overrightarrow{PQ}=\langle -2,1,-2\rangle$, so parametric equations are $x=2-2t$, $y=-4+t$, $z=7-2t$.

1. $\mathbf{F}(x,y)=\frac{1}{2}\mathbf{i}+\frac{1}{2}\mathbf{j}$

All vectors point in the same direction and have length $\frac{\sqrt{2}}{2}$.

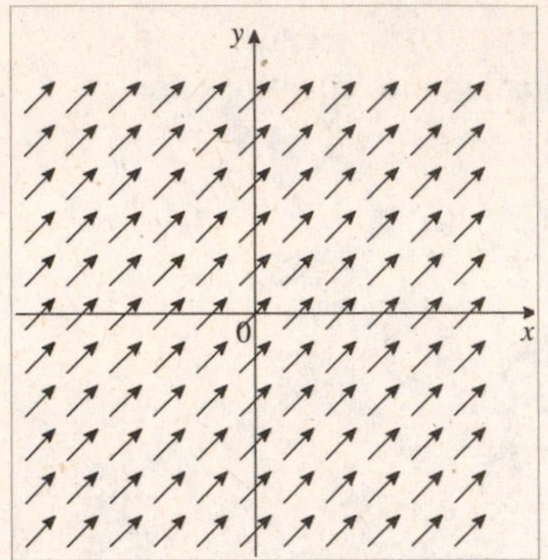

3. $\mathbf{F}(x,y)=y\mathbf{i}+\frac{1}{2}\mathbf{j}$

The vectors point to the left for $y<0$ and to the right for $y>0$.

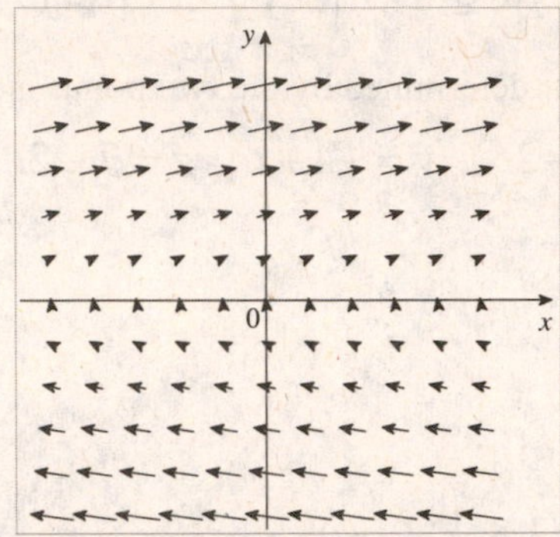

5. $\mathbf{F}(x,y)=\dfrac{y\mathbf{i}+x\mathbf{j}}{\sqrt{x^2+y^2}}$

The length of the vector $\dfrac{y\mathbf{i}+x\mathbf{j}}{\sqrt{x^2+y^2}}$ is 1.

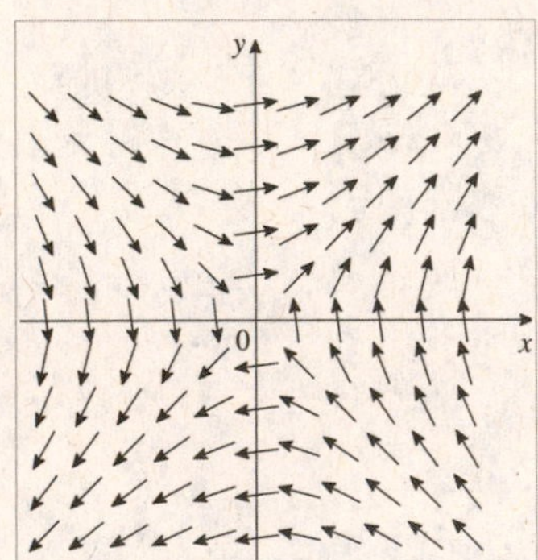

7. $\mathbf{F}(x,y,z)=\mathbf{j}$

All vectors in this field are parallel to the y-axis and have length 1.

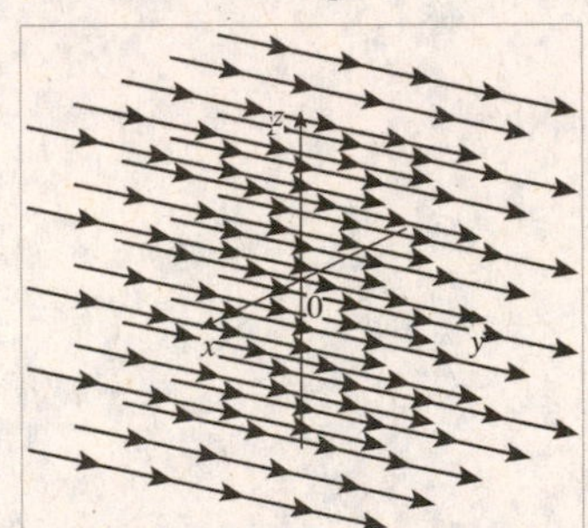

9. $\mathbf{F}(x,y,z)=z\mathbf{j}$

At each point (x,y,z), $\mathbf{F}(x,y,z)$ is a vector of length $|z|$. For $z>0$, all point in the direction of the positive y-axis while for $z<0$, all are in the direction of the negative y-axis.

11. $\mathbf{F}(x,y)=\langle y,x\rangle$ corresponds to graph II, because in the first quadrant all the vectors have positive x- and y-components, in the second quadrant all vectors have positive x-components and negative y-components, in the third quadrant all vectors have negative x- and y-components, and in the fourth quadrant all vectors have negative x-components and positive y-components.

13. $\mathbf{F}(x, y) = \langle x-2, y+1 \rangle$ corresponds to graph I because the vectors are independent of y (vectors along vertical lines are identical) and, as we move to the right, both the x - and the y -components get larger.

15. $\mathbf{F}(x, y, z) = \mathbf{i} + 2\mathbf{j} + 3\mathbf{k}$ corresponds to graph IV, since all vectors have identical length and direction.

17. $\mathbf{F}(x, y, z) = x\mathbf{i} + y\mathbf{j} + 3\mathbf{k}$ corresponds to graph III; the projection of each vector onto the xy -plane is $x\mathbf{i} + y\mathbf{j}$, which points away from the origin, and the vectors point generally upward because their z -components are all 3 .

19.

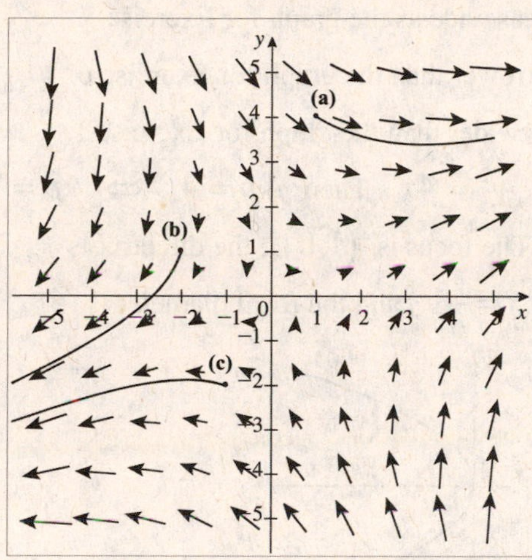

7 Conic Sections

7.1 Parabolas

1. A parabola is the set of all points in the plane equidistant from a fixed point called the *focus* and a fixed line called the *directrix* of the parabola.

3. The graph of the equation $y^2 = 4px$ is a parabola with focus $F(p,0)$ and directrix $x = -p$. So the graph of $y^2 = 12x$ is a parabola with focus $F(3,0)$ and directrix $x = -3$.

5. $y^2 = 2x$ is Graph III, which opens to the right and is not as wide as the graph for Exercise 5.

7. $x^2 = -6y$ is Graph II, which opens downward and is narrower than the graph for Exercise 6.

9. $y^2 - 8x = 0$ is Graph VI, which opens to the right and is wider than the graph for Exercise 1.

11. $x^2 = 9y$. Then $4p = 9 \Leftrightarrow p = \frac{9}{4}$. The focus is $\left(0, \frac{9}{4}\right)$, the directrix is $y = -\frac{9}{4}$, and the focal diameter is 9.

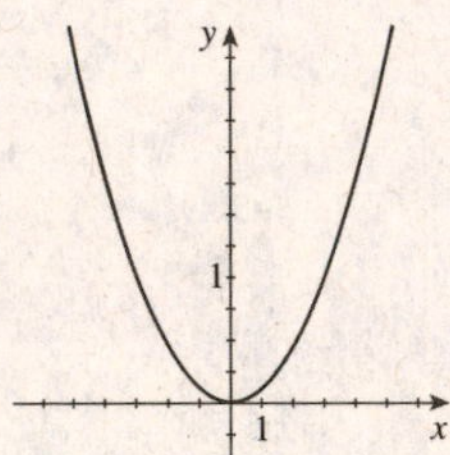

13. $y^2 = 4x$. Then $4p = 4 \Leftrightarrow p = 1$. The focus is $(1,0)$, the directrix is $x = -1$, and the focal diameter is 4.

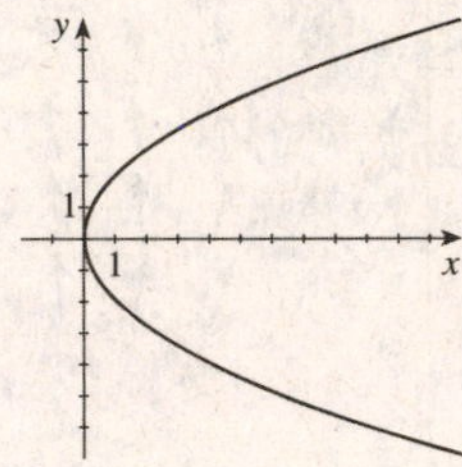

15. $y = 5x^2 \Leftrightarrow x^2 = \frac{1}{5}y$. Then $4p = \frac{1}{5} \Leftrightarrow p = \frac{1}{20}$. The focus is $\left(0, \frac{1}{20}\right)$, the directrix is $y = -\frac{1}{20}$, and the focal diameter is $\frac{1}{5}$.

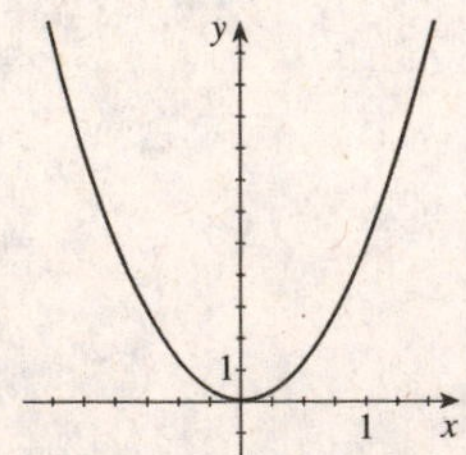

17. $x = -8y^2 \Leftrightarrow y^2 = -\frac{1}{8}x$. Then $4p = -\dfrac{1}{8} \Leftrightarrow p = -\frac{1}{32}$. The focus is $\left(-\frac{1}{32}, 0\right)$, the directrix is $x = \frac{1}{32}$, and the focal diameter is $\frac{1}{8}$.

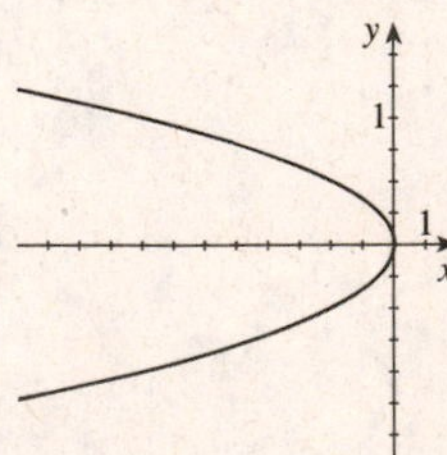

19. $x^2 + 6y = 0 \Leftrightarrow x^2 = -6y$. Then $4p = -6 \Leftrightarrow p = -\frac{3}{2}$. The focus is $\left(0, -\frac{3}{2}\right)$, the directrix is $y = \frac{3}{2}$, and the focal diameter is 6.

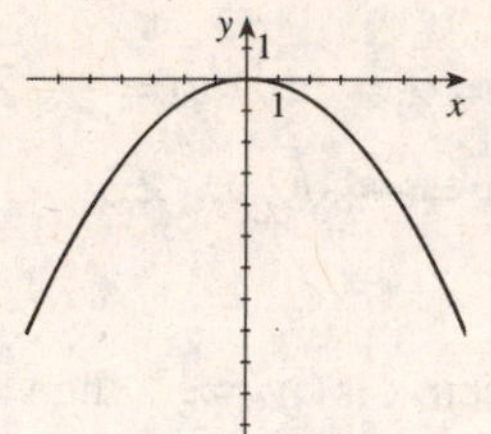

21. $5x + 3y^2 = 0 \Leftrightarrow y^2 = -\frac{5}{3}x$. Then $4p = -\frac{5}{3} \Leftrightarrow p = -\frac{5}{12}$. The focus is $\left(-\frac{5}{12}, 0\right)$, the directrix is $x = \frac{5}{12}$, and the focal diameter is $\frac{5}{3}$.

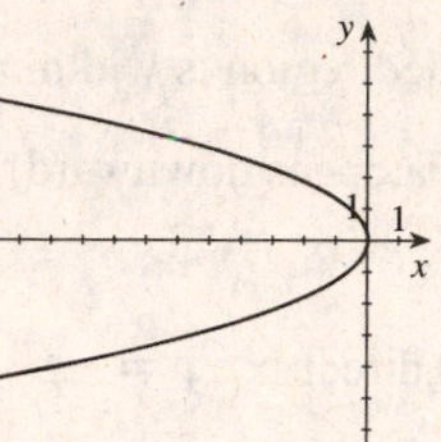

23. $x^2 = 16y$

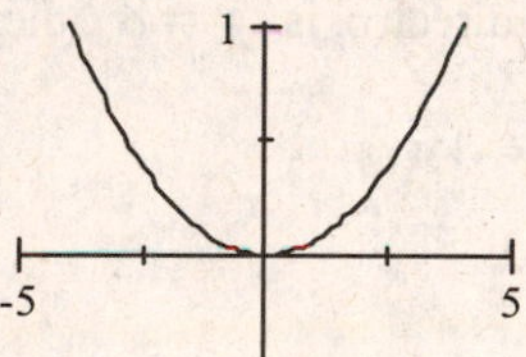

25. $y^2 = -\frac{1}{3}x$

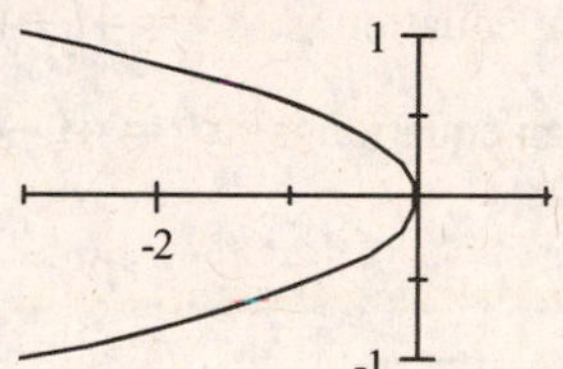

27. $4x + y^2 = 0$

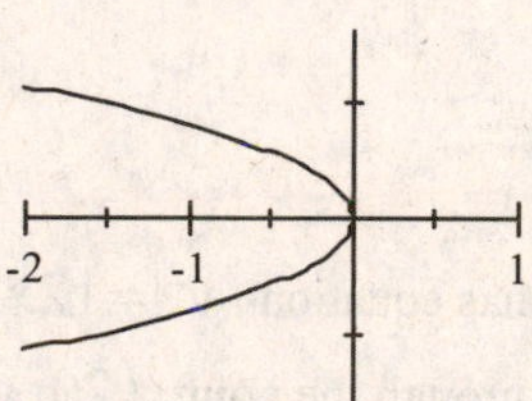

29. Since the focus is $(0,2)$, $p = 2 \Leftrightarrow 4p = 8$. Hence, an equation of the parabola is $x^2 = 8y$.

31. Since the focus is $(-8,0)$, $p = -8 \Leftrightarrow 4p = -32$. Hence, an equation of the parabola is $y^2 = -32x$.

33. Since the directrix is $x = 2$, $p = -2 \Leftrightarrow 4p = -8$. Hence, an equation of the parabola is $y^2 = -8x$.

35. Since the directrix is $y = -10$, $p = 10 \Leftrightarrow 4p = 40$. Hence, an equation of the parabola is $x^2 = 40y$.

37. The focus is on the positive x-axis, so the parabola opens horizontally with $2p = 2 \Leftrightarrow 4p = 4$. So an equation of the parabola is $y^2 = 4x$.

39. Since the parabola opens upward with focus 5 units from the vertex, the focus is $(5,0)$. So $p = 5 \Leftrightarrow 4p = 20$. Thus an equation of the parabola is $x^2 = 20y$.

41. $p = 2 \Leftrightarrow 4p = 8$. Since the parabola opens upward, its equation is $x^2 = 8y$.

43. $p=4 \quad \Leftrightarrow \quad 4p=16$. Since the parabola opens to the left, its equation is $y^2=-16x$.

45. The focal diameter is $4p=\frac{3}{2}+\frac{3}{2}=3$. Since the parabola opens to the left, its equation is $y^2=-3x$.

47. The equation of the parabola has the form $y^2=4px$. Since the parabola passes through the point $(4,-2)$, $(-2)^2=4p(4) \quad \Leftrightarrow \quad 4p=1$, and so an equation is $y^2=x$.

49. The area of the shaded region is width $\times$ height $=4p\cdot p=8$, and so $p^2=2 \quad \Leftrightarrow \quad p=-\sqrt{2}$ (because the parabola opens downward). Therefore, an equation is $x^2=4py=-4\sqrt{2}y \quad \Leftrightarrow$ $x^2=-4\sqrt{2}y$.

51. **(a)** A parabola with directrix $y=-p$ has equation $x^2=4py$. If the directrix is $y=\frac{1}{2}$, then $p=-\frac{1}{2}$, so an equation is $x^2=4\left(-\frac{1}{2}\right)y \quad \Leftrightarrow \quad x^2=-2y$. If the directrix is $y=1$, then $p=-1$, so an equation is $x^2=4(-1)y \quad \Leftrightarrow \quad x^2=-4y$. If the directrix is $y=4$, then $p=-4$, so an equation is $x^2=4(-4)y \quad \Leftrightarrow \quad x^2=-16y$. If the directrix is $y=8$, then $p=-8$, so an equation is $x^2=4(-8)y \quad \Leftrightarrow \quad x^2=-32y$.

(b)

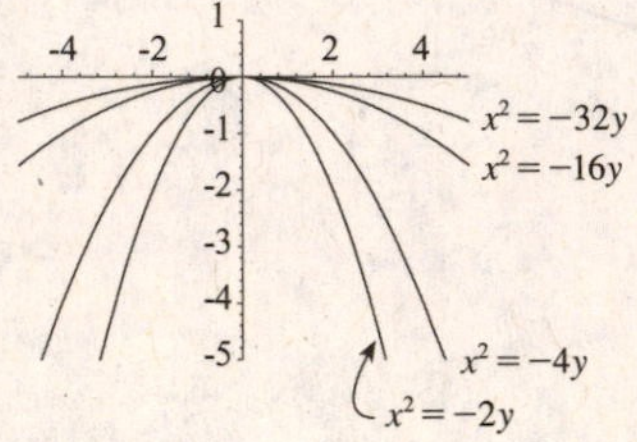

As the directrix moves further from the vertex, the parabolas get flatter.

53. **(a)** Since the focal diameter is 12 cm, $4p=12$. Hence, the parabola has equation $y^2=12x$.

(b) At a point 20 cm horizontally from the vertex, the parabola passes through the point $(20,y)$, and hence from part (a), $y^2=12(20) \quad \Leftrightarrow \quad y^2=240 \quad \Leftrightarrow \quad y=\pm 4\sqrt{15}$. Thus, $|CD|=8\sqrt{15}\approx 31$ cm.

55. With the vertex at the origin, the top of one tower will be at the point $(300,150)$. Inserting this point into the equation $x^2=4py$ gives $(300)^2=4p(150) \quad \Leftrightarrow \quad 90000=600p \quad \Leftrightarrow \quad p=150$. So an equation of the parabolic part of the cables is $x^2=4(150)y \quad \Leftrightarrow \quad x^2=600y$.

57. Many answers are possible: satellite dish TV antennas, sound surveillance equipment, solar collectors for hot water heating or electricity generation, bridge pillars, etc.

7.2 Ellipses

1. An ellipse is the set of all points in the plane for which the *sum* of the distances from two fixed points F_1 and F_2 is constant. The points F_1 and F_2 are called the *foci* of the ellipse.

3. The graph of the equation $\frac{x^2}{b^2}+\frac{y^2}{a^2}=1$ with $a>b>0$ is an ellipse with vertices $(0,a)$ and $(0,-a)$ and foci $(0,\pm c)$, where $c=\sqrt{a^2-b^2}$. So the graph of $\frac{x^2}{4^2}+\frac{y^2}{5^2}=1$ is an ellipse with vertices $(0,5)$ and $(0,-5)$ and foci $(0,3)$ and $(0,-3)$.

5. $\frac{x^2}{16}+\frac{y^2}{4}=1$ is Graph II. The major axis is horizontal and the vertices are $(\pm 4,0)$.

7. $4x^2+y^2=4$ is Graph I. The major axis is vertical and the vertices are $(0,\pm 2)$.

9. $\frac{x^2}{25}+\frac{y^2}{9}=1$. This ellipse has $a=5$, $b=3$, and so $c^2=a^2-b^2=16 \iff c=4$. The vertices are $(\pm 5,0)$, the foci are $(\pm 4,0)$, the eccentricity is $e=\frac{c}{a}=\frac{4}{5}=0.8$, the length of the major axis is $2a=10$, and the length of the minor axis is $2b=6$.

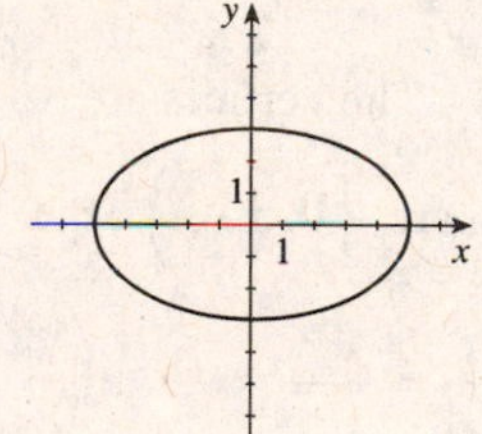

11. $9x^2+4y^2=36 \iff \frac{x^2}{4}+\frac{y^2}{9}=1$. This ellipse has $a=3$, $b=2$, and so $c^2=9-4=5 \iff c=\sqrt{5}$. The vertices are $(0,\pm 3)$, the foci are $(0,\pm\sqrt{5})$, the eccentricity is $e=\frac{c}{a}=\frac{\sqrt{5}}{3}$, the length of the major axis is $2a=6$, and the length of the minor axis is $2b=4$.

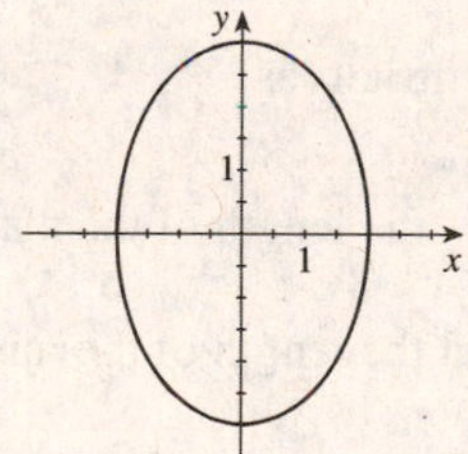

13. $x^2 + 4y^2 = 16 \Leftrightarrow \dfrac{x^2}{16} + \dfrac{y^2}{4} = 1$. This ellipse has $a = 4$, $b = 2$, and so $c^2 = 16 - 4 = 12 \Leftrightarrow c = 2\sqrt{3}$. The vertices are $(\pm 4, 0)$, the foci are $(\pm 2\sqrt{3}, 0)$, the eccentricity is $e = \dfrac{c}{a} = \dfrac{2\sqrt{3}}{4} = \frac{\sqrt{3}}{2}$, the length of the major axis is $2a = 8$, and the length of the minor axis is $2b = 4$.

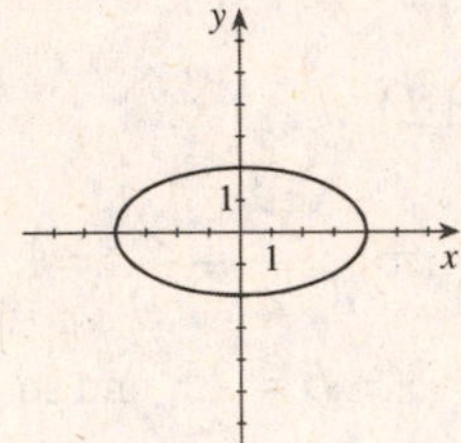

15. $2x^2 + y^2 = 3 \Leftrightarrow \dfrac{x^2}{\frac{3}{2}} + \dfrac{y^2}{3} = 1$. This ellipse has $a = \sqrt{3}$, $b = \sqrt{\frac{3}{2}}$, and so $c^2 = 3 - \frac{3}{2} = \frac{3}{2} \Leftrightarrow c = \sqrt{\frac{3}{2}} = \frac{\sqrt{6}}{2}$. The vertices are $(0, \pm\sqrt{3})$, the foci are $(0, \pm\frac{\sqrt{6}}{2})$, the eccentricity is $e = \dfrac{c}{a} = \dfrac{\frac{\sqrt{6}}{2}}{\sqrt{3}} = \frac{\sqrt{2}}{2}$, the length of the major axis is $2a = 2\sqrt{3}$, and the length of the minor axis is $2b = 2 \cdot \frac{\sqrt{6}}{2} = \sqrt{6}$.

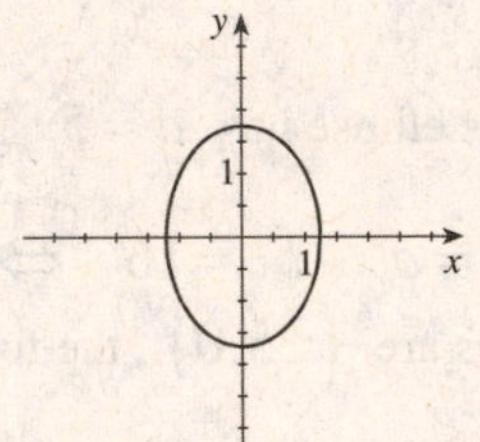

17. $x^2 + 4y^2 = 1 \Leftrightarrow \dfrac{x^2}{1} + \dfrac{y^2}{\frac{1}{4}} = 1$. This ellipse has $a = 1$, $b = \frac{1}{2}$, and so $c^2 = 1 - \frac{1}{4} = \frac{3}{4} \Leftrightarrow c = \frac{\sqrt{3}}{2}$. The vertices are $(\pm 1, 0)$, the foci are $(\pm\frac{\sqrt{3}}{2}, 0)$, the eccentricity is $e = \dfrac{c}{a} = \frac{\sqrt{3}/2}{1} = \frac{\sqrt{3}}{2}$, the length of the major axis is $2a = 2$, and the length of the minor axis is $2b = 1$.

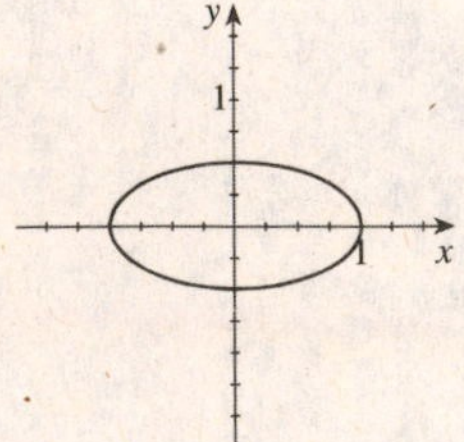

19. $\frac{1}{2}x^2 + \frac{1}{8}y^2 = \frac{1}{4} \Leftrightarrow 2x^2 + \frac{1}{2}y^2 = 1 \Leftrightarrow \dfrac{x^2}{\frac{1}{2}} + \dfrac{y^2}{2} = 1$. This ellipse has $a = \sqrt{2}$, $b = \frac{1}{\sqrt{2}}$, and so $c^2 = 2 - \frac{1}{2} = \frac{3}{2} \Leftrightarrow c = \sqrt{\frac{3}{2}} = \frac{\sqrt{6}}{2}$. The vertices are $(0, \pm\sqrt{2})$, the foci are $(0, \pm\frac{\sqrt{6}}{2})$, the eccentricity is $e = \dfrac{c}{a} = \dfrac{\frac{\sqrt{6}}{2}}{\sqrt{2}} = \frac{\sqrt{3}}{2}$, the length of the major axis is $2a = 2\sqrt{2}$, and length of the minor axis is $2b = \sqrt{2}$.

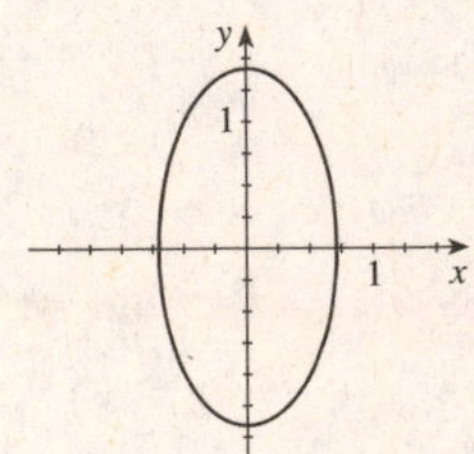

21. $y^2 = 1 - 2x^2 \Leftrightarrow 2x^2 + y^2 = 1 \Leftrightarrow \dfrac{x^2}{\frac{1}{2}} + \dfrac{y^2}{1} = 1$. This ellipse has $a = 1$, $b = \frac{\sqrt{2}}{2}$, and so $c^2 = 1 - \frac{1}{2} = \frac{1}{2} \Leftrightarrow c = \frac{\sqrt{2}}{2}$. The vertices are $(0, \pm 1)$, the foci are $\left(0, \pm\frac{\sqrt{2}}{2}\right)$, the eccentricity is $e = \dfrac{c}{a} = \frac{1/\sqrt{2}}{1} = \frac{\sqrt{2}}{2}$, the length of the major axis is $2a = 2$, and the length of the minor axis is $2b = \sqrt{2}$.

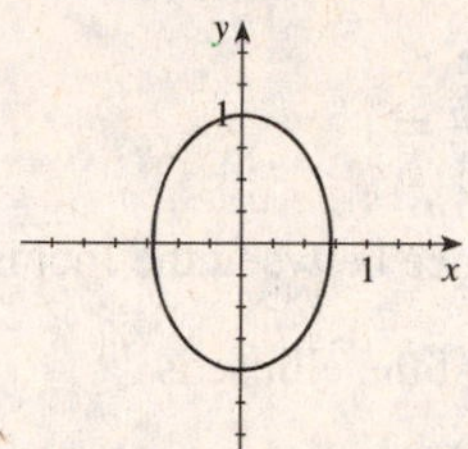

23. This ellipse has a horizontal major axis with $a = 5$ and $b = 4$, so an equation is $\dfrac{x^2}{(5)^2} + \dfrac{y^2}{(4)^2} = 1 \Leftrightarrow \dfrac{x^2}{25} + \dfrac{y^2}{16} = 1$.

25. This ellipse has a vertical major axis with $c = 2$ and $b = 2$. So $a^2 = c^2 + b^2 = 2^2 + 2^2 = 8 \Leftrightarrow a = 2\sqrt{2}$. So an equation is $\dfrac{x^2}{(2)^2} + \dfrac{y^2}{\left(2\sqrt{2}\right)^2} = 1 \Leftrightarrow \dfrac{x^2}{4} + \dfrac{y^2}{8} = 1$.

27. This ellipse has a horizontal major axis with $a = 16$, so an equation of the ellipse is of the form $\dfrac{x^2}{16^2} + \dfrac{y^2}{b^2} = 1$. Substituting the point $(8, 6)$ into the equation, we get $\frac{64}{256} + \dfrac{36}{b^2} = 1 \Leftrightarrow \dfrac{36}{b^2} = 1 - \frac{1}{4} \Leftrightarrow \dfrac{36}{b^2} = \dfrac{3}{4} \Leftrightarrow b^2 = \dfrac{4(36)}{3} = 48$. Thus, an equation of the ellipse is $\dfrac{x^2}{256} + \dfrac{y^2}{48} = 1$.

29. $\dfrac{x^2}{25} + \dfrac{y^2}{20} = 1 \Leftrightarrow \dfrac{y^2}{20} = 1 - \dfrac{x^2}{25} \Leftrightarrow y^2 = 20 - \dfrac{4x^2}{5} \Rightarrow y = \pm\sqrt{20 - \dfrac{4x^2}{5}}$.

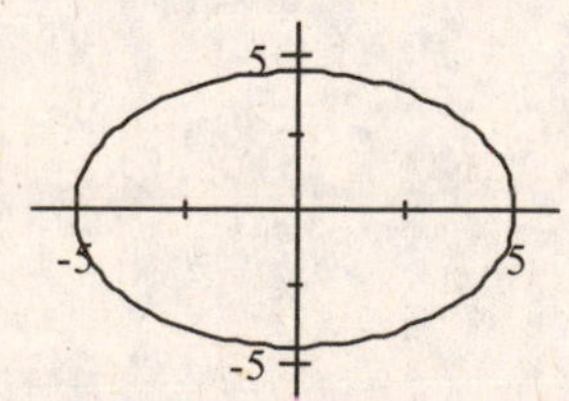

31. $6x^2 + y^2 = 36 \Leftrightarrow y^2 = 36 - 6x^2 \Rightarrow y = \pm\sqrt{36 - 6x^2}$.

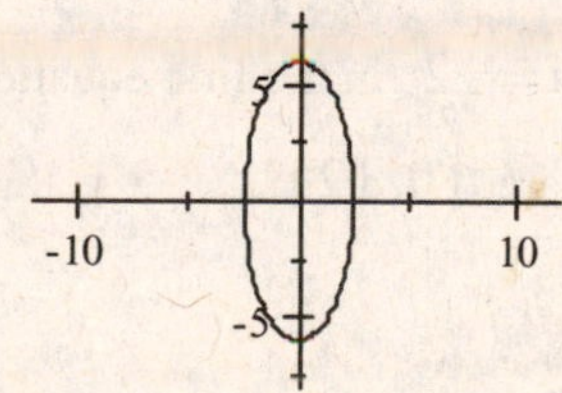

33. The foci are $(\pm 4, 0)$, and the vertices are $(\pm 5, 0)$. Thus, $c = 4$ and $a = 5$, and so $b^2 = 25 - 16 = 9$. Therefore, an equation of the ellipse is $\frac{x^2}{25} + \frac{y^2}{9} = 1$.

35. The length of the major axis is $2a = 4 \Leftrightarrow a = 2$, the length of the minor axis is $2b = 2 \Leftrightarrow b = 1$, and the foci are on the y-axis. Therefore, an equation of the ellipse is $x^2 + \frac{y^2}{4} = 1$.

37. The foci are $(0, \pm 2)$, and the length of the minor axis is $2b = 6 \Leftrightarrow b = 3$. Thus, $a^2 = 4 + 9 = 13$. Since the foci are on the y-axis, an equation is $\frac{x^2}{9} + \frac{y^2}{13} = 1$.

39. The endpoints of the major axis are $(\pm 10, 0) \Leftrightarrow a = 10$, and the distance between the foci is $2c = 6 \Leftrightarrow c = 3$. Therefore, $b^2 = 100 - 9 = 91$, and so an equation of the ellipse is $\frac{x^2}{100} + \frac{y^2}{91} = 1$.

41. The length of the major axis is 10, so $2a = 10 \Leftrightarrow a = 5$, and the foci are on the x-axis, so the form of the equation is $\frac{x^2}{25} + \frac{y^2}{b^2} = 1$. Since the ellipse passes through $(\sqrt{5}, 2)$, we know that $\frac{(\sqrt{5})^2}{25} + \frac{(2)^2}{b^2} = 1 \Leftrightarrow \frac{5}{25} + \frac{4}{b^2} = 1 \Leftrightarrow \frac{4}{b^2} = \frac{4}{5} \Leftrightarrow b^2 = 5$, and so an equation is $\frac{x^2}{25} + \frac{y^2}{5} = 1$.

43. Since the foci are $(\pm 1.5, 0)$, we have $c = \frac{3}{2}$. Since the eccentricity is $0.8 = \frac{c}{a}$, we have $a = \frac{\frac{3}{2}}{\frac{4}{5}} = \frac{15}{8}$, and so $b^2 = \frac{225}{64} - \frac{9}{4} = \frac{225 - 16 \cdot 9}{64} = \frac{81}{64}$. Therefore, an equation of the ellipse is $\frac{x^2}{(15/8)^2} + \frac{y^2}{81/64} = 1 \Leftrightarrow \frac{64x^2}{225} + \frac{64y^2}{81} = 1$.

45. $\begin{cases} 4x^2 + y^2 = 4 \\ 4x^2 + 9y^2 = 36 \end{cases}$

Subtracting the first equation from the second gives $8y^2 = 32 \Leftrightarrow y^2 = 4 \Leftrightarrow y = \pm 2$. Substituting $y = \pm 2$ in the first equation gives $4x^2 + (\pm 2)^2 = 4 \Leftrightarrow x = 0$, and so the points of intersection are $(0, \pm 2)$.

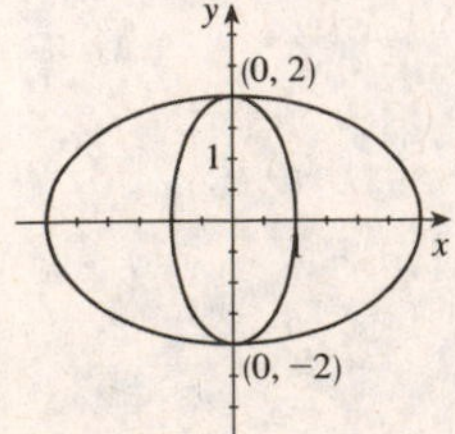

47. $\begin{cases} 100x^2 + 25y^2 = 100 \\ x^2 + \dfrac{y^2}{9} = 1 \end{cases}$

Dividing the first equation by 100 gives $x^2 + \dfrac{y^2}{4} = 1$. Subtracting this equation from the second equation gives $\dfrac{y^2}{9} - \dfrac{y^2}{4} = 0 \Leftrightarrow \left(\frac{1}{9} - \frac{1}{4}\right)y^2 = 0 \Leftrightarrow y = 0$. Substituting $y = 0$ in the second equation gives $x^2 + (0)^2 = 1 \Leftrightarrow x = \pm 1$, and so the points of intersection are $(\pm 1, 0)$.

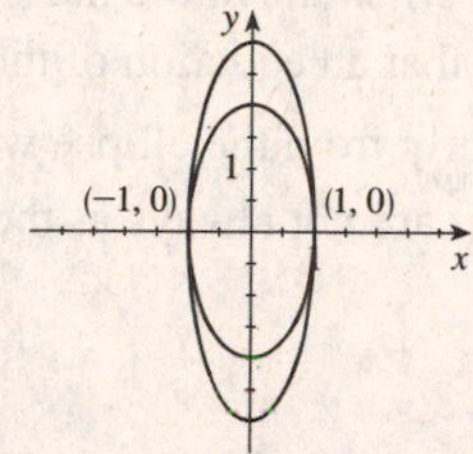

49. (a) $x^2 + ky^2 = 100 \Leftrightarrow ky^2 = 100 - x^2 \Leftrightarrow y = \pm \dfrac{1}{k}\sqrt{100 - x^2}$. For the top half, we graph $y = \dfrac{1}{k}\sqrt{100 - x^2}$ for $k = 4$, 10, 25, and 50.

(b) This family of ellipses have common major axes and vertices, and the eccentricity increases as k increases.

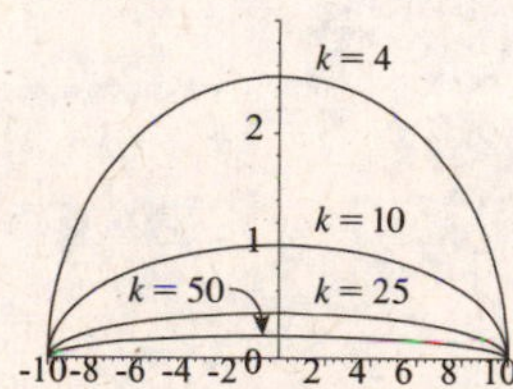

51. Using the perihelion, $a - c = 147{,}000{,}000$, while using the aphelion, $a + c = 153{,}000{,}000$. Adding, we have $2a = 300{,}000{,}000 \Leftrightarrow a = 150{,}000{,}000$. So $b^2 = a^2 - c^2 = \left(150 \times 10^6\right)^2 - \left(3 \times 10^6\right)^2 = 22{,}491 \times 10^{12} = 2.2491 \times 10^{16}$. Thus, an equation of the orbit is $\dfrac{x^2}{2.2500 \times 10^{16}} + \dfrac{y^2}{2.2491 \times 10^{16}} = 1$.

53. Using the perilune, $a - c = 1075 + 68 = 1143$, and using the apolune, $a + c = 1075 + 195 = 1270$. Adding, we get $2a = 2413 \Leftrightarrow a = 1206.5$. So $c = 1270 - 1206.5 \Leftrightarrow c = 63.5$. Therefore, $b^2 = (1206.5)^2 - (63.5)^2 = 1{,}451{,}610$.

Since $a^2 \approx 1{,}455{,}642$, an equation of Apollo 11's orbit is $\dfrac{x^2}{1{,}455{,}642} + \dfrac{y^2}{1{,}451{,}610} = 1$.

55. From the diagram, $a = 40$ and $b = 20$, and so an equation of the ellipse whose top half is the window is $\frac{x^2}{1600} + \frac{y^2}{400} = 1$. Since the ellipse passes through the point $(25, h)$, by substituting, we have $\frac{25^2}{1600} + \frac{h^2}{400} = 1 \Leftrightarrow 625 + 4y^2 = 1600 \Leftrightarrow y = \frac{\sqrt{975}}{2} = \frac{5\sqrt{39}}{2} \approx 15.61$ in. Therefore, the window is approximately 15.6 inches high at the specified point.

57. We start with the flashlight perpendicular to the wall; this shape is a circle. As the angle of elevation increases, the shape of the light changes to an ellipse. When the flashlight is angled so that the outer edge of the light cone is parallel to the wall, the shape of the light is a parabola. Finally, as the angle of elevation increases further, the shape of the light is hyperbolic.

59. The shape drawn on the paper is almost, but not quite, an ellipse. For example, when the bottle has radius 1 unit and the compass legs are set 1 unit apart, then it can be shown that an equation of the resulting curve is $1 + y^2 = 2\cos x$. The graph of this curve differs very slightly from the ellipse with the same major and minor axis. This example shows that in mathematics, things are not always as they appear to be.

7.3 Hyperbolas

1. A hyperbola is the set of all points in the plane for which the *difference* of the distances from two fixed point F_1 and F_2 is constant. The points F_1 and F_2 are called the *foci* of the hyperbola.

3. The graph of the equation $\frac{y^2}{a^2}-\frac{x^2}{b^2}=1$ with $a>0$, $b>0$ is a hyperbola with vertices $(0,a)$ and $(0,-a)$ and foci $(0,\pm c)$, where $c=\sqrt{a^2+b^2}$. So the graph of $\frac{y^2}{4^2}-\frac{x^2}{3^2}=1$ is a hyperbola with vertices $(0,4)$ and $(0,-4)$ and foci $(0,5)$ and $(0,-5)$.

5. $\frac{x^2}{4}-y^2=1$ is Graph III, which opens horizontally and has vertices at $(\pm 2,0)$.

7. $16y^2-x^2=144$ is Graph II, which pens vertically and has vertices at $(0,\pm 3)$.

9. The hyperbola $\frac{x^2}{4}-\frac{y^2}{16}=1$ has $a=2$, $b=4$, and $c^2=16+4 \Rightarrow c=2\sqrt{5}$. The vertices are $(\pm 2,0)$, the foci are $(\pm 2\sqrt{5},0)$, and the asymptotes are $y=\pm\frac{4}{2}x \Leftrightarrow y=\pm 2x$.

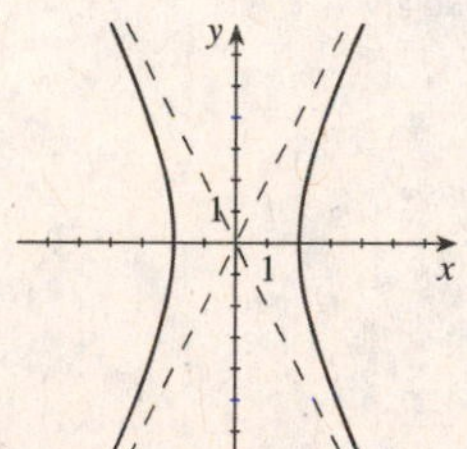

11. The hyperbola $\frac{y^2}{1}-\frac{x^2}{25}=1$ has $a=1$, $b=5$, and $c^2=1+25=26 \Rightarrow c=\sqrt{26}$. The vertices are $(0,\pm 1)$, the foci are $(0,\pm\sqrt{26})$, and the asymptotes are $y=\pm\frac{1}{5}x$.

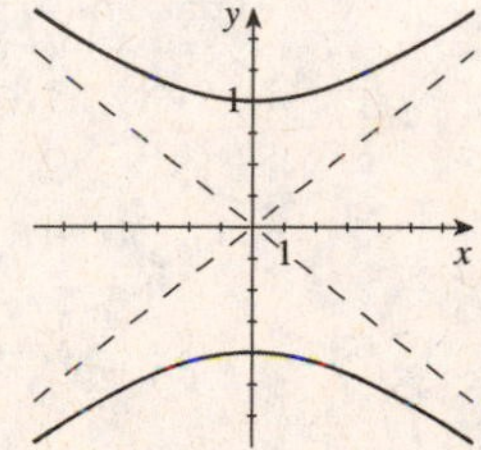

13. The hyperbola $x^2-y^2=1$ has $a=1$, $b=1$, and $c^2=1+1=2 \Rightarrow c=\sqrt{2}$. The vertices are $(\pm 1,0)$, the foci are $(\pm\sqrt{2},0)$, and the asymptotes are $y=\pm x$.

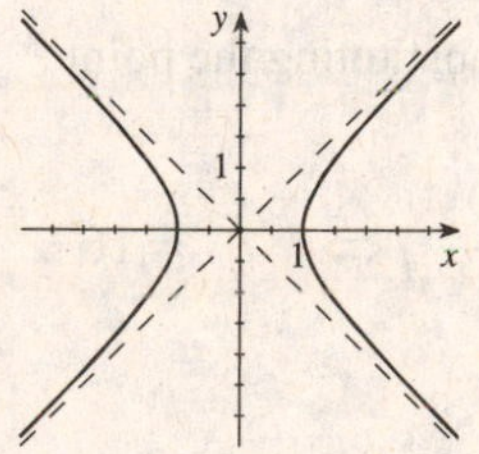

15. The hyperbola $25y^2 - 9x^2 = 225 \Leftrightarrow \frac{y^2}{9} - \frac{x^2}{25} = 1$ has $a = 3$, $b = 5$, and $c^2 = 25 + 9 = 34 \Rightarrow c = \sqrt{34}$. The vertices are $(0, \pm 3)$, the foci are $(0, \pm\sqrt{34})$, and the asymptotes are $y = \pm\frac{3}{5}x$.

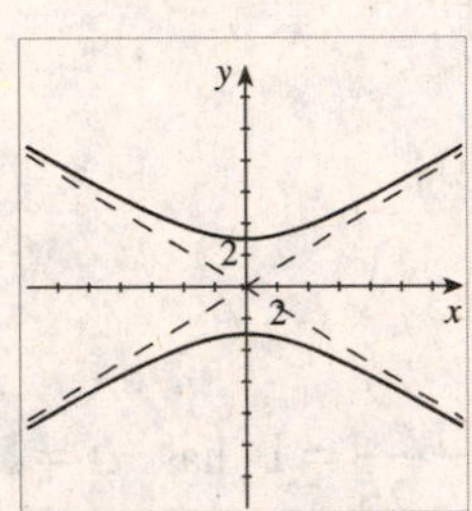

17. The hyperbola $x^2 - 4y^2 - 8 = 0 \Leftrightarrow \frac{x^2}{8} - \frac{y^2}{2} = 1$ has $a = \sqrt{8}$, $b = \sqrt{2}$, and $c^2 = 8 + 2 = 10 \Rightarrow c = \sqrt{10}$. The vertices are $(\pm 2\sqrt{2}, 0)$, the foci are $(\pm\sqrt{10}, 0)$, and the asymptotes are $y = \pm\frac{\sqrt{2}}{\sqrt{8}}x = \pm\frac{1}{2}x$.

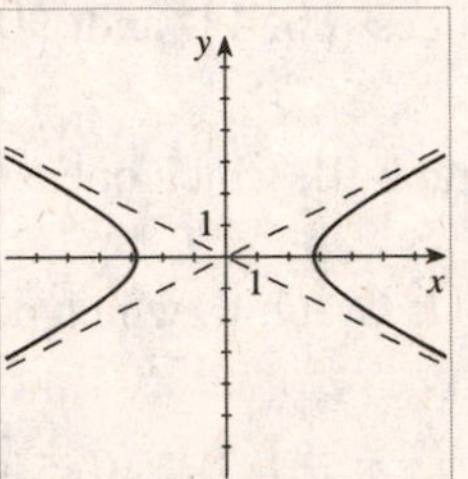

19. The hyperbola $4y^2 - x^2 = 1 \Leftrightarrow \frac{y^2}{\frac{1}{4}} - x^2 = 1$ has $a = \frac{1}{2}$, $b = 1$, and $c^2 = \frac{1}{4} + 1 = \frac{5}{4} \Rightarrow c = \frac{\sqrt{5}}{2}$. The vertices are $(0, \pm\frac{1}{2})$, the foci are $(0, \pm\frac{\sqrt{5}}{2})$, and the asymptotes are $y = \pm\frac{1/2}{1}x = \pm\frac{1}{2}x$.

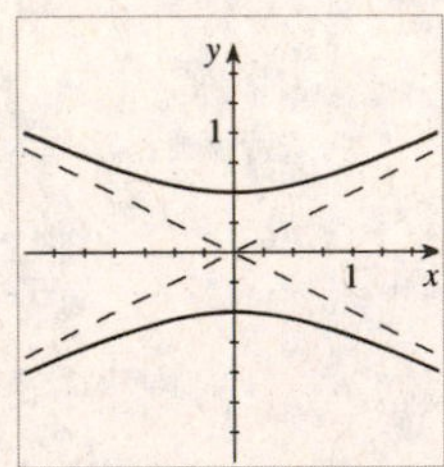

21. From the graph, the foci are $(\pm 4, 0)$, and the vertices are $(\pm 2, 0)$, so $c = 4$ and $a = 2$. Thus, $b^2 = 16 - 4 = 12$, and since the vertices are on the x-axis, an equation of the hyperbola is $\frac{x^2}{4} - \frac{y^2}{12} = 1$.

23. From the graph, the vertices are $(0, \pm 4)$, the foci are on the y-axis, and the hyperbola passes through the point $(3, -5)$. So the equation is of the form $\frac{y^2}{16} - \frac{x^2}{b^2} = 1$. Substituting the point $(3, -5)$, we have $\frac{(-5)^2}{16} - \frac{(3)^2}{b^2} = 1 \Leftrightarrow \frac{25}{16} - 1 = \frac{9}{b^2} \Leftrightarrow \frac{9}{16} = \frac{9}{b^2} \Leftrightarrow b^2 = 16$.

Thus, an equation of the hyperbola is $\frac{y^2}{16} - \frac{x^2}{16} = 1$.

25. The vertices are $(\pm 3,0)$, so $a=3$. Since the asymptotes are $y=\pm\frac{1}{2}x=\pm\frac{b}{a}x$, we have $\frac{b}{3}=\frac{1}{2}$ $\Leftrightarrow$ $b=\frac{3}{2}$. Since the vertices are on the x-axis, an equation is $\frac{x^2}{3^2}-\frac{y^2}{(3/2)^2}=1$ $\Leftrightarrow$ $\frac{x^2}{9}-\frac{4y^2}{9}=1$.

27. $x^2-2y^2=8$ $\Leftrightarrow$ $2y^2=x^2-8$ $\Leftrightarrow$ $y^2=\frac{1}{2}x^2-4$ $\Rightarrow$ $y=\pm\sqrt{\frac{1}{2}x^2-4}$

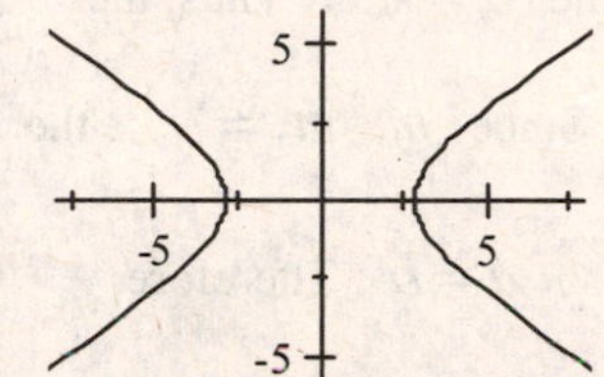

29. $\frac{y^2}{2}-\frac{x^2}{6}=1$ $\Leftrightarrow$ $\frac{y^2}{2}=\frac{x^2}{6}+1$ $\Leftrightarrow$ $y^2=\frac{x^2}{3}+2$ $\Rightarrow$ $y=\pm\sqrt{\frac{x^2}{3}+2}$

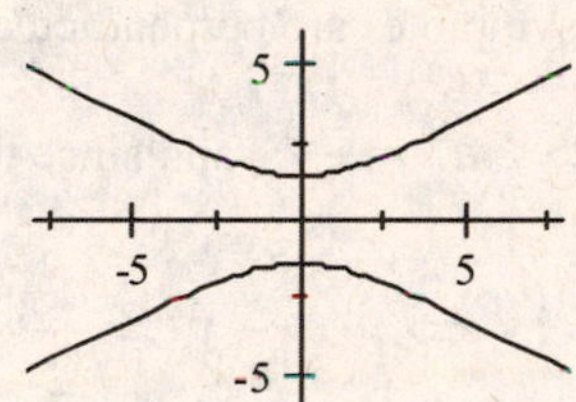

31. The foci are $(\pm 5,0)$ and the vertices are $(\pm 3,0)$, so $c=5$ and $a=3$. Then $b^2=25-9=16$, and since the vertices are on the x-axis, an equation of the hyperbola is $\frac{x^2}{9}-\frac{y^2}{16}=1$.

33. The foci are $(0,\pm 2)$ and the vertices are $(0,\pm 1)$, so $c=2$ and $a=1$. Then $b^2=4-1=3$, and since the vertices are on the y-axis, an equation is $y^2-\frac{x^2}{3}=1$.

35. The vertices are $(\pm 1,0)$ and the asymptotes are $y=\pm 5x$, so $a=1$. The asymptotes are $y=\pm\frac{b}{a}x$, so $\frac{b}{1}=5$ $\Leftrightarrow$ $b=5$. Therefore, an equation of the hyperbola is $x^2-\frac{y^2}{25}=1$.

37. The foci are $(0,\pm 8)$, and the asymptotes are $y=\pm\frac{1}{2}x$, so $c=8$. The asymptotes are $y=\pm\frac{a}{b}x$, so $\frac{a}{b}=\frac{1}{2}$ and $b=2a$. Since $a^2+b^2=c^2=64$, we have $a^2+4a^2=64$ $\Leftrightarrow$ $a^2=\frac{64}{5}$ and $b^2=4a^2=\frac{256}{5}$. Thus, an equation of the hyperbola is $\frac{y^2}{64/5}-\frac{x^2}{256/5}=1$ $\Leftrightarrow$ $\frac{5y^2}{64}-\frac{5x^2}{256}=1$.

39. The asymptotes of the hyperbola are $y = \pm x$, so $b = a$. Since the hyperbola passes through the point $(5,3)$, its foci are on the x-axis, and an equation is of the form, $\frac{x^2}{a^2} - \frac{y^2}{a^2} = 1$, so it follows that $\frac{25}{a^2} - \frac{9}{a^2} = 1 \Leftrightarrow a^2 = 16 = b^2$. Therefore, an equation of the hyperbola is $\frac{x^2}{16} - \frac{y^2}{16} = 1$.

41. The foci are $(\pm 5, 0)$, and the length of the transverse axis is 6, so $c = 5$ and $2a = 6 \Leftrightarrow a = 3$. Thus, $b^2 = 25 - 9 = 16$, and an equation is $\frac{x^2}{9} - \frac{y^2}{16} = 1$.

43. **(a)** The hyperbola $x^2 - y^2 = 5 \Leftrightarrow \frac{x^2}{5} - \frac{y^2}{5} = 1$ has $a = \sqrt{5}$ and $b = \sqrt{5}$. Thus, the asymptotes are $y = \pm x$, and their slopes are $m_1 = 1$ and $m_2 = -1$. Since $m_1 \cdot m_2 = -1$, the asymptotes are perpendicular.

(b) Since the asymptotes are perpendicular, they must have slopes ± 1, so $a = b$. Therefore, $c^2 = 2a^2 \Leftrightarrow a^2 = \frac{c^2}{2}$, and since the vertices are on the x-axis, an equation is

$$\frac{x^2}{\frac{1}{2}c^2} - \frac{y^2}{\frac{1}{2}c^2} = 1 \Leftrightarrow x^2 - y^2 = \frac{c^2}{2}.$$

45. $\sqrt{(x+c)^2 + y^2} - \sqrt{(x-c)^2 + y^2} = \pm 2a$. Let us consider the positive case only. Then $\sqrt{(x+c)^2 + y^2} = 2a + \sqrt{(x-c)^2 + y^2}$, and squaring both sides gives

$$x^2 + 2cx + c^2 + y^2 = 4a^2 + 4a\sqrt{(x-c)^2 + y^2} + x^2 - 2cx + c^2 + y^2 \Leftrightarrow$$

$4a\sqrt{(x-c)^2 + y^2} = 4cx - 4a^2$. Dividing by 4 and squaring both sides gives

$$a^2\left(x^2 - 2cx + c^2 + y^2\right) = c^2x^2 - 2a^2cx + a^4 \Leftrightarrow$$

$$a^2x^2 - 2a^2cx + a^2c^2 + a^2y^2 = c^2x^2 - 2a^2cx + a^4 \Leftrightarrow a^2x^2 + a^2c^2 + a^2y^2 = c^2x^2 + a^4.$$

Rearranging the order, we have $c^2x^2 - a^2x^2 - a^2y^2 = a^2c^2 - a^4 \Leftrightarrow$ $\left(c^2 - a^2\right)x^2 - a^2y^2 = a^2\left(c^2 - a^2\right)$. The negative case gives the same result.

47. (a) From the equation, we have $a^2 = k$ and $b^2 = 16 - k$. Thus, $c^2 = a^2 + b^2 = k + 16 - k = 16$ $\Rightarrow$ $c = \pm 4$. Thus the foci of the family of hyperbolas are $(0, \pm 4)$.

(b) $\dfrac{y^2}{k} - \dfrac{x^2}{16-k} = 1$ $\Leftrightarrow$ $y^2 = k\left(1 + \dfrac{x^2}{16-k}\right)$ $\Rightarrow$ $y = \pm\sqrt{k + \dfrac{kx^2}{16-k}}$. For the top branch, we graph $y = \sqrt{k + \dfrac{kx^2}{16-k}}$, $k = 1$, 4, 8, 12. As k increases, the asymptotes get steeper and the vertices move further apart.

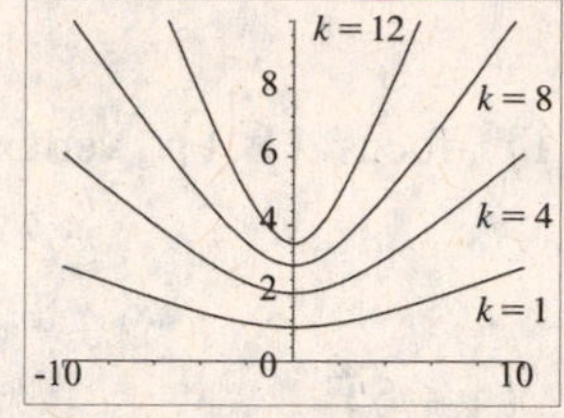

49. Since the asymptotes are perpendicular, $a = b$. Also, since the sun is a focus and the closest distance is 2×10^9, it follows that $c - a = 2 \times 10^9$. Now $c^2 = a^2 + b^2 = 2a^2$, and so $c = \sqrt{2}a$. Thus, $\sqrt{2}a - a = 2 \times 10^9$ $\Rightarrow$ $a = \dfrac{2 \times 10^9}{\sqrt{2} - 1}$ and $a^2 = b^2 = \dfrac{4 \times 10^{18}}{3 - 2\sqrt{2}} \approx 2.3 \times 10^{19}$. Therefore, an equation of the hyperbola is $\dfrac{x^2}{2.3 \times 10^{19}} - \dfrac{y^2}{2.3 \times 10^{19}} = 1$ $\Leftrightarrow$ $x^2 - y^2 = 2.3 \times 10^{19}$.

51. Some possible answers are: as cross-sections of nuclear power plant cooling towers, or as reflectors for camouflaging the location of secret installations.

7.4 Shifted Conics

1. **(a)** If we replace x by $x-3$ the graph of the equation is shifted to the *right* by 3 units. If we replace x by $x+3$ the graph is shifted to the *left* by 3 units.
(b) If we replace y by $y-1$ the graph of the equation is shifted *upward* by 1 unit. If we replace y by $y+1$ the graph is shifted *downward* by 1 unit.

3. $\frac{x^2}{5^2}+\frac{y^2}{4^2}=1$, from left to right: vertex $(-5,0)$, focus $(-3,0)$, focus $(3,0)$, vertex $(5,0)$.
$\frac{(x-3)^2}{5^2}+\frac{(y-1)^2}{4^2}=1$, from left to right: vertex $(-2,1)$, focus $(0,1)$, focus $(6,1)$, vertex $(8,1)$.

5. The ellipse $\frac{(x-2)^2}{9}+\frac{(y-1)^2}{4}=1$ is obtained from the ellipse $\frac{x^2}{9}+\frac{y^2}{4}=1$ by shifting it 2 units to the right and 1 unit upward. So $a=3$, $b=2$, and $c=\sqrt{9-4}=\sqrt{5}$. The center is $(2,1)$, the foci are $\left(2\pm\sqrt{5},1\right)$, the vertices are $(2\pm 3,1)=(-1,1)$ and $(5,1)$, the length of the major axis is $2a=6$, and the length of the minor axis is $2b=4$.

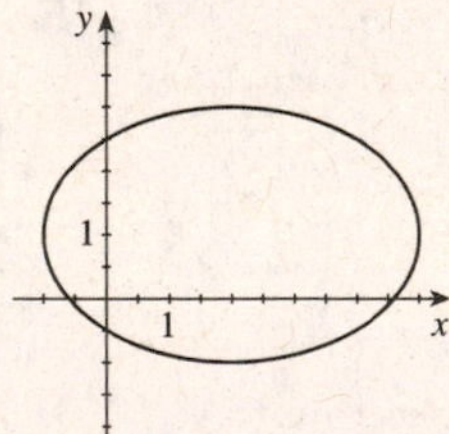

7. The ellipse $\frac{x^2}{9}+\frac{(y+5)^2}{25}=1$ is obtained from the ellipse $\frac{x^2}{9}+\frac{y^2}{25}=1$ by shifting it 5 units downward. So $a=5$, $b=3$, and $c=\sqrt{25-9}=4$. The center is $(0,-5)$, the foci are $(0,-5\pm 4)=(0,-9)$ and $(0,-1)$, the vertices are $(0,-5\pm 5)=(0,-10)$ and $(0,0)$, the length of the major axis is $2a=10$, and the length of the minor axis is $2b=6$.

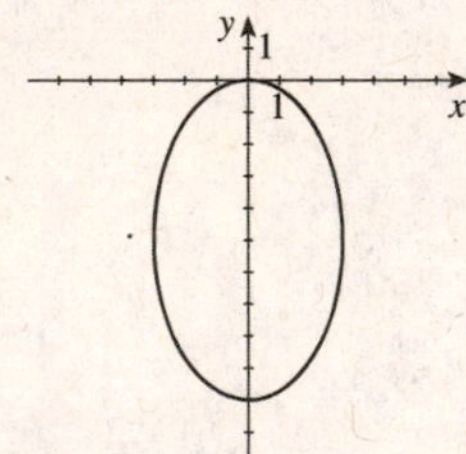

9. The parabola $(x-3)^2 = 8(y+1)$ is obtained from the parabola $x^2 = 8y$ by shifting it 3 units to the right and 1 unit down. So $4p = 8 \Leftrightarrow p = 2$. The vertex is $(3,-1)$, the focus is $(3,-1+2) = (3,1)$, and the directrix is $y = -1-2 = -3$.

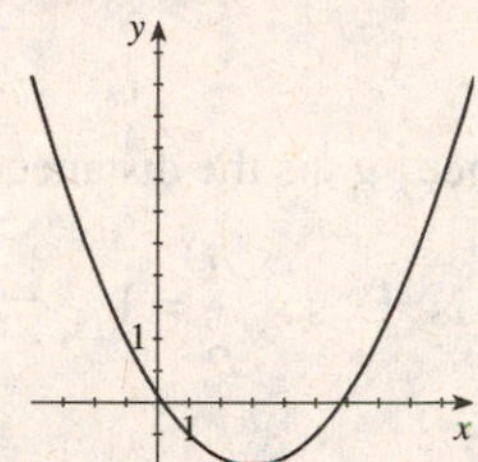

11. The parabola $-4\left(x+\frac{1}{2}\right)^2 = y \Leftrightarrow \left(x+\frac{1}{2}\right)^2 = -\frac{1}{4}y$ is obtained from the parabola $x^2 = -\frac{1}{4}y$ by shifting it $\frac{1}{2}$ unit to the left. So $4p = -\frac{1}{4} \Leftrightarrow p = -\frac{1}{16}$. The vertex is $\left(-\frac{1}{2},0\right)$, the focus is $\left(-\frac{1}{2},0-\frac{1}{16}\right) = \left(-\frac{1}{2},-\frac{1}{16}\right)$, and the directrix is $y = 0+\frac{1}{16} = \frac{1}{16}$.

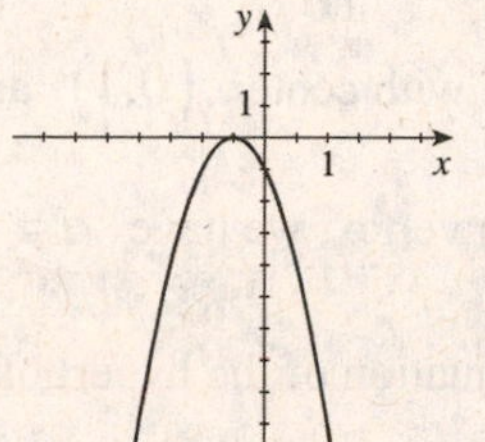

13. The hyperbola $\dfrac{(x+1)^2}{9} - \dfrac{(y-3)^2}{16} = 1$ is obtained from the hyperbola $\dfrac{x^2}{9} - \dfrac{y^2}{16} = 1$ by shifting it 1 unit to the left and 3 units up. So $a = 3$, $b = 4$, and $c = \sqrt{9+16} = 5$. The center is $(-1,3)$, the foci are $(-1\pm 5,3) = (-6,3)$ and $(4,3)$, the vertices are $(-1\pm 3,3) = (-4,3)$ and $(2,3)$, and the asymptotes are $(y-3) = \pm\frac{4}{3}(x+1) \Leftrightarrow y = \pm\frac{4}{3}(x+1)+3 \Leftrightarrow 3y = 4x+13$ and $3y = -4x+5$.

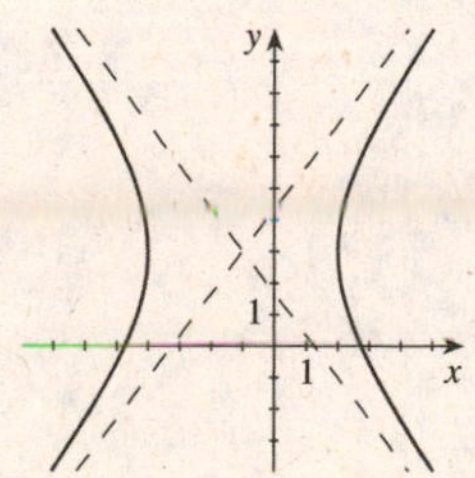

15. The hyperbola $y^2 - \dfrac{(x+1)^2}{4} = 1$ is obtained from the hyperbola $y^2 - \dfrac{x^2}{4} = 1$ by shifting it 1 unit to the left. So $a = 1$, $b = 2$, and $c = \sqrt{1+4} = \sqrt{5}$. The center is $(-1,0)$, the foci are $\left(-1,\pm\sqrt{5}\right) = \left(-1,-\sqrt{5}\right)$ and $\left(-1,\sqrt{5}\right)$, the vertices are $(-1,\pm 1) = (-1,-1)$ and $(-1,1)$, and the asymptotes are $y = \pm\frac{1}{2}(x+1) \Leftrightarrow y = \frac{1}{2}(x+1)$ and $y = -\frac{1}{2}(x+1)$.

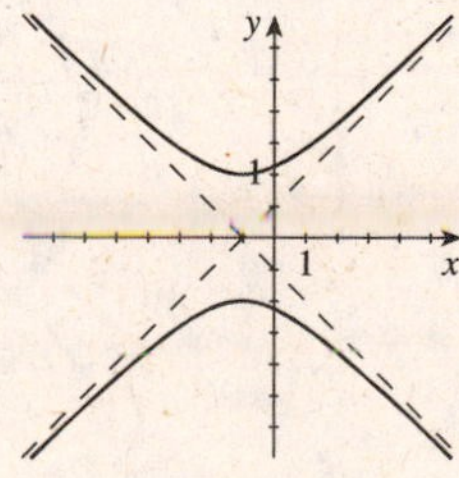

17. This is a parabola that opens down with its vertex at $(0,4)$, so its equation is of the form $x^2 = a(y-4)$. Since $(1,0)$ is a point on this parabola, we have $(1)^2 = a(0-4)$ $\Leftrightarrow$ $1 = -4a$ $\Leftrightarrow$ $a = -\frac{1}{4}$. Thus, an equation is $x^2 = -\frac{1}{4}(y-4)$.

19. This is an ellipse with the major axis parallel to the x -axis, with one vertex at $(0,0)$, the other vertex at $(10,0)$, and one focus at $(8,0)$. The center is at $\left(\frac{0+10}{2},0\right) = (5,0)$, $a = 5$, and $c = 3$ (the distance from one focus to the center). So $b^2 = a^2 - c^2 = 25 - 9 = 16$. Thus, an equation is $\dfrac{(x-5)^2}{25} + \dfrac{y^2}{16} = 1$.

21. This is a hyperbola with center $(0,1)$ and vertices $(0,0)$ and $(0,2)$. Since a is the distance form the center to a vertex, we have $a = 1$. The slope of the given asymptote is 1, so $\dfrac{a}{b} = 1$ $\Leftrightarrow$ $b = 1$. Thus, an equation of the hyperbola is $(y-1)^2 - x^2 = 1$.

23. $y^2 = 4(x+2y)$ $\Leftrightarrow$ $y^2 - 8y = 4x$ $\Leftrightarrow$ $y^2 - 8y + 16 = 4x + 16$ $\Leftrightarrow$ $(y-4)^2 = 4(x+4)$. This is a parabola with $4p = 4$ $\Leftrightarrow$ $p = 1$. The vertex is $(-4,4)$, the focus is $(-4+1,4) = (-3,4)$, and the directrix is $x = -4 - 1 = -5$.

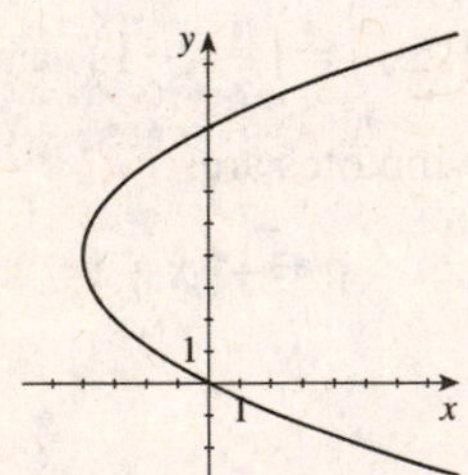

25. $x^2 - 4y^2 - 2x + 16y = 20$ $\Leftrightarrow$ $(x^2 - 2x + 1) - 4(y^2 - 4y + 4) = 20 + 1 - 16$ $\Leftrightarrow$ $(x-1)^2 - 4(y-2)^2 = 5$ $\Leftrightarrow$ $\dfrac{(x-1)^2}{5} - \dfrac{(y-2)^2}{\frac{5}{4}} = 1$. This is a hyperbola with $a = \sqrt{5}$, $b = \frac{1}{2}\sqrt{5}$, and $c = \sqrt{5 + \frac{5}{4}} = \frac{5}{2}$. The center is $(1,2)$, the foci are $\left(1 \pm \frac{5}{2},2\right) = \left(-\frac{3}{2},2\right)$ and $\left(\frac{7}{2},2\right)$, the vertices are $\left(1 \pm \sqrt{5},2\right)$, and the asymptotes are $y - 2 = \pm\frac{1}{2}(x-1)$ $\Leftrightarrow$ $y = \pm\frac{1}{2}(x-1) + 2$ $\Leftrightarrow$ $y = \frac{1}{2}x + \frac{3}{2}$ and $y = -\frac{1}{2}x + \frac{5}{2}$.

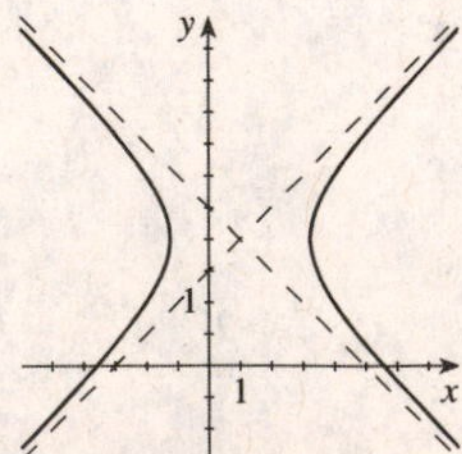

26. $x^2+6x+12y+9=0 \Leftrightarrow$ $x^2+6x+9=-12y \Leftrightarrow$ $(x+3)^2=-12y$. This is a parabola with $4p=-12 \Leftrightarrow p=-3$. The vertex is $(-3,0)$, the focus is $(-3,-3)$, and the directrix is $y=3$.

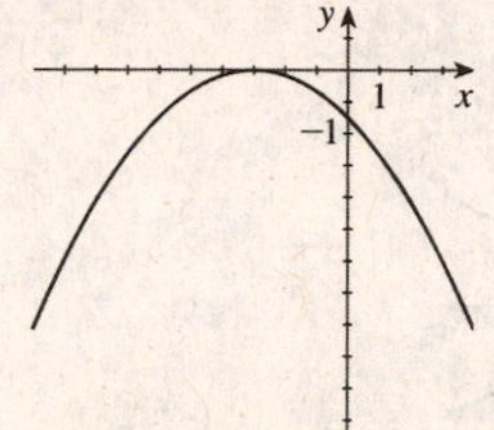

27. $4x^2+25y^2-24x+250y+561=0$ $\Leftrightarrow$ $4(x^2-6x+9)+25(y^2+10y+25)=-561+36+625$ $\Leftrightarrow 4(x-3)^2+25(y+5)^2=100$ $\Leftrightarrow \dfrac{(x-3)^2}{25}+\dfrac{(y+5)^2}{4}=1$. This is an ellipse with $a=5$, $b=2$, and $c=\sqrt{25-4}=\sqrt{21}$. The center is $(3,-5)$, the foci are $(3\pm\sqrt{21},-5)$, the vertices are $(3\pm5,-5)=(-2,-5)$ and $(8,-5)$, the length of the major axis is $2a=10$, and the length of the minor axis is $2b=4$.

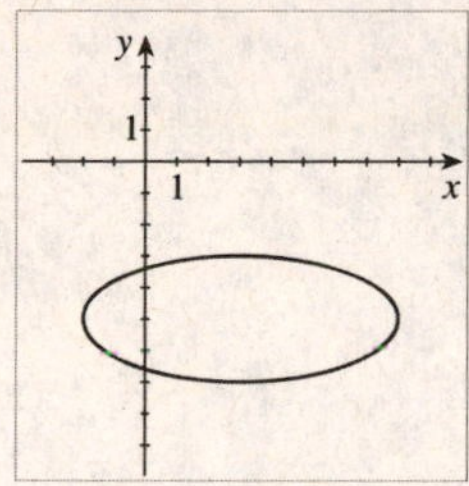

29. $16x^2 - 9y^2 - 96x + 288 = 0 \;\Leftrightarrow$
$16\left(x^2 - 6x\right) - 9y^2 + 288 = 0 \;\Leftrightarrow$
$16\left(x^2 - 6x + 9\right) - 9y^2 = 144 - 288$
$\Leftrightarrow\; 16\left(x-3\right)^2 - 9y^2 = -144 \;\Leftrightarrow$
$\dfrac{y^2}{16} - \dfrac{\left(x-3\right)^2}{9} = 1$. This is a hyperbola with $a = 4$, $b = 3$, and $c = \sqrt{16+9} = 5$. The center is $\left(3,0\right)$, the foci are $\left(3,\pm 5\right)$, the vertices are $\left(3,\pm 4\right)$, and the asymptotes are $y = \pm\frac{4}{3}\left(x-3\right) \;\Leftrightarrow\; y = \frac{4}{3}x - 4$ and $y = 4 - \frac{4}{3}x$.

31. $x^2 + 16 = 4\left(y^2 + 2x\right) \;\Leftrightarrow$
$x^2 - 8x - 4y^2 + 16 = 0 \;\Leftrightarrow$
$\left(x^2 - 8x + 16\right) - 4y^2 = -16 + 16 \;\Leftrightarrow$
$4y^2 = \left(x-4\right)^2 \;\Leftrightarrow\; y = \pm\frac{1}{2}\left(x-4\right)$.
Thus, the conic is degenerate, and its graph is the pair of lines $y = \frac{1}{2}\left(x-4\right)$ and $y = -\frac{1}{2}\left(x-4\right)$.

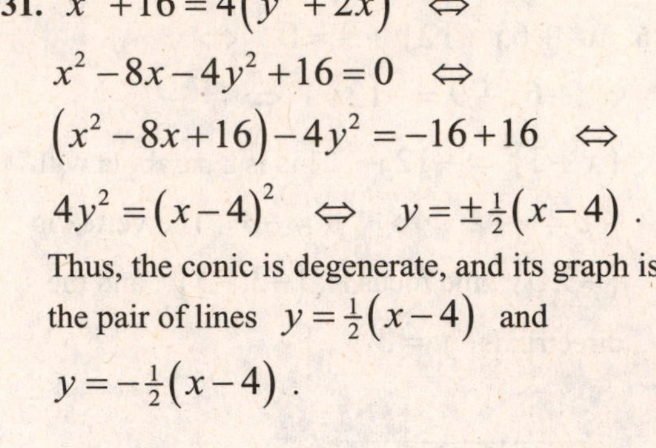

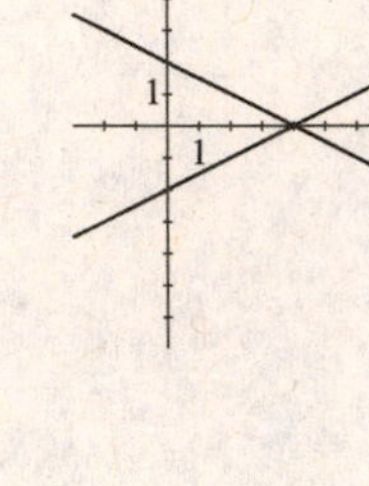

33. $3x^2 + 4y^2 - 6x - 24y + 39 = 0 \;\Leftrightarrow$
$3\left(x^2 - 2x\right) + 4\left(y^2 - 6y\right) = -39 \;\Leftrightarrow$
$3\left(x^2 - 2x + 1\right) + 4\left(y^2 - 6y + 9\right) = -39 + 3 + 36$
$\Leftrightarrow\; 3\left(x-1\right)^2 + 4\left(y-3\right)^2 = 0 \;\Leftrightarrow$
$x = 1$ and $y = 3$. This is a degenerate conic whose graph is the point $\left(1,3\right)$.

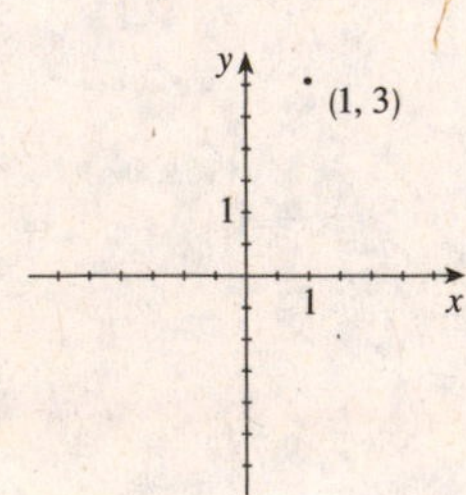

35. $2x^2 - 4x + y + 5 = 0 \;\Leftrightarrow$
$y = -2x^2 + 4x - 5$.

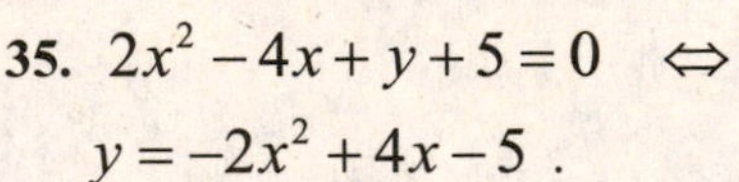

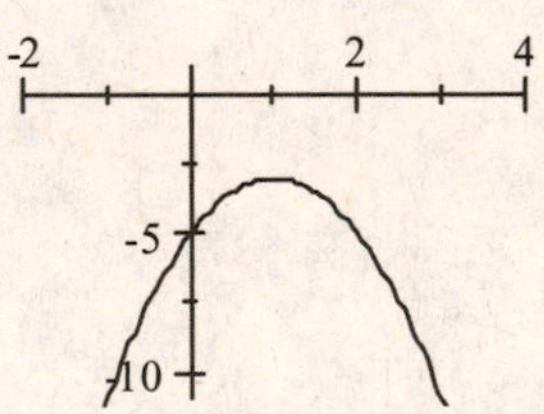

37. $9x^2+36=y^2+36x+6y \quad \Leftrightarrow \quad x^2-36x+36=y^2+6y \quad \Leftrightarrow$

$9x^2-36x+45=y^2+6y+9 \quad \Leftrightarrow \quad 9\left(x^2-4x+5\right)=\left(y+3\right)^2 \quad \Leftrightarrow$

$y+3=\pm\sqrt{9\left(x^2-4x+5\right)} \quad \Leftrightarrow \quad y=-3\pm3\sqrt{x^2-4x+5}$

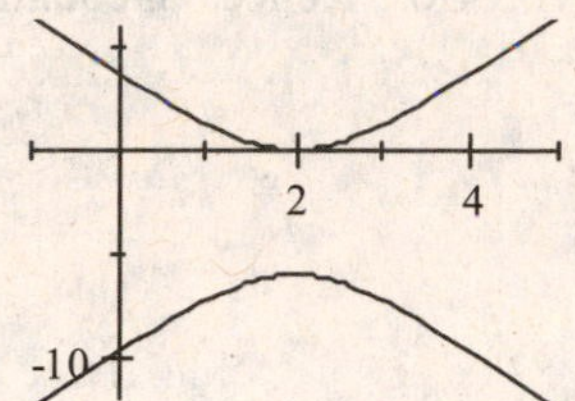

39. $4x^2+y^2+4\left(x-2y\right)+F=0 \quad \Leftrightarrow \quad 4\left(x^2+x\right)+\left(y^2-8y\right)=-F \quad \Leftrightarrow$

$4\left(x^2+x+\frac{1}{4}\right)+\left(y^2-8y+16\right)=16+1-F \quad \Leftrightarrow \quad 4\left(x+\frac{1}{2}\right)^2+\left(y-1\right)^2=17-F$

(a) For an ellipse, $17-F>0 \quad \Leftrightarrow \quad F<17$.

(b) For a single point, $17-F=0 \quad \Leftrightarrow \quad F=17$.

(c) For the empty set, $17-F<0 \quad \Leftrightarrow \quad F>17$.

41. (a) $x^2=4p\left(y+p\right)$, for $p=-2$, $-\frac{3}{2}$, -1 , $-\frac{1}{2}$, $\frac{1}{2}$, 1 , $\frac{3}{2}$, 2 .

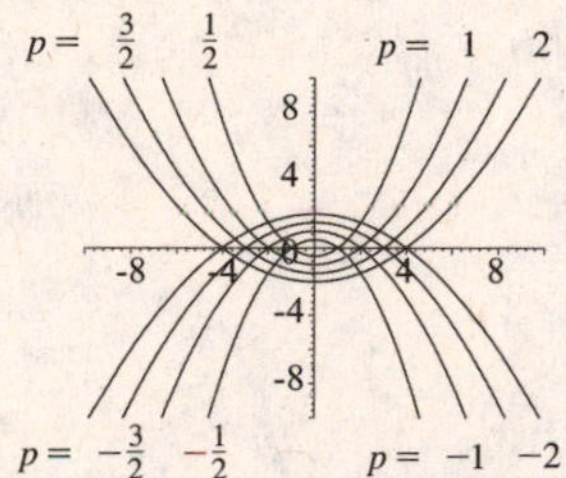

(b) The graph of $x^2=4p\left(y+p\right)$ is obtained by shifting the graph of $x^2=4py$ vertically $-p$ units so that the vertex is at $\left(0,-p\right)$. The focus of $x^2=4py$ is at $\left(0,p\right)$, so this point is also shifted $-p$ units vertically to the point $\left(0,p-p\right)=\left(0,0\right)$. Thus, the focus is located at the origin.**(c)** The parabolas become narrower as the vertex moves toward the origin.

43. Since the height of the satellite above the earth varies between 140 and 440, the length of the major axis is $2a = 140 + 2(3960) + 440 = 8500 \Leftrightarrow a = 4250$. Since the center of the earth is at one focus, we have $a - c = (\text{earth radius}) + 140 = 3960 + 140 = 4100 \Leftrightarrow$ $c = a - 4100 = 4250 - 4100 = 150$. Thus, the center of the ellipse is $(-150, 0)$. So $b^2 = a^2 - c^2 = 4250^2 - 150^2 = 18,062,500 - 22500 = 18,040,000$. Hence, an equation is $\dfrac{(x+150)^2}{18,062,500} + \dfrac{y^2}{18,040,000} = 1$.

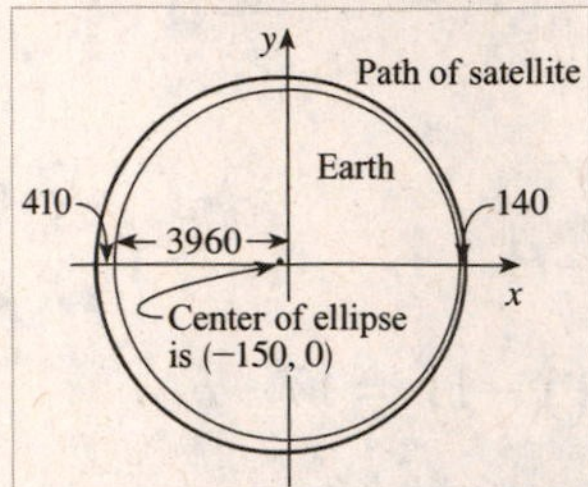

44. (a) We assume that $(0,1)$ is the focus closer to the vertex $(0,0)$, as shown in the figure in the text. Then the center of the ellipse is $(0,a)$ and $1=a-c$. So $c=a-1$ and $(a-1)^2=a^2-b^2$ $\Leftrightarrow$ $a^2-2a+1=a^2-b^2$ $\Leftrightarrow$ $b^2=2a-1$. Thus, the equation is $\dfrac{x^2}{2a-1}+\dfrac{(y-a)^2}{a^2}=1$. If we choose $a=2$, then we get $\dfrac{x^2}{3}+\dfrac{(y-2)^2}{4}=1$. If we choose $a=5$, then we get $\dfrac{x^2}{9}+\dfrac{(y-5)^2}{25}=1$. (Answers will vary, depending on your choices of $a>1$.)

(b) Since a vertex is at $(0,0)$ and a focus is at $(0,1)$, we must have $c-a=1$ ($a>0$), and the center of the hyperbola is $(0,-a)$. So $c=a+1$ and $(a+1)^2=a^2+b^2$ $\Leftrightarrow$ $a^2+2a+1=a^2+b^2$ $\Leftrightarrow$ $b^2=2a+1$. Thus the equation is $\dfrac{(y+a)^2}{a^2}-\dfrac{x^2}{2a+1}=1$. If we let $a=2$, then we get $\dfrac{(y+2)^2}{4}-\dfrac{x^2}{5}=1$. If we let $a=5$, then we get $\dfrac{(y+5)^2}{25}-\dfrac{x^2}{11}=1$. (Answers will vary, depending on your choices of a.)

(c) Since the vertex is at $(0,0)$ and the focus is at $(0,1)$, we must have $p=1$. So $(x-0)^2=4(1)(y-0)$ $\Leftrightarrow$ $x^2=4y$, and there is no other possible parabola.

(d) Graphs will vary, depending on the choices of a in parts (a) and (b).

(e) The ellipses are inside the parabola and the hyperbolas are outside the parabola. All touch at the origin.

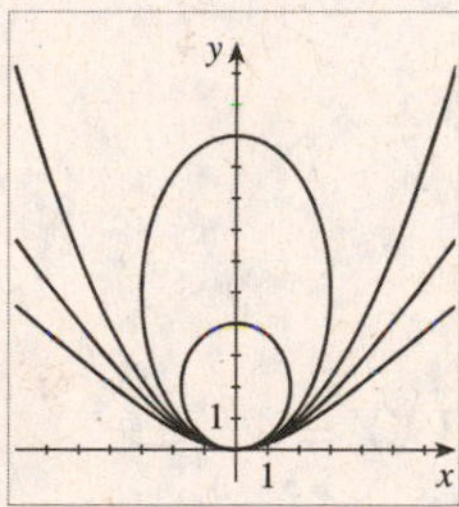

7.5 Rotation of Axes

1. If the x - and y -axes are rotated through an acute angle φ to produce the new X - and Y -axes, then the xy -coordinates (x,y) and the XY -coordinates (X,Y) of a point P in the plane are related by the formulas $x = X\cos\varphi - Y\sin\varphi$, $y = X\sin\varphi + Y\cos\varphi$, $X = x\cos\varphi + y\sin\varphi$, and $Y = -x\sin\varphi + y\cos\varphi$.

3. $(x,y)=(1,1)$, $\varphi = 45^\circ$. Then $X = x\cos\varphi + y\sin\varphi = 1\cdot\frac{1}{\sqrt{2}} + 1\cdot\frac{1}{\sqrt{2}} = \sqrt{2}$ and $Y = -x\sin\varphi + y\cos\varphi = -1\cdot\frac{1}{\sqrt{2}} + 1\cdot\frac{1}{\sqrt{2}} = 0$. Therefore, the XY -coordinates of the given point are $(X,Y) = (\sqrt{2},0)$.

5. $(x,y) = (3,-\sqrt{3})$, $\varphi = 60^\circ$. Then $X = x\cos\varphi + y\sin\varphi = 3\cdot\frac{1}{2} - \sqrt{3}\cdot\frac{\sqrt{3}}{2} = 0$ and $Y = -x\sin\varphi + y\cos\varphi = -3\cdot\frac{\sqrt{3}}{2} - \sqrt{3}\cdot\frac{1}{2} = -2\sqrt{3}$. Therefore, the XY -coordinates of the given point are $(X,Y) = (0,-2\sqrt{3})$.

7. $(x,y) = (0,2)$, $\varphi = 55^\circ$. Then $X = x\cos\varphi + y\sin\varphi = 0\cos 55^\circ + 2\sin 55^\circ \approx 1.6383$ and $Y = -x\sin\varphi + y\cos\varphi = -0\sin 55^\circ + 2\cos 55^\circ \approx 1.1472$. Therefore, the XY -coordinates of the given point are approximately $(X,Y) = (1.6383, 1.1472)$.

9. $x^2 - 3y^2 = 4$, $\varphi = 60^\circ$. Then $x = X\cos 60^\circ - Y\sin 60^\circ = \frac{1}{2}X - \frac{\sqrt{3}}{2}Y$ and $y = X\sin 60^\circ + Y\cos 60^\circ = \frac{\sqrt{3}}{2}X + \frac{1}{2}Y$. Substituting these values into the equation, we get

$$\left(\tfrac{1}{2}X - \tfrac{\sqrt{3}}{2}Y\right)^2 - 3\left(\tfrac{\sqrt{3}}{2}X + \tfrac{1}{2}Y\right)^2 = 4 \Leftrightarrow$$

$$\frac{X^2}{4} - \frac{\sqrt{3}XY}{2} + \frac{3Y^2}{4} - 3\left(\frac{3X^2}{4} + \frac{\sqrt{3}XY}{2} + \frac{Y^2}{4}\right) = 4 \Leftrightarrow$$

$$\frac{X^2}{4} - \frac{9}{4}X^2 + \frac{3Y^2}{4} - \frac{3Y^2}{4} - \frac{\sqrt{3}XY}{2} - \frac{3\sqrt{3}XY}{2} = 4 \Leftrightarrow -2X^2 - 2\sqrt{3}XY = 4 \Leftrightarrow$$

$$X^2 + \sqrt{3}XY = -2 .$$

11. $x^2 - y^2 = 2y$, $\varphi = \cos^{-1}\left(\frac{3}{5}\right)$. So $\cos\varphi = \frac{3}{5}$ and $\sin\varphi = \frac{4}{5}$. Then

$$(X\cos\varphi - Y\sin\varphi)^2 - (X\sin\varphi + Y\cos\varphi)^2 = 2(X\sin\varphi + Y\cos\varphi) \Leftrightarrow$$

$$\left(\tfrac{3}{5}X - \tfrac{4}{5}Y\right)^2 - \left(\tfrac{4}{5}X + \tfrac{3}{5}Y\right)^2 = 2\left(\tfrac{4}{5}X + \tfrac{3}{5}Y\right) \Leftrightarrow$$

$$\frac{9X^2}{25} - \frac{24XY}{25} + \frac{16Y^2}{25} - \frac{16X^2}{25} - \frac{24XY}{25} - \frac{9Y^2}{25} = \frac{8X}{5} + \frac{6Y}{5} \Leftrightarrow$$

$$-\frac{7X^2}{25} - \frac{48XY}{25} + \frac{7Y^2}{25} - \frac{8X}{5} - \frac{6Y}{5} = 0 \Leftrightarrow 7Y^2 - 48XY - 7X^2 - 40X - 30Y = 0 .$$

13. $x^2+2\sqrt{3}xy-y^2=4$, $\varphi=30^\circ$. Then $x=X\cos 30^\circ-Y\sin 30^\circ=\frac{\sqrt{3}}{2}X-\frac{1}{2}Y=\frac{1}{2}\left(\sqrt{3}X-Y\right)$ and $y=X\ \ \sin 30^\circ+Y$ $\cos 30^\circ=\frac{1}{2}X+\frac{\sqrt{3}}{2}Y=\frac{1}{2}\left(X+\sqrt{3}Y\right)$. Substituting these values into the equation, we get

$$\left[\tfrac{1}{2}\left(\sqrt{3}X-Y\right)\right]^2+2\sqrt{3}\left[\tfrac{1}{2}\left(\sqrt{3}X-Y\right)\right]\left[\tfrac{1}{2}\left(X+\sqrt{3}Y\right)\right]-\left[\tfrac{1}{2}\left(X+\sqrt{3}Y\right)\right]^2=4 \quad\Leftrightarrow$$

$$\left(\sqrt{3}X-Y\right)^2+2\sqrt{3}\left(\sqrt{3}X-Y\right)\left(X+\sqrt{3}Y\right)-\left(X+\sqrt{3}Y\right)^2=16 \quad\Leftrightarrow$$

$$\left(3X^2-2\sqrt{3}XY+Y^2\right)+\left(6X^2+4\sqrt{3}XY-6Y^2\right)-\left(X^2+2\sqrt{3}XY+3Y^2\right)=16 \quad\Leftrightarrow\quad 8X^2$$

$-8Y^2=16 \quad\Leftrightarrow\quad \dfrac{X^2}{2}-\dfrac{Y^2}{2}=1$. This is a hyperbola.

15. (a) $xy=8 \quad\Leftrightarrow\quad 0x^2+xy+0y^2=8$. So $A=0$, $B=1$, and $C=0$, and so the discriminant is $B^2-4AC=1^2-4(0)(0)=1$. Since the discriminant is positive, the equation represents a hyperbola.

(b) $\cot 2\varphi=\dfrac{A-C}{B}=0 \quad\Rightarrow\quad 2\varphi=90^\circ \quad\Leftrightarrow\quad \varphi=45^\circ$. Therefore, $x=\frac{\sqrt{2}}{2}X-\frac{\sqrt{2}}{2}Y$ and $y=\frac{\sqrt{2}}{2}X+\frac{\sqrt{2}}{2}Y$. After substitution, the original equation becomes

$$\left(\tfrac{\sqrt{2}}{2}X-\tfrac{\sqrt{2}}{2}Y\right)\left(\tfrac{\sqrt{2}}{2}X+\tfrac{\sqrt{2}}{2}Y\right)=8 \quad\Leftrightarrow\quad \frac{(X-Y)(X+Y)}{2}=8 \quad\Leftrightarrow\quad \frac{X^2}{16}-\frac{Y^2}{16}=1$$

. This is a hyperbola with $a=4$, $b=4$, and $c=4\sqrt{2}$. Hence, the vertices are $V(\pm 4,0)$ and the foci are $F\left(\pm 4\sqrt{2},0\right)$.

(c)

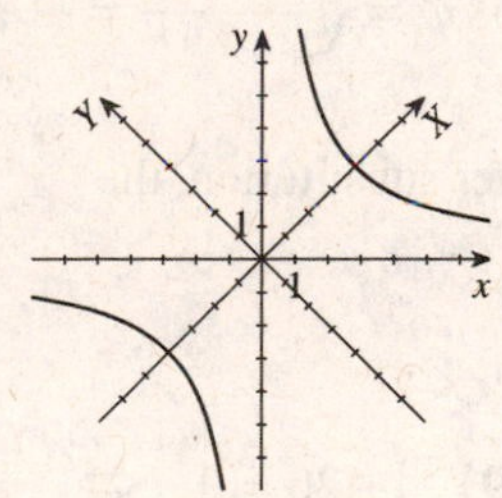

17. (a) $x^2+2\sqrt{3}xy-y^2+2=0$. So $A=1$, $B=2\sqrt{3}$, and $C=-1$, and so the discriminant is $B^2-4AC=\left(2\sqrt{3}\right)^2-4(1)(-1)>0$. Since the discriminant is positive, the equation represents a hyperbola.

(b) $\cot 2\varphi=\dfrac{A-C}{B}=\dfrac{1+1}{2\sqrt{3}}=\dfrac{1}{\sqrt{3}}$ $\Rightarrow$ $2\varphi=60^\circ$ $\Leftrightarrow$ $\varphi=30^\circ$. Therefore, $x=\frac{\sqrt{3}}{2}X-\frac{1}{2}Y$ and $y=\frac{1}{2}X+\frac{\sqrt{3}}{2}Y$. After substitution, the original equation becomes

$$\left(\tfrac{\sqrt{3}}{2}X-\tfrac{1}{2}Y\right)^2+2\sqrt{3}\left(\tfrac{\sqrt{3}}{2}X-\tfrac{1}{2}Y\right)\left(\tfrac{1}{2}X+\tfrac{\sqrt{3}}{2}Y\right)-\left(\tfrac{1}{2}X+\tfrac{\sqrt{3}}{2}Y\right)^2+2=0 \quad\Leftrightarrow$$

$$\tfrac{3}{4}X^2-\tfrac{\sqrt{3}}{2}XY+\tfrac{1}{4}Y^2+\tfrac{\sqrt{3}}{2}\left(\sqrt{3}X^2+2XY-\sqrt{3}Y^2\right)-\tfrac{1}{4}X^2-\tfrac{\sqrt{3}}{2}XY-\tfrac{3}{4}Y^2+2=0 \quad\Leftrightarrow$$

$$X^2\left(\tfrac{3}{4}+\tfrac{3}{2}-\tfrac{1}{4}\right)+XY\left(-\tfrac{\sqrt{3}}{2}+\sqrt{3}-\tfrac{\sqrt{3}}{2}\right)+Y^2\left(\tfrac{1}{4}-\tfrac{3}{2}-\tfrac{3}{4}\right)=-2 \quad\Leftrightarrow\quad 2X^2-2Y^2=-2 \quad\Leftrightarrow$$

$$Y^2-X^2=1 .$$

(c)

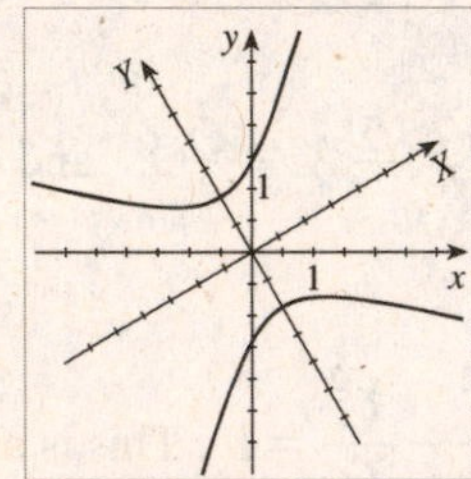

19. (a) $11x^2-24xy+4y^2+20=0$. So $A=11$, $B=-24$, and $C=4$, and so the discriminant is $B^2-4AC=(-24)^2-4(11)(4)>0$. Since the discriminant is positive, the equation represents a hyperbola.

(b) $\cot 2\varphi=\dfrac{A-C}{B}=\dfrac{11-4}{-24}=-\dfrac{7}{24}$ $\Rightarrow$ $\cos 2\varphi=-\frac{7}{25}$. Therefore, $\cos\varphi=\sqrt{\frac{1+(-7/25)}{2}}=\frac{3}{5}$ and $\sin\varphi=\sqrt{\frac{1-(-7/25)}{2}}=\frac{4}{5}$. Hence, $x=\dfrac{3X}{5}-\frac{4}{5}Y$ and $y=\frac{4}{5}X+\frac{3}{5}Y$. After substitution, the original equation becomes

$$11\left(\tfrac{3}{5}X-\tfrac{4}{5}Y\right)^2-24\left(\tfrac{3}{5}X-\tfrac{4}{5}Y\right)\left(\tfrac{4}{5}X+\tfrac{3}{5}Y\right)+4\left(\tfrac{4}{5}X+\tfrac{3}{5}Y\right)^2+20=0 \quad\Leftrightarrow$$

$$\tfrac{11}{25}\left(9X^2-24XY+16Y^2\right)-\tfrac{24}{25}\left(12X^2-7XY-12Y^2\right)+\tfrac{4}{25}\left(16X^2+24XY+9Y^2\right)+20=0 \quad\Leftrightarrow$$

$$X^2(99-288+64)+XY(-264+168+96)+Y^2(176+288+36)=-500 \quad\Leftrightarrow$$

$$-125X^2+500Y^2=-500 \quad\Leftrightarrow\quad \tfrac{1}{4}X^2-Y^2=1 .$$

(c)

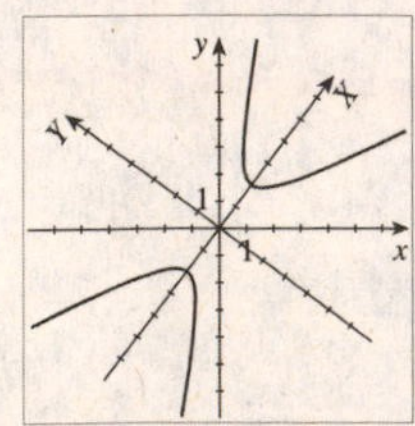

Since $\cos 2\varphi=-\frac{7}{25}$, we have $2\varphi\approx 106.26^\circ$, so $\varphi\approx 53^\circ$.

21. (a) $\sqrt{3}x^2 + 3xy = 3$. So $A = \sqrt{3}$, $B = 3$, and $C = 0$, and so the discriminant is $B^2 - 4AC = (3)^2 - 4\left(\sqrt{3}\right)(0) = 9$. Since the discriminant is positive, the equation represents a hyperbola.

(b) $\cot 2\varphi = \dfrac{A-C}{B} = \dfrac{1}{\sqrt{3}} \Rightarrow 2\varphi = 60^\circ \Leftrightarrow \varphi = 30^\circ$. Therefore, $x = \frac{\sqrt{3}}{2}X - \frac{1}{2}Y$ and $y = \frac{1}{2}X + \frac{\sqrt{3}}{2}Y$. After substitution, the equation becomes

$$\sqrt{3}\left(\tfrac{\sqrt{3}}{2}X - \tfrac{1}{2}Y\right)^2 + 3\left(\tfrac{\sqrt{3}}{2}X - \tfrac{1}{2}Y\right)\left(\tfrac{1}{2}X + \tfrac{\sqrt{3}}{2}Y\right) = 3 \Leftrightarrow$$

$$\tfrac{\sqrt{3}}{4}\left(3X^2 - 2\sqrt{3}XY + Y^2\right) + \tfrac{3}{4}\left(\sqrt{3}X^2 + 2XY - \sqrt{3}Y^2\right) = 3 \Leftrightarrow$$

$$X^2\left(\tfrac{3\sqrt{3}}{4} + \tfrac{3\sqrt{3}}{4}\right) + XY\left(\tfrac{-6}{4} + \tfrac{6}{4}\right) + Y^2\left(\tfrac{\sqrt{3}}{4} - \tfrac{3\sqrt{3}}{4}\right) = 3 \Leftrightarrow \tfrac{3\sqrt{3}}{2}X^2 - \tfrac{\sqrt{3}}{2}Y^2 = 3 \Leftrightarrow$$

$\frac{\sqrt{3}}{2}X^2 - \frac{1}{2\sqrt{3}}Y^2 = 1$. This is a hyperbola with $a = \sqrt{\frac{2}{\sqrt{3}}}$ and $b = \sqrt{2\sqrt{3}}$.

(c)

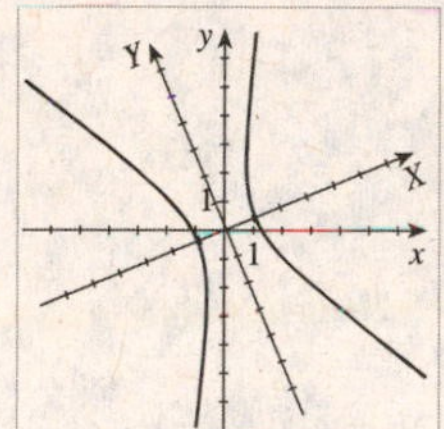

23. (a) $x^2 + 2xy + y^2 + x - y = 0$. So $A = 1$, $B = 2$, and $C = 1$, and so the discriminant is $B^2 - 4AC = 2^2 - 4(1)(1) = 0$. Since the discriminant is zero, the equation represents a parabola.

(b) $\cot 2\varphi = \dfrac{A-C}{B} = 0 \Rightarrow 2\varphi = 90^\circ \Leftrightarrow \varphi = 45^\circ$. Therefore, $x = \frac{\sqrt{2}}{2}X - \frac{\sqrt{2}}{2}Y$ and $y = \frac{\sqrt{2}}{2}X + \frac{\sqrt{2}}{2}Y$. After substitution, the original equation becomes

$$\left(\frac{\sqrt{2}}{2}X - \frac{\sqrt{2}}{2}Y\right) + 2\left(\frac{\sqrt{2}}{2}X - \frac{\sqrt{2}}{2}Y\right)\left(\frac{\sqrt{2}}{2} + \frac{\sqrt{2}}{2}Y\right) + \left(\frac{\sqrt{2}}{2}X + \frac{\sqrt{2}}{2}Y\right)^2$$

$$+ \left(\frac{\sqrt{2}}{2}X - \frac{\sqrt{2}}{2}Y\right) - \left(\frac{\sqrt{2}}{2}X + \frac{\sqrt{2}}{2}Y\right) = 0$$

$$\Leftrightarrow \quad \frac{1}{2}X^2 - XY + \frac{1}{2}Y^2 + X^2 - Y^2 + \frac{1}{2}X^2 + XY + Y^2 - \sqrt{2}Y = 0 \quad \Leftrightarrow \quad 2X^2 - \sqrt{2}Y = 0$$

$\Leftrightarrow \quad X^2 = \frac{\sqrt{2}}{2}Y$. This is a parabola with $4p = \frac{1}{\sqrt{2}}$ and hence the focus is $F\left(0, \frac{1}{4\sqrt{2}}\right)$.

(c)

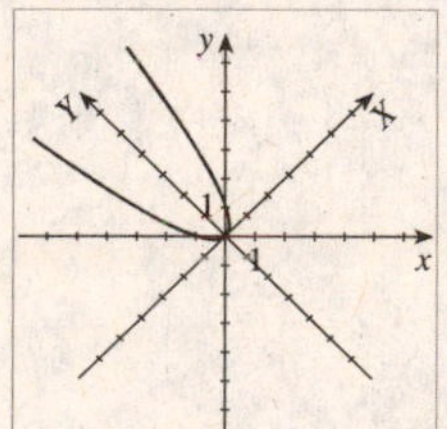

25. (a) $2\sqrt{3}x^2 - 6xy + \sqrt{3}x + 3y = 0$. So $A = 2\sqrt{3}$, $B = -6$, and $C = 0$, and so the discriminant is $B^2 - 4AC = (-6)^2 - 4(2\sqrt{3})(0) = 36$. Since the discriminant is positive, the equation represents a hyperbola.

(b) $\cot 2\varphi = \dfrac{A-C}{B} = \dfrac{2\sqrt{3}}{-6} = -\dfrac{1}{\sqrt{3}} \Rightarrow 2\varphi = 120^\circ \Leftrightarrow \varphi = 60^\circ$. Therefore, $x = \frac{1}{2}X - \frac{\sqrt{3}}{2}Y$ and $y = \frac{\sqrt{3}}{2}X + \frac{1}{2}Y$, and substituting gives

$$2\sqrt{3}\left(\tfrac{1}{2}X - \tfrac{\sqrt{3}}{2}Y\right)^2 - 6\left(\tfrac{1}{2}X - \tfrac{\sqrt{3}}{2}Y\right)\left(\tfrac{\sqrt{3}}{2}X + \tfrac{1}{2}Y\right) + \sqrt{3}\left(\tfrac{1}{2}X - \tfrac{\sqrt{3}}{2}Y\right) + 3\left(\tfrac{\sqrt{3}}{2}X + \tfrac{1}{2}Y\right) = 0 \Leftrightarrow$$

$$\tfrac{\sqrt{3}}{2}\left(X^2 - 2\sqrt{3}XY + 3Y^2\right) - \tfrac{3}{2}\left(\sqrt{3}X^2 - 2XY - \sqrt{3}Y^2\right) + \tfrac{\sqrt{3}}{2}\left(X - \sqrt{3}Y\right) + \tfrac{3}{2}\left(\sqrt{3}X + Y\right) = 0$$

$$\Leftrightarrow \quad X^2\left(\tfrac{\sqrt{3}}{2} - \tfrac{3\sqrt{3}}{2}\right) + X\left(\tfrac{\sqrt{3}}{2} + \tfrac{3\sqrt{3}}{2}\right) + XY(-3+3) + Y^2\left(\tfrac{3\sqrt{3}}{2} + \tfrac{3\sqrt{3}}{2}\right) + Y\left(-\tfrac{3}{2} + \tfrac{3}{2}\right) = 0 \Leftrightarrow$$

$$-\sqrt{3}X^2 + 2\sqrt{3}X + 3\sqrt{3}Y^2 = 0 \Leftrightarrow -X^2 + 2X + 3Y^2 = 0 \Leftrightarrow$$

$3Y^2 - \left(X^2 - 2X + 1\right) = -1 \Leftrightarrow (X-1)^2 - 3Y^2 = 1$. This is a hyperbola with $a = 1$, $b = \frac{\sqrt{3}}{3}$, $c = \sqrt{1 + \frac{1}{3}} = \frac{2}{\sqrt{3}}$, and $C(1,0)$.

(c)

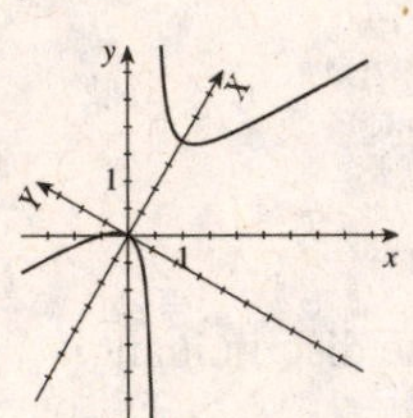

27. (a) $52x^2+72xy+73y^2=40x-30y+75$. So $A=52$, $B=72$, and $C=73$, and so the discriminant is $B^2-4AC=(72)^2-4(52)(73)=-10{,}000$. Since the discriminant is decidedly negative, the equation represents an ellipse.

(b) $\cot 2\varphi=\dfrac{A-C}{B}=\dfrac{52-73}{72}=-\dfrac{7}{24}$. Therefore, as in Exercise 19(b), we get $\cos\varphi=\frac{3}{5}$, $\sin\varphi=\frac{4}{5}$, and $x=\frac{3}{5}X-\frac{4}{5}Y$, $y=\frac{4}{5}X+\frac{3}{5}Y$. By substitution,

$$52\left(\frac{3}{5}X-\frac{4}{5}Y\right)^2+72\left(\frac{3}{5}X-\frac{4}{5}Y\right)\left(\frac{4}{5}X+\frac{3}{5}Y\right)$$
$$+73\left(\frac{4}{5}X+\frac{3}{5}Y\right)=40\left(\frac{3}{5}X-\frac{4}{5}Y\right)-30\left(\frac{4}{5}X+\frac{3}{5}Y\right)+75$$

$$\Leftrightarrow\ \frac{52}{25}\left(9X^2-24XY+16Y2\right)+\frac{72}{25}\left(12X^2-7XY-12Y^2\right)$$
$$+\frac{73}{25}\left(16X^2+24XY+9Y^2\right)=24X-32Y-24X-18Y+75$$

$\Leftrightarrow\ 468X^2+832Y^2+864X^2-864Y^2+1168X^2+657Y^2=-1250Y+1875\ \Leftrightarrow$ $2500X^2+625Y^2+1250Y=1875\ \Leftrightarrow\ 100X^2+25Y^2+50Y=75\ \Leftrightarrow$ $X^2+\frac{1}{4}(Y+1)^2=1$. This is an ellipse with $a=2$, $b=1$, $c=\sqrt{4-1}=\sqrt{3}$, and center $C(0,-1)$.

(c)

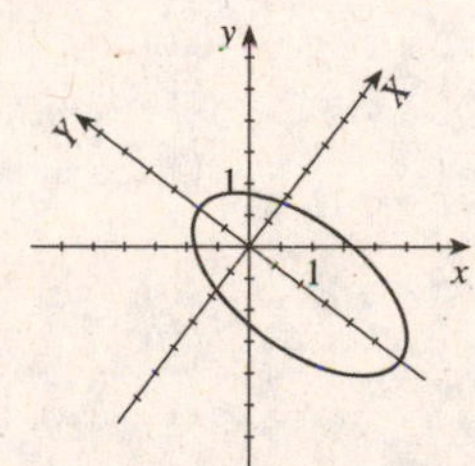

Since $\cos 2\varphi=-\frac{7}{25}$, we have $2\varphi=\cos^{-1}\left(-\frac{7}{25}\right)\approx 106.26^\circ$ and so $\varphi\approx 53^\circ$.

29. (a) The discriminant is $B^2-4AC=(-4)^2+4(2)(2)=0$. Since the discriminant is 0, the equation represents a parabola.

(b) $2x^2-4xy+2y^2-5x-5=0\ \Leftrightarrow\ 2y^2-4xy=-2x^2+5x+5\ \Leftrightarrow$ $2\left(y^2-2xy\right)=-2x^2+5x+5\ \Leftrightarrow\ 2\left(y^2-2xy+x^2\right)=-2x^2+5x+5+2x^2\ \Leftrightarrow$ $2(y-x)^2=5x+5\ \Leftrightarrow\ (y-x)^2=\frac{5}{2}x+\frac{5}{2}\Rightarrow y-x=\pm\sqrt{\frac{5}{2}x+\frac{5}{2}}\ \Leftrightarrow\ y=x\pm\sqrt{\frac{5}{2}x+\frac{5}{2}}$

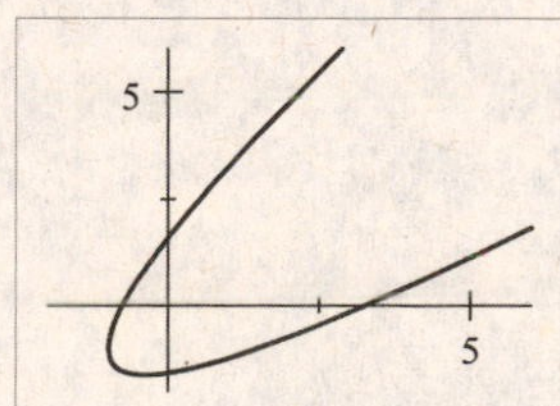

31. (a) The discriminant is $B^2 - 4AC = 10^2 + 4(6)(3) = 28 > 0$. Since the discriminant is positive, the equation represents a hyperbola.

(b) $6x^2 + 10xy + 3y^2 - 6y = 36 \quad \Leftrightarrow \quad 3y^2 + 10xy - 6y = 36 - 6x^2 \quad \Leftrightarrow$

$3y^2 + 2(5x-3)y = 36 - 6x^2 \quad \Leftrightarrow \quad y^2 + 2\left(\frac{5}{3}x - 1\right)y = 12 - 2x^2 \quad \Leftrightarrow$

$y^2 + 2\left(\frac{5}{3}x-1\right)y + \left(\frac{5}{3}x-1\right)^2 = \left(\frac{5}{3}x-1\right)^2 + 12 - 2x^2 \quad \Leftrightarrow$

$\left[y + \left(\frac{5}{3}x-1\right)\right]^2 = \frac{25}{9}x^2 - \frac{10}{3}x + 1 + 12 - 2x^2 \quad \Leftrightarrow \quad \left[y + \left(\frac{5}{3}x-1\right)\right]^2 = \frac{7}{9}x^2 - \frac{10}{3}x + 13 \quad \Leftrightarrow$

$y + \left(\frac{5}{3}x-1\right) = \pm\sqrt{\frac{7}{9}x^2 - \frac{10}{3}x + 13} \quad \Leftrightarrow \quad y = -\frac{5}{3}x + 1 \pm \sqrt{\frac{7}{9}x^2 - \frac{10}{3}x + 13}$

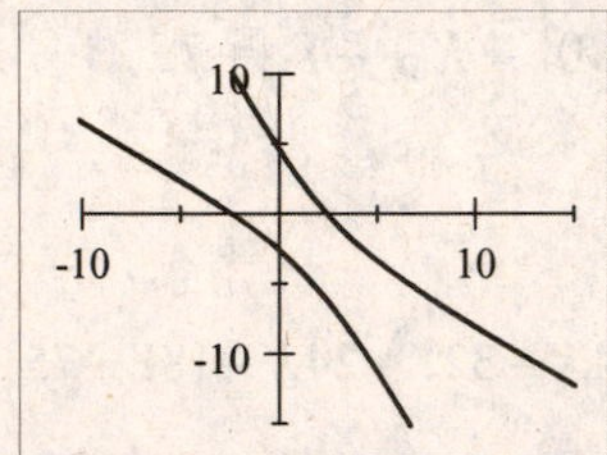

33. (a) $7x^2+48xy-7y^2-200x-150y+600=0$. Then $A=7$, $B=48$, and $C=-7$, and so the discriminant is $B^2-4AC=(48)^2-4(7)(7)>0$. Since the discriminant is positive, the equation represents a hyperbola. We now find the equation in terms of XY -coordinates. We have $\cot 2\varphi=\dfrac{A-C}{B}=\dfrac{7}{24}$ $\Rightarrow$ $\cos\varphi=\frac{4}{5}$ and $\sin\varphi=\frac{3}{5}$. Therefore, $x=\frac{4}{5}X-\frac{3}{5}Y$ and $y=\frac{3}{5}X+\frac{4}{5}Y$, and substitution gives

$$7\left(\tfrac{4}{5}X-\tfrac{3}{5}Y\right)^2+48\left(\tfrac{4}{5}X-\tfrac{3}{5}Y\right)\left(\tfrac{3}{5}X+\tfrac{4}{5}Y\right)-7\left(\tfrac{3}{5}X+\tfrac{4}{5}Y\right)^2 -200\left(\tfrac{4}{5}X-\tfrac{3}{5}Y\right)-150\left(\tfrac{3}{5}X+\tfrac{4}{5}Y\right)+600=0 \Leftrightarrow$$

$$\frac{7}{25}\left(16X^2-24XY+9Y^2\right)+\frac{48}{25}\left(12X^2+7XY-12Y^2\right) -\frac{7}{25}\left(9X^2+24XY+16Y^2\right)-160X+120Y-90X-120Y+600=0$$

$$\Leftrightarrow \quad 112X^2-168XY+63Y^2+576X^2+336XY-576Y^2-63X^2 -168XY-112Y^2-6250X+15{,}000=0 \Leftrightarrow$$

$25X^2-25Y^2-250X+600=0$ $\Leftrightarrow$ $25\left(X^2-10X+25\right)-25Y^2=-600+625$ $\Leftrightarrow$ $(X-5)^2-Y^2=1$. This is a hyperbola with $a=1$, $b=1$, $c=\sqrt{1+1}=\sqrt{2}$, and center $C(5,0)$.

(b) In the XY -plane, the center is $C(5,0)$, the vertices are $V(5\pm1,0)=V_1(4,0)$ and $V_2(6,0)$, and the foci are $F\left(5\pm\sqrt{2},0\right)$. In the xy -plane, the center is $C\left(\frac{4}{5}\cdot5-\frac{3}{5}\cdot0,\frac{3}{5}\cdot5+\frac{4}{5}\cdot0\right)=C(4,3)$, the vertices are $V_1\left(\frac{4}{5}\cdot4-\frac{3}{5}\cdot0,\frac{3}{5}\cdot4+\frac{4}{5}\cdot0\right)=V_1\left(\frac{16}{5},\frac{12}{5}\right)$ and $V_2\left(\frac{4}{5}\cdot6-\frac{3}{5}\cdot0,\frac{3}{5}\cdot6+\frac{4}{5}\cdot0\right)=V_2\left(\frac{24}{5},\frac{18}{5}\right)$, and the foci are $F_1\left(4+\frac{4}{5}\sqrt{2},3+\frac{3}{5}\sqrt{2}\right)$ and $F_2\left(4-\frac{4}{5}\sqrt{2},3-\frac{3}{5}\sqrt{2}\right)$.

(c) In the XY -plane, the equations of the asymptotes are $Y=X-5$ and $Y=-X+5$. In the xy -plane, these equations become $-x\cdot\frac{3}{5}+y\cdot\frac{4}{5}=x\cdot\frac{4}{5}+y\cdot\frac{3}{5}-5$ $\Leftrightarrow$ $7x-y-25=0$. Similarly, $-x\cdot\frac{3}{5}+y\cdot\frac{4}{5}=-x\cdot\frac{4}{5}-y\cdot\frac{3}{5}+5$ $\Leftrightarrow$ $x+7y-25=0$.

35. We use the hint and eliminate Y by adding: $x = X\cos\varphi - Y\sin\varphi \iff$ $x\cos\varphi = X\cos^2\varphi - Y\sin\varphi\cos\varphi$ and $y = X\sin\varphi + Y\cos\varphi \iff$ $y\sin\varphi = X\sin^2\varphi + Y\sin\varphi\cos\varphi$, and adding these two equations gives $x\cos\varphi + y$ $\sin\varphi = X\left(\cos^2\varphi + \sin^2\varphi\right) \iff x\cos\varphi + y\sin\varphi = X$. In a similar manner, we eliminate X by subtracting:
$x = X\cos\varphi - Y\sin\varphi \iff -x\sin\varphi = -X\cos\varphi\sin\varphi + Y\sin^2\varphi$ and $y = X\sin\varphi + Y\cos\varphi \iff y\cos\varphi = X\sin\varphi\cos\varphi + Y\cos^2\varphi$, so $-x\sin\varphi + y$ $\cos\varphi = Y\left(\cos^2\varphi + \sin^2\varphi\right) \iff -x \quad \sin\varphi + y \quad \cos\varphi = Y$. Thus, $X = x$ $\cos\varphi + y\sin\varphi$ and $Y = -x \quad \sin\varphi + y\cos\varphi$.

37. $Z = \begin{bmatrix} x \\ y \end{bmatrix}$, $Z' = \begin{bmatrix} X \\ Y \end{bmatrix}$, and $R = \begin{bmatrix} \cos\varphi & -\sin\varphi \\ \sin\varphi & \cos\varphi \end{bmatrix}$.

Thus $Z = RZ' \iff \begin{bmatrix} x \\ y \end{bmatrix} = \begin{bmatrix} \cos\varphi & -\sin\varphi \\ \sin\varphi & \cos\varphi \end{bmatrix}\begin{bmatrix} X \\ Y \end{bmatrix} = \begin{bmatrix} X\cos\varphi - Y\sin\varphi Y \\ X\sin\varphi + Y\cos\varphi \end{bmatrix}$. Equating the entries in this matrix equation gives the first pair of rotation of axes formulas. Now

$R^{-1} = \dfrac{1}{\cos^2\varphi + \sin^2\varphi}\begin{bmatrix} \cos\varphi & \sin\varphi \\ -\sin\varphi & \cos\varphi \end{bmatrix} = \begin{bmatrix} \cos\varphi & \sin\varphi \\ -\sin\varphi & \cos\varphi \end{bmatrix}$ and so $Z' = R^{-1}Z \iff$

$\begin{bmatrix} X \\ Y \end{bmatrix} = \begin{bmatrix} \cos\varphi & \sin\varphi \\ -\sin\varphi & \cos\varphi \end{bmatrix}\begin{bmatrix} x \\ y \end{bmatrix} = \begin{bmatrix} x\cos\varphi + y\sin\varphi \\ -x\sin\varphi + y\cos\varphi \end{bmatrix}$. Equating the entries in this matrix equation gives the second pair of rotation of axes formulas.

39. Let P be the point (x_1, y_1) and Q be the point (x_2, y_2) and let $P'(X_1, Y_1)$ and $Q'(X_2, Y_2)$ be the images of P and Q under the rotation of φ . So $X_1 = x_1\cos\varphi + y_1\sin\varphi, Y_1 = -x_1\sin\varphi + y_1\cos\varphi$, $X_2 = x_2\cos\varphi + y_2\sin\varphi$, and $Y_2 = -x_2\sin\varphi + y_2\cos\varphi$. Thus $d(P', Q') = \sqrt{(X_2 - X_1)^2 + (Y_2 - Y_1)^2}$, where

$$(X_2 - X_1)^2 = \left[(x_2\cos\varphi + y_2\sin\varphi) - (x_1\cos\varphi + y_1\sin\varphi)\right]^2 = \left[(x_2 - x_1)\cos\varphi + (y_2 - y_1)\sin\varphi\right]^2$$
$$= (x_2 - x_1)^2\cos^2\varphi + (x_2 - x_1)(y_2 - y_1)\sin\varphi\cos\varphi + (y_2 - y_1)^2\sin^2\varphi$$

and
$$(Y_2 - Y_1)^2 = \left[(-x_2\sin\varphi + y_2\cos\varphi) - (-x_1\sin\varphi + y_1\cos\varphi)\right]^2 = \left[-(x_2 - x_1)\sin\varphi + (y_2 - y_1)\cos\varphi\right]^2$$
$$= (x_2 - x_1)^2\sin^2\varphi - (x_2 - x_1)(y_2 - y_1)\sin\varphi\cos\varphi + (y_2 - y_1)^2\cos^2\varphi$$

So

$$\begin{aligned}(X_2 - X_1)^2 + (Y_2 - Y_1)^2 &= (x_2 - x_1)^2\cos^2\varphi + (x_2 - x_1)(y_2 - y_1)\sin\varphi\cos\varphi + (y_2 - y_1)^2\sin^2\varphi \\ &\quad + (x_2 - x_1)^2\sin^2\varphi - (x_2 - x_1)(y_2 - y_1)\sin\varphi\cos\varphi + (y_2 - y_1)^2\cos^2\varphi \\ &= (x_2 - x_1)^2\cos^2\varphi + (y_2 - y_1)^2\sin^2\varphi + (x_2 - x_1)^2\sin^2\varphi + (y_2 - y_1)^2\cos^2\varphi \\ &= (x_2 - x_1)^2\left(\cos^2\varphi + \sin^2\varphi\right) + (y_2 - y_1)^2\left(\sin^2\varphi + \cos^2\varphi\right) = (x_2 - x_1)^2 + (y_2 - y_1)^2\end{aligned}$$

Putting these equations together gives

$$d(P', Q') = \sqrt{(X_2 - X_1)^2 + (Y_2 - Y_1)^2} = \sqrt{(x_2 - x_1)^2 + (y_2 - y_1)^2} = d(P, Q).$$

7.6 Polar Equations of Conics

1. All conics can be described geometrically using a fixed point F called the *focus* and a fixed line ℓ called the *directrix*. For a fixed positive number e the set of all points P satisfying $\dfrac{\text{distance from } P \text{ to } F}{\text{distance from } P \text{ to } \ell} = e$ is a *conic section*. If $e=1$ the conic is a *parabola*, if $e<1$ the conic is an *ellipse*, and if $e>1$ the conic is a *hyperbola*. The number e is called the *eccentricity* of the conic.

3. Substituting $e=\frac{2}{3}$ and $d=3$ into the general equation of a conic with vertical directrix, we get $r=\dfrac{\frac{2}{3}\cdot 3}{1+\frac{2}{3}\cos\theta} \Leftrightarrow r=\dfrac{6}{3+2\cos\theta}$.

5. Substituting $e=1$ and $d=2$ into the general equation of a conic with horizontal directrix, we get $r=\dfrac{1\cdot 2}{1+\sin\theta} \Leftrightarrow r=\dfrac{2}{1+\sin\theta}$.

7. $r=5\sec\theta \Leftrightarrow r\cos\theta=5 \Leftrightarrow x=5$. So $d=5$ and $e=4$ gives $r=\dfrac{4\cdot 5}{1+4\cos\theta} \Leftrightarrow r=\dfrac{20}{1+4\cos\theta}$.

9. Since this is a parabola whose focus is at the origin and vertex at $(5,\pi/2)$, the directrix must be $y=10$. So $d=10$ and $e=1$ gives $r=\dfrac{1\cdot 10}{1+\sin\theta}=\dfrac{10}{1+\sin\theta}$.

11. $r=\dfrac{6}{1+\cos\theta}$ is Graph II. The eccentricity is 1, so this is a parabola. When $\theta=0$, we have $r=3$ and when $\theta=\frac{\pi}{2}$, we have $r=6$.

13. $r=\dfrac{3}{1-2\sin\theta}$ is Graph VI. $e=2$, so this is a hyperbola. When $\theta=0$, $r=3$, and when $\theta=\pi$, $r=3$.

15. $r=\dfrac{12}{3+2\sin\theta}$ is Graph IV. $r=\dfrac{4}{1+\frac{2}{3}\sin\theta}$, so $e=\frac{2}{3}$ and this is an ellipse. When $\theta=0$, $r=4$, and when $\theta=\pi$, $r=4$.

17. (a) The equation $r = \dfrac{4}{1-\sin\theta}$ has $e=1$ and $d=4$, so it represents a parabola.

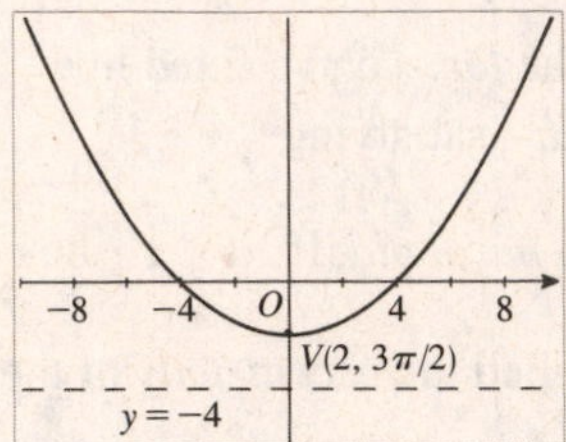

(b) Because the equation is of the form $r = \dfrac{ed}{1-e\sin\theta}$, the directrix is parallel to the polar axis and has equation $y=-4$. The vertex is $\left(2, \frac{3\pi}{2}\right)$.

19. (a) The equation $r = \dfrac{5}{3+3\cos\theta} = \dfrac{\frac{5}{3}}{1+\cos\theta}$ has $e=1$ and $d=\frac{5}{3}$, so it represents a parabola.

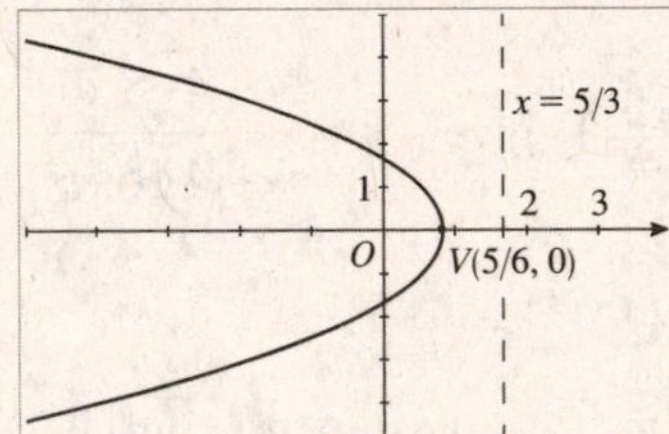

(b) Because the equation is of the form $r = \dfrac{ed}{1+e\cos\theta}$, the directrix is parallel to the polar axis and has equation $x=d=\frac{5}{3}$. The vertex is $\left(\frac{5}{6}, 0\right)$.

21. (a) The equation $r = \dfrac{4}{2-\cos\theta} = \dfrac{2}{1-\frac{1}{2}\cos\theta}$ has $e=\frac{1}{2}<1$, so it represents an ellipse.

(b) Because the equation is of the form $r = \dfrac{ed}{1-e\cos\theta}$ with $d=4$, the directrix is vertical and has equation $x=-4$. Thus, the vertices are $V_1(4,0)$ and $V_2\left(\frac{4}{3},\pi\right)$. **(c)** The length of the major axis is $2a = |V_1V_2| = 4+\frac{4}{3} = \frac{16}{3}$ and the center is at the midpoint of V_1V_2, $\left(\frac{4}{3},0\right)$. The minor axis has length $2b$ where $b^2 = a^2-c^2 = a^2-(ae)^2 = \left(\frac{8}{3}\right)^2 - \left(\frac{8}{3}\cdot\frac{1}{2}\right)^2 = \frac{16}{3}$, so $2b = 2\cdot\sqrt{\frac{16}{3}} = \frac{8\sqrt{3}}{3} \approx 4.62$.

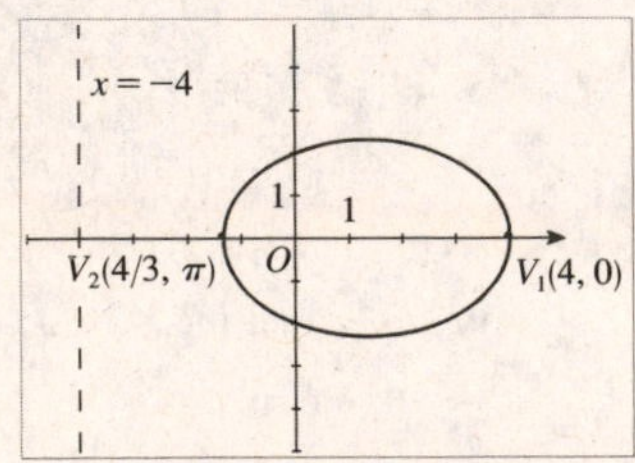

23. **(a)** The equation $r = \dfrac{12}{4+3\sin\theta} = \dfrac{3}{1+\frac{3}{4}\sin\theta}$ has $e = \frac{3}{4} < 1$, so it represents an ellipse.

(b) Because the equation is of the form $r = \dfrac{ed}{1+e\sin\theta}$ with $d = 4$, the directrix is horizontal and has equation $y = 4$. Thus, the vertices are $V_1\left(\frac{12}{7}, \frac{\pi}{2}\right)$ and $V_2\left(12, \frac{3\pi}{2}\right)$.

(c) The length of the major axis is $2a = |V_1V_2| = \frac{12}{7} + 12 = \frac{96}{7}$ and the center is at the midpoint of V_1V_2, $\left(\frac{36}{7}, \frac{3\pi}{2}\right)$. The minor axis has length $2b$ where

$b^2 = a^2 - c^2 = a^2 - (ae)^2 = \left(\frac{48}{7}\right)^2 - \left(\frac{48}{7}\cdot\frac{3}{4}\right)^2 = \frac{144}{7}$, so $2b = 2\cdot\sqrt{\frac{144}{7}} = \frac{24\sqrt{7}}{7} \approx 9.07$.

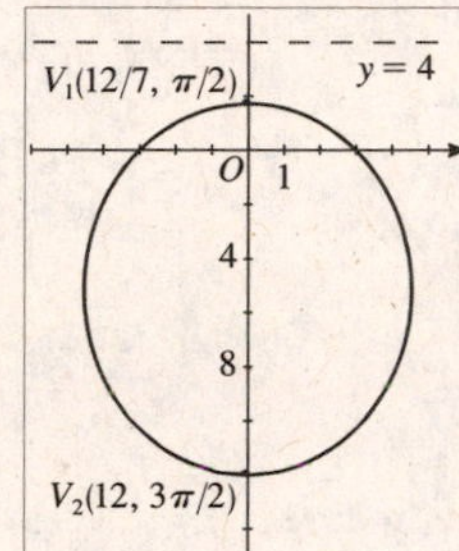

25. **(a)** The equation $r = \dfrac{8}{1+2\cos\theta}$ has $e = 2 > 1$, so it represents a hyperbola.

(b) Because the equation has the form $r = \dfrac{ed}{1+\cos\theta}$ with $d = 4$, the transverse axis is horizontal and the directrix has equation $x = 4$. The vertices are $V_1\left(\frac{8}{3}, 0\right)$ and $V_2(-8, \pi) = (8, 0)$.

(c) The center is the midpoint of V_1V_2, $\left(\frac{16}{3}, 0\right)$. To sketch the central box and the asymptotes, we find a and b. The length of the transverse axis is $2a = \frac{16}{3}$, and so $a = \frac{8}{3}$, and

$b^2 = c^2 - a^2 = (ae)^2 - a^2 = \left(\frac{8}{3}\cdot 2\right)^2 - \left(\frac{8}{3}\right)^2 = \frac{64}{3}$, so $b = \sqrt{\frac{64}{3}} = \frac{8\sqrt{3}}{3} \approx 4.62$.

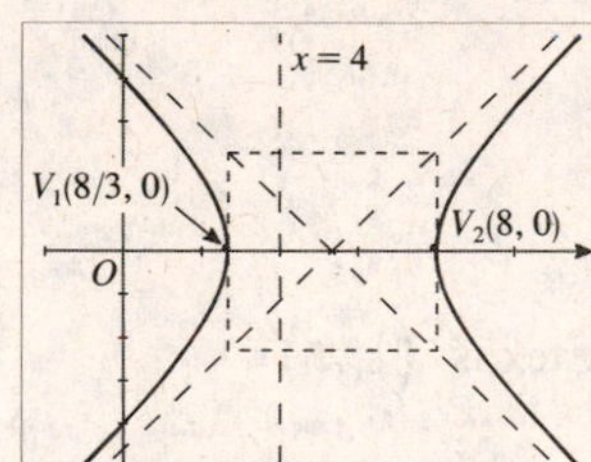

27. (a) The equation $r = \dfrac{20}{2-3\sin\theta} = \dfrac{10}{1-\frac{3}{2}\sin\theta}$ has $e = \frac{3}{2} > 1$, so it represents a hyperbola.

(b) Because the equation has the form $r = \dfrac{ed}{1+\cos\theta}$ with $d = \frac{20}{3}$, the transverse axis is vertical and the directrix has equation $y = -\frac{20}{3}$. The vertices are $V_1\left(-20,\frac{\pi}{2}\right) = \left(20,\frac{3\pi}{2}\right)$ and $V_2\left(4,\frac{3\pi}{2}\right)$.

(c) The center is the midpoint of V_1V_2 , $\left(12,\frac{3\pi}{2}\right)$. To sketch the central box and the asymptotes, we find a and b . The length of the transverse axis is $2a = 16$, and so $a = 8$, and $b^2 = c^2 - a^2 = (ae)^2 - a^2 = \left(8\cdot\frac{3}{2}\right)^2 - 8^2 = 80$, so $b = \sqrt{80} = 4\sqrt{5} \approx 8.94$.

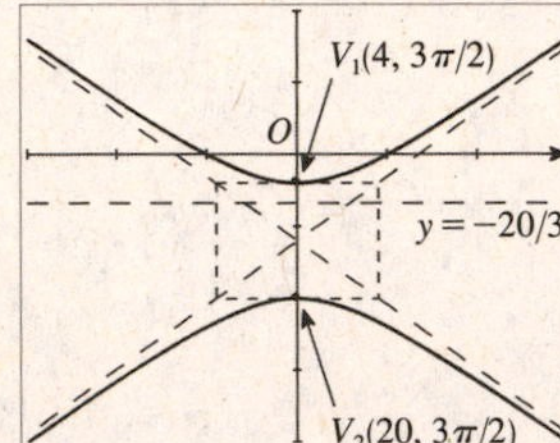

29. (a) $r = \dfrac{4}{1+3\cos\theta}$ $\Rightarrow$ $e = 3$, so the conic is a hyperbola.

(b) The vertices occur where $\theta = 0$ and $\theta = \pi$. Now $\theta = 0 \Rightarrow r = \dfrac{4}{1+3\cos 0} = 1$, and $\theta = \pi \Rightarrow r = \dfrac{4}{1+3\cos\pi} = \dfrac{4}{-2} = -2$. Thus the vertices are $(1,0)$ and $(-2,\pi)$.

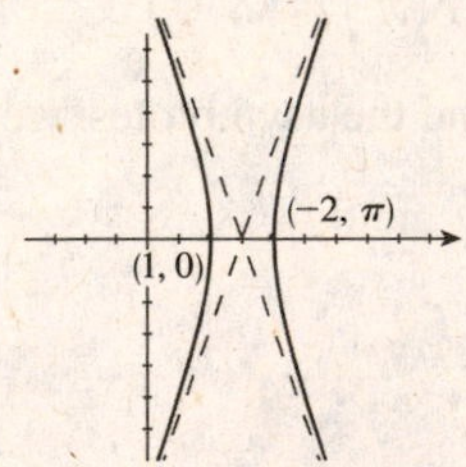

31. (a) $r = \dfrac{2}{1-\cos\theta}$ $\Rightarrow$ $e = 1$, so the conic is a parabola.

(b) Substituting $\theta = \pi$, we have $r = \dfrac{2}{1-\cos\pi} = \frac{2}{2} = 1$. Thus the vertex is $(1,\pi)$.

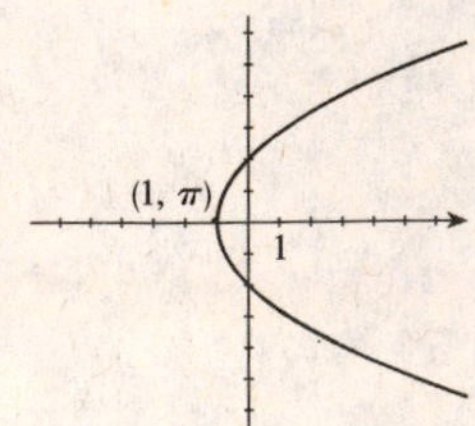

33. (a) $r=\dfrac{6}{2+\sin\theta} \iff r=\dfrac{\frac{1}{2}\cdot 6}{1+\frac{1}{2}\sin\theta} \implies e=\frac{1}{2}$, so the conic is an ellipse.

(b) The vertices occur where $\theta=\frac{\pi}{2}$ and $\theta=\frac{3\pi}{2}$. Now $\theta=\frac{\pi}{2} \implies r=\dfrac{6}{2+\sin\frac{\pi}{2}}=\dfrac{6}{3}=2$ and

$\theta=\frac{3\pi}{2} \implies r=\dfrac{6}{2+\sin\frac{3\pi}{2}}=\dfrac{6}{1}=6$. Thus, the vertices are $\left(2,\frac{\pi}{2}\right)$ and $\left(6,\frac{3\pi}{2}\right)$.

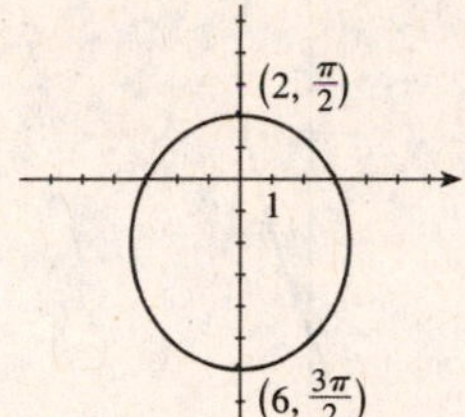

35. (a) $r=\dfrac{7}{2-5\sin\theta} \iff$

$r=\dfrac{\frac{7}{2}}{1-\frac{5}{2}\sin\theta} \implies e=\frac{5}{2}$, so the conic is a hyperbola.

(b) The vertices occur where $\theta=\frac{\pi}{2}$ and

$\theta=\frac{3\pi}{2}$. $r=\dfrac{7}{2-5\sin\frac{\pi}{2}}=\frac{7}{-3}=-\frac{7}{3}$ and

$\theta=\frac{3\pi}{2} \implies r=\dfrac{7}{2-5\sin\frac{3\pi}{2}}=\frac{7}{7}=1$.

Thus, the vertices are $\left(-\frac{7}{3},\frac{\pi}{2}\right)$ and $\left(1,\frac{3\pi}{2}\right)$.

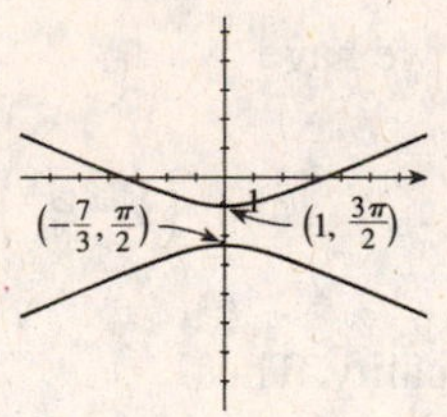

37. (a) $r=\dfrac{1}{4-3\cos\theta}=\dfrac{\frac{1}{4}}{1-\frac{3}{4}\cos\theta} \implies$

$e=\frac{3}{4}$, so the conic is an ellipse. The vertices occur where $\theta=0$ and $\theta=\pi$.

Now $\theta=0 \implies r=\dfrac{1}{4-3\cos 0}=1$

and $\theta=\pi \implies r=\dfrac{1}{4-3\cos\pi}=\frac{1}{7}$.

Thus, the vertices are $(1,0)$ and $\left(\frac{1}{7},\pi\right)$.

We have $d=\frac{1}{3}$, so the directrix is $x=-\frac{1}{3}$.

(b) If the ellipse is rotated through $\frac{\pi}{3}$, the equation of the resulting conic is

$$r=\frac{1}{4-3\cos\left(\theta-\frac{\pi}{3}\right)} .$$

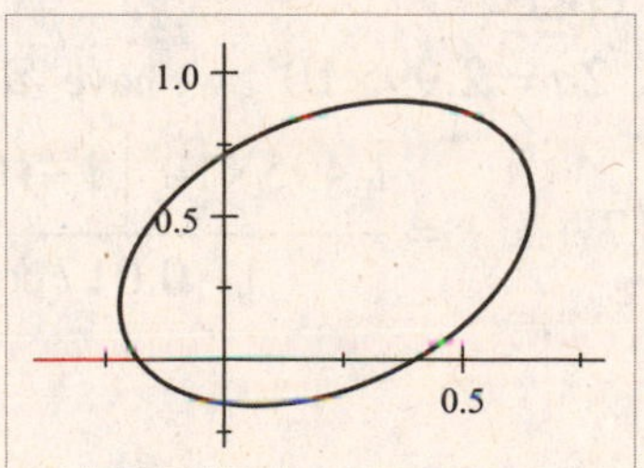

39. (a) $r=\dfrac{2}{1+\sin\theta}$ $\Rightarrow$ $e=1$, so the conic is a parabola. Substituting $\theta=\frac{\pi}{2}$, we have $r=t\dfrac{2}{1+\sin\frac{\pi}{2}}=1$, so the vertex is $\left(1,\frac{\pi}{2}\right)$. Because $d=2$, the directrix is $y=2$.

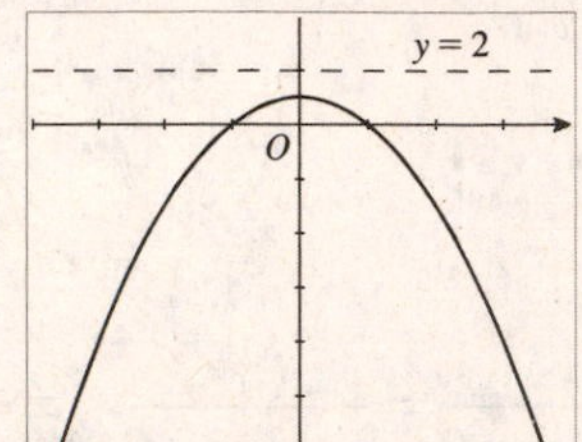

(b) If the ellipse is rotated through $\theta=-\frac{\pi}{4}$, the equation of the resulting conic is $r=\dfrac{2}{1+\sin\left(\theta+\frac{\pi}{4}\right)}$.

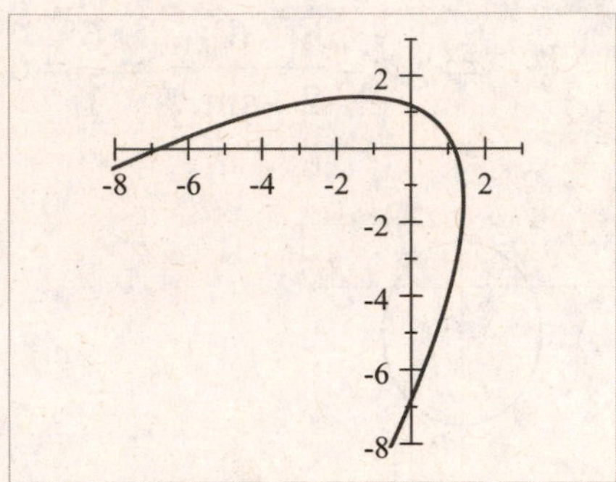

41. The ellipse is nearly circular when e is close to 0 and becomes more elongated as $e\to 1^-$. At $e=1$, the curve becomes a parabola.

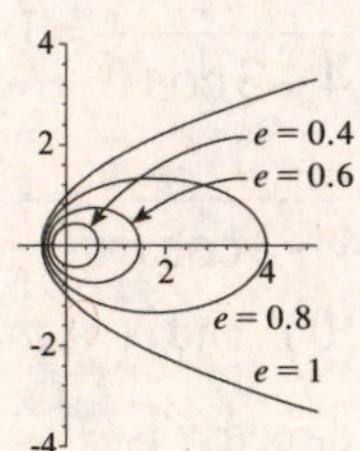

43. (a) Since the polar form of an ellipse with directrix $x=-d$ is $r=\dfrac{ed}{1-e\cos\theta}$ we need to show that $ed=a\left(1-e^2\right)$. From the proof of the Equivalent Description of Conics we have $a^2=\dfrac{e^2d^2}{\left(1-e^2\right)^2}$. Since the conic is an ellipse, $e<1$ and so the quantities a , d , and $\left(1-e^2\right)$ are all positive. Thus we can take the square roots of both sides and maintain equality. Thus $a^2=\dfrac{e^2d^2}{\left(1-e^2\right)^2}$ $\Leftrightarrow$ $a=\dfrac{ed}{1-e^2}$ $\Leftrightarrow$ $ed=a\left(1-e^2\right)$. As a result, $r=\dfrac{ed}{1-e\cos\theta}$ $\Leftrightarrow$ $r=\dfrac{a\left(1-e^2\right)}{1-e\cos\theta}$.

(b) Since $2a=2.99\times10^8$ we have $a=1.495\times10^8$, so a polar equation for the earth's orbit (using $e\approx0.017$) is $r=\dfrac{1.495\times10^8\left[1-(0.017)^2\right]}{1-0.017\cos\theta}\approx\dfrac{1.49\times10^8}{1-0.017\cos\theta}$.

45. From Exercise 44, we know that at perihelion $r = 4.43\times10^9 = a(1-e)$ and at aphelion $r = 7.37\times10^9 = a(1+e)$. Dividing these equations gives $\dfrac{7.37\times10^9}{4.43\times10^9} = \dfrac{a(1+e)}{a(1-e)}$ $\Leftrightarrow$

$1.664 = \dfrac{1+e}{1-e}$ $\Leftrightarrow$ $1.664(1-e) = 1+e$ $\Leftrightarrow$ $1.664 - 1 = e + 1.664$ $\Leftrightarrow$

$0.664 = 2.664e$ $\Leftrightarrow$ $e = \dfrac{0.664}{2.664} \approx 0.25$.

47. The r -coordinate of the satellite will be its distance from the focus (the center of the earth). From the r -coordinate we can easily calculate the height of the satellite.

1. $y^2 = 4x$. This is a parabola with $4p = 4$ $\Leftrightarrow$ $p = 1$. The vertex is $(0,0)$, the focus is $(1,0)$, and the directrix is $x = -1$.

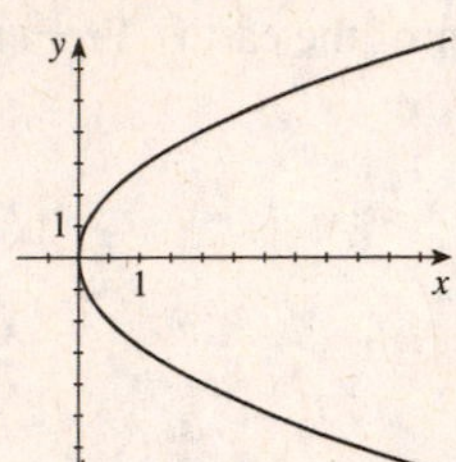

3. $x^2 + 8y = 0$ $\Leftrightarrow$ $x^2 = -8y$. This is a parabola with $4p = -8$ $\Leftrightarrow$ $p = -2$. The vertex is $(0,0)$, the focus is $(0,-2)$, and the directrix is $y = 2$.

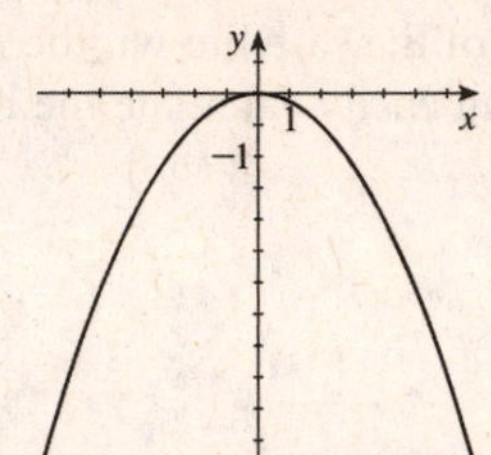

5. $x - y^2 + 4y - 2 = 0$ $\Leftrightarrow$

$x - (y^2 - 4y + 4) - 2 = -4$ $\Leftrightarrow$

$x - (y-2)^2 = -2$ $\Leftrightarrow$

$(y-2)^2 = x + 2$. This is a parabola with $4p = 1$ $\Leftrightarrow$ $p = \frac{1}{4}$. The vertex is $(-2,2)$, the focus is $\left(-2+\frac{1}{4},2\right) = \left(-\frac{7}{4},2\right)$, and the directrix is $x = -2 - \frac{1}{4} = -\frac{9}{4}$.

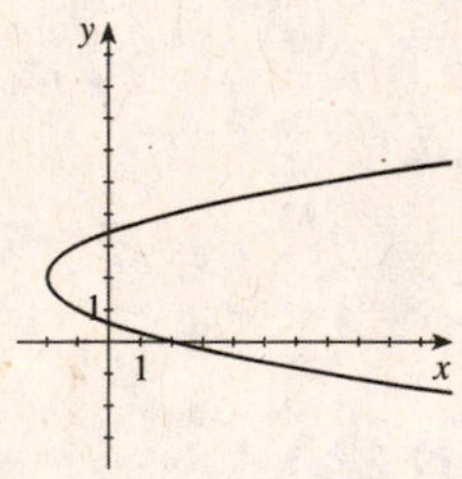

7. $\frac{1}{2}x^2 + 2x = 2y + 4$ $\Leftrightarrow$

$x^2 + 4x = 4y + 8$ $\Leftrightarrow$

$x^2 + 4x + 4 = 4y + 8$ $\Leftrightarrow$

$(x+2)^2 = 4(y+3)$. This is a parabola with $4p = 4$ $\Leftrightarrow$ $p = 1$. The vertex is $(-2,-3)$, the focus is $(-2,-3+1) = (-2,-2)$, and the directrix is $y = -3 - 1 = -4$.

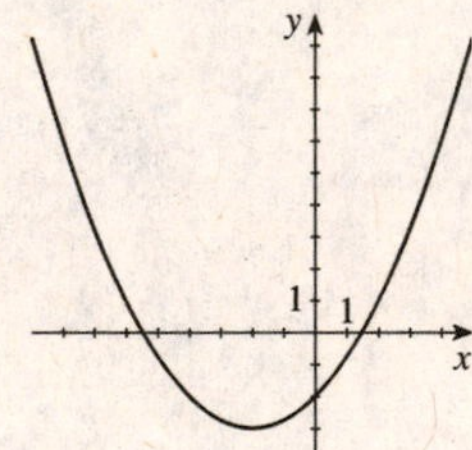

9. $\dfrac{x^2}{9}+\dfrac{y^2}{25}=1$. This is an ellipse with $a=5$, $b=3$, and $c=\sqrt{25-9}=4$. The center is $(0,0)$, the vertices are $(0,\pm5)$, the foci are $(0,\pm4)$, the length of the major axis is $2a=10$, and the length of the minor axis is $2b=6$.

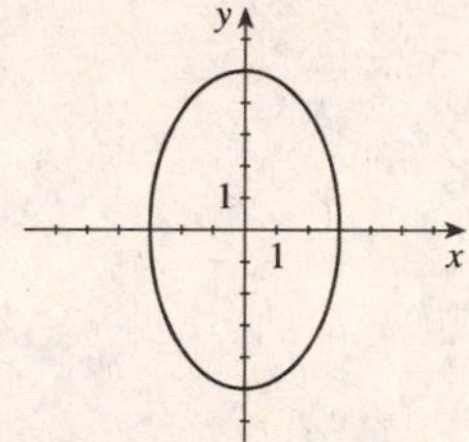

11. $x^2+4y^2=16 \Leftrightarrow \dfrac{x^2}{16}+\dfrac{y^2}{4}=1$. This is an ellipse with $a=4$, $b=2$, and $c=\sqrt{16-4}=2\sqrt{3}$. The center is $(0,0)$, the vertices are $(\pm4,0)$, the foci are $\left(\pm2\sqrt{3},0\right)$, the length of the major axis is $2a=8$, and the length of the minor axis is $2b=4$.

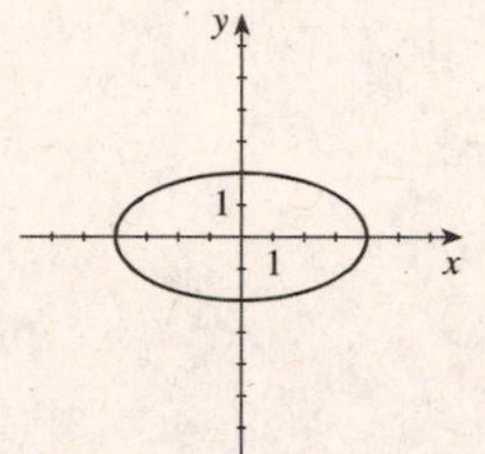

13. $\dfrac{(x-3)^2}{9}+\dfrac{y^2}{16}=1$. This is an ellipse with $a=4$, $b=3$, and $c=\sqrt{16-9}=\sqrt{7}$. The center is $(3,0)$, the vertices are $(3,\pm4)$, the foci are $\left(3,\pm\sqrt{7}\right)$, the length of the major axis is $2a=8$, and the length of the minor axis is $2b=6$.

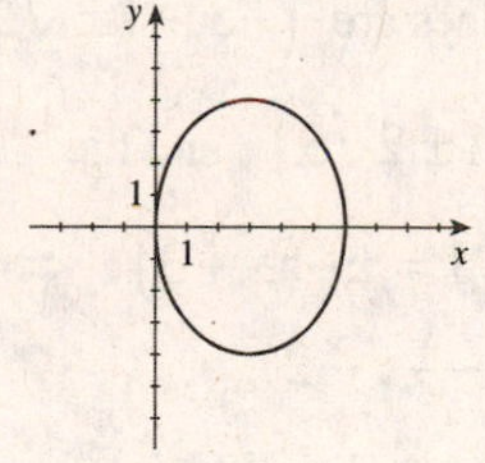

15. $4x^2+9y^2=36y \Leftrightarrow$
$4x^2+9\left(y^2-4y+4\right)=36 \Leftrightarrow$
$4x^2+9(y-2)^2=36 \Leftrightarrow$
$\dfrac{x^2}{9}+\dfrac{(y-2)^2}{4}=1$. This is an ellipse with $a=3$, $b=2$, and $c=\sqrt{9-4}=\sqrt{5}$. The center is $(0,2)$, the vertices are $(\pm3,2)$, the foci are $\left(\pm\sqrt{5},2\right)$, the length of the major axis is $2a=6$, and the length of the minor axis is $2b=4$.

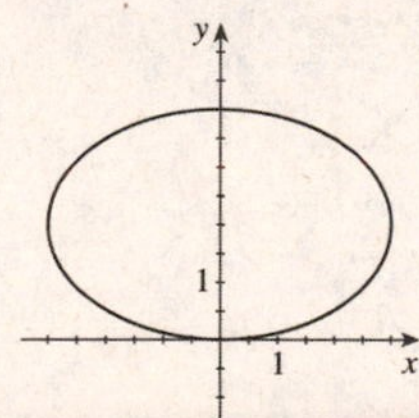

17. $-\frac{x^2}{9}+\frac{y^2}{16}=1 \Leftrightarrow \frac{y^2}{16}-\frac{x^2}{9}=0$. This is a hyperbola with $a=4$, $b=3$, and $c=\sqrt{16+9}=\sqrt{25}=5$. The center is $(0,0)$, the vertices are $(0,\pm 4)$, the foci are $(0,\pm 5)$, and the asymptotes are $y=\pm\frac{4}{3}x$.

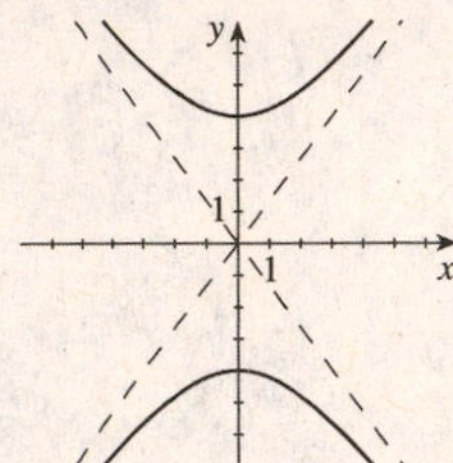

19. $x^2-2y^2=16 \Leftrightarrow \frac{x^2}{16}-\frac{y^2}{8}=1$. This is a hyperbola with $a=4$, $b=2\sqrt{2}$, and $c=\sqrt{16+8}=\sqrt{24}=2\sqrt{6}$. The center is $(0,0)$, the vertices are $(\pm 4,0)$, the foci are $(\pm 2\sqrt{6},0)$, and the asymptotes are $y=\pm\frac{2\sqrt{2}}{4}x \Leftrightarrow y=\pm\frac{1}{\sqrt{2}}x$.

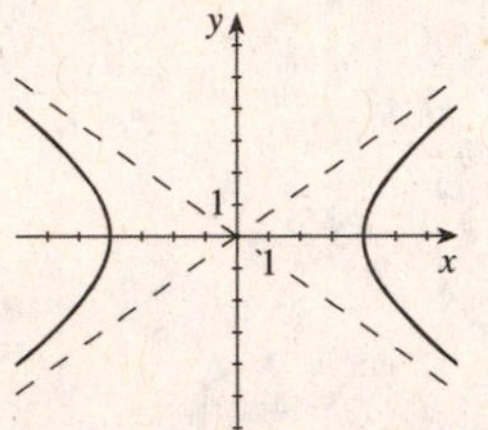

21. $\frac{(x+4)^2}{16}-\frac{y^2}{16}=1$. This is a hyperbola with $a=4$, $b=4$ and $c=\sqrt{16+16}=4\sqrt{2}$. The center is $(-4,0)$, the vertices are $(-4\pm 4,0)$ which are $(-8,0)$ and $(0,0)$, the foci are $(-4\pm 4\sqrt{2},0)$, and the asymptotes are $y=\pm(x+4)$.

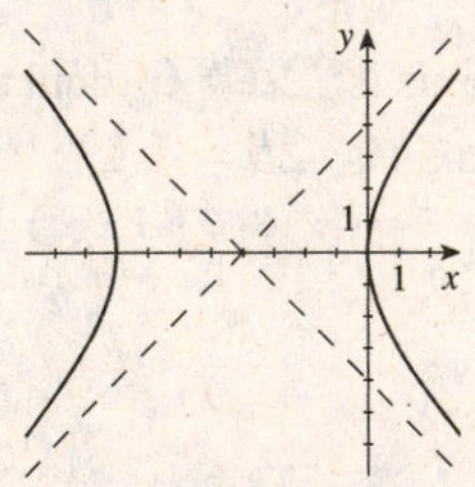

23. $9y^2+18y=x^2+6x+18 \Leftrightarrow$

$9(y^2+2y+1)=(x^2+6x+9)+9-9+18$

$\Leftrightarrow 9(y+1)^2-(x+3)^2=18 \Leftrightarrow$

$\frac{(y+1)^2}{2}-\frac{(x+3)^2}{18}=1$. This is a hyperbola with $a=\sqrt{2}$, $b=3\sqrt{2}$, and $c=\sqrt{2+18}=2\sqrt{5}$. The center is $(-3,-1)$, the vertices are $(-3,-1\pm\sqrt{2})$, the foci are $(-3,-1\pm 2\sqrt{5})$, and the asymptotes are $y+1=\pm\frac{1}{3}(x+3) \Leftrightarrow y=\frac{1}{3}x$ and $y=-\frac{1}{3}x-2$.

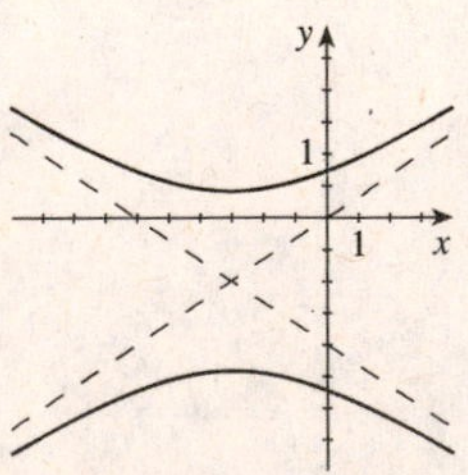

25. This is a parabola that opens to the right with its vertex at $(0,0)$ and the focus at $(2,0)$. So $p=2$, and the equation is $y^2=4(2)x \Leftrightarrow y^2=8x$.

27. From the graph, the center is $(0,0)$, and the vertices are $(0,-4)$ and $(0,4)$. Since a is the distance from the center to a vertex, we have $a=4$. Because one focus is $(0,5)$, we have $c=5$, and since $c^2=a^2+b^2$, we have $25=16+b^2 \Leftrightarrow b^2=9$. Thus an equation of the hyperbola is $\dfrac{y^2}{16}-\dfrac{x^2}{9}=1$.

29. From the graph, the center of the ellipse is $(4,2)$, and so $a=4$ and $b=2$. The equation is

$$\frac{(x-4)^2}{4^2}+\frac{(y-2)^2}{2^2}=1 \Leftrightarrow \frac{(x-4)^2}{16}+\frac{(y-2)^2}{4}=1.$$

31. $\dfrac{x^2}{12}+y=1 \Leftrightarrow \dfrac{x^2}{12}=-(y-1) \Leftrightarrow x^2=-12(y-1)$. This is a parabola with $4p=-12$ $\Leftrightarrow p=-3$. The vertex is $(0,1)$ and the focus is $(0,1-3)=(0,-2)$.

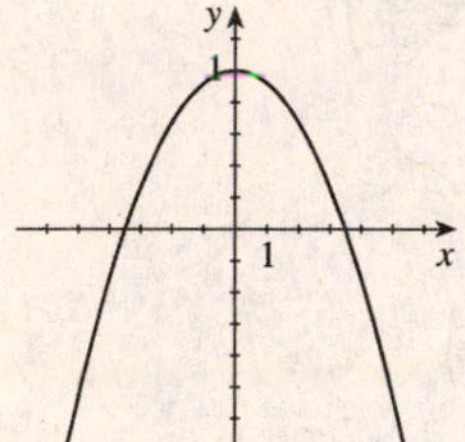

33. $x^2-y^2+144=0 \Leftrightarrow$ $\dfrac{y^2}{144}-\dfrac{x^2}{144}=1$. This is a hyperbola with $a=12$, $b=12$, and $c=\sqrt{144+144}=12\sqrt{2}$. The vertices are $(0,\pm12)$ and the foci are $(0,\pm12\sqrt{2})$.

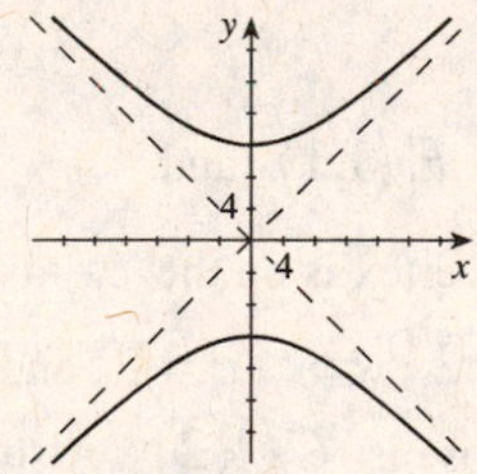

35. $4x^2+y^2=8(x+y) \Leftrightarrow$

$4(x^2-2x)+(y^2-8y)=0 \Leftrightarrow$

$4(x^2-2x+1)+(y^2-8y+16)=4+16$

$\Leftrightarrow 4(x-1)^2+(y-4)^2=20 \Leftrightarrow$

$\dfrac{(x-1)^2}{5}+\dfrac{(y-4)^2}{20}=1$. This is an ellipse with $a=2\sqrt{5}$, $b=\sqrt{5}$, and $c=\sqrt{20-5}=\sqrt{15}$. The vertices are $(1,4\pm2\sqrt{5})$ and the foci are $(1,4\pm\sqrt{15})$.

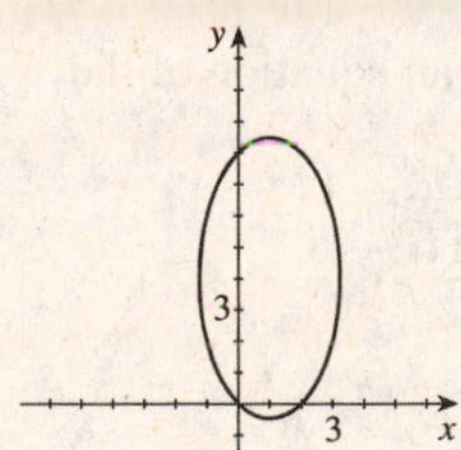

37. $x = y^2 - 16y \Leftrightarrow$
$x + 64 = y^2 - 16y + 64 \Leftrightarrow$
$(y-8)^2 = x + 64$. This is a parabola with $4p = 1 \Leftrightarrow p = \frac{1}{4}$. The vertex is $(-64, 8)$ and the focus is $\left(-64 + \frac{1}{4}, 8\right) = \left(-\frac{255}{4}, 8\right)$.

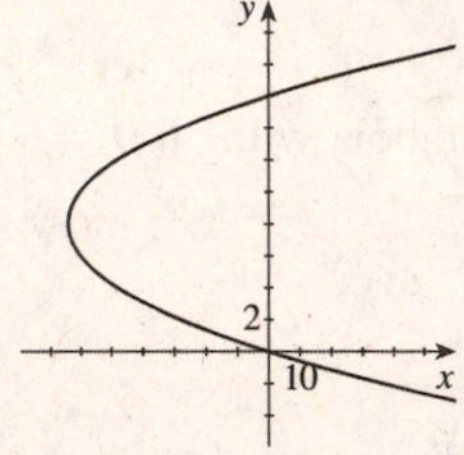

39. $2x^2 - 12x + y^2 + 6y + 26 = 0 \Leftrightarrow$
$2(x^2 - 6x) + (y^2 + 6y) = -26 \Leftrightarrow$
$2(x^2 - 6x + 9) + (y^2 + 6y + 9) = -26 + 18 + 9$
$\Leftrightarrow 2(x-3)^2 + (y+3)^2 = 1 \Leftrightarrow$
$\dfrac{(x-3)^2}{\frac{1}{2}} + (y+3)^2 = 1$. This is an ellipse with $a = 1$, $b = \frac{\sqrt{2}}{2}$, and $c = \sqrt{1 - \frac{1}{2}} = \frac{\sqrt{2}}{2}$. The vertices are $(3, -3 \pm 1) = (3, -4)$ and $(3, -2)$, and the foci are $\left(3, -3 \pm \frac{\sqrt{2}}{2}\right)$.

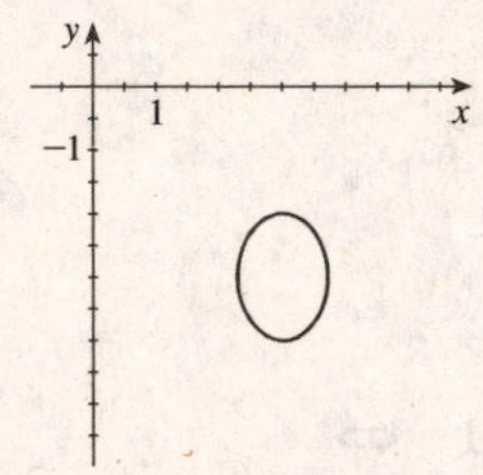

41. $9x^2 + 8y^2 - 15x + 8y + 27 = 0 \Leftrightarrow$
$9\left(x^2 - \frac{5}{3}x + \frac{25}{36}\right) + 8\left(y^2 + y + \frac{1}{4}\right) = -27 + \frac{25}{4} + 2$
$\Leftrightarrow 9\left(x - \frac{5}{6}\right)^2 + 8\left(y + \frac{1}{2}\right)^2 = -\frac{75}{4}$.
However, since the left-hand side of the equation is greater than or equal to 0, there is no point that satisfies this equation. The graph is empty.

43. The parabola has focus $(0,1)$ and directrix $y = -1$. Therefore, $p = 1$ and so $4p = 4$. Since the focus is on the y-axis and the vertex is $(0,0)$, an equation of the parabola is $x^2 = 4y$.

45. The hyperbola has vertices $(0, \pm 2)$ and asymptotes $y = \pm \frac{1}{2}x$. Therefore, $a = 2$, and the foci are on the y-axis. Since the slopes of the asymptotes are $\pm \frac{1}{2} = \pm \dfrac{a}{b}$ $\Leftrightarrow b = 2a = 4$, an equation of the hyperbola is $\dfrac{y^2}{4} - \dfrac{x^2}{16} = 1$.

47. The ellipse has foci $F_1(1,1)$ and $F_2(1,3)$, and one vertex is on the x-axis. Thus, $2c = 3 - 1 = 2 \Leftrightarrow c = 1$, and so the center of the ellipse is $C(1,2)$. Also, since one vertex is on the x-axis, $a = 2 - 0 = 2$, and thus $b^2 = 4 - 1 = 3$. So an equation of the ellipse is
$\dfrac{(x-1)^2}{3} + \dfrac{(y-2)^2}{4} = 1$.

49. The ellipse has vertices $V_1(7,12)$ and $V_2(7,-8)$ and passes through the point $P(1,8)$. Thus, $2a=12-(-8)=20 \Leftrightarrow a=10$, and the center is $\left(7,\dfrac{-8+12}{2}\right)=(7,2)$. Thus an equation of the ellipse has the form $\dfrac{(x-7)^2}{b^2}+\dfrac{(y-2)^2}{100}=1$. Since the point $P(1,8)$ is on the ellipse, $\dfrac{(1-7)^2}{b^2}+\dfrac{(8-2)^2}{100}=1 \Leftrightarrow 3600+36b^2=100b^2 \Leftrightarrow 64b^2=3600 \Leftrightarrow b^2=\frac{225}{4}$.

Therefore, an equation of the ellipse is $\dfrac{(x-7)^2}{225/4}+\dfrac{(y-2)^2}{100}=1 \Leftrightarrow \dfrac{4(x-7)^2}{225}+\dfrac{(y-2)^2}{100}=1$.

51. The length of the major axis is $2a=186{,}000{,}000 \Leftrightarrow a=93{,}000{,}000$. The eccentricity is $e=c/a=0.017$, and so $c=0.017(93{,}000{,}000)=1{,}581{,}000$.

(a) The earth is closest to the sun when the distance is $a-c=93{,}000{,}000-1{,}581{,}000=91{,}419{,}000$.

(b) The earth is furthest from the sun when the distance is $a+c=93{,}000{,}000+1{,}581{,}000=94{,}581{,}000$.

53. (a) The graphs of $\dfrac{x^2}{16+k^2}+\dfrac{y^2}{k^2}=1$ for $k=1$, 2, 4, and 8 are shown in the figure.

(b) $c^2=(16+k^2)-k^2=16 \Rightarrow c=\pm 4$. Since the center is $(0,0)$, the foci of each of the ellipses are $(\pm 4,0)$.

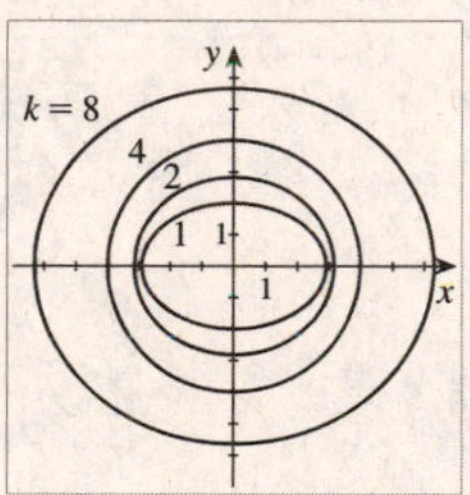

55. (a) $x^2+4xy+y^2=1$. Then $A=1$, $B=4$, and $C=1$, so the discriminant is $4^2-4(1)(1)=12$. Since the discriminant is positive, the equation represents a hyperbola.

(b) $\cot 2\varphi=\dfrac{A-C}{B}=\dfrac{1-1}{4}=0 \;\Rightarrow\; 2\varphi=90^\circ \;\Leftrightarrow\; \varphi=45^\circ$. Therefore, $x=\frac{\sqrt{2}}{2}X-\frac{\sqrt{2}}{2}Y$ and $y=\frac{\sqrt{2}}{2}X+\frac{\sqrt{2}}{2}Y$. Substituting into the original equation gives

$$\left(\tfrac{\sqrt{2}}{2}X-\tfrac{\sqrt{2}}{2}Y\right)^2+4\left(\tfrac{\sqrt{2}}{2}X-\tfrac{\sqrt{2}}{2}Y\right)\left(\tfrac{\sqrt{2}}{2}X+\tfrac{\sqrt{2}}{2}Y\right)+\left(\tfrac{\sqrt{2}}{2}X+\tfrac{\sqrt{2}}{2}Y\right)^2=1 \;\Leftrightarrow$$

$$\tfrac{1}{2}\left(X^2-2XY+Y^2\right)+2\left(X^2+XY-XY-Y^2\right)+\tfrac{1}{2}\left(X^2+2XY+Y^2\right)=1 \;\Leftrightarrow\; 3X^2-Y^2=1 \;\Leftrightarrow\; 3X^2-Y^2=1.$$

This is a hyperbola with $a=\frac{1}{\sqrt{3}}$, $b=1$, and $c=\sqrt{\frac{1}{3}+1}=\frac{2}{\sqrt{3}}$. Therefore, the hyperbola has vertices $V\left(\pm\frac{1}{\sqrt{3}},0\right)$ and foci $F\left(\pm\frac{2}{\sqrt{3}},0\right)$, in XY-coordinates.

(c)

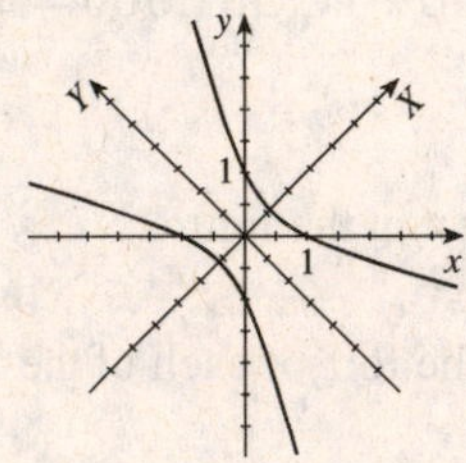

57. (a) $7x^2-6\sqrt{3}xy+13y^2-4\sqrt{3}x-4y=0$. Then $A=7$, $B=-6\sqrt{3}$, and $C=13$, so the discriminant is $\left(-6\sqrt{3}\right)^2-4(7)(13)=-256$. Since the discriminant is negative, the equation represents an ellipse.

(b) $\cot 2\varphi=\dfrac{A-C}{B}=\dfrac{7-13}{-6\sqrt{3}}=\dfrac{1}{\sqrt{3}} \Rightarrow 2\varphi=60^\circ \Leftrightarrow \varphi=30^\circ$. Therefore, $x=\frac{\sqrt{3}}{2}X-\frac{1}{2}Y$ and $y=\frac{1}{2}X+\frac{\sqrt{3}}{2}Y$. Substituting into the original equation gives

$$7\left(\frac{\sqrt{3}}{2}X-\frac{1}{2}Y\right)^2-6\sqrt{3}\left(\frac{\sqrt{3}}{2}X-\frac{1}{2}Y\right)\left(\frac{1}{2}X+\frac{\sqrt{3}}{2}Y\right)$$
$$+13\left(\frac{1}{2}X+\frac{\sqrt{3}}{2}Y\right)^2-4\sqrt{3}\left(\frac{\sqrt{3}}{2}X-\frac{1}{2}Y\right)-4\left(\frac{1}{2}X+\frac{\sqrt{3}}{2}Y\right)=0 \Leftrightarrow$$

$$\frac{7}{4}\left(3X^2-2\sqrt{3}XY+Y^2\right)-\frac{3\sqrt{3}}{2}\left(\sqrt{3}X^2+3XY-XY-\sqrt{3}Y^2\right)$$
$$+\frac{13}{4}\left(X^2+2\sqrt{3}XY+3Y^2\right)-6X+2\sqrt{3}Y-2X-2\sqrt{3}Y=0 \Leftrightarrow$$
$$X^2\left(\tfrac{21}{4}-\tfrac{9}{2}+\tfrac{13}{4}\right)-8X+Y^2\left(\tfrac{7}{4}+\tfrac{9}{2}+\tfrac{39}{4}\right)=0 \Leftrightarrow 4X^2-8X+16Y^2=0 \Leftrightarrow$$

$4\left(X^2-2X+1\right)+16Y^2=4 \Leftrightarrow (X-1)^2+4Y^2=1$. This ellipse has $a=1$, $b=\frac{1}{2}$, and $c=\sqrt{1-\frac{1}{4}}=\frac{1}{2}\sqrt{3}$. Therefore, the vertices are $V(1\pm1,0)=V_1(0,0)$ and $V_2(2,0)$ and the foci are $F\left(1\pm\frac{1}{2}\sqrt{3},0\right)$.

(c)

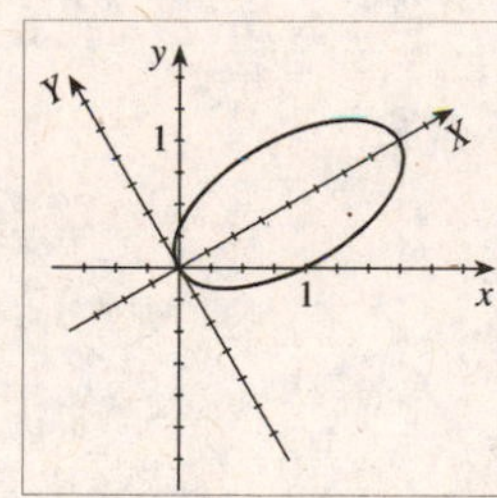

59. $5x^2+3y^2=60 \Leftrightarrow 3y^2=60-5x^2 \Leftrightarrow y^2=20-\frac{5}{3}x^2$. This conic is an ellipse.

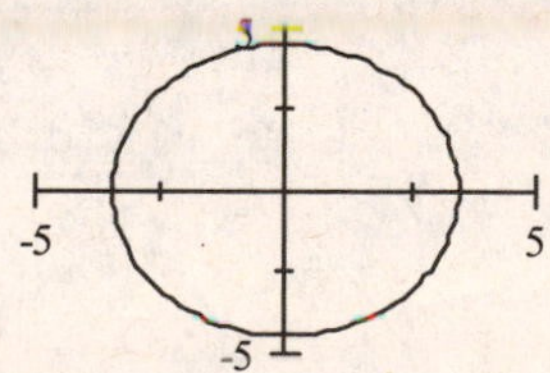

61. $6x+y^2-12y=30 \Leftrightarrow y^2-12y=30-6x \Leftrightarrow y^2-12y+36=66-6x \Leftrightarrow$ $(y-6)^2=66-6x \Leftrightarrow y-6=\pm\sqrt{66-6x} \Leftrightarrow y=6\pm\sqrt{66-6x}$. This conic is a parabola.

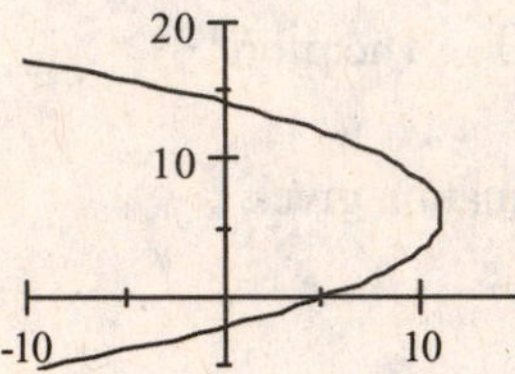

63. (a) $r=\dfrac{1}{1-\cos\theta} \Rightarrow e=1$. Therefore, this is a parabola.

(b)

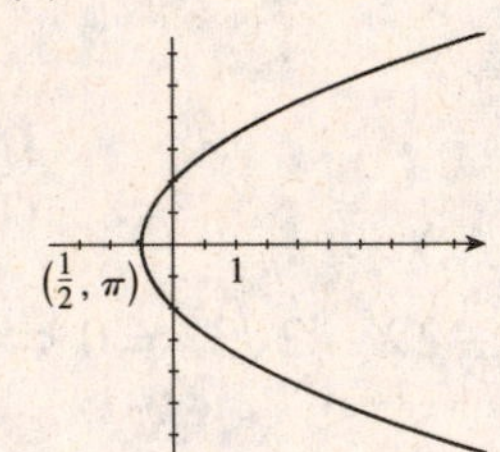

65. (a) $r=\dfrac{4}{1+2\sin\theta} \Rightarrow e=2$. Therefore, this is a hyperbola.

(b)

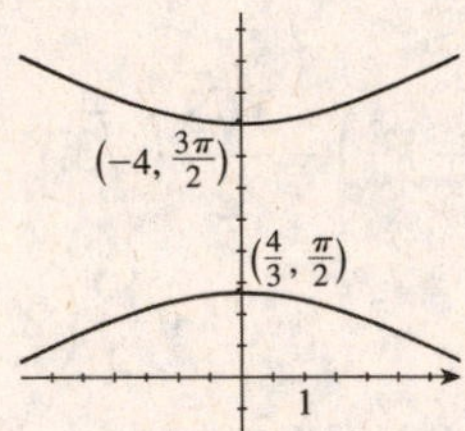

1. $x^2 = -12y$. This is a parabola with $4p = -12 \Leftrightarrow p = -3$. The focus is $(0,-3)$ and the directrix is $y = 3$.

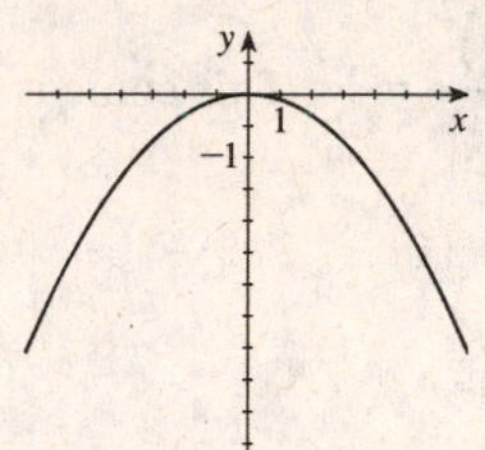

2. $\dfrac{x^2}{16} + \dfrac{y^2}{4} = 1$. This is an ellipse with $a = 4$, $b = 2$, and $c = \sqrt{16-4} = 2\sqrt{3}$. The vertices are $(\pm 4, 0)$, the foci are $(\pm 2\sqrt{3}, 0)$, the length of the major axis is $2a = 8$, and the length of the minor axis is $2b = 4$.

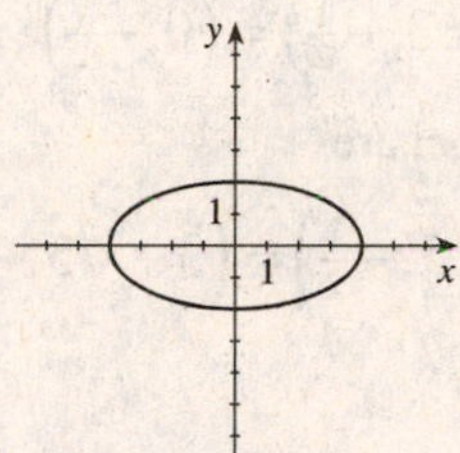

3. $\dfrac{y^2}{9} - \dfrac{x^2}{16} = 1$. This is a hyperbola with $a = 3$, $b = 4$, and $c = \sqrt{9+16} = 5$. The vertices are $(0, \pm 3)$, the foci are $(0, \pm 5)$, and the asymptotes are $y = \pm \frac{3}{4} x$.

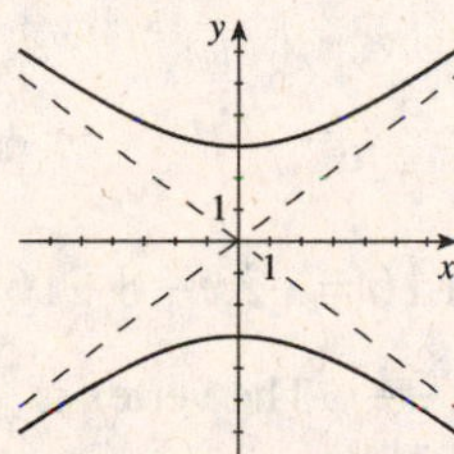

4. This is a parabola that opens to the left with its vertex at $(0,0)$. So its equation is of the form $y^2 = 4px$ with $p < 0$. Substituting the point $(-4,2)$, we have $2^2 = 4p(-4) \Leftrightarrow 4 = -16p \Leftrightarrow p = -\frac{1}{4}$. So an equation is $y^2 = 4\left(-\frac{1}{4}\right)x \Leftrightarrow y^2 = -x$.

5. This is an ellipse tangent to the x-axis at $(0,0)$ and with one vertex at the point $(4,3)$. The center is $(0,3)$, and $a = 4$ and $b = 3$. Thus the equation is $\dfrac{x^2}{16} + \dfrac{(y-3)^2}{9} = 1$.

6. This a hyperbola with a horizontal transverse axis, vertices at $(1,0)$ and $(3,0)$, and foci at $(0,0)$ and $(4,0)$. Thus the center is $(2,0)$, and $a = 3 - 2 = 1$ and $c = 4 - 2 = 2$. Thus $b^2 = 2^2 - 1^2 = 3$. So an equation is $\dfrac{(x-2)^2}{1^2} - \dfrac{y^2}{3} = 1 \Leftrightarrow (x-2)^2 - \dfrac{y^2}{3} = 1$.

7. $16x^2+36y^2-96x+36y+9=0 \iff$

$16(x^2-6x)+36(y^2+y)=-9 \iff$

$16(x^2-6x+9)+36(y^2+y+\frac{1}{4})=-9+144+9$

$\iff 16(x-3)^2+36(y+\frac{1}{2})^2=144$

$\iff \dfrac{(x-3)^2}{9}+\dfrac{(y+\frac{1}{2})^2}{4}=1$. This is an ellipse with $a=3$, $b=2$, and $c=\sqrt{9-4}=\sqrt{5}$. The center is $(3,-\frac{1}{2})$, the vertices are $(3\pm 3,-\frac{1}{2})=(0,-\frac{1}{2})$ and $(6,-\frac{1}{2})$, and the foci are $(h\pm c,k)=(3\pm\sqrt{5},-\frac{1}{2})=(3+\sqrt{5},-\frac{1}{2})$ and $(3-\sqrt{5},-\frac{1}{2})$.

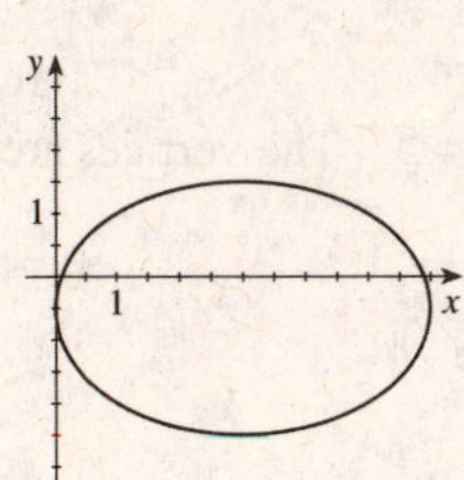

8. $9x^2-8y^2+36x+64y=164 \iff$

$9(x^2+4x)-8(y^2-8y)=164 \iff$

$9(x^2+4x+4)-8(y^2-8y+16)=164+36-128$

$\iff 9(x+2)^2-8(y-4)^2=72 \iff$

$\dfrac{(x+2)^2}{8}-\dfrac{(y-4)^2}{9}=1$. This conic is a hyperbola.

9. $2x+y^2+8y+8=0 \iff y^2+8y+16=-2x-8+16 \iff (y+4)^2=-2(x-4)$. This is a parabola with $4p=-2 \iff p=-\frac{1}{2}$. The vertex is $(4,-4)$ and the focus is $(4-\frac{1}{2},-4)=(\frac{7}{2},-4)$.

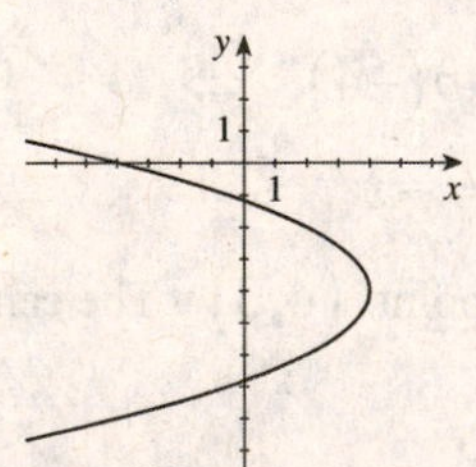

10. The hyperbola has foci $(0,\pm 5)$ and asymptotes $y=\pm\frac{3}{4}x$. Since the foci are $(0,\pm 5)$, $c=5$, the foci are on the y-axis, and the center is $(0,0)$. Also, since $y=\pm\frac{3}{4}x=\pm\dfrac{a}{b}x$, it follows that $\dfrac{a}{b}=\dfrac{3}{4} \iff a=\frac{3}{4}b$. Then $c^2=5^2=25=a^2+b^2=(\frac{3}{4}b)^2+b^2=\frac{25}{16}b^2 \iff b^2=16$, and by substitution, $a=\frac{3}{4}(4)=3$. Therefore, an equation of the hyperbola is $\dfrac{y^2}{9}-\dfrac{x^2}{16}=1$.

11. The parabola has focus $(2,4)$ and directrix the x -axis ($y=0$). Therefore, $2p=4-0=4$ $\Leftrightarrow$ $p=2$ $\Leftrightarrow$ $4p=8$, and the vertex is $(2,4-p)=(2,2)$. Hence, an equation of the parabola is $(x-2)^2=8(y-2)$ $\Leftrightarrow$ $x^2-4x+4=8y-16$ $\Leftrightarrow$ $x^2-4x-8y+20=0$.

12. We place the vertex of the parabola at the origin, so the parabola contains the points $(3,\pm3)$, and the equation is of the form $y^2=4px$. Substituting the point $(3,3)$, we get $3^2=4p(3)$ $\Leftrightarrow$ $9=12p$ $\Leftrightarrow$ $p=\frac{3}{4}$. So the focus is $\left(\frac{3}{4},0\right)$, and we should place the light bulb $\frac{3}{4}$ inch from the vertex.

13. (a) $5x^2+4xy+2y^2=18$. Then $A=5$, $B=4$, and $C=2$, so the discriminant is $(4)^2-4(5)(2)=-24$. Since the discriminant is negative, the equation represents an ellipse.

(b) $\cot 2\varphi=\dfrac{A-C}{B}=\dfrac{5-2}{4}=\dfrac{3}{4}$. Thus, $\cos 2\varphi=\frac{3}{5}$ and so $\cos\varphi=\sqrt{\frac{1+(3/5)}{2}}=\frac{2\sqrt{5}}{5}$, $\sin\varphi=\sqrt{\frac{1-(3/5)}{2}}=\frac{\sqrt{5}}{5}$. It follows that $x=\frac{2\sqrt{5}}{5}X-\frac{\sqrt{5}}{5}Y$ and $y=\frac{\sqrt{5}}{5}X+\frac{2\sqrt{5}}{5}Y$. By substitution,

$$5\left(\tfrac{2\sqrt{5}}{5}X-\tfrac{\sqrt{5}}{5}Y\right)^2+4\left(\tfrac{2\sqrt{5}}{5}X-\tfrac{\sqrt{5}}{5}Y\right)\left(\tfrac{\sqrt{5}}{5}X+\tfrac{2\sqrt{5}}{5}Y\right)+2\left(\tfrac{\sqrt{5}}{5}X+\tfrac{2\sqrt{5}}{5}Y\right)^2=18 \quad\Leftrightarrow$$

$$4X^2-4XY+Y^2+\tfrac{4}{5}\left(2X^2+4XY-XY-2Y^2\right)+\tfrac{2}{5}\left(X^2+4XY+4Y^2\right)=18 \quad\Leftrightarrow$$

$$X^2\left(4+\tfrac{8}{5}+\tfrac{2}{5}\right)+XY\left(-4+\tfrac{12}{5}+\tfrac{8}{5}\right)+Y^2\left(1-\tfrac{8}{5}+\tfrac{4}{5}\right)=18 \quad\Leftrightarrow\quad 6X^2+Y^2=18 \quad\Leftrightarrow$$

$\dfrac{X^2}{3}+\dfrac{Y^2}{18}=1$. This is an ellipse with $a=3\sqrt{2}$ and $b=\sqrt{3}$.

(c)

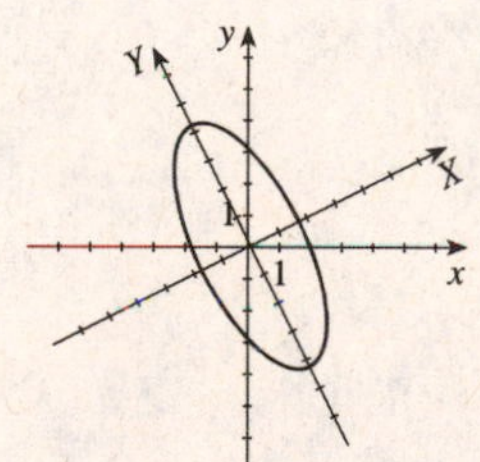

Since $\cos 2\varphi=\frac{3}{5}$ we have $2\varphi=\cos^{-1}\frac{3}{5}\approx 53.13^\circ$, so $\varphi\approx 27^\circ$.

(d) In XY -coordinates, the vertices are $V\left(0,\pm3\sqrt{2}\right)$. Therefore, in xy -coordinates, the vertices are $x=-\frac{3\sqrt{2}}{\sqrt{5}}$ and $y=\frac{6\sqrt{2}}{\sqrt{5}}$ $\Rightarrow$ $V_1\left(-\frac{3\sqrt{2}}{\sqrt{5}},\frac{6\sqrt{2}}{\sqrt{5}}\right)$, and $x=\frac{3\sqrt{2}}{\sqrt{5}}$ and $y=-\frac{6\sqrt{2}}{\sqrt{5}}$ $\Rightarrow$ $V_2\left(\frac{3\sqrt{2}}{\sqrt{5}},-\frac{6\sqrt{2}}{\sqrt{5}}\right)$.

14. (a) Since the focus of this conic is the origin and the directrix is $x = 2$, the equation has the form $r = \dfrac{ed}{1 + e\cos\theta}$. Subsituting $e = \frac{1}{2}$ and $d = 2$ we get $r = \dfrac{1}{1 + \frac{1}{2}\cos\theta} \Leftrightarrow r = \dfrac{2}{2 + \cos\theta}$.

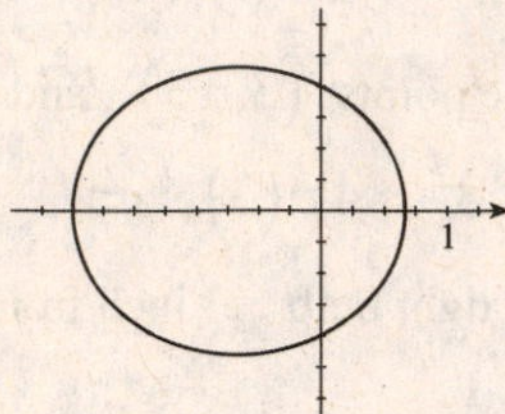

(b) $r = \dfrac{3}{2 - \sin\theta} \Leftrightarrow r = \dfrac{\frac{3}{2}}{1 - \frac{1}{2}\sin\theta}$. So $e = \frac{1}{2}$ and the conic is an ellipse.

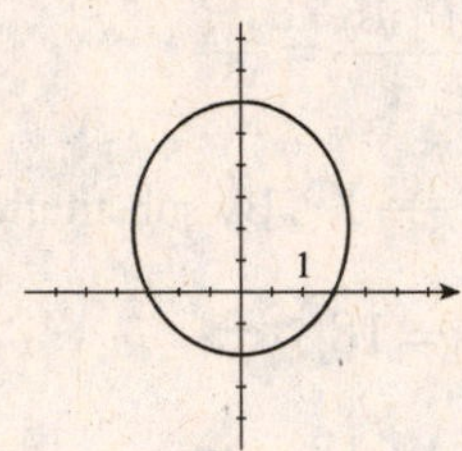

1. Answers will vary.

5. (a) The tangent line passes though the point (a,a^2) , so an equation is $y-a^2=m(x-a)$.

(b) Because the tangent line intersects the parabola at only the one point (a,a^2) , the system $\begin{cases} y-a^2=m(x-a) \\ y=x^2 \end{cases}$ has only one solution, namely $x=a$, $y=a^2$.

(c) $\begin{cases} y-a^2=m(x-a) \\ y=x^2 \end{cases} \Leftrightarrow \begin{cases} y=a^2+m(x-a) \\ y=x^2 \end{cases} \Leftrightarrow a^2+m(x-a)=x^2 \Leftrightarrow$ $x^2-mx+am-a^2=0$. This quadratic has discriminant $(-m)^2-4(1)(am-a^2)=m^2-4am+4a^2=(m-2a)^2$. Setting this equal to 0 , we find $m=2a$.

(d) An equation of the tangent line is $y-a^2=2a(x-a) \Leftrightarrow y=a^2+2ax-2a^2 \Leftrightarrow$ $y=2ax-a^2$.

8 Exponential and Logarithmic Functions

8.1 Exponential Functions

1. The function $f(x) = 5^x$ is an exponential function with base 5 ; $f(-2) = 5^{-2} = \frac{1}{25}$, $f(0) = 5^0 = 1$, $f(2) = 5^2 = 25$, and $f(6) = 5^6 = 15,625$.

3. (a) To obtain the graph of $g(x) = 2^x - 1$ we start with the graph of $f(x) = 2^x$ and shift it *downward* 1 unit.

(b) To obtain the graph of $h(x) = 2^{x-1}$ we start with the graph of $f(x) = 2^x$ and shift it to the *right* 1 unit.

5. $f(x) = 4^x$; $f(0.5) = 2$, $f(\sqrt{2}) \approx 7.103$, $f(\pi) \approx 77.880$, $f\left(\frac{1}{3}\right) \approx 1.587$.

Note: In the first printing of the text, the problem statement asks for $f(-\pi) \approx 0.0128$ instead of $f(\pi)$.

7. $g(x) = \left(\frac{2}{3}\right)^{x-1}$; $g(1.3) \approx 0.885$, $g(\sqrt{5}) \approx 0.606$, $g(2\pi) \approx 0.117$, $g\left(-\frac{1}{2}\right) \approx 1.837$

9. $f(x) = 2^x$

x	y
-4	$\frac{1}{16}$
-2	$\frac{1}{4}$
0	1
2	4
4	16

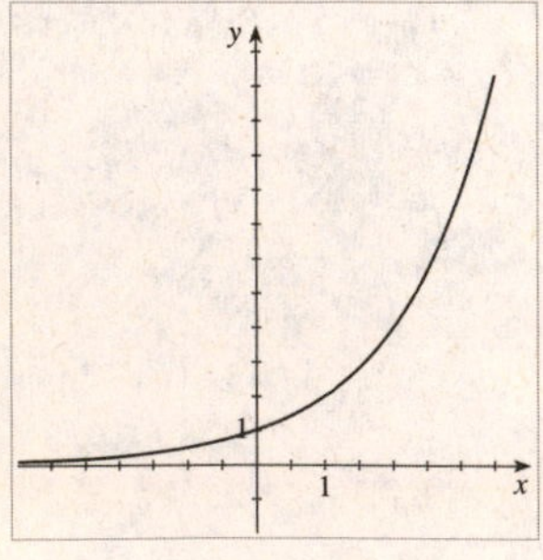

11. $f(x) = \left(\frac{1}{3}\right)^x$

x	y
-2	9
-1	3
0	1
1	$\frac{1}{3}$
2	$\frac{1}{9}$

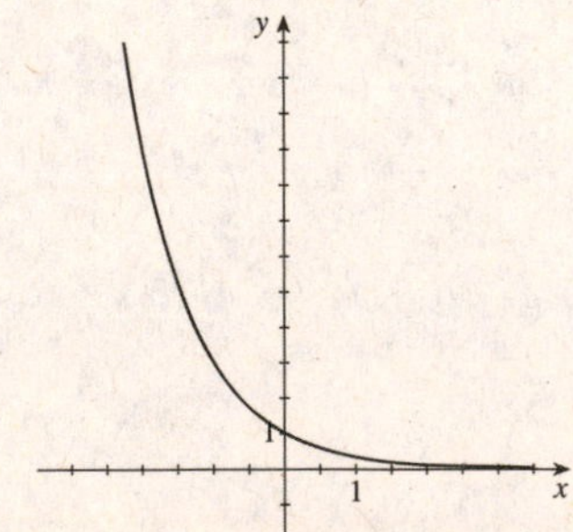

13. $g(x) = 3(1.3)^x$

x	y
-2	1.775
-1	2.308
0	3.0
1	3.9
2	5.07
3	6.591
4	8.568

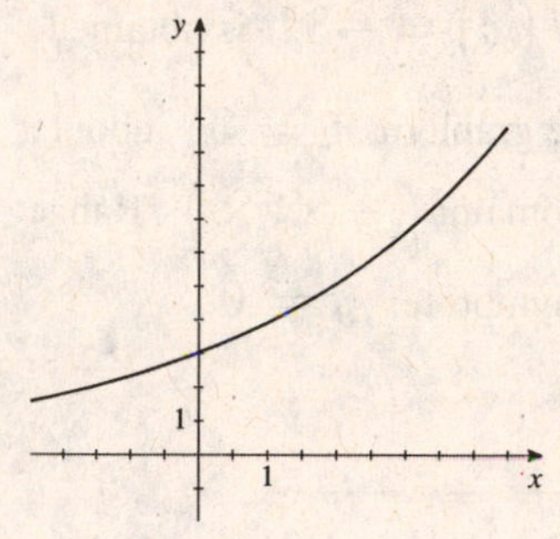

15. $f(x) = 2^x$ and $g(x) = 2^{-x}$

17. $f(x) = 4^x$ and $g(x) = 7^x$.

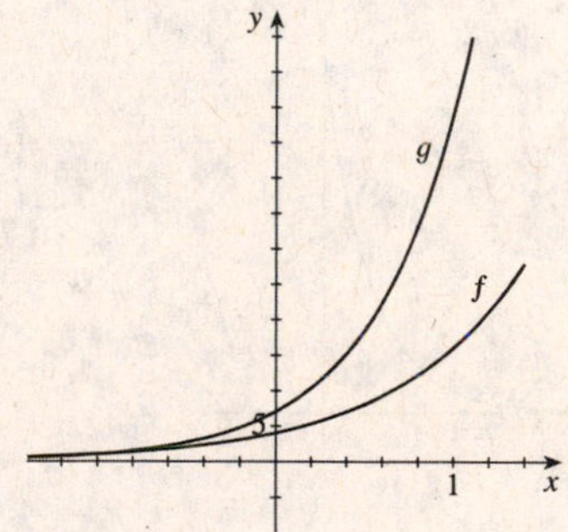

19. From the graph, $f(2) = a^2 = 9$, so $a = 3$. Thus $f(x) = 3^x$.

21. From the graph, $f(2) = a^2 = \frac{1}{16}$, so $a = \frac{1}{4}$. Thus $f(x) = \left(\frac{1}{4}\right)^x$.

23. The graph of $f(x) = 5^{x+1}$ is obtained from that of $y = 5^x$ by shifting 1 unit to the left, so it has graph II.

25. The graph of $f(x) = -3^x$ is obtained by reflecting the graph of $y = 3^x$ about the x -axis. Domain: $(-\infty,\infty)$. Range: $(-\infty,0)$. Asymptote: $y = 0$.

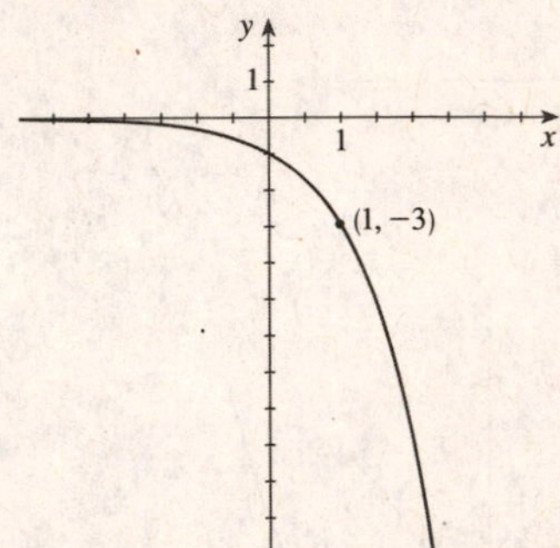

27. $g(x) = 2^x - 3$. The graph of g is obtained by shifting the graph of $y = 2^x$ downward 3 units. Domain: $(-\infty,\infty)$. Range: $(-3,\infty)$. Asymptote: $y = -3$.

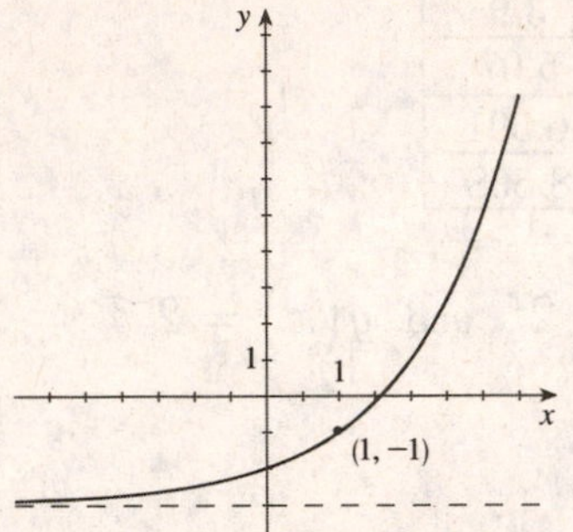

29. $h(x) = 4 + \left(\frac{1}{2}\right)^x$. The graph of h is obtained by shifting the graph of $y = \left(\frac{1}{2}\right)^x$ upward 4 units. Domain: $(-\infty,\infty)$. Range: $(4,\infty)$. Asymptote: $y = 4$.

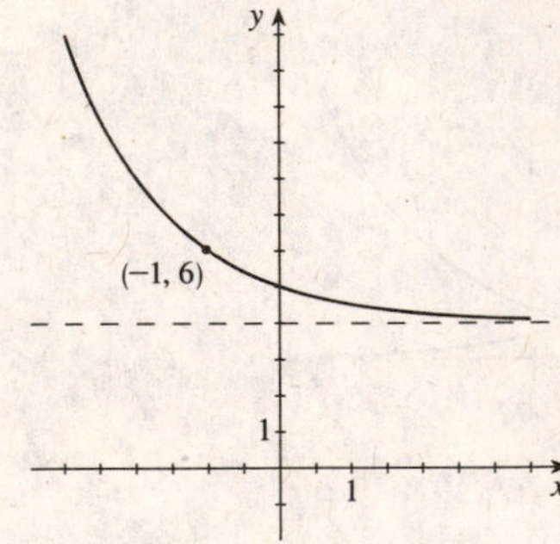

31. $f(x) = 10^{x+3}$. The graph of f is obtained by shifting the graph of $y = 10^x$ to the left 3 units. Domain: $(-\infty,\infty)$. Range: $(0,\infty)$. Asymptote: $y = 0$.

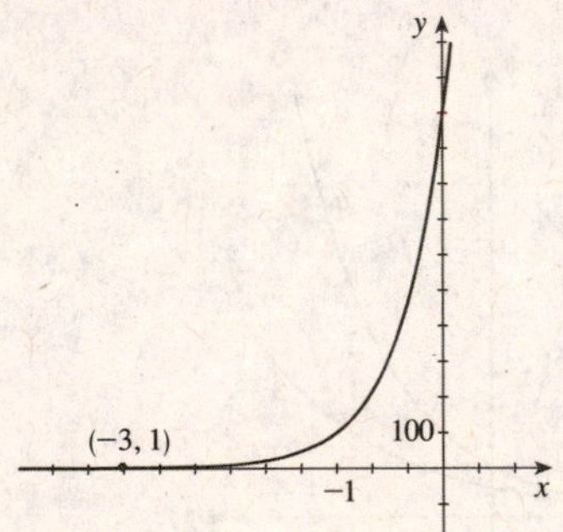

33. $y = 5^{-x} + 1$. The graph of y is obtained by reflecting the graph of $y = 5^x$ about the x -axis and then shifting upward 1 unit. Domain: $(-\infty,\infty)$. Range: $(1,\infty)$. Asymptote: $y = 1$.

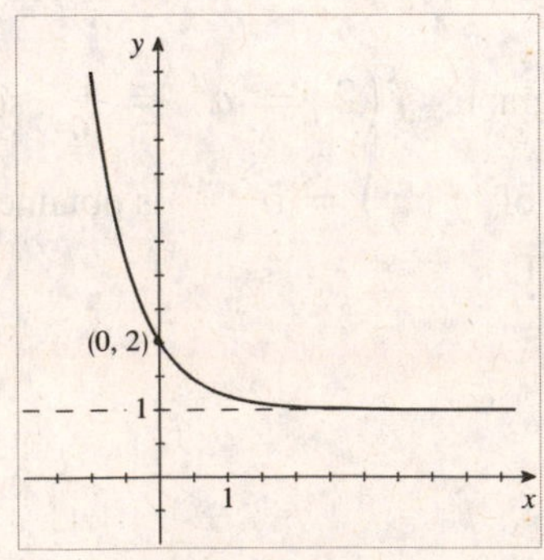

35. $y = 3 - 10^{x-1} = -10^{x-1} + 3$. The graph of y is obtained by reflecting the graph of $y = 10^x$ about the y-axis, then shifting to the right 1 unit and upward 3 units. Domain: $(-\infty, \infty)$. Range: $(-\infty, 3)$. Asymptote: $y = 3$.

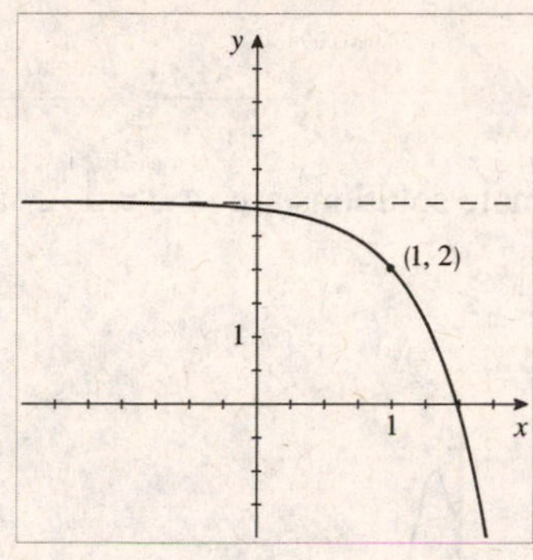

37. (a)

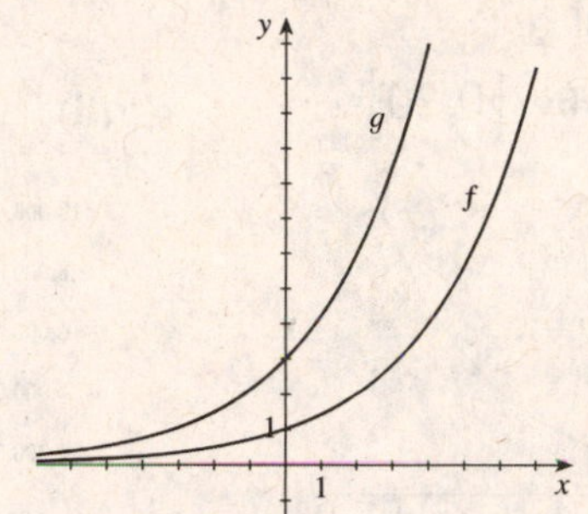

(b) Since $g(x) = 3(2^x) = 3f(x)$ and $f(x) > 0$, the height of the graph of $g(x)$ is always three times the height of the graph of $f(x) = 2^x$, so the graph of g is steeper than the graph of f.

39.

x	$f(x) = x^3$	$g(x) = 3^x$
0	0	1
1	1	3
2	8	9
3	27	27
4	64	81
5	125	243
6	216	729
7	343	2187
8	512	6561
9	729	19,683
10	1000	59,049
15	3375	14,348,907
20	8000	3,486,784,401

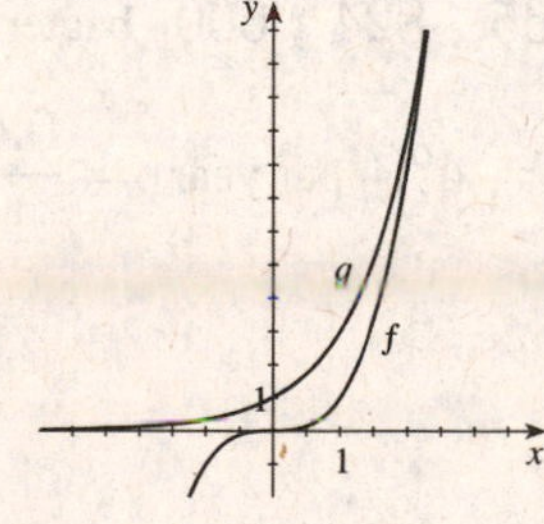

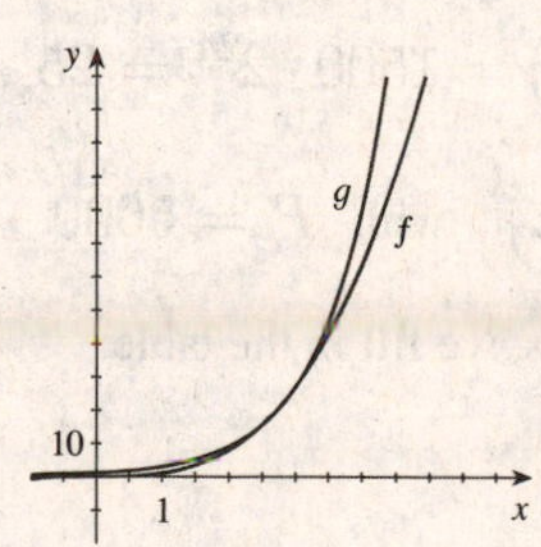

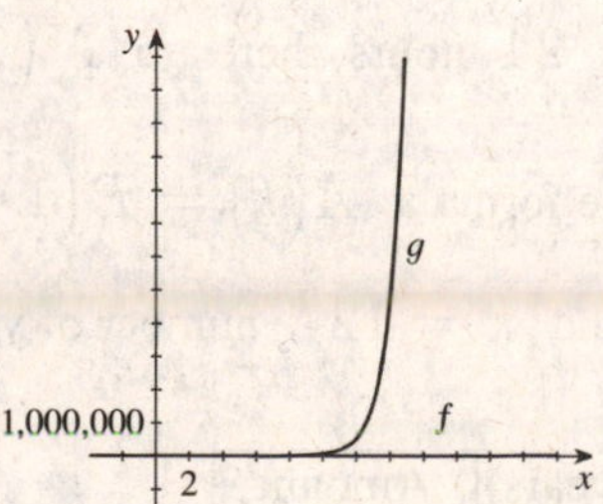

41. **(a)** From the graphs below, we see that the graph of f ultimately increases much more quickly than the graph of g.

(i) $[0,5]$ by $[0,20]$ **(ii)** $[0,25]$ by $[0,10^7]$ **(iii)** $[0,50]$ by $[0,10^8]$

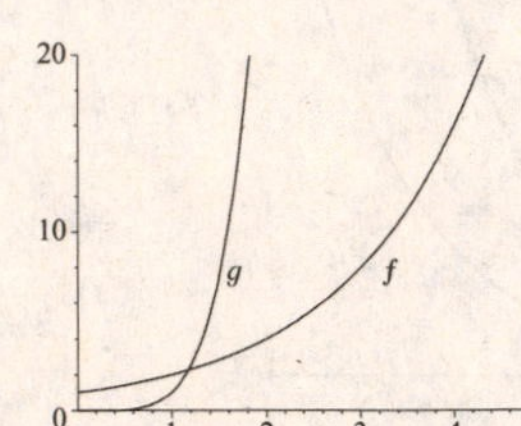

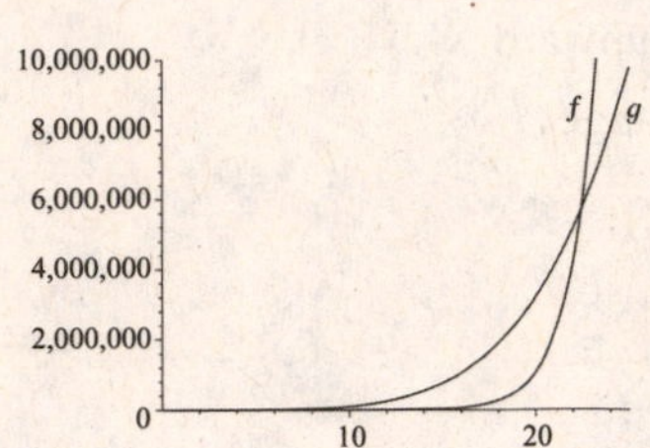

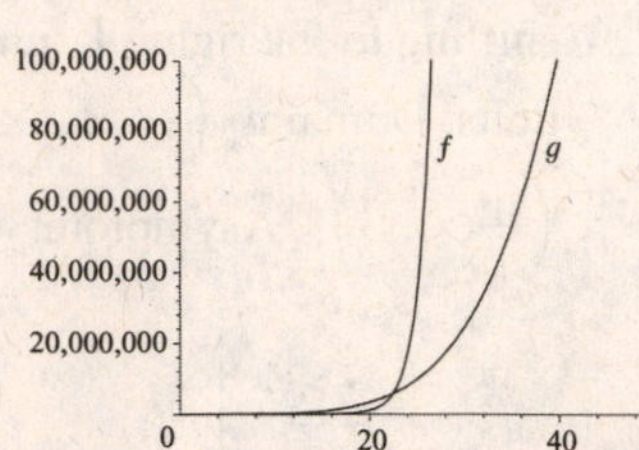

(b) From the graphs in parts (a)(i) and (a)(ii), we see that the approximate solutions are $x \approx 1.2$ and $x \approx 22.4$.

43.

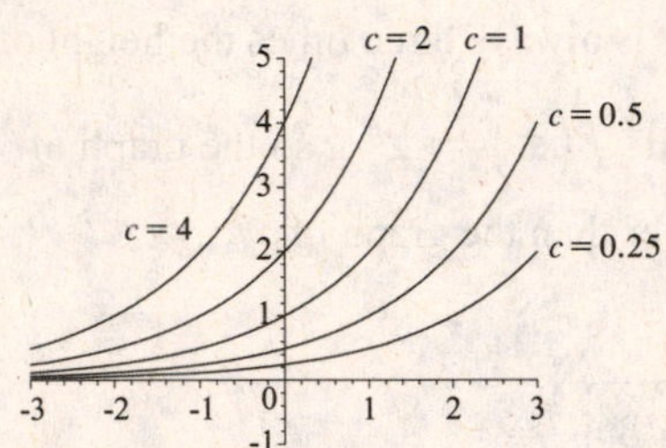

The larger the value of c, the more rapidly the graph of $f(x) = c2^x$ increases. Also notice that the graphs are just shifted horizontally 1 unit. This is because of our choice of c; each c in this exercise is of the form 2^k. So $f(x) = 2^k \cdot 2^x = 2^{x+k}$.

45. $y = 10^{x-x^2}$

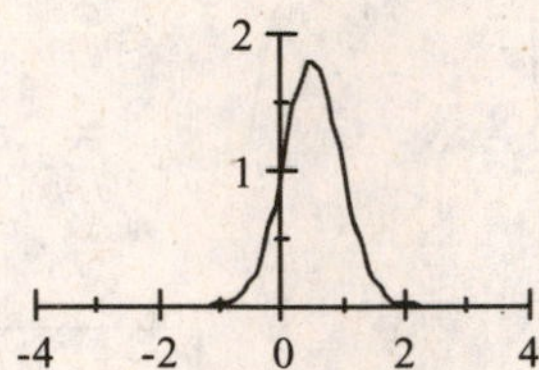

(a) From the graph, we see that the function is increasing on $(-\infty, 0.50]$ and decreasing on $[0.50, \infty)$. **(b)** From the graph, we see that the range is approximately $(0, 1.78]$.

47. **(a)** After 1 hour, there are $1500 \cdot 2 = 3000$ bacteria. After 2 hours, there are $(1500 \cdot 2) \cdot 2 = 6000$ bacteria. After 3 hours, there are $(1500 \cdot 2 \cdot 2) \cdot 2 = 12,000$ bacteria. We see that after t hours, there are $N(t) = 1500 \cdot 2^t$ bacteria.

(b) After 24 hours, there are $N(24) = 1500 \cdot 2^{24} = 25,165,824,000$ bacteria.

49. Using the formula $A(t) = P(1+i)^k$ with $P = 5000$, $i = 4\%$ per year $= \dfrac{0.04}{12}$ per month, and $k = 12 \cdot$ number of years, we fill in the table:

Time (years)	Amount
1	\$5203.71
2	\$5415.71
3	\$5636.36
4	\$5865.99
5	\$6104.98
6	\$6353.71

51. $P = 10{,}000$, $r = 0.03$, and $n = 2$. So

$A(t) = 10{,}000\left(1 + \frac{0.03}{2}\right)^{2t} = 10{,}000 \cdot 1.015^{2t}$.

(a) $A(5) = 10000 \cdot 1.015^{10} \approx 11\ ,\ 605.41$, and so the value of the investment is $\$11\ ,$ 605.41 .

(b) $A(10) = 10000 \cdot 1.015^{20} \approx 13\ ,\ 468.55$, and so the value of the investment is $\$13\ ,$ 468.55 .

(c) $A(15) = 10000 \cdot 1.015^{30} \approx 15\ ,\ 630.80$, and so the value of the investment is $\$15\ ,$ 630.80 .

53. $P = 500$, $r = 0.0375$, and $n = 4$. So $A(t) = 500\left(1 + \frac{0.0375}{4}\right)^{4t}$.

(a) $A(1) = 500\left(1 + \frac{0.0375}{4}\right)^{4} \approx 519.02$, and so the value of the investment is $\$519.02$.

(b) $A(2) = 500\left(1 + \frac{0.0375}{4}\right)^{8} \approx 538.75$, and so the value of the investment is $\$538.75$.

(c) $A(10) = 500\left(1 + \frac{0.0375}{4}\right)^{40} \approx 726.23$, and so the value of the investment is $\$726.23$.

55. We must solve for P in the equation $10000 = P\left(1 + \frac{0.09}{2}\right)^{2(3)} = P(1.045)^{6} \Leftrightarrow$ $10000 = 1.3023P \Leftrightarrow P = 7678.96$. Thus, the present value is $\$7\ ,\ 678.96$.

57. $r_{\text{APY}} = \left(1 + \dfrac{r}{n}\right)^{n} - 1$. Here $r = 0.08$ and $n = 12$, so

$r_{\text{APY}} = \left(1 + \dfrac{0.08}{12}\right)^{12} - 1 \approx (1.0066667)^{12} - 1 \approx 0.083000$. Thus, the annual percentage yield is about 8.3% .

59. (a) In this case the payment is $\$1$ million.

(b) In this case the total pay is $2 + 2^2 + 2^3 + \cdots + 2^{30} > 2^{30}$ cents $= \$10{,}737{,}418.24$. Since this is much more than method (a), method (b) is more profitable.

8.2 The Natural Exponential Function

1. The function $f(x) = e^x$ is called the *natural* exponential function. The number e is approximately equal to 2.71828.

3. $h(x) = e^x$; $h(3) = 20.086$, $h(0.23) = 1.259$, $h(1) = 2.718$, $h(-2) = 0.135$

5. $f(x) = 3e^x$

x	y
-2	0.41
-1	1.10
-0.5	1.82
0	3
0.5	4.95
1	8.15
2	22.17

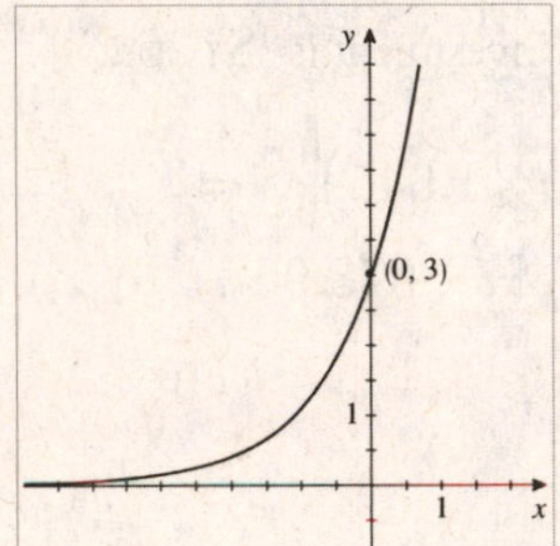

7. $y = -e^x$. The graph of $y = -e^x$ is obtained from the graph of $y = e^x$ by reflecting it about the x -axis. Domain: $(-\infty, \infty)$. Range: $(-\infty, 0)$. Asymptote: $y = 0$.

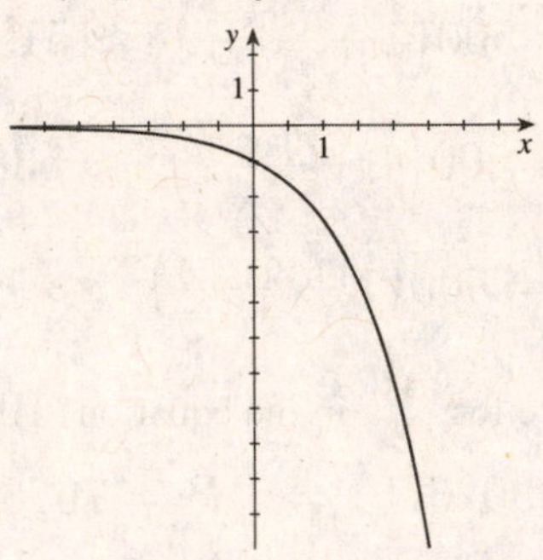

9. $y = e^{-x} - 1$. The graph of $y = e^{-x} - 1$ is obtained from the graph of $y = e^x$ by reflecting it about the y -axis then shifting downward 1 unit. Domain: $(-\infty, \infty)$. Range: $(-1, \infty)$. Asymptote: $y = -1$.

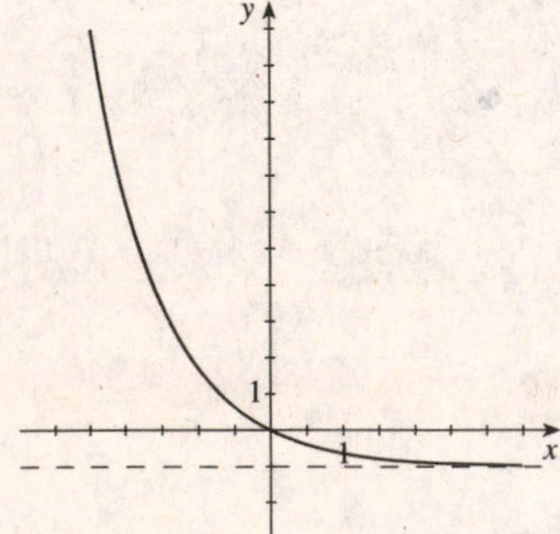

11. $y = e^{x-2}$. The graph of $y = e^{x-2}$ is obtained from the graph of $y = e^x$ by shifting it to the right 2 units. Domain: $(-\infty, \infty)$. Range: $(0, \infty)$. Asymptote: $y = 0$.

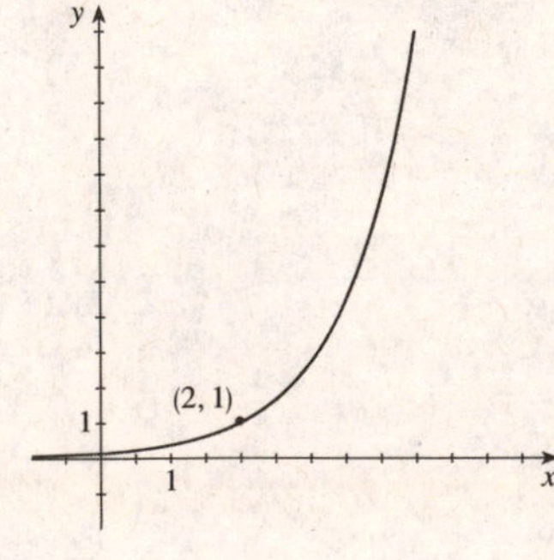

13. $h(x) = e^{x+1} - 3$. The graph of h is obtained from the graph of $y = e^x$ by shifting it to the left 1 unit and downward 3 units. Domain: $(-\infty, \infty)$. Range: $(-3, \infty)$. Asymptote: $y = -3$.

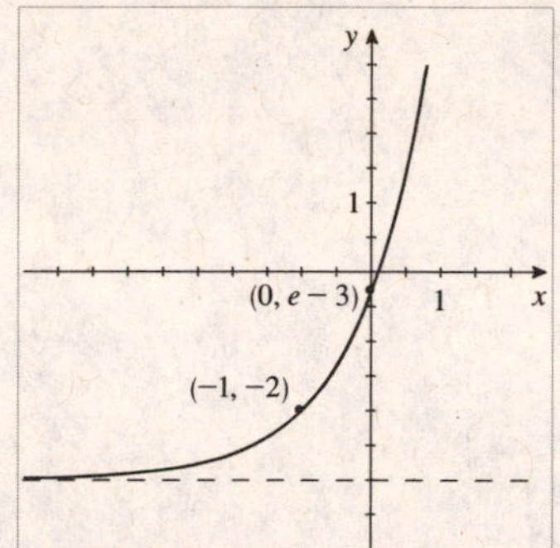

15. (a)

y = cosh x

$y = \frac{1}{2}e^{-x}$

$y = \frac{1}{2}e^{x}$

(b)

$$\begin{aligned}\cosh(-x) &= \frac{e^{-x} + e^{-(-x)}}{2} \\ &= \frac{e^{-x} + e^{x}}{2} \\ &= \frac{e^{x} + e^{-x}}{2} = \cosh x\end{aligned}$$

17. (a)

a = 1, a = 1.5, a = 0.5, a = 2

(b) As a increases the curve $y = \frac{a}{2}\left(e^{x/a} + e^{-x/a}\right)$ flattens out and the y intercept increases.

19. $g(x) = e^x + e^{-3x}$. The graph of $g(x)$ is shown in the viewing rectangle $[-4, 4]$ by $[0, 20]$. From the graph, we see that there is a local minimum of approximately 1.75 when $x \approx 0.27$.

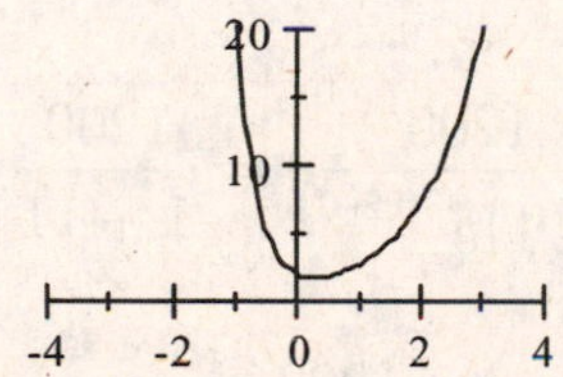

21. $m(t) = 13e^{-0.015t}$

(a) $m(0) = 13$ kg.

(b)

$$m(45) = 13e^{-0.015(45)} = 13e^{-0.675} = 6.619$$

kg. Thus the mass of the radioactive substance after 45 days is about 6.6 kg.

23. $v(t) = 80\left(1 - e^{-0.2t}\right)$ **(a)**

$v(0) = 80\left(1 - e^{0}\right) = 80(1 - 1) = 0$

.**(b)**

$v(5) = 80\left(1 - e^{-0.2(5)}\right) \approx 80(0.632) = 50.57$

ft/s. So the velocity after 5 s is about 50.6 ft/s.

$v(10) = 80\left(1 - e^{-0.2(10)}\right) \approx 80(0.865) = 69.2$

ft/s. So the velocity after 10 s is about 69.2 ft/s.**(d)** The terminal velocity is 80 ft/s.

(c)

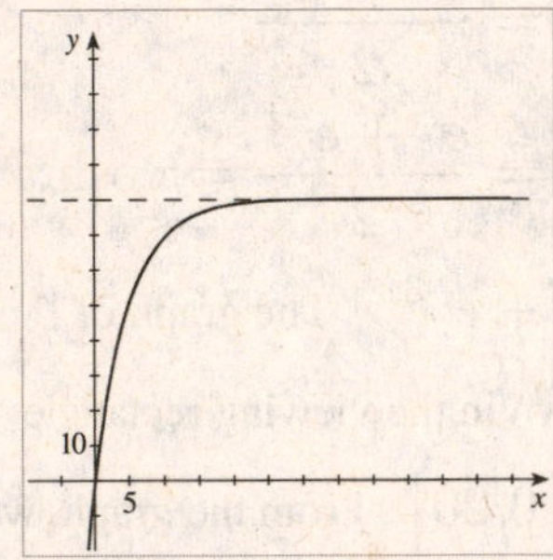

25. $P(t) = \dfrac{1200}{1 + 11e^{-0.2t}}$

(a)

$$P(0) = \frac{1200}{1 + 11e^{-0.2(0)}} = \frac{1200}{1 + 11} = 100$$

.

(b) $P(10) = \dfrac{1200}{1 + 11e^{-0.2(10)}} \approx 482$.

$P(20) = \dfrac{1200}{1 + 11e^{-0.2(20)}} \approx 999$.

$P(30) = \dfrac{1200}{1 + 11e^{-0.2(30)}} \approx 1168$.

(c) As $t \to \infty$ we have $e^{-0.2t} \to 0$,

so $P(t) \to \dfrac{1200}{1 + 0} = 1200$. The graph shown confirms this.

27. $P(t) = \dfrac{73.2}{6.1 + 5.9e^{-0.02t}}$

(a) In the year 2200, $t = 2200 - 2000 = 200$, and the population is predicted to be

$P(200) = \dfrac{73.2}{6.1 + 5.9e^{-0.02(200)}} \approx 11.79$ billion. In 2300 , $t = 300$, and

$P(300) = \dfrac{73.2}{6.1 + 5.9e^{-0.02(300)}} \approx 11.97$ billion.**(c)** As t increases, the denominator approaches 6.1 , so according to this model, the world population approaches $\frac{73.2}{6.1} = 12$ billion people.

(b)

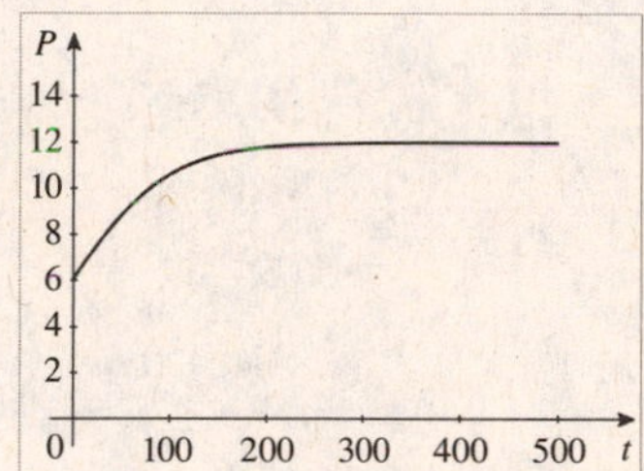

29. Using the formula $A(t) = Pe^{rt}$ with $P = 7000$ and $r = 3\% = 0.03$, we fill in the table:

Time (years)	Amount
1	$7213.18
2	$7432.86
3	$7659.22
4	$7892.48
5	$8132.84
6	$8380.52

31. We use the formula $A(t) = Pe^{rt}$ with $P = 2000$ and $r = 3.5\% = 0.035$

(a) $A(2) = 2000e^{0.035 \cdot 2} \approx \$2145.02$

(b) $A(42) = 2000e^{0.035 \cdot 4} \approx \$2300.55$

(c) $A(12) = 2000e^{0.035 \cdot 12} \approx \$3043.92$

33. (a) Using the formula $A(t) = P(1+i)^k$ with $P = 600$, $i = 2.5\%$ per year $= 0.025$, and $k = 10$, we calculate $A(10) = 600(1.025)^{10} \approx \$768.05$.

(b) Here $i = \frac{0.025}{2}$ semiannually and $k = 10 \cdot 2 = 20$, so

$A(10) = 600\left(1 + \frac{0.025}{2}\right)^{20} \approx \$769.22$.

(c) Here $i = 2.5\%$ per year $= \frac{0.025}{4}$ quarterly and $k = 10 \cdot 4 = 40$, so

$A(10) = 600\left(1 + \frac{0.025}{4}\right)^{40} \approx \$769.82$.

(d) Using the formula $A(t) = Pe^{rt}$ with $P = 600$, $r = 2.5\% = 0.025$, and $t = 10$, we have $A(10) = 600e^{0.025 \cdot 10} \approx \$770.42$.

35. *Investment 1:* After 1 year, a $\$100$ investment grows to

$A(1) = 100\left(1 + \frac{0.025}{2}\right)^{2} \approx 102.52$.

Investment 2: After 1 year, a $\$100$ investment grows to

$A(1) = 100\left(1 + \frac{0.0225}{4}\right)^{4} = 102.27$.

Investment 3: After 1 year, a $\$100$ investment grows to $A(1) = 100e^{0.02} \approx 102.02$.

We see that Investment 1 yields the highest return.

37. (a) $A(t) = Pe^{rt} = 5000e^{0.09t}$

(b)

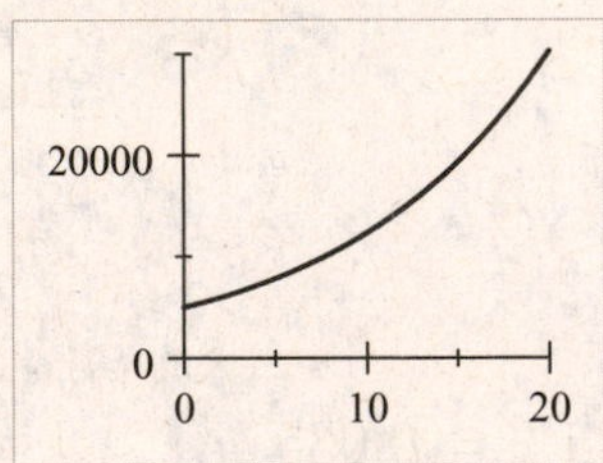

(c) $A(t) = 25,000$ when $t \approx 17.88$ years.

8.3 Logarithmic Functions

1. $\log x$ is the exponent to which the base 10 must be raised in order to get x.

x	10^3	10^2	10^1	10^0	10^{-1}	10^{-2}	10^{-3}	$10^{1/2}$
$\log x$	3	2	1	0	-1	-2	-3	$1/2$

3. (a) $5^3 = 125$, so $\log_5 125 = 3$.

(b) $\log_5 25 = 2$, so $5^2 = 25$.

5.

Logarithmic form	Exponential form
$\log_8 8 = 1$	$8^1 = 8$
$\log_8 64 = 2$	$8^2 = 64$
$\log_8 4 = \frac{2}{3}$	$8^{2/3} = 4$
$\log_8 512 = 3$	$8^3 = 512$
$\log_8 \frac{1}{8} = -1$	$8^{-1} = \frac{1}{8}$
$\log_8 \frac{1}{64} = -2$	$8^{-2} = \frac{1}{64}$

7. (a) $5^2 = 25$

(b) $5^0 = 1$

9. (a) $8^{1/3} = 2$

(b) $2^{-3} = \frac{1}{8}$

11. (a) $e^x = 5$

(b) $e^5 = y$

13. (a) $\log_5 125 = 3$

(b) $\log_{10} 0.0001 = -4$

15. (a) $\log_8 \frac{1}{8} = -1$

(b) $\log_2 \left(\frac{1}{8}\right) = -3$

17. (a) $\ln 2 = x$

(b) $\ln y = 3$

19. (a) $\log_3 3 = 1$

(b) $\log_3 1 = \log_3 3^0 = 0$

(c) $\log_3 3^2 = 2$

21. (a) $\log_6 36 = \log_6 6^2 = 2$

(b) $\log_9 81 = \log_9 9^2 = 2$

(c) $\log_7 7^{10} = 10$

23. (a) $\log_3 \left(\frac{1}{27}\right) = \log_3 3^{-3} = -3$

(b) $\log_{10} \sqrt{10} = \log_{10} 10^{1/2} = \frac{1}{2}$

(c) $\log_5 0.2 = \log_5 \left(\frac{1}{5}\right) = \log_5 5^{-1}$

$= -1$

25. (a) $2^{\log_2 37} = 37$

(b) $3^{\log_3 8} = 8$

(c) $e^{\ln \sqrt{5}} = \sqrt{5}$

27. (a) $\log_8 0.25 = \log_8 8^{-2/3} = -\frac{2}{3}$

(b) $\ln e^4 = 4$

(c) $\ln\left(\frac{1}{e}\right) = \ln e^{-1} = -1$

29. (a) $\log_2 x = 5 \;\Leftrightarrow\; x = 2^5 = 32$

(b) $x = \log_2 16 = \log_2 2^4 = 4$

31. (a) $x = \log_3 243 = \log_3 3^5 = 5$

(b) $\log_3 x = 3 \quad \Leftrightarrow \quad x = 3^3 = 27$

33. (a) $\log_{10} x = 2 \quad \Leftrightarrow$

$x = 10^2 = 100$

(b) $\log_5 x = 2 \quad \Leftrightarrow \quad x = 5^2 = 25$

35. (a) $\log_x 16 = 4 \quad \Leftrightarrow \quad x^4 = 16 \quad \Leftrightarrow \quad x = 2$

(b) $\log_x 8 = \frac{3}{2} \quad \Leftrightarrow \quad x^{3/2} = 8$

$\Leftrightarrow \quad x = 8^{2/3} = 4$

37. (a) $\log 2 \approx 0.3010$

(b) $\log 35.2 \approx 1.5465$

(c) $\log\left(\frac{2}{3}\right) \approx -0.1761$

39. (a) $\ln 5 \approx 1.6094$

(b) $\ln 25.3 \approx 3.2308$

(c) $\ln\left(1 + \sqrt{3}\right) \approx 1.0051$

41. 

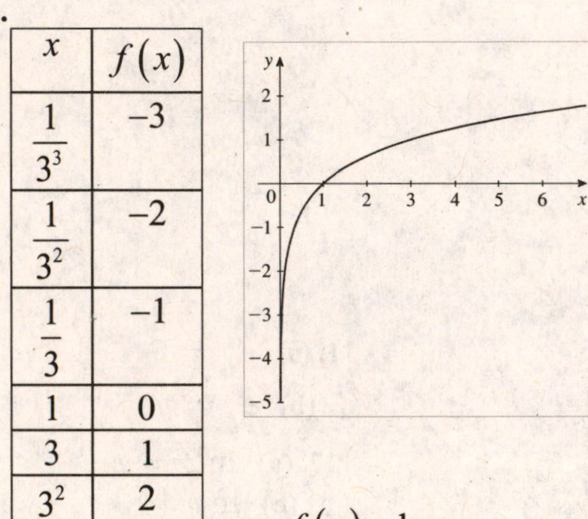

x	$f(x)$
$\frac{1}{3^3}$	-3
$\frac{1}{3^2}$	-2
$\frac{1}{3}$	-1
1	0
3	1
3^2	2

$f(x) = \log_3 x$

43.

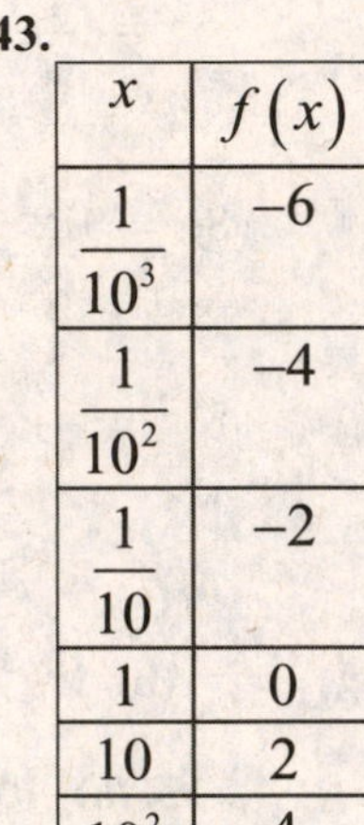

x	$f(x)$
$\frac{1}{10^3}$	-6
$\frac{1}{10^2}$	-4
$\frac{1}{10}$	-2
1	0
10	2
10^2	4

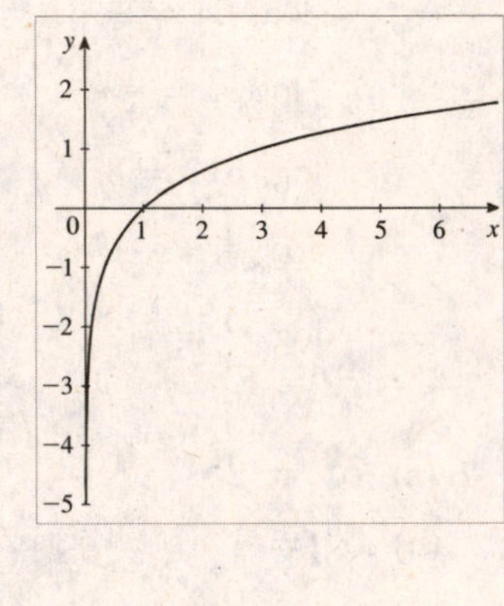

$f(x) = 2\log x$

45. Since the point $\left(5,1\right)$ is on the graph, we have $1 = \log_a 5 \quad \Leftrightarrow \quad a^1 = 5$. Thus the function is $y = \log_5 x$.

47. Since the point $\left(3,\frac{1}{2}\right)$ is on the graph, we have $\frac{1}{2} = \log_a 3 \quad \Leftrightarrow \quad a^{1/2} = 3 \quad \Leftrightarrow \quad a = 9$. Thus the function is $y = \log_9 x$.

49. I

51. The graph of $y = \log_4 x$ is obtained from the graph of $y = 4^x$ by reflecting it about the line $y = x$.

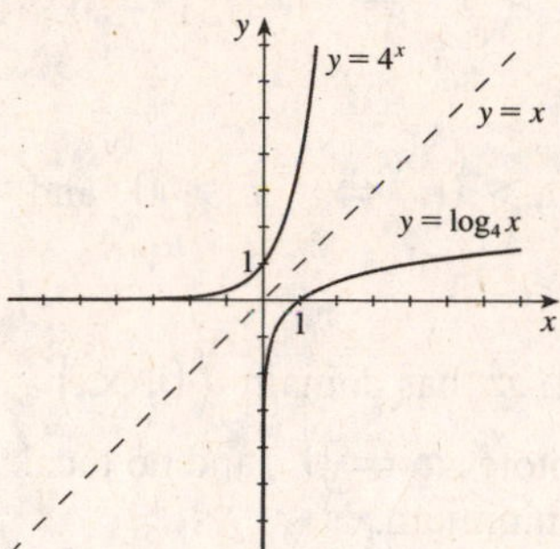

53. $f(x) = \log_2(x-4)$. The graph of f is obtained from the graph of $y = \log_2 x$ by shifting it to the right 4 units. Domain: $(4, \infty)$. Range: $(-\infty, \infty)$. Vertical asymptote: $x = 4$.

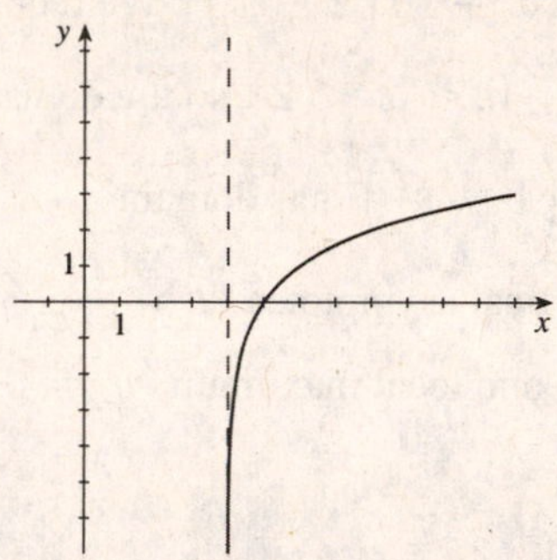

55. $g(x) = \log_5(-x)$. The graph of g is obtained from the graph of $y = \log_5 x$ by reflecting it about the y-axis. Domain: $(-\infty, 0)$. Range: $(-\infty, \infty)$. Vertical asymptote: $x = 0$.

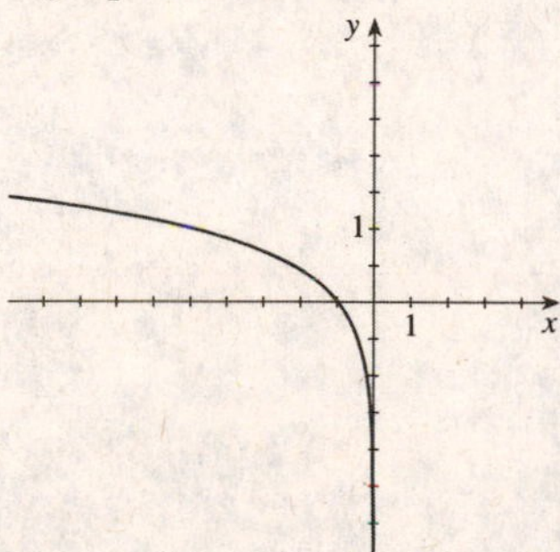

57. $y = 2 + \log_3 x$. The graph of $y = 2 + \log_3 x$ is obtained from the graph of $y = \log_3 x$ by shifting it upward 2 units. Domain: $(0, \infty)$. Range: $(-\infty, \infty)$. Vertical asymptote: $x = 0$.

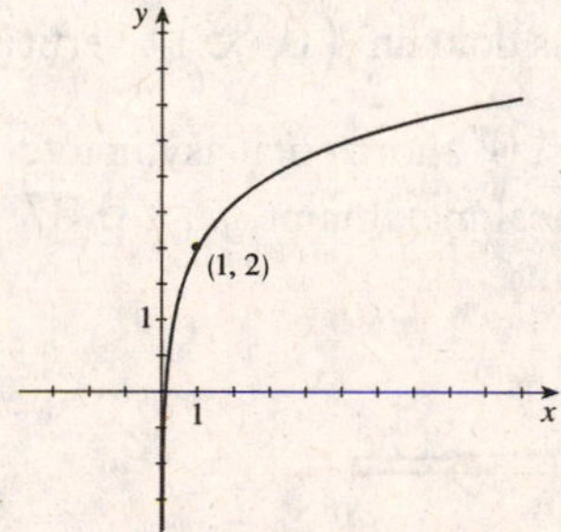

59. $y = 1 - \log_{10} x$. The graph of $y = 1 - \log_{10} x$ is obtained from the graph of $y = \log_{10} x$ by reflecting it about the x-axis, and then shifting it upward 1 unit. Domain: $(0, \infty)$. Range: $(-\infty, \infty)$. Vertical asymptote: $x = 0$.

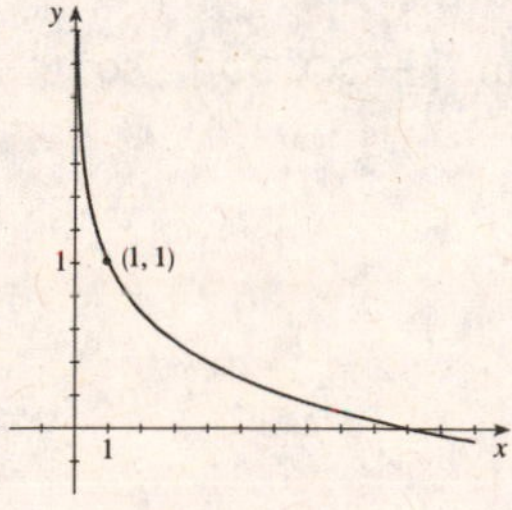

61. $y = |\ln x|$. The graph of $y = |\ln x|$ is obtained from the graph of $y = \ln x$ by reflecting the part of the graph for $0 < x < 1$ about the x-axis. Domain: $(0, \infty)$. Range: $[0, \infty)$. Vertical asymptote: $x = 0$.

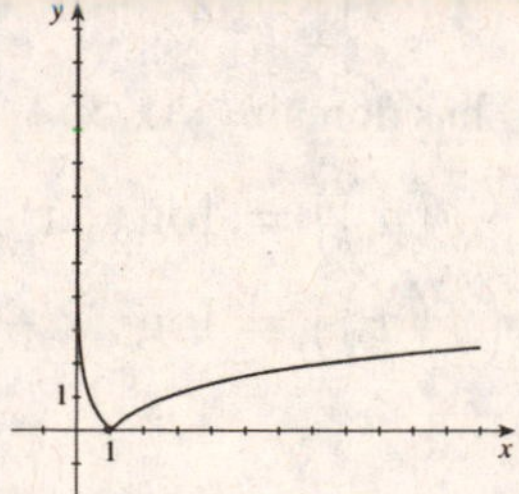

63. $f(x) = \log_{10}(x+3)$. We require that $x+3>0 \Leftrightarrow x>-3$, so the domain is $(-3,\infty)$.

65. $g(x) = \log_3(x^2-1)$. We require that $x^2-1>0 \Leftrightarrow x^2>1 \Rightarrow x<-1$ or $x>1$, so the domain is $(-\infty,-1)\cup(1,\infty)$.

67. $h(x) = \ln x + \ln(2-x)$. We require that $x>0$ and $2-x>0 \Leftrightarrow x>0$ and $x<2 \Leftrightarrow 0<x<2$, so the domain is $(0,2)$.

69. $y = \log_{10}(1-x^2)$ has domain $(-1,1)$, vertical asymptotes $x=-1$ and $x=1$, and local maximum $y=0$ at $x=0$.

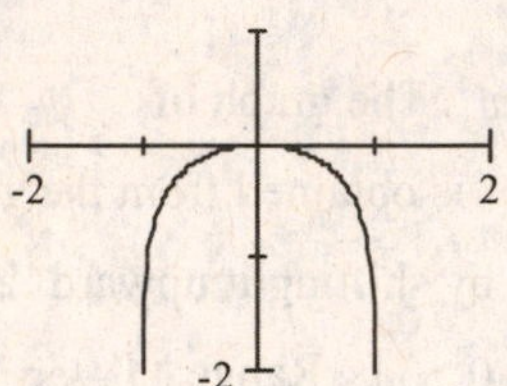

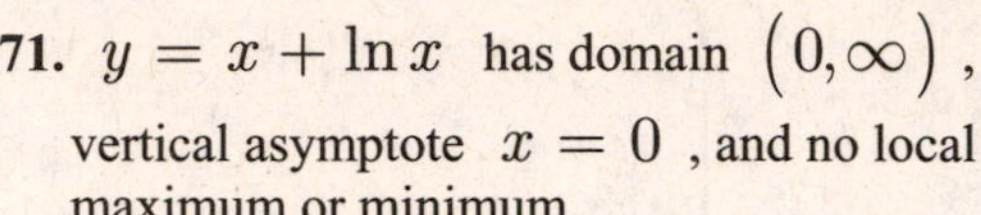

71. $y = x + \ln x$ has domain $(0,\infty)$, vertical asymptote $x=0$, and no local maximum or minimum.

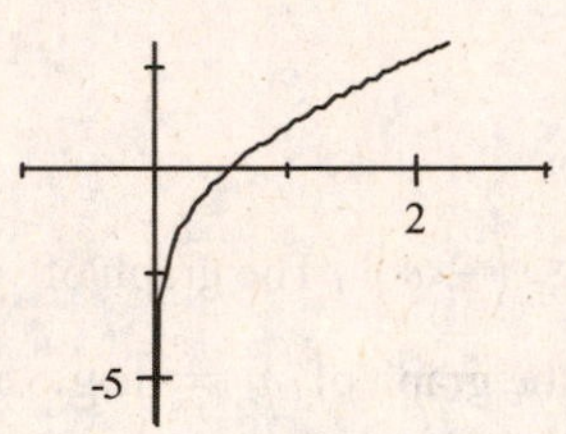

73. $y = \dfrac{\ln x}{x}$ has domain $(0,\infty)$, vertical asymptote $x=0$, horizontal asymptote $y=0$, and local maximum $y\approx 0.37$ at $x\approx 2.72$.

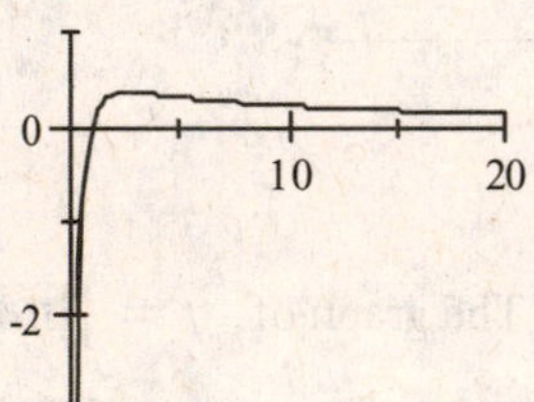

75. $f(x) = 2^x$ and $g(x) = x+1$ both have domain $(-\infty,\infty)$, so $(f\circ g)(x) = f(g(x)) = 2^{g(x)} = 2^{x+1}$ with domain $(-\infty,\infty)$ and $(g\circ f)(x) = g(f(x)) = 2^x+1$ with domain $(-\infty,\infty)$.

77. $f(x) = \log_2 x$ has domain $(0,\infty)$ and $g(x) = x-2$ has domain $(-\infty,\infty)$, so $(f\circ g)(x) = f(g(x)) = \log_2(x-2)$ with domain $(2,\infty)$ and $(g\circ f)(x) = g(f(x)) = \log_2 x - 2$ with domain $(0,\infty)$.

79. The graph of $g(x) = \sqrt{x}$ grows faster than the graph of $f(x) = \ln x$.

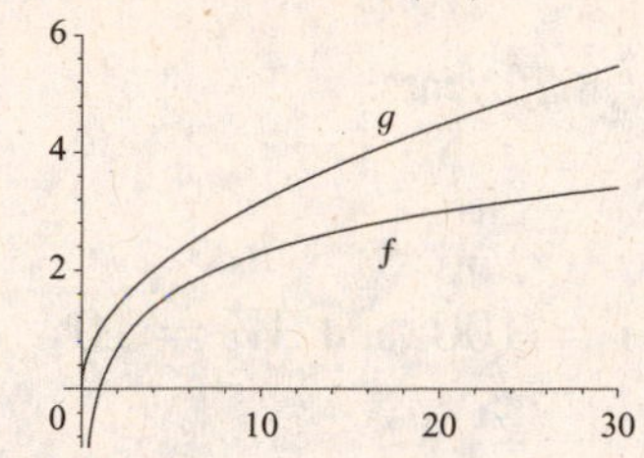

81. (a)

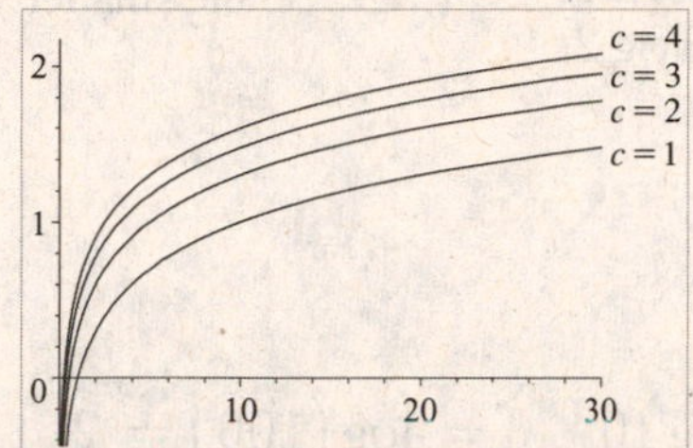

(b) Notice that $f(x) = \log(cx) = \log c + \log x$, so as c increases, the graph of $f(x) = \log(cx)$ is shifted upward $\log c$ units.

83. (a) $f(x) = \log_2(\log_{10} x)$. Since the domain of $\log_2 x$ is the positive real numbers, we have: $\log_{10} x > 0 \Leftrightarrow x > 10^0 = 1$. Thus the domain of $f(x)$ is $(1, \infty)$.

(b) $y = \log_2(\log_{10} x) \Leftrightarrow 2^y = \log_{10} x \Leftrightarrow 10^{2^y} = x$. Thus $f^{-1}(x) = 10^{2^x}$.

85. (a) $f(x) = \dfrac{2^x}{1+2^x}$. $y = \dfrac{2^x}{1+2^x} \Leftrightarrow y + y2^x = 2^x \Leftrightarrow$

$y = 2^x - y2^x = 2^x(1-y) \Leftrightarrow 2^x = \dfrac{y}{1-y} \Leftrightarrow x = \log_2\left(\dfrac{y}{1-y}\right)$. Thus

$f^{-1}(x) = \log_2\left(\dfrac{x}{1-x}\right)$.

(b) $\dfrac{x}{1-x} > 0$. Solving this using the methods from Chapter 1, we start with the endpoints, 0 and 1.

Interval	$(-\infty, 0)$	$(0,1)$	$(1,\infty)$
Sign of x Sign of $1-x$	$-$ $+$	$+$ $+$	$+$ $-$
Sign of $\dfrac{x}{1-x}$	$-$	$+$	$-$

Thus the domain of $f^{-1}(x)$ is $(0,1)$.

87. Using $D = 0.73D_0$ we have $A = -8267\ln\left(\dfrac{D}{D_0}\right) = -8267\ln 0.73 \approx 2601$ years.

89. When $r = 6\%$ we have $t = \dfrac{\ln 2}{0.06} \approx 11.6$ years. When $r = 7\%$ we have $t = \dfrac{\ln 2}{0.07} \approx 9.9$ years. And when $r = 8\%$ we have $t = \dfrac{\ln 2}{0.08} \approx 8.7$ years.

91. Using $A = 100$ and $W = 5$ we find the ID to be $\dfrac{\log(2A\,/\,W)}{\log 2} = \dfrac{\log(2 \cdot 100\,/\,5)}{\log 2} = \dfrac{\log 40}{\log 2} \approx 5.32$. Using $A = 100$ and $W = 10$ we find the ID to be $\dfrac{\log(2A\,/\,W)}{\log 2} = \dfrac{\log(2 \cdot 100\,/\,10)}{\log 2} = \dfrac{\log 20}{\log 2} \approx 4.32$. So the smaller icon is $\dfrac{5.23}{4.32} \approx 1.23$ times harder.

93. $\log\left(\log 10^{100}\right) = \log 100 = 2$

$$\log\left(\log\left(\log 10^{\text{googol}}\right)\right) = \log\left(\log\left(\text{googol}\right)\right) = \log\left(\log 10^{100}\right) = \log\left(100\right) = 2$$

95. The numbers between 1000 and 9999 (inclusive) each have 4 digits, while $\log 1000 = 3$ and $\log 10{,}000 = 4$. Since $\left[\left[\log x\right]\right] = 3$ for all integers x where $1000 \le x < 10\,,000$, the number of digits is $\left[\left[\log x\right]\right] + 1$. Likewise, if x is an integer where $10^{n-1} \le x < 10^{n}$, then x has n digits and $\left[\left[\log x\right]\right] = n - 1$. Since $\left[\left[\log x\right]\right] = n - 1 \Leftrightarrow n = \left[\left[\log x\right]\right] + 1$, the number of digits in x is $\left[\left[\log x\right]\right] + 1$.

8.4 Laws of Logarithms

1. The logarithm of a product of two numbers is the same as the *sum* of the logarithms of these numbers. So $\log_5(25 \cdot 125) = \log_5 25 + \log_5 125 = 2 + 3 = 5$.

3. The logarithm of a number raised to a power is the same as the power *times* the logarithm of the number. So $\log_5(25^{10}) = 10 \cdot \log_5 25 = 10 \cdot 2 = 20$.

5. Most calculators can find logarithms with base 10 and base e. To find logarithms with different bases we use the *change of base* formula. To find $\log_7 12$ we write

$$\log_7 12 = \frac{\log 12}{\log 7} \approx \frac{1.079}{0.845} \approx 1.277.$$

7. $\log_3 \sqrt{27} = \log_3 3^{3/2} = \frac{3}{2}$

9. $\log 4 + \log 25 = \log(4 \cdot 25) = \log 100 = 2$

11. $\log_4 192 - \log_4 3 = \log_4 \frac{192}{3} = \log_4 64 = \log_4 4^3 = 3$

13. $\log_2 6 - \log_2 15 + \log_2 20 = \log_2 \frac{6}{15} + \log_2 20 = \log_2\left(\frac{2}{5} \cdot 20\right) = \log_2 8 = \log_2 2^3 = 3$

15. $\log_4 16^{100} = \log_4 \left(4^2\right)^{100} = \log_4 4^{200} = 200$

17.
$$\log\left(\log 10^{10{,}000}\right) = \log(10{,}000 \log 10) = \log(10{,}000 \cdot 1) = \log(10{,}000) = \log 10^4 = 4 \log 10 = 4$$

19. $\log_2 2x = \log_2 2 + \log_2 x = 1 + \log_2 x$

21. $\log_2[x(x-1)] = \log_2 x + \log_2(x-1)$

23. $\log 6^{10} = 10 \log 6$

25. $\log_2\left(AB^2\right) = \log_2 A + \log_2 B^2 = \log_2 A + 2\log_2 B$

27. $\log_3\left(x\sqrt{y}\right) = \log_3 x + \log_3 \sqrt{y} = \log_3 x + \frac{1}{2}\log_3 y$

29. $\log_5 \sqrt[3]{x^2+1} = \frac{1}{3}\log_5\left(x^2+1\right)$

31. $\ln\sqrt{ab} = \frac{1}{2}\ln ab = \frac{1}{2}(\ln a + \ln b)$

33. $\log\left(\frac{x^3y^4}{z^6}\right) = \log\left(x^3y^4\right) - \log z^6 = 3\log x + 4\log y - 6\log z$

35. $\log_2\left(\frac{x\left(x^2+1\right)}{\sqrt{x^2-1}}\right) = \log_2 x + \log_2\left(x^2+1\right) - \frac{1}{2}\log_2\left(x^2-1\right)$

37. $\ln\left(x\sqrt{\frac{y}{z}}\right) = \ln x + \frac{1}{2}\ln\left(\frac{y}{z}\right) = \ln x + \frac{1}{2}(\ln y - \ln z)$

39. $\log \sqrt[4]{x^2+y^2} = \frac{1}{4}\log\left(x^2+y^2\right)$

41. $\log\sqrt{\dfrac{x^2+4}{(x^2+1)(x^3-7)^2}} = \frac{1}{2}\log\dfrac{x^2+4}{(x^2+1)(x^3-7)^2}$

$= \frac{1}{2}\left[\log(x^2+4) - \log(x^2+1)(x^3-7)^2\right]$

$= \frac{1}{2}\left[\log(x^2+4) - \log(x^2+1) - 2\log(x^3-7)\right]$

43. $\ln\dfrac{x^3\sqrt{x-1}}{3x+4} = \ln\left(x^3\sqrt{x-1}\right) - \ln(3x+4) = 3\ln x + \frac{1}{2}\ln(x-1) - \ln(3x+4)$

45. $\log_3 5 + 5\log_3 2 = \log_3 5 + \log_3 2^5 = \log_3\left(5\cdot 2^5\right) = \log_3 160$

47. $\log_2 A + \log_2 B - 2\log_2 C = \log_2(AB) - \log_2\left(C^2\right) = \log_2\left(\dfrac{AB}{C^2}\right)$

49. $4\log x - \frac{1}{3}\log(x^2+1) + 2\log(x-1) = \log x^4 - \log\sqrt[3]{x^2+1} + \log(x-1)^2$

$= \log\left(\dfrac{x^4}{\sqrt[3]{x^2+1}}\right) + \log(x-1)^2$

$= \log\left(\dfrac{x^4(x-1)^2}{\sqrt[3]{x^2+1}}\right)$

51. $\ln 5 + 2\ln x + 3\ln(x^2+5) = \ln(5x^2) + \ln(x^2+5)^3 = \ln\left[5x^2(x^2+5)^3\right]$

53. $\frac{1}{3}\log(x+2)^3 + \frac{1}{2}\left[\log x^4 - \log(x^2-x-6)^2\right] = 3\cdot\frac{1}{3}\log(x+2) + \frac{1}{2}\log\dfrac{x^4}{(x^2-x-6)^2}$

$= \log(x+2) + \log\left(\dfrac{x^4}{[(x-3)(x+2)]^2}\right)^{1/2} = \log(x+2) + \log\dfrac{x^2}{(x-3)(x+2)}$

$= \log\dfrac{x^2(x+2)}{(x-3)(x+2)} = \log\dfrac{x^2}{x-3}$

55. $\log_2 5 = \dfrac{\log 5}{\log 2} \approx 2.321928$

57. $\log_3 16 = \dfrac{\log 16}{\log 3} \approx 2.523719$

59. $\log_7 2.61 = \dfrac{\log 2.61}{\log 7} \approx 0.493008$

61. $\log_4 125 = \dfrac{\log 125}{\log 4} \approx 3.482892$

63. $\log_3 x = \dfrac{\log_e x}{\log_e 3} = \dfrac{\ln x}{\ln 3} = \dfrac{1}{\ln 3}\ln x$. The graph of $y = \dfrac{1}{\ln 3}\ln x$ is shown in the viewing rectangle $[-1,4]$ by $[-3,2]$.

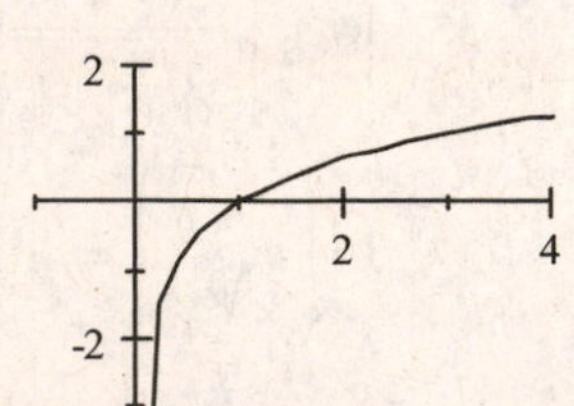

65. $\log e = \dfrac{\ln e}{\ln 10} = \dfrac{1}{\ln 10}$

67. $$-\ln\left(x - \sqrt{x^2 - 1}\right) = \ln\left(\frac{1}{x - \sqrt{x^2 - 1}}\right) = \ln\left(\frac{1}{x - \sqrt{x^2 - 1}} \cdot \frac{x + \sqrt{x^2 - 1}}{x + \sqrt{x^2 - 1}}\right)$$
$$= \ln\left(\frac{x + \sqrt{x^2 - 1}}{x^2 - \left(x^2 - 1\right)}\right)$$
$$= \ln\left(x + \sqrt{x^2 - 1}\right)$$

69. (a) $\log P = \log c - k \log W \Leftrightarrow \log P = \log c - \log W^k \Leftrightarrow \log P = \log\left(\dfrac{c}{W^k}\right)$

$\Leftrightarrow P = \dfrac{c}{W^k}$.

(b) Using $k = 2.1$ and $c = 8000$, when $W = 2$ we have $P = \dfrac{8000}{2^{2.1}} \approx 1866$ and when $W = 10$ we have $P = \dfrac{8000}{10^{2.1}} \approx 64$.

71. (a) $M = -2.5 \log\left(B / B_0\right) = -2.5 \log B + 2.5 \log B_0$.

(b) Suppose B_1 and B_2 are the brightness of two stars such that $B_1 < B_2$ and let M_1 and M_2 be their respective magnitudes. Since $\log$ is an increasing function, we have $\log B_1 < \log B_2$. Then $\log B_1 < \log B_2 \Leftrightarrow \log B_1 - \log B_0 < \log B_2 - \log B_0 \Leftrightarrow \log\left(B_1 / B_0\right) < \log\left(B_2 / B_0\right) \Leftrightarrow -2.5 \log\left(B_1 / B_0\right) > -2.5 \log\left(B_2 / B_0\right) \Leftrightarrow M_1 > M_2$. Thus the brighter star has less magnitudes.

(c) Let B_1 be the brightness of the star Albiero. Then $100B_1$ is the brightness of Betelgeuse, and its magnitude is

$$\begin{aligned} M &= -2.5\log\left(100B_1 / B_0\right) = -2.5\left[\log 100 + \log\left(B_1 / B_0\right)\right] = -2.5\left[2 + \log\left(B_1 / B_0\right)\right] \\ &= -5 - 2.5\log\left(B_1 / B_0\right) \\ &= -5 + \text{ magnitude of Albiero} \end{aligned}$$

73. The error is on the first line: $\log 0.1 < 0$, so $2 \log 0.1 < \log 0.1$.

8.5 Exponential and Logarithmic Equations

1. (a) First we isolate e^x to get the equivalent equation $e^x = 25$.

(b) Next, we take the natural logarithm of each side to get the equivalent equation $x = \ln 25$.

(c) Now we use a calculator to find $x \approx 3.219$.

3. $10^x = 25 \Leftrightarrow \log 10^x = \log 25 \Leftrightarrow x \log 10 = \log 25 \Leftrightarrow x \approx 1.398$

5. $e^{-2x} = 7 \Leftrightarrow \ln e^{-2x} = \ln 7 \Leftrightarrow -2x \ln e = \ln 7 \Leftrightarrow -2x = \ln 7 \Leftrightarrow$
$x = -\frac{1}{2}\ln 7 \approx -0.9730$

7. $2^{1-x} = 3 \Leftrightarrow \log 2^{1-x} = \log 3 \Leftrightarrow (1-x)\log 2 = \log 3 \Leftrightarrow 1 - x = \dfrac{\log 3}{\log 2}$
$\Leftrightarrow x = 1 - \dfrac{\log 3}{\log 2} \approx -0.5850$

9. $3e^x = 10 \Leftrightarrow e^x = \frac{10}{3} \Leftrightarrow x = \ln\left(\frac{10}{3}\right) \approx 1.2040$

11. $e^{1-4x} = 2 \Leftrightarrow 1 - 4x = \ln 2 \Leftrightarrow -4x = -1 + \ln 2 \Leftrightarrow$
$x = \dfrac{1 - \ln 2}{4} = 0.0767$

13. $4 + 3^{5x} = 8 \Leftrightarrow 3^{5x} = 4 \Leftrightarrow \log 3^{5x} = \log 4 \Leftrightarrow 5x \log 3 = \log 4 \Leftrightarrow$
$5x = \dfrac{\log 4}{\log 3} \Leftrightarrow x = \dfrac{\log 4}{5 \log 3} \approx 0.2524$

15. $8^{0.4x} = 5 \Leftrightarrow \log 8^{0.4x} = \log 5 \Leftrightarrow 0.4x \log 8 = \log 5 \Leftrightarrow 0.4x = \dfrac{\log 5}{\log 8} \Leftrightarrow$
$x = \dfrac{\log 5}{0.4 \log 8} \approx 1.9349$

17. $5^{-x/100} = 2 \Leftrightarrow \log 5^{-x/100} = \log 2 \Leftrightarrow -\dfrac{x}{100}\log 5 = \log 2 \Leftrightarrow$
$x = -\dfrac{100 \log 2}{\log 5} \approx -43.0677$

19. $e^{2x+1} = 200 \Leftrightarrow 2x + 1 = \ln 200 \Leftrightarrow 2x = -1 + \ln 200 \Leftrightarrow$
$x = \dfrac{-1 + \ln 200}{2} \approx 2.1492$

21. $5^x = 4^{x+1} \Leftrightarrow \log 5^x = \log 4^{x+1} \Leftrightarrow x \log 5 = (x+1)\log 4 = x \log 4 + \log 4$
$\Leftrightarrow x \log 5 - x \log 4 = \log 4 \Leftrightarrow x(\log 5 - \log 4) = \log 4 \Leftrightarrow$
$x = \dfrac{\log 4}{\log 5 - \log 4} \approx 6.2126$

23. $2^{3x+1} = 3^{x-2} \Leftrightarrow \log 2^{3x+1} = \log 3^{x-2} \Leftrightarrow (3x+1)\log 2 = (x-2)\log 3 \Leftrightarrow$ $3x\log 2 + \log 2 = x\log 3 - 2\log 3 \Leftrightarrow 3x\log 2 - x\log 3 = -\log 2 - 2\log 3 \Leftrightarrow$ $x(3\log 2 - \log 3) = -(\log 2 + 2\log 3) \Leftrightarrow s = -\dfrac{\log 2 + 2\log 3}{3\log 2 - \log 3} \approx -2.9469$

25. $\dfrac{50}{1+e^{-x}} = 4 \Leftrightarrow 50 = 4 + 4e^{-x} \Leftrightarrow 46 = 4e^{-x} \Leftrightarrow 11.5 = e^{-x} \Leftrightarrow$ $\ln 11.5 = -x \Leftrightarrow x = -\ln 11.5 \approx -2.4423$

27. $100(1.04)^{2t} = 300 \Leftrightarrow 1.04^{2t} = 3 \Leftrightarrow \log 1.04^{2t} = \log 3 \Leftrightarrow$ $2t\log 1.04 = \log 3 \Leftrightarrow t = \dfrac{\log 3}{2\log 1.04} \approx 14.0055$

29. $e^{2x} - 3e^x + 2 = 0 \Leftrightarrow (e^x - 1)(e^x - 2) = 0 \Rightarrow e^x - 1 = 0$ or $e^x - 2 = 0$. If $e^x - 1 = 0$, then $e^x = 1 \Leftrightarrow x = \ln 1 = 0$. If $e^x - 2 = 0$, then $e^x = 2 \Leftrightarrow$ $x = \ln 2 \approx 0.6931$. So the solutions are $x = 0$ and $x \approx 0.6931$.

31. $e^{4x} + 4e^{2x} - 21 = 0 \Leftrightarrow (e^{2x} + 7)(e^{2x} - 3) = 0 \Rightarrow e^{2x} = -7$ or $e^{2x} = 3$. Now $e^{2x} = -7$ has no solution, since $e^{2x} > 0$ for all x . But we can solve $e^{2x} = 3 \Leftrightarrow$ $2x = \ln 3 \Leftrightarrow x = \frac{1}{2}\ln 3 \approx 0.5493$. So the only solution is $x \approx 0.5493$.

33. $x^2 2^x - 2^x = 0 \Leftrightarrow 2^x(x^2 - 1) = 0 \Rightarrow 2^x = 0$ (never) or $x^2 - 1 = 0$. If $x^2 - 1 = 0$, then $x^2 = 1 \Rightarrow x = \pm 1$. So the only solutions are $x = \pm 1$.

35. $4x^3e^{-3x} - 3x^4e^{-3x} = 0 \Leftrightarrow x^3e^{-3x}(4 - 3x) = 0 \Rightarrow x = 0$ or $e^{-3x} = 0$ (never) or $4 - 3x = 0$. If $4 - 3x = 0$, then $3x = 4 \Leftrightarrow x = \frac{4}{3}$. So the solutions are $x = 0$ and $x = \frac{4}{3}$.

37. $\ln x = 10 \Leftrightarrow x = e^{10} \approx 22,026$

39. $\log x = -2 \Leftrightarrow x = 10^{-2} = 0.01$

41. $\log(3x + 5) = 2 \Leftrightarrow 3x + 5 = 10^2 = 100 \Leftrightarrow 3x = 95 \Leftrightarrow$ $x = \frac{95}{3} \approx 31.6667$

43. $4 - \log(3 - x) = 3 \Leftrightarrow \log(3 - x) = 1 \Leftrightarrow 3 - x = 10 \Leftrightarrow x = -7$

45. $\log_2 3 + \log_2 x = \log_2 5 + \log_2(x - 2) \Leftrightarrow \log_2(3x) = \log_2(5x - 10) \Leftrightarrow$ $3x = 5x - 10 \Leftrightarrow 2x = 10 \Leftrightarrow x = 5$

47. $\log x + \log(x - 1) = \log(4x) \Leftrightarrow \log[x(x - 1)] = \log(4x) \Leftrightarrow x^2 - x = 4x$ $\Leftrightarrow x^2 - 5x = 0 \Leftrightarrow x(x - 5) = 0 \Rightarrow x = 0$ or $x = 5$. So the possible solutions are $x = 0$ and $x = 5$. However, when $x = 0$, $\log x$ is undefined. Thus the only solution is $x = 5$.

49. $\log_5(x+1) - \log_5(x-1) = 2 \Leftrightarrow \log_5\left(\dfrac{x+1}{x-1}\right) = 2 \Leftrightarrow \dfrac{x+1}{x-1} = 5^2 \Leftrightarrow$

$x+1 = 25x - 25 \Leftrightarrow 24x = 26 \Leftrightarrow x = \frac{13}{12}$

51. $\log_2 x + \log_2(x-3) = 2 \Leftrightarrow \log_2[x(x-3)] = 2 \Leftrightarrow x^2 - 3x = 2^2 \Leftrightarrow$

$x^2 - 3x - 4 = 0 \Leftrightarrow (x-4)(x+1) \Leftrightarrow x = -1$ or $x = 4$. Since

$\log(-1-3) = \log(-4)$ is undefined, the only solution is $x = 4$.

53. $\log_9(x-5) + \log_9(x+3) = 1 \Leftrightarrow \log_9[(x-5)(x+3)] = 1 \Leftrightarrow$

$(x-5)(x+3) = 9^1 \Leftrightarrow x^2 - 2x - 24 = 0 \Leftrightarrow (x-6)(x+4) = 0 \Rightarrow$

$x = 6$ or -4. However, $x = -4$ is inadmissible, so $x = 6$ is the only solution.

55. $\log(x+3) = \log x + \log 3 \Leftrightarrow \log(x+3) = \log(3x) \Leftrightarrow x+3 = 3x \Leftrightarrow$

$2x = 3 \Leftrightarrow x = \frac{3}{2}$

57. $2^{2/\log_5 x} = \frac{1}{16} \Leftrightarrow \log_2 2^{2/\log_5 x} = \log_2\left(\frac{1}{16}\right) \Leftrightarrow \dfrac{2}{\log_5 x} = -4 \Leftrightarrow \log_5 x = -\frac{1}{2}$

$\Leftrightarrow x = 5^{-1/2} = \frac{1}{\sqrt{5}} \approx 0.4472$

59. $\ln x = 3 - x \Leftrightarrow$
$\ln x + x - 3 = 0$. Let
$f(x) = \ln x + x - 3$. We need to solve
the equation $f(x) = 0$. From the graph
of f, we get $x \approx 2.21$.

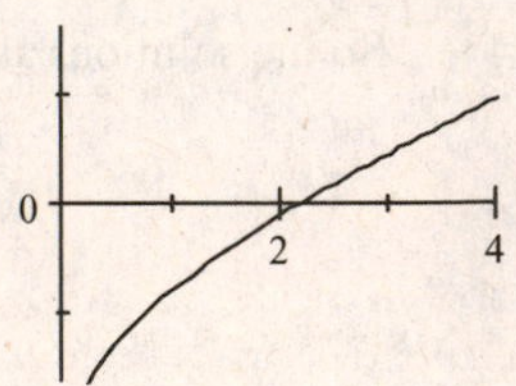

61. $x^3 - x = \log_{10}(x+1) \Leftrightarrow$
$x^3 - x - \log_{10}(x+1) = 0$. Let
$f(x) = x^3 - x - \log_{10}(x+1)$. We
need to solve the equation $f(x) = 0$.
From the graph of f, we get $x = 0$ or
$x \approx 1.14$.

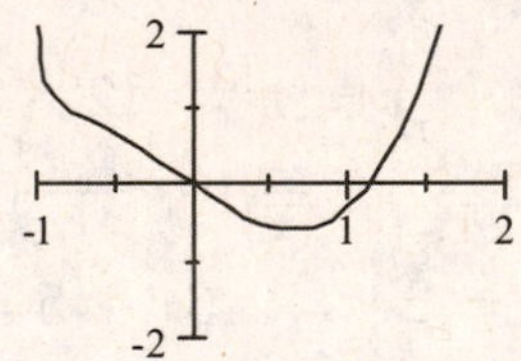

63. $e^x = -x \quad \Leftrightarrow \quad e^x + x = 0$. Let $f(x) = e^x + x$. We need to solve the equation $f(x) = 0$. From the graph of f, we get $x \approx -0.57$.

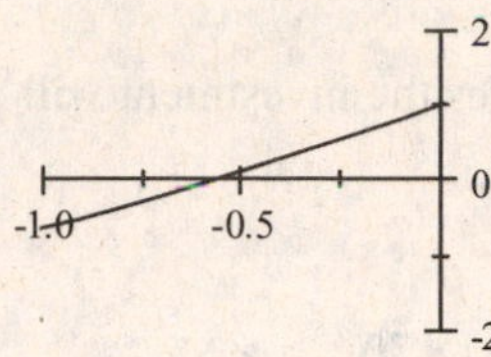

65. $4^{-x} = \sqrt{x} \quad \Leftrightarrow \quad 4^{-x} - \sqrt{x} = 0$. Let $f(x) = 4^{-x} - \sqrt{x}$. We need to solve the equation $f(x) = 0$. From the graph of f, we get $x \approx 0.36$.

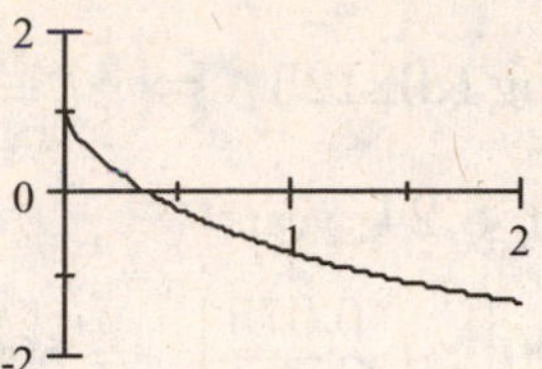

67. $\log(x-2) + \log(9-x) < 1 \quad \Leftrightarrow \quad \log[(x-2)(9-x)] < 1 \quad \Leftrightarrow$ $\log(-x^2 + 11x - 18) < 1 \quad \Rightarrow \quad -x^2 + 11x - 18 < 10^1 \quad \Leftrightarrow \quad 0 < x^2 - 11x + 28$ $\Leftrightarrow \quad 0 < (x-7)(x-4)$. Also, since the domain of a logarithm is positive we must have $0 < -x^2 + 11x - 18 \quad \Leftrightarrow \quad 0 < (x-2)(9-x)$. Using the methods from Chapter 1 with the endpoints 2, 4, 7, 9 for the intervals, we make the following table:

Interval	$(-\infty,2)$	$(2,4)$	$(4,7)$	$(7,9)$	$(9,\infty)$
Sign of $x-7$	$-$	$-$	$-$	$+$	$+$
Sign of $x-4$	$-$	$-$	$+$	$+$	$+$
Sign of $x-2$	$-$	$+$	$+$	$+$	$+$
Sign of $9-x$	$+$	$+$	$+$	$+$	$-$
Sign of $(x-7)(x-4)$	$+$	$+$	$-$	$+$	$+$
Sign of $(x-2)(9-x)$	$-$	$+$	$+$	$+$	$-$

Thus the solution is $(2,4) \cup (7,9)$.

69. $2 < 10^x < 5 \quad \Leftrightarrow \quad \log 2 < x < \log 5 \quad \Leftrightarrow \quad 0.3010 < x < 0.6990$. Hence the solution to the inequality is approximately the interval $(0.3010, 0.6990)$.

71. To find the inverse of $f(x) = 2^{2x}$, we set $y = f(x)$ and solve for x. $y = 2^{2x} \quad \Leftrightarrow$ $\ln y = \ln(2^{2x}) = 2x \ln 2 \quad \Leftrightarrow \quad x = \dfrac{\ln y}{2\ln 2}$. Interchange x and y: $y = \dfrac{\ln x}{2\ln 2}$. Thus, $f^{-1}(x) = \dfrac{\ln x}{2\ln 2}$.

73. To find the inverse of $f(x) = \log_2(x-1)$, we set $y = f(x)$ and solve for x. $y = \log_2(x-1) \quad \Leftrightarrow \quad 2^y = 2^{\log_2(x-1)} = x - 1 \quad \Leftrightarrow \quad x = 2^y + 1$. Interchange x and y: $y = 2^x + 1$. Thus, $f^{-1}(x) = 2^x + 1$.

75. (a) $A(3) = 5000\left(1 + \frac{0.085}{4}\right)^{4(3)} = 5000\left(1.02125^{12}\right) = 6435.09$. Thus the amount after 3 years is $\$6{,}435.09$.

(b) $10000 = 5000\left(1 + \frac{0.085}{4}\right)^{4t} = 5000\left(1.02125^{4t}\right) \Leftrightarrow 2 = 1.02125^{4t} \Leftrightarrow$

$\log 2 = 4t \log 1.02125 \Leftrightarrow t = \frac{\log 2}{4 \log 1.02125} \approx 8.24$ years. Thus the investment will double in about 8.24 years.

77. $8000 = 5000\left(1 + \frac{0.075}{4}\right)^{4t} = 5000\left(1.01875^{4t}\right) \Leftrightarrow 1.6 = 1.01875^{4t} \Leftrightarrow$

$\log 1.6 = 4t \log 1.01875 \Leftrightarrow t = \frac{\log 1.6}{4 \log 1.01875} \approx 6.33$ years. The investment will increase to $\$8000$ in approximately 6 years and 4 months.

79. $2 = e^{0.085t} \Leftrightarrow \ln 2 = 0.085t \Leftrightarrow t = \frac{\ln 2}{0.085} \approx 8.15$ years. Thus the investment will double in about 8.15 years.

81. $15e^{-0.087t} = 5 \Leftrightarrow e^{-0.087t} = \frac{1}{3} \Leftrightarrow -0.087t = \ln\left(\frac{1}{3}\right) = -\ln 3 \Leftrightarrow$

$t = \frac{\ln 3}{0.087} \approx 12.6277$. So only 5 grams remain after approximately 13 days.

83. (a) $P(3) = \frac{10}{1 + 4e^{-0.8(3)}} = 7.337$, so there are approximately 7337 fish after 3 years.

(b) We solve for t . $\frac{10}{1 + 4e^{-0.8t}} = 5 \Leftrightarrow 1 + 4e^{-0.8t} = \frac{10}{5} = 2 \Leftrightarrow 4e^{-0.8t} = 1$

$\Leftrightarrow e^{-0.8t} = 0.25 \Leftrightarrow -0.8t = \ln 0.25 \Leftrightarrow t = \frac{\ln 0.25}{-0.8} = 1.73$. So the population will reach 5000 fish in about 1 year and 9 months.

85. (a) $\ln\left(\frac{P}{P_0}\right) = -\frac{h}{k} \Leftrightarrow \frac{P}{P_0} = e^{-h/k} \Leftrightarrow P = P_0 e^{-h/k}$. Substituting $k = 7$ and $P_0 = 100$ we get $P = 100e^{-h/7}$.

(b) When $h = 4$ we have $P = 100e^{-4/7} \approx 56.47$ kPa.

87. (a) $I = \frac{60}{13}\left(1 - e^{-13t/5}\right) \Leftrightarrow \frac{13}{60}I = 1 - e^{-13t/5} \Leftrightarrow e^{-13t/5} = 1 - \frac{13}{60}I \Leftrightarrow$

$-\frac{13}{5}t = \ln\left(1 - \frac{13}{60}I\right) \Leftrightarrow t = -\frac{5}{13}\ln\left(1 - \frac{13}{60}I\right)$.

(b) Substituting $I = 2$, we have $t = -\frac{5}{13}\ln\left[1 - \frac{13}{60}(2)\right] \approx 0.218$ seconds.

89. Since $9^1 = 9$, $9^2 = 81$, and $9^3 = 729$, the solution of $9^x = 20$ must be between 1 and 2 (because 20 is between 9 and 81), whereas the solution to $9^x = 100$ must be between 2 and 3 (because 100 is between 81 and 729).

91. (a) $(x-1)^{\log(x-1)} = 100(x-1) \Leftrightarrow \log\left((x-1)^{\log(x-1)}\right) = \log\left(100(x-1)\right)$

$\Leftrightarrow$

$[\log(x-1)]\log(x-1) = \log 100 + \log(x-1) \Leftrightarrow$

$[\log(x-1)]^2 - \log(x-1) - 2 = 0 \Leftrightarrow [\log(x-1)-2][\log(x-1)+1] = 0$.

Thus either $\log(x-1) = 2 \Leftrightarrow x = 101$ or $\log(x-1) = -1 \Leftrightarrow x = \frac{11}{10}$.

(b) $\log_2 x + \log_4 x + \log_8 x = 11 \Leftrightarrow \log_2 x + \log_2 \sqrt{x} + \log_2 \sqrt[3]{x} = 11 \Leftrightarrow$

$\log_2\left(x\sqrt{x}\sqrt[3]{x}\right) = 11 \Leftrightarrow \log_2\left(x^{11/6}\right) = 11 \Leftrightarrow \frac{11}{6}\log_2 x = 11 \Leftrightarrow \log_2 x = 6$

$\Leftrightarrow x = 2^6 = 64$

(c) $4^x - 2^{x+1} = 3 \Leftrightarrow (2^x)^2 - 2(2^x) - 3 = 0 \Leftrightarrow (2^x - 3)(2^x + 1) = 0 \Leftrightarrow$

either $2^x = 3 \Leftrightarrow x = \dfrac{\ln 3}{\ln 2}$ or $2^x = -1$, which has no real solution. So $x = \dfrac{\ln 3}{\ln 2}$ is the only real solution.

1. (a) Here $n_0 = 10$ and $a = 1.5$ hours, so $n(t) = 10 \cdot 2^{t/1.5} = 10 \cdot 2^{2t/3}$.

(b) After 35 hours, there will be $n(35) = 10 \cdot 2^{2(35)/3} \approx 1.06 \times 10^8$ bacteria.

(c) $n(t) = 10 \cdot 2^{2t/3} = 10,000 \Leftrightarrow 2^{2t/3} = 1000 \Leftrightarrow \ln(2^{2t/3}) = \ln 1000 \Leftrightarrow$

$\frac{2t}{3}\ln 2 = \ln 1000 \Leftrightarrow t = \frac{3}{2}\frac{\ln 1000}{\ln 2} \approx 14.9$, so the bacteria count will reach $10,000$ in about 14.9 hours.

3. (a) A model for the squirrel population is $n(t) = n_0 \cdot 2^{t/6}$. We are given that $n(30) = 100,000$, so $n_0 \cdot 2^{30/6} = 100,000$ $\Leftrightarrow n_0 = \frac{100{,}000}{2^5} = 3125$. Initially, there were approximately 3125 squirrels.

(b) In 10 years, we will have $t = 40$, so the population will be $n(40) = 3125 \cdot 2^{40/6} \approx 317,480$ squirrels.

(c)

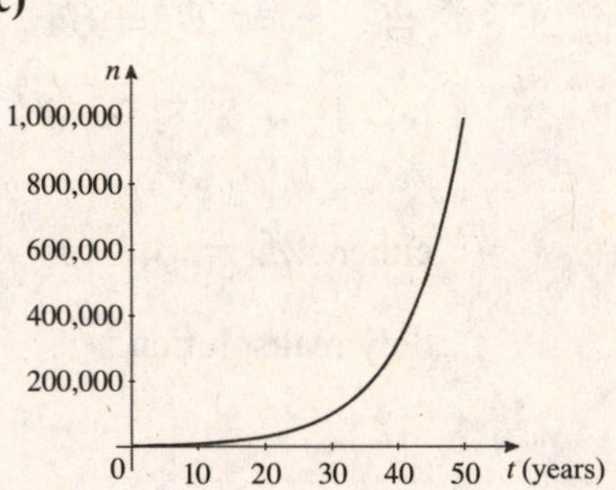

5. (a) $r = 0.08$ and $n(0) = 18,000$. Thus the population is given by the formula $n(t) = 18,000e^{0.08t}$.

(b) $t = 2013 - 2005 = 8$. Then we have $n(8) = 18,000e^{0.08(8)} = 18000e^{0.64} \approx 34,137$. Thus there should be $34,137$ foxes in the region by the year 2013 .

(c)

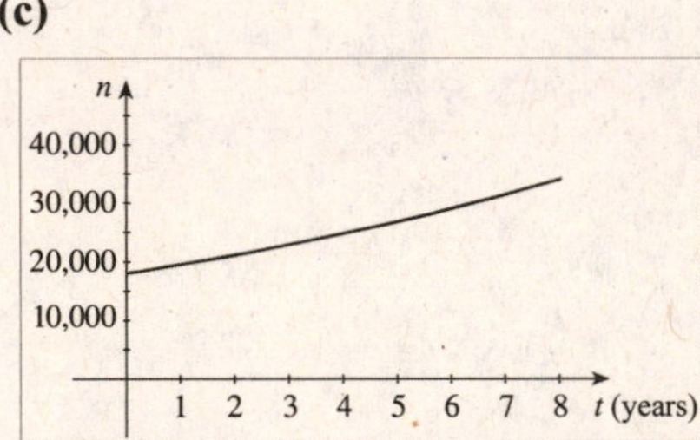

7. $n(t) = n_0 e^{rt}$; $n_0 = 110$ million, $t = 2020 - 1995 = 25$.

(a) $r = 0.03; n(25) = 110{,}000,000e^{0.03(25)} = 110{,}000{,}000e^{0.75} \approx 232,870,000$.

Thus at a 3% growth rate, the projected population will be approximately 233 million people by the year 2020 .

(b) $r = 0.02; n(25) = 110{,}000,000e^{0.02(25)} = 110{,}000{,}000e^{0.50} \approx 181,359,340$.

Thus at a 2% growth rate, the projected population will be approximately 181 million people by the year 2020 .

9. (a) The doubling time is 18 years and the initial population is $112\ ,\ 000$, so a model is $n\left(t\right) = 112\ ,\ 000 \cdot 2^{t/18}$.

(b) We need to find the relative growth rate r . Since the population is $2 \cdot 112\ ,\ 000 = 224\ ,\ 000$ when $t = 18$, we have $224\ ,\ 000 = 112\ ,\ 000e^{18r} \Leftrightarrow 2 = e^{18r} \Leftrightarrow \ln 2 = 18r \Leftrightarrow r = \frac{\ln 2}{18} \approx 0.0385$. Thus, a model is $n\left(t\right) = 112\ ,\ 000e^{0.0385t}$.

(c)

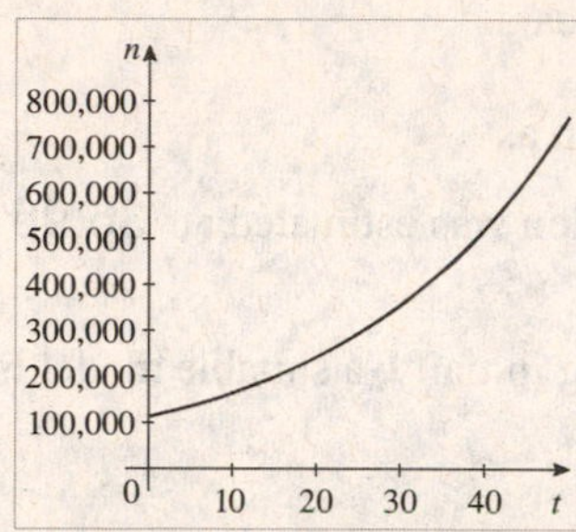

(d) Using the model in part (a), we solve the equation $n\left(t\right) = 112\ ,\ 000 \cdot 2^{t/18} = 500\ ,\ 000 \Leftrightarrow 2^{t/18} = \frac{125}{28} \Leftrightarrow \ln 2^{t/18} = \ln \frac{125}{28} \Leftrightarrow \frac{t}{18} \ln 2 = \ln \frac{125}{28} \Leftrightarrow t = \dfrac{18 \ln \frac{125}{28}}{\ln 2} \approx 38.85$. Therefore, the population should reach $500\ ,\ 000$ in the year 2045.

11. (a) The deer population in 2003 was $20{,}000$.

(b) Using the model $n\left(t\right) = 20\ ,\ 000e^{rt}$ and the point $\left(4, 31000\right)$, we have $31\ ,\ 000 = 20\ ,\ 000e^{4r} \Leftrightarrow 1.55 = e^{4r} \Leftrightarrow 4r = \ln 1.55 \Leftrightarrow r = \frac{1}{4} \ln 1.55 \approx 0.1096$. Thus $n\left(t\right) = 20\ ,\ 000e^{0.1096t}$

(c) $n\left(8\right) = 20\ ,\ 000e^{0.1096\left(8\right)} \approx 48\ ,\ 218$, so the projected deer population in 2011 is about $48\ ,\ 000$.

(d) $100\ ,\ 000 = 20\ ,\ 000e^{0.1096t} \Leftrightarrow 5 = e^{0.1096t} \Leftrightarrow 0.1096t = \ln 5 \Leftrightarrow t = \dfrac{\ln 5}{0.1096} \approx 14.63$. Since $2003 + 14.63 = 2017.63$, the deer population will reach $100{,}000$ during the year 2017.

13. (a) Using the formula $n\left(t\right) = n_0 e^{rt}$ with $n_0 = 8600$ and $n\left(1\right) = 10000$, we solve for r, giving $10000 = n\left(1\right) = 8600e^{r} \Leftrightarrow \frac{50}{43} = e^{r} \Leftrightarrow r = \ln\left(\frac{50}{43}\right) \approx 0.1508$. Thus $n\left(t\right) = 8600e^{0.1508t}$.

(b) $n\left(2\right) = 8600e^{0.1508\left(2\right)} \approx 11627$. Thus the number of bacteria after two hours is about $11\ ,\ 600$.

(c) $17200 = 8600e^{0.1508t} \Leftrightarrow 2 = e^{0.1508t} \Leftrightarrow 0.1508t = \ln 2 \Leftrightarrow t = \dfrac{\ln 2}{0.1508} \approx 4.596$. Thus the number of bacteria will double in about 4.6 hours.

15. (a) Calculating dates relative to 1990 gives $n_0 = 29.76$ and $n(10) = 33.87$. Then $n(10) = 29.76e^{10r} = 33.87 \quad\Leftrightarrow\quad e^{10r} = \frac{33.87}{29.76} \approx 1.1381 \quad\Leftrightarrow\quad 10r = \ln 1.1381$ $\Leftrightarrow\quad r = \frac{1}{10}\ln 1.1381 \approx 0.012936$. Thus $n(t) = 29.76e^{0.012936t}$ million people.

(b) $2(29.76) = 29.76e^{0.012936t} \quad\Leftrightarrow\quad 2 = e^{0.012936t} \quad\Leftrightarrow\quad \ln 2 = 0.012936t \quad\Leftrightarrow$

$t = \dfrac{\ln 2}{0.012936} \approx 53.58$, so the population doubles in about 54 years.

(c) $t = 2010 - 1990 = 20$, so our model gives the 2010 population as $n(20) \approx 29.76e^{0.012936(20)} \approx 38.55$ million. The actual population was estimated at 36.96 million in 2009.

17. (a) Because the half-life is 1600 years and the sample weighs 22 mg initially, a suitable model is $m(t) = 22 \cdot 2^{-t/1600}$.

(b) From the formula for radioactive decay, we have $m(t) = m_0 e^{-rt}$, where $m_0 = 22$ and $r = \dfrac{\ln 2}{h} = \dfrac{\ln 2}{1600} \approx 0.000433$. Thus, the amount after t years is given by $m(t) = 22e^{-0.000433t}$.

(c) $m(4000) = 22e^{-0.000433(4000)} \approx 3.89$, so the amount after 4000 years is about 4 mg.

(d) We have to solve for t in the equation $18 = 22 \quad e^{-0.000433t}$. This gives $18 = 22e^{-0.000433t} \quad\Leftrightarrow\quad \frac{9}{11} = e^{-0.000433t} \quad\Leftrightarrow\quad -0.000433t = \ln\left(\frac{9}{11}\right) \quad\Leftrightarrow$

$t = \dfrac{\ln\left(\frac{9}{11}\right)}{-0.000433} \approx 463.4$, so it takes about 463 years.

19. By the formula in the text, $m(t) = m_0 e^{-rt}$ where $r = \dfrac{\ln 2}{h}$, so $m(t) = 50e^{-[(\ln 2)/28]t}$.

We need to solve for t in the equation $32 = 50e^{-[(\ln 2)/28]t}$. This gives $e^{-[(\ln 2)/28]t} = \frac{32}{50}$

$\Leftrightarrow\quad -\dfrac{\ln 2}{28}t = \ln\left(\frac{32}{50}\right) \quad\Leftrightarrow\quad t = -\dfrac{28}{\ln 2} \cdot \ln\left(\frac{32}{50}\right) \approx 18.03$, so it takes about 18 years.

21. By the formula for radioactive decay, we have $m(t) = m_0 e^{-rt}$, where $r = \dfrac{\ln 2}{h}$, in other words $m(t) = m_0 e^{-[(\ln 2)/h]t}$. In this exercise we have to solve for h in the equation

$200 = 250e^{-[(\ln 2)/h]\cdot 48} \quad\Leftrightarrow\quad 0.8 = e^{-[(\ln 2)/h]\cdot 48} \quad\Leftrightarrow\quad \ln(0.8) = -\dfrac{\ln 2}{h} \cdot 48 \quad\Leftrightarrow$

$h = -\dfrac{\ln 2}{\ln 0.8} \cdot 48 \approx 149.1$ hours. So the half-life is approximately 149 hours.

23. By the formula in the text, $m(t) = m_0 e^{-[(\ln 2)/h]\cdot t}$, so we have $0.65 = 1 \cdot e^{-[(\ln 2)/5730]\cdot t}$ $\Leftrightarrow$ $\ln(0.65) = -\dfrac{\ln 2}{5730} t$ $\Leftrightarrow$ $t = -\dfrac{5730 \ln 0.65}{\ln 2} \approx 3561$. Thus the artifact is about 3560 years old.

25. (a) $T(0) = 65 + 145e^{-0.05(0)} = 65 + 145 = 210°$ F.

(b) $T(10) = 65 + 145e^{-0.05(10)} \approx 152.9$. Thus the temperature after 10 minutes is about $153°$ F.

(c) $100 = 65 + 145e^{-0.05t}$ $\Leftrightarrow$ $35 = 145e^{-0.05t}$ $\Leftrightarrow$ $0.2414 = e^{-0.05t}$ $\Leftrightarrow$ $\ln 0.2414 = -0.05t$ $\Leftrightarrow$ $t = -\dfrac{\ln 0.2414}{0.05} \approx 28.4$. Thus the temperature will be $100°$ F in about 28 minutes.

27. Using Newton's Law of Cooling, $T(t) = T_s + D_0 e^{-kt}$ with $T_s = 75$ and $D_0 = 185 - 75 = 110$. So $T(t) = 75 + 110e^{-kt}$.

(a) Since $T(30) = 150$, we have $T(30) = 75 + 110e^{-30k} = 150$ $\Leftrightarrow$ $110e^{-30k} = 75$ $\Leftrightarrow$ $e^{-30k} = \frac{15}{22}$ $\Leftrightarrow$ $-30k = \ln\left(\frac{15}{22}\right)$ $\Leftrightarrow$ $k = -\frac{1}{30}\ln\left(\frac{15}{22}\right)$. Thus we have $T(45) = 75 + 110e^{(45/30)\ln(15/22)} \approx 136.9$, and so the temperature of the turkey after 45 minutes is about $137°$ F.

(b) The temperature will be $100°$ F when $75 + 110e^{(t/30)\ln(15/22)} = 100$ $\Leftrightarrow$ $e^{(t/30)\ln(15/22)} = \dfrac{25}{110} = \frac{5}{22}$ $\Leftrightarrow$ $\left(\dfrac{t}{30}\right)\ln\left(\frac{15}{22}\right) = \ln\left(\frac{5}{22}\right)$ $\Leftrightarrow$ $t = 30\dfrac{\ln\left(\frac{5}{22}\right)}{\ln\left(\frac{15}{22}\right)} \approx 116.1$. So the temperature will be $100°$ F after 116 minutes.

29. (a) pH $= -\log[\mathrm{H}^+] = -\log(5.0 \times 10^{-3}) \approx 2.3$

(b) pH $= -\log[\mathrm{H}^+] = -\log(3.2 \times 10^{-4}) \approx 3.5$

(c) pH $= -\log[\mathrm{H}^+] = -\log(5.0 \times 10^{-9}) \approx 8.3$

31. (a) pH $= -\log[\mathrm{H}^+] = 3.0$ $\Leftrightarrow$ $[\mathrm{H}^+] = 10^{-3}$ M

(b) pH $= -\log[\mathrm{H}^+] = 6.5$ $\Leftrightarrow$ $[\mathrm{H}^+] = 10^{-6.5} \approx 3.2 \times 10^{-7}$ M

33. $4.0 \times 10^{-7} \le [\mathrm{H}^+] \le 1.6 \times 10^{-5}$ $\Leftrightarrow$ $\log(4.0 \times 10^{-7}) \le \log[\mathrm{H}^+] \le \log(1.6 \times 10^{-5})$ $\Leftrightarrow$ $-\log(4.0 \times 10^{-7}) \ge$ pH $\ge -\log(1.6 \times 10^{-5})$ $\Leftrightarrow$ $6.4 \ge$ pH ≥ 4.8. Therefore the range of pH readings for cheese is approximately 4.8 to 6.4.

35. Let I_0 be the intensity of the smaller earthquake and I_1 the intensity of the larger earthquake. Then $I_1 = 20I_0$. Notice that $M_0 = \log\left(\frac{I_0}{S}\right) = \log I_0 - \log S$ and

$M_1 = \log\left(\frac{I_1}{S}\right) = \log\left(\frac{20I_0}{S}\right) = \log 20 + \log I_0 - \log S$. Then

$M_1 - M_0 = \log 20 + \log I_0 - \log S - \log I_0 + \log S = \log 20 \approx 1.3$. Therefore the magnitude is 1.3 times larger.

37. Let the subscript A represent the Alaska earthquake and S represent the San Francisco earthquake. Then $M_A = \log\left(\frac{I_A}{S}\right) = 8.6 \Leftrightarrow I_A = S \cdot 10^{8.6}$; also, $M_S = \log\left(\frac{I_S}{S}\right) = 8.3 \Leftrightarrow$

$I_S = S \cdot 10^{8.6}$. So $\frac{I_A}{I_S} = \frac{S \cdot 10^{8.6}}{S \cdot 10^{8.3}} = 10^{0.3} \approx 1.995$, and hence the Alaskan earthquake was roughly twice as intense as the San Francisco earthquake.

39. Let the subscript M represent the Mexico City earthquake, and T represent the Tangshan earthquake. We have $\frac{I_T}{I_M} = 1.26 \Leftrightarrow$

$\log 1.26 = \log \frac{I_T}{I_M} = \log \frac{I_T / S}{I_M / S} = \log \frac{I_T}{S} - \log \frac{I_M}{S} = M_T - M_M$. Therefore

$M_T = M_M + \log 1.26 \approx 8.1 + 0.1 = 8.2$. Thus the magnitude of the Tangshan earthquake was roughly 8.2.

41. $\beta = 10\log\left(\frac{I}{I_0}\right) = 10\log\left(\frac{2.0 \times 10^{-5}}{1.0 \times 10^{-12}}\right) = 10\log\left(2 \times 10^7\right) = 10\left(\log 2 + \log 10^7\right)$
$= 10\left(\log 2 + 7\right) \approx 73$

Therefore the intensity level was 73 dB.

43. (a) $\beta_1 = 10 \log\left(\frac{I_1}{I_0}\right)$ and $I_1 = \frac{k}{d_1^2} \Leftrightarrow$

$\beta_1 = 10\log\left(\frac{k}{d_1^2 I_0}\right) = 10\left[\log\left(\frac{k}{I_0}\right) - 2\log d_1\right] = 10\log\left(\frac{k}{I_0}\right) - 20\log d_1$. Similarly,

$\beta_2 = 10\log\left(\frac{k}{I_0}\right) - 20\log d_2$. Substituting the expression for β_1 gives

$$\beta_2 = 10\log\left(\frac{k}{I_0}\right) - 20\log d_1 + 20\log d_1 - 20\log d_2 = \beta_1 + 20\log d_1 - 20\log d_2 = \beta_1 + 20\log\left(\frac{d_1}{d_2}\right)$$

(b) $\beta_1 = 120$, $d_1 = 2$, and $d_2 = 10$. Then

$\beta_2 = \beta_1 + 20\log\left(\frac{d_1}{d_2}\right) = 120 + 20\log\left(\frac{2}{10}\right) = 120 + 20\log 0.2 \approx 106$, and so the intensity level at 10 m is approximately 106 dB.

8.7 Modeling Harmonic Motion

1. $y = ke^{-a} \sin wt$

3. (a) $k = 2$, $c = 1.5$, and $f = 3$ $\Rightarrow$ $\omega = 6\pi$, so we have $y = 2e^{-1.5t} \cos 6\pi t$.

(b)

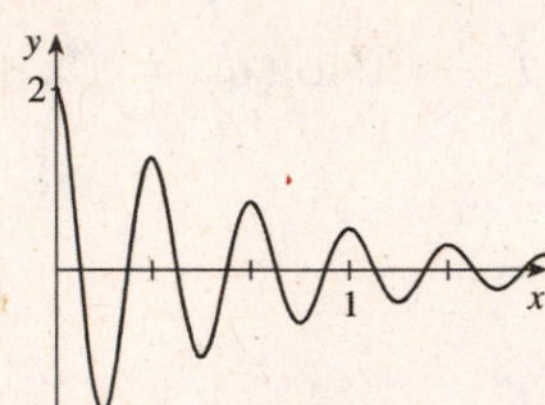

5. (a) $k = 100$, $c = 0.05$, and $p = 4$ $\Rightarrow$ $\omega = \frac{\pi}{2}$, so we have $y = 100e^{-0.05t} \cos \frac{\pi}{2} t$.

(b)

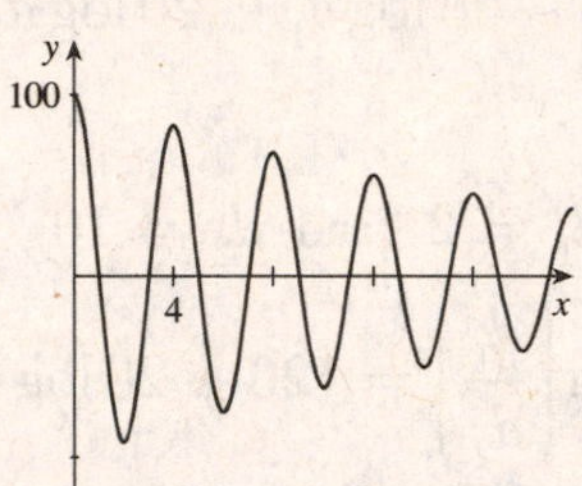

7. (a) $k = 7$, $c = 10$, and $p = \frac{\pi}{6}$ $\Rightarrow$ $\omega = 12$, so we have $y = 7e^{-10t} \sin 12t$.

(b)

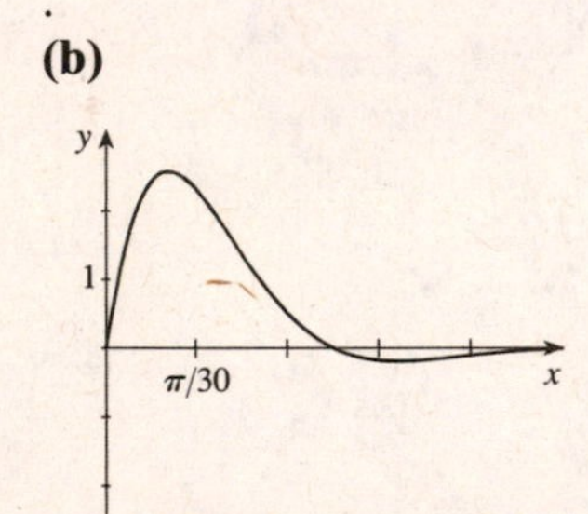

9. (a) $k = 0.3$, $c = 0.2$, and $f = 20$ $\Rightarrow$ $\omega = 40\pi$, so we have $y = 0.3e^{-0.2t} \sin 40\pi t$.

(b)

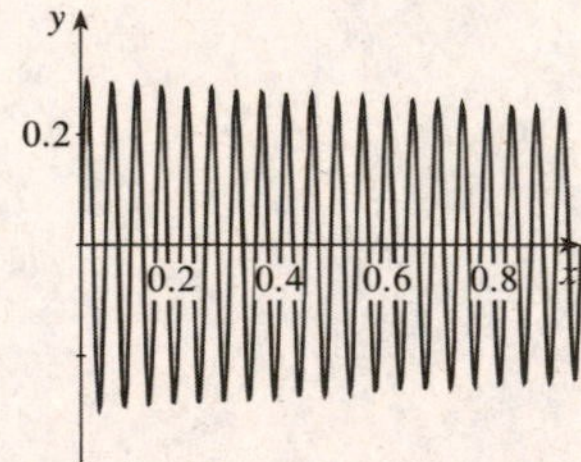

11. Since $y = 0$ and $t = 0$ we use the model $y = ke^{-ct} \sin(wt)$. From the graph the period is 2π so $y = ke^{-ct} \sin t$. To find k and c we substitute the given points $\left(\frac{\pi}{2}, 5\right)$ and $\left(\frac{9\pi}{2}, 1\right)$ into the equation and solve for k. $5 = ke^{-\frac{\pi}{2}c} \sin \frac{\pi}{2} \Leftrightarrow 5 = ke^{-\frac{\pi}{2}c} \Leftrightarrow 5e^{\frac{\pi}{2}c} = k$ and

$1 = ke^{-\frac{9\pi}{2}c} \sin \frac{9\pi}{2} \Leftrightarrow 1 = ke^{-\frac{9\pi}{2}c} \Leftrightarrow e^{\frac{9\pi}{2}c} = k$. Equating these we get

$5e^{\frac{\pi}{2}c} = e^{\frac{9\pi}{2}c} \Leftrightarrow 5 = \dfrac{e^{\frac{9\pi}{2}c}}{e^{\frac{\pi}{2}c}} = e^{4\pi c} \Leftrightarrow \ln\ 5 = 4\pi c \Leftrightarrow c = \dfrac{\ln 5}{4\pi} \approx 0.128$. Using $k = e^{\frac{9\pi}{2}c}$ we have

$k = e^{\frac{9\pi}{2}(0.128)} \approx 6.114$. Hence the model is $y = 6.114e^{-0.128t} \sin t$.

13. Since $y \neq 0$ when $t = 0$ we use the model $y = ke^{-ct}\cos(wt)$. Since $y = k$ when $t = 0$ we get $k = 100$. From the graph the period is 8 so $w = \dfrac{2\pi}{8} = \dfrac{\pi}{4}$ and the model is $y = 100e^{-ct}\cos\left(\dfrac{\pi}{4}t\right)$.

To find c we substitute the given point (8, 40) into the equation and solve. $40 = 100e^{-8c}\cos\left(\dfrac{\pi}{4}\cdot 8\right) = 100e^{-8c} \Leftrightarrow 0.4 = e^{-8c} \Leftrightarrow -8c = \ln 0.4 \Leftrightarrow c = \dfrac{\ln 0.4}{-8} \approx 0.115$. Thus the model is $y = 100e^{-0.115t}\cos\left(\dfrac{\pi}{4}t\right)$

15. $k = 1$, $c = 0.9$, and $\dfrac{\omega}{2\pi} = \frac{1}{2} \Leftrightarrow \omega = \pi$. Since $f(0) = 0$, $f(t) = e^{-0.9t}\sin \pi t$.

17. $\dfrac{ke^{-ct}}{ke^{-c(t+3)}} = 4 \Leftrightarrow$

$e^{-ct+c(t+3)} = 4 \Leftrightarrow e^{3c} = 4 \Leftrightarrow 3c = \ln 4 \Leftrightarrow c = \frac{1}{3}\ln 4 \approx 0.46$.

8 Review

1. $f(x) = 5^x$; $f(-1.5) \approx 0.0894$, $f(\sqrt{2}) \approx 9.739$, $f(2.5) \approx 55.902$

3. $g(x) = 4 \cdot \left(\frac{2}{3}\right)^{x-2}$; $g(-0.7) \approx 11.954$, $g(e) \approx 2.989$, $g(\pi) \approx 2.518$

5. $f(x) = 2^{-x+1}$. Domain $(-\infty, \infty)$, range $(0, \infty)$, asymptote $y = 0$.

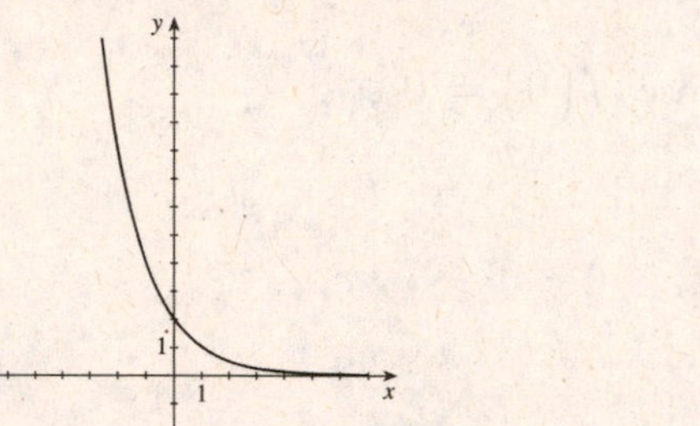

7. $g(x) = 3 + 2^x$. Domain $(-\infty, \infty)$, range $(3, \infty)$, asymptote $y = 3$.

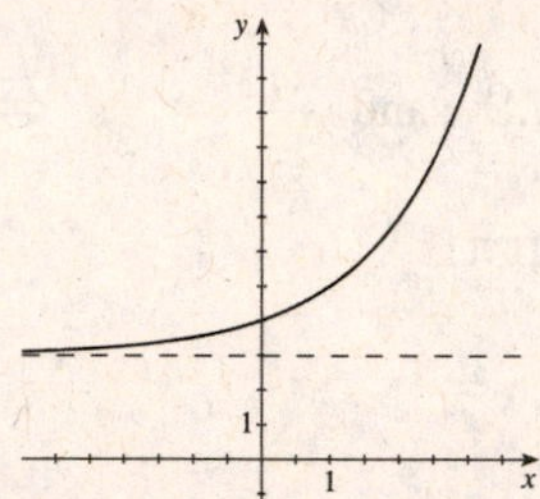

9. $f(x) = \log_3(x-1)$. Domain $(1, \infty)$, range $(-\infty, \infty)$, asymptote $x = 1$.

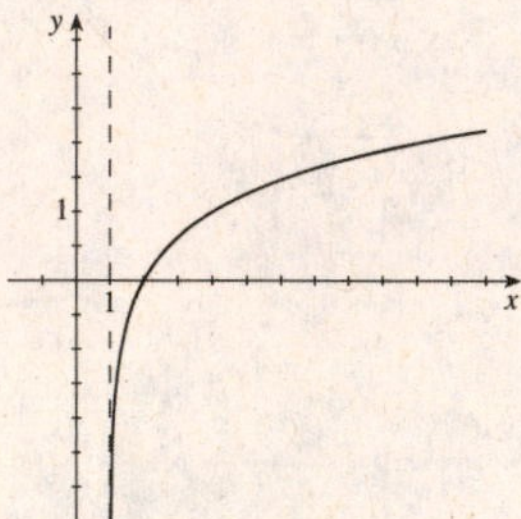

11. $f(x) = 2 - \log_2 x$. Domain $(0, \infty)$, range $(-\infty, \infty)$, asymptote $x = 0$.

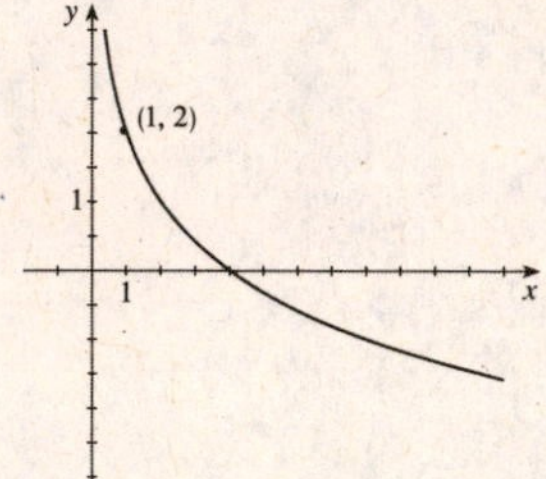

13. $F(x) = e^x - 1$. Domain $(-\infty, \infty)$, range $(-1, \infty)$, asymptote $y = -1$.

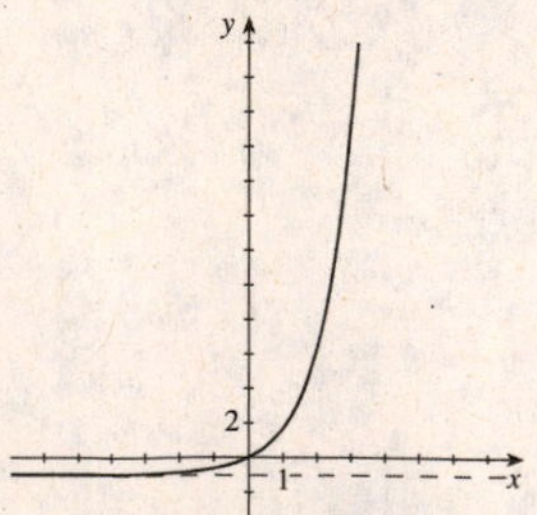

15. $g(x) = 2 \ln x$. Domain $(0, \infty)$, range $(-\infty, \infty)$, asymptote $x = 0$.

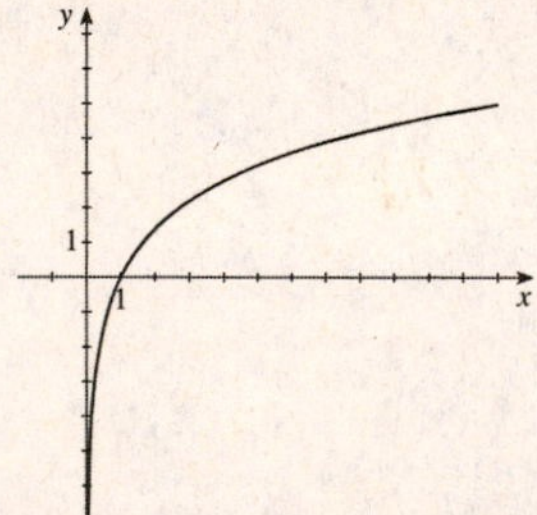

17. $f(x) = 10^{x^2} + \log(1 - 2x)$. Since $\log u$ is defined only for $u > 0$, we require $1 - 2x > 0 \Leftrightarrow -2x > -1 \Leftrightarrow x < \frac{1}{2}$, and so the domain is $\left(-\infty, \frac{1}{2}\right)$.

19. $h(x)=\ln(x^2-4)$. We must have $x^2-4>0$ (since $\ln y$ is defined only for $y>0$)

$\Leftrightarrow \quad x^2-4>0 \quad \Leftrightarrow \quad (x-2)(x+2)>0$. The endpoints of the intervals are -2 and 2.

Interval	$(-\infty,-2)$	$(-2,2)$	$(2,\infty)$
Sign of $x-2$	$-$	$-$	$+$
Sign of $x+2$	$-$	$+$	$+$
Sign of $(x-2)(x+2)$	$+$	$-$	$+$

Thus the domain is $(-\infty,-2)\cup(2,\infty)$.

21. $\log_2 1024=10 \quad \Leftrightarrow \quad 2^{10}=1024$

23. $\log x=y \quad \Leftrightarrow \quad 10^y=x$

25. $2^6=64 \quad \Leftrightarrow \quad \log_2 64=6$

27. $10^x=74 \Leftrightarrow \log_{10}74=x$

$\Leftrightarrow \log 74=x$

29. $\log_2 128=\log_2\left(2^7\right)=7$

31. $10^{\log 45}=45$

33. $\ln\left(e^6\right)=6$

35. $\log_3\frac{1}{27}=\log_3 3^{-3}=-3$

37. $\log_5\sqrt{5}=\log_5 5^{1/2}=\frac{1}{2}$

39.

$\log 25+\log 4=\log(25\cdot 4)=\log 10^2=2$

41. $\log_2\left(16^{23}\right)=\log_2\left(2^4\right)^{23}=\log_2 2^{92}=92$

43. $\log_8 6-\log_8 3+\log_8 2=\log_8\left(\frac{6}{3}\cdot 2\right)=\log_8 4=\log_8 8^{2/3}=\frac{2}{3}$

45. $\log\left(AB^2C^3\right)=\log A+2\log B+3\log C$

47. $\ln\sqrt{\dfrac{x^2-1}{x^2+1}}=\frac{1}{2}\ln\left(\dfrac{x^2-1}{x^2+1}\right)=\frac{1}{2}\left[\ln\left(x^2-1\right)-\ln\left(x^2+1\right)\right]$

49. $\log_5\left(\dfrac{x^2(1-5x)^{3/2}}{\sqrt{x^3-x}}\right)=\log_5 x^2(1-5x)^{3/2}-\log_5\sqrt{x\left(x^2-1\right)}$

$$=2\log_5 x+\frac{3}{2}\log_5(1-5x)-\frac{1}{2}\log_5\left(x^3-x\right)$$

51. $\log 6+4\log 2=\log 6+\log 2^4=\log\left(6\cdot 2^4\right)=\log 96$

53. $\frac{3}{2}\ \log_2(x-y)-2$

$$\log_2\left(x^2+y^2\right)=\log_2(x-y)^{3/2}-\log_2\left(x^2+y^2\right)^2=\log_2\left(\frac{(x-y)^{3/2}}{\left(x^2+y^2\right)^2}\right)$$

55. $\log(x-2)+\log(x+2)-\frac{1}{2}$

$$\log(x^2+4)=\log[(x-2)(x+2)]-\log\sqrt{x^2+4}=\log\left(\frac{x^2-4}{\sqrt{x^2+4}}\right)$$

57. $3^{2x-7}=27 \Leftrightarrow 3^{2x-7}=3^3 \Leftrightarrow 2x-7=3 \Leftrightarrow 2x=10 \Leftrightarrow x=5$

59. $2^{3x-5}=7 \Leftrightarrow \log_2(2^{3x-5})=\log_2 7 \Leftrightarrow 3x-5=\log_2 7 \Leftrightarrow$

$x=\frac{1}{3}(\log_2 7+5)$. Using the Change of Base Formula, we have $\log_2 7=\frac{\log 7}{\log 2}\approx 2.807$, so

$x\approx\frac{1}{3}(2.807+5)\approx 2.602$.

61. $4^{1-x}=3^{2x+5} \Leftrightarrow \log 4^{1-x}=\log 3^{2x+5} \Leftrightarrow (1-x)\log 4=(2x+5)\log 3 \Leftrightarrow$

$\log 4-5\log 3=2x\log 3+x\log 4 \Leftrightarrow x(\log 3+\log 4)=\log 4-5\log 3 \Leftrightarrow$

$$x=\frac{\log 4-5\log 3}{2\log 3+\log 4}\approx -1.146$$

63. $x^2e^{2x}+2xe^{2x}=8e^{2x} \Leftrightarrow e^{2x}(x^2+2x-8)=0 \Leftrightarrow x^2+2x-8=0$ (since $e^{2x}\neq 0$) $\Leftrightarrow (x+4)(x-2)=0 \Leftrightarrow x=-4$ or $x=2$

65. $\log_2(1-x)=4 \Leftrightarrow 1-x=2^4 \Leftrightarrow x=1-16=-15$

67. $\log_8(x+5)-\log_8(x-2)=1 \Leftrightarrow \log_8\frac{x+5}{x-2}=1 \Leftrightarrow \frac{x+5}{x-2}=8 \Leftrightarrow$

$x+5=8(x-2) \Leftrightarrow x+5=8x-16 \Leftrightarrow 7x=21 \Leftrightarrow x=3$

69. $5^{-2x/3}=0.63 \Leftrightarrow \frac{-2x}{3}\log 5=\log 0.63 \Leftrightarrow x=-\frac{3\log 0.63}{2\log 5}\approx 0.430618$

71. $5^{2x+1}=3^{4x-1} \Leftrightarrow (2x+1)\log 5=(4x-1)\log 3 \Leftrightarrow$

$2x\log 5+\log 5=4x\log 3-\log 3 \Leftrightarrow x(2\log 5-4\log 3)=-\log 3-\log 5 \Leftrightarrow$

$$x=\frac{\log 3+\log 5}{4\log 3-2\log 5}\approx 2.303600$$

73. $y=e^{x/(x+2)}$. Vertical asymptote $x=-2$, horizontal asymptote $y=2.72$, no maximum or minimum.

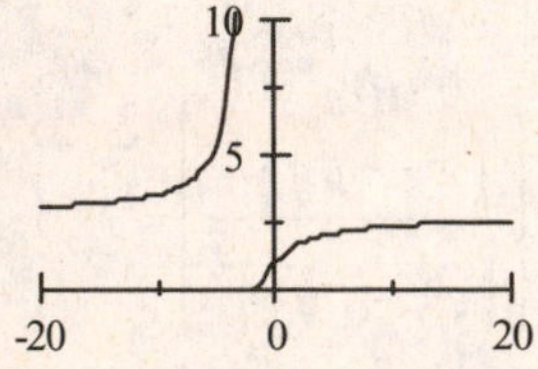

75. $y=\log(x^3-x)$. Vertical asymptotes $x=-1$, $x=0$, $x=1$, no horizontal asymptote, local maximum of about -0.41 when $x\approx -0.58$.

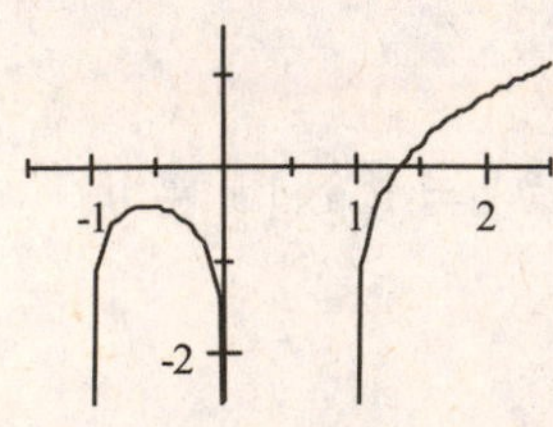

77. $3\log x = 6 - 2x$. We graph $y = 3\log x$ and $y = 6 - 2x$ in the same viewing rectangle. The solution occurs where the two graphs intersect. From the graphs, we see that the solution is $x \approx 2.42$.

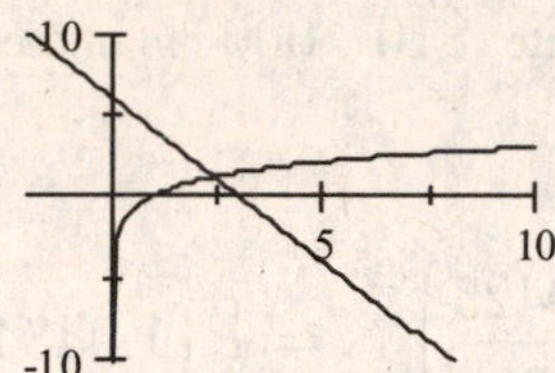

79. $\ln x > x - 2$. We graph the function $f(x) = \ln x - x + 2$, and we see that the graph lies above the x -axis for $0.16 < x < 3.15$. So the approximate solution of the given inequality is $0.16 < x < 3.15$.

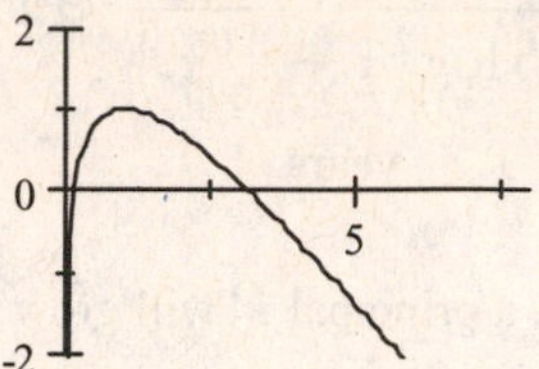

81. $f(x) = e^x - 3e^{-x} - 4x$. We graph the function $f(x)$, and we see that the function is increasing on $(-\infty,0]$ and $[1.10,\infty)$ and that it is decreasing on $[0,1.10]$.

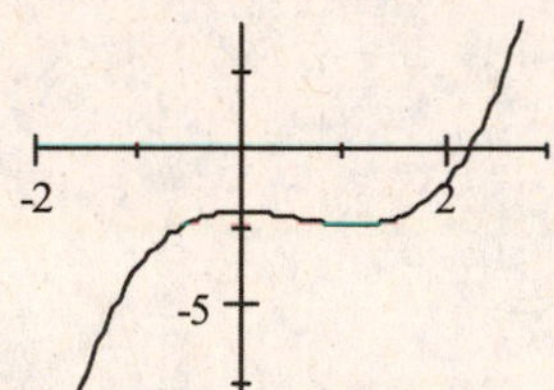

83. $\log_4 15 = \dfrac{\log 15}{\log 4} = 1.953445$

85. $\log_9 0.28 = \dfrac{\log 0.28}{\log 9} \approx -0.579352$

86. $\log_{100} 250 = \dfrac{\log 250}{\log 100} \approx 1.198970$

87. Notice that $\log_4 258 > \log_4 256 = \log_4 4^4 = 4$ and so $\log_4 258 > 4$. Also $\log_5 620 < \log_5 625 = \log_5 5^4 = 4$ and so $\log_5 620 < 4$. Then $\log_4 258 > 4 > \log_5 620$ and so $\log_4 258$ is larger.

89. $P = 12{,}000$, $r = 0.10$, and $t = 3$. Then $A = P\left(1+\dfrac{r}{n}\right)^{nt}$.

(a) For $n = 2$, $A = 12{,}000\left(1+\frac{0.10}{2}\right)^{2(3)} = 12{,}000\left(1.05^6\right) \approx \$16,081.15$.

(b) For $n = 12$, $A = 12{,}000\left(1+\frac{0.10}{12}\right)^{12(3)} \approx \$16,178.18$.

(c) For $n = 365$, $A = 12{,}000\left(1+\frac{0.10}{365}\right)^{365(3)} \approx \$16,197.64$.

(d) For $n = \infty$, $A = Pe^{rt} = 12,000e^{0.10(3)} \approx \$16,198.31$.

91. We use the formula $A = P\left(1+\frac{r}{n}\right)^{nt}$ with $P = 100\ ,\ 000$, $r = 0.052$, $n = 365$, and $A = 100\ ,\ 000 + 10\ ,\ 000 = 110\ ,\ 000$, and solve for t : $110\ ,\ 000 = 100\ ,\ 000\left(1+\frac{0.052}{365}\right)^{365t} \Leftrightarrow 1.1 = \left(1+\frac{0.052}{365}\right)^{365t} \Leftrightarrow \log 1.1 = 365t \log\left(1+\frac{0.052}{365}\right)$

$\Leftrightarrow t = \dfrac{\log 1.1}{365 \log\left(1+\frac{0.052}{365}\right)} \approx 1.833$. The account will accumulate $\$10\ ,\ 000$ in interest in approximately 1.8 years.

93. After one year, a principal P will grow to the amount $A = P\left(1+\dfrac{0.0425}{365}\right)^{365} = P\left(1.04341\right)$.

The formula for simple interest is $A = P\left(1+r\right)$. Comparing, we see that $1+r = 1.04341$, so $r = 0.04341$. Thus the annual percentage yield is 4.341% .

95. (a) Using the model $n\left(t\right) = n_0 e^{rt}$, with $n_0 = 30$ and $r = 0.15$, we have the formula $n\left(t\right) = 30e^{0.15t}$.

(b) $n\left(4\right) = 30e^{0.15(4)} \approx 55$.

(c) $500 = 30e^{0.15t} \Leftrightarrow \frac{50}{3} = e^{0.15t} \Leftrightarrow 0.15t = \ln\left(\frac{50}{3}\right) \Leftrightarrow$

$t = \dfrac{1}{0.15}\ln\left(\frac{50}{3}\right) \approx 18.76$. So the stray cat population will reach 500 in about 19 years.

97. (a) From the formula for radioactive decay, we have $m\left(t\right) = 10e^{-rt}$, where $r = -\dfrac{\ln 2}{2.7 \times 10^5}$.

So after 1000 years the amount remaining is

$m\left(1000\right) = 10 \cdot e^{\left[-\ln 2/\left(2.7\times 10^5\right)\right]\cdot 1000} = 10e^{-(\ln 2)/\left(2.7\times 10^2\right)} = 10e^{-(\ln 2)/270} \approx 9.97$.

Therefore the amount remaining is about 9.97 mg.

(b) We solve for t in the equation $7 = 10e^{-\left[\ln 2/\left(2.7\times 10^5\right)\right]\cdot t}$. We have

$7 = 10e^{-\left[\ln 2/\left(2.7\times 10^5\right)\right]\cdot t} \Leftrightarrow 0.7 = e^{-\left[\ln 2/\left(2.7\times 10^5\right)\right]\cdot t} \Leftrightarrow \ln 0.7 = -\dfrac{\ln 2}{2.7 \times 10^5} \cdot t$

$\Leftrightarrow t = -\dfrac{\ln 0.7}{\ln 2} \cdot 2.7 \times 10^5 \approx 138\ ,\ 934.75$. Thus it takes about $139\ ,\ 000$ years.

99. (a) From the formula for radioactive decay, $r = \frac{\ln 2}{1590} \approx 0.0004359$ and $n(t) = 150 \cdot e^{-0.0004359t}$.

(b) $n(1000) = 150 \cdot e^{-0.0004359 \cdot 1000} \approx 97.00$, and so the amount remaining is about 97.00 mg.

(c) Find t so that $50 = 150 \cdot e^{-0.0004359t}$. We have $50 = 150 \cdot e^{-0.0004359t}$ $\Leftrightarrow$ $\frac{1}{3} = e^{-0.0004359t}$ $\Leftrightarrow$ $t = -\frac{1}{0.0004359} \ln\left(\frac{1}{3}\right) \approx 2520$. Thus only 50 mg remain after about 2520 years.

101. (a) Using $n_0 = 1500$ and $n(5) = 3200$ in the formula $n(t) = n_0 e^{rt}$, we have $3200 = n(5) = 1500e^{5r}$ $\Leftrightarrow$ $e^{5r} = \frac{32}{15}$ $\Leftrightarrow$ $5r = \ln\left(\frac{32}{15}\right)$ $\Leftrightarrow$ $r = \frac{1}{5}\ln\left(\frac{32}{15}\right) \approx 0.1515$. Thus $n(t) = 1500 \cdot e^{0.1515t}$.

(b) We have $t = 1999 - 1988 = 11$ so $n(11) = 1500e^{0.1515 \cdot 11} \approx 7940$. Thus in 1999 the bird population should be about 7940.

103. $\left[\mathrm{H}^+\right] = 1.3 \times 10^{-8}$ M. Then pH $= -\log\left[\mathrm{H}^+\right] = -\log\left(1.3 \times 10^{-8}\right) \approx 7.9$, and so fresh egg whites are basic.

105. Let I_0 be the intensity of the smaller earthquake and I_1 be the intensity of the larger earthquake. Then $I_1 = 35I_0$. Since $M = \log\left(\frac{I}{S}\right)$, we have $M_0 = \log\left(\frac{I_0}{S}\right) = 6.5$ and

$$M_1 = \log\left(\frac{I_1}{S}\right) = \log\left(\frac{35I_0}{S}\right) = \log 35 + \log\left(\frac{I_0}{S}\right) = \log 35 + M_0 = \log 35 + 6.5 \approx 8.04$$

. So the magnitude on the Richter scale of the larger earthquake is approximately 8.0.

8 Test

1. (a)

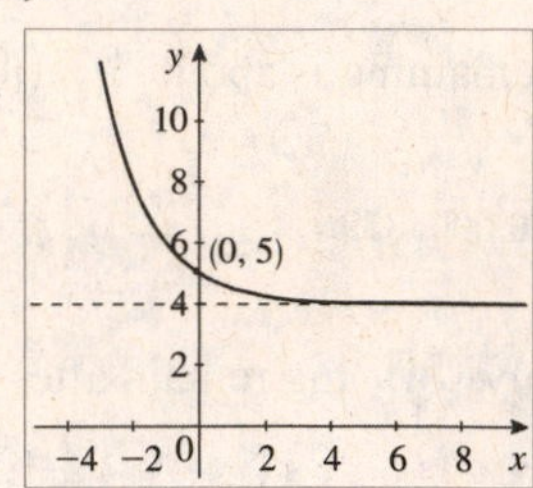

$f(x) = 2^{-x} + 4$ has domain $(-\infty, \infty)$, range $(4, \infty)$, and horizontal asymptote $y = 4$.

(b)

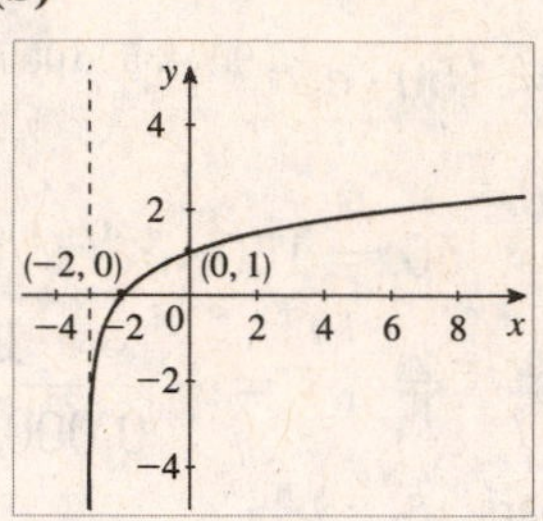

has domain $(-3, \infty)$, range $(-\infty, \infty)$, and vertical asymptote $x = -3$.

2. (a) $6^{2x} = 25 \Leftrightarrow \log_6 6^{2x} = \log_6 25 \Leftrightarrow 2x = \log_6 25$

(b) $\ln A = 3 \Leftrightarrow e^{\ln A} = e^3 \Leftrightarrow A = e^3$

3. (a) $10^{\log 36} = 36$ **(b)** $\ln e^3 = 3$ **(c)** $\log_3 \sqrt{27} = \log_3 \left(3^3\right)^{1/2} = \log_3 3^{3/2} = \frac{3}{2}$

(d) $\log_2 80 - \log_2 10 = \log_2 \left(\frac{80}{10}\right) = \log_2 8 = \log_2 2^3 = 3$

(e) $\log_8 4 = \log_8 8^{2/3} = \frac{2}{3}$ **(f)** $\log_6 4 + \log_6 9 = \log_6 (4 \cdot 9) = \log_6 6^2 = 2$

4. $\log \sqrt[3]{\dfrac{x+2}{x^4\left(x^2+4\right)}} = \dfrac{1}{3}\log\left(\dfrac{x+2}{x^4\left(x^2+4\right)}\right) = \frac{1}{3}\left[\log(x+2) - \left(4\log x + \log\left(x^2+4\right)\right)\right]$

$= \frac{1}{3}\log(x+2) - \frac{4}{3}\log x - \frac{1}{3}\log\left(x^2+4\right)$

5. $\ln x - 2\ln\left(x^2+1\right) + \frac{1}{2}\ln\left(3-x^4\right) = \ln\left(x\sqrt{3-x^4}\right) - \ln\left(x^2+1\right)^2 = \ln\left(\dfrac{x\sqrt{3-x^4}}{\left(x^2+1\right)^2}\right)$

6. (a) $2^{x-1} = 10 \Leftrightarrow \log 2^{x-1} = \log 10 = 1 \Leftrightarrow (x-1)\log 2 = 1 \Leftrightarrow$

$x - 1 = \dfrac{1}{\log 2} \Leftrightarrow x = 1 + \dfrac{1}{\log 2} \approx 4.32$

(b) $5\ln(3-x) = 4 \Leftrightarrow \ln(3-x) = \frac{4}{5} \Leftrightarrow e^{\ln(3-x)} = e^{4/5} \Leftrightarrow 3 - x = e^{4/5}$

$\Leftrightarrow x = 3 - e^{4/5} \approx 0.77$

(c) $10^{x+3} = 6^{2x} \Leftrightarrow \log 10^{x+3} = \log 6^{2x} \Leftrightarrow x + 3 = 2x\log 6 \Leftrightarrow$

$2x\log 6 - x = 3 \Leftrightarrow x(2\log 6 - 1) = 3 \Leftrightarrow x = \dfrac{3}{2\log 6 - 1} \approx 5.39$

(d) $\log_2(x+2) + \log_2(x-1) = 2 \Leftrightarrow \log_2((x+2)(x-1)) = 2 \Leftrightarrow$

$x^2 + x - 2 = 2^2 \Leftrightarrow x^2 + x - 6 = 0 \Leftrightarrow (x+3)(x-2) = 0 \Rightarrow x = -3$ or $x = 2$. However, both logarithms are undefined at $x = -3$, so the only solution is $x = 2$.

7. (a) From the formula for population growth, we have $8000 = 1000e^{r\cdot 1} \Leftrightarrow 8 = e^r \Leftrightarrow r = \ln 8 \approx 2.07944$. Thus $n(t) = 1000e^{2.07944t}$.

(b) $n(1.5) = 1000e^{2.07944(1.5)} \approx 22,627$

(c) $15000 = 1000e^{2.07944t} \Leftrightarrow 15 = e^{2.07944t} \Leftrightarrow \ln 15 = 2.07944t \Leftrightarrow t = \dfrac{\ln 15}{2.07944} \approx 1.3$. Thus the population will reach $15{,}000$ after approximately 1.3 hours.

(d)

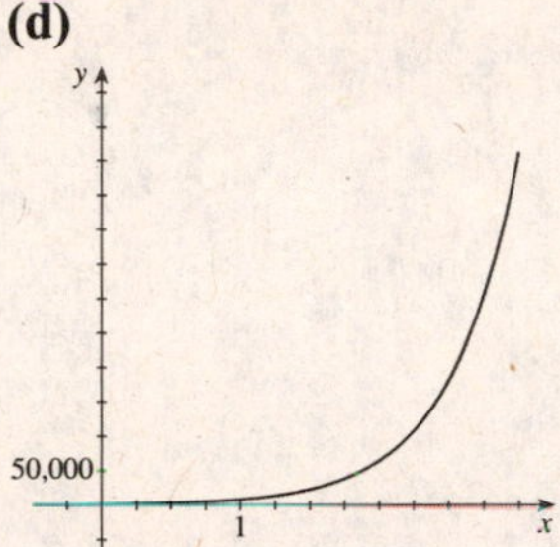

8. (a) $A(t) = 12{,}000\left(1 + \frac{0.056}{12}\right)^{12t}$, where t is in years.

(b) $A(t) = 12{,}000\left(1 + \frac{0.056}{365}\right)^{365t}$. So $A(3) = 12{,}000\left(1 + \dfrac{0.056}{365}\right)^{365(3)} = \$14{,}195.06$

(c) $A(t) = 12{,}000\left(1 + \frac{0.056}{2}\right)^{2t} = 12{,}000(1.028)^{2t}$. So $20{,}000 = 12,000(1.028)^{2t}$ $\Leftrightarrow 1.6667 = 1.028^{2t} \Leftrightarrow \ln 1.6667 = \ln 1.028^{2t} \Leftrightarrow \ln 1.6667 = 2t \ln 1.028$ $\Leftrightarrow t = \frac{\ln 1.6667}{2 \ln 1.028} \approx 9.25$ years.

9. (a) The initial mass is $m_0 = 3$ and the half-life is $h = 10$, so using the radioactive decay model with $m_0 = 3$ and $r = \frac{\ln 2}{h} = \frac{\ln 2}{10}$, we have $A(t) = 3e^{-[(\ln 2)/10]t} \approx 3e^{-0.069t}$.

(b) After 1 minute $= 60$ seconds, the amount remaining is $A(60) = 3e^{-0.069(60)} \approx 0.048$ g.

(c) We solve $3e^{-0.069t} = 10^{-6} \Leftrightarrow e^{-0.069t} = \dfrac{10^{-6}}{3} \Leftrightarrow \ln e^{-0.069t} = \ln\left(\dfrac{10^{-6}}{3}\right) \Leftrightarrow$ $-0.069t = \ln\left(\dfrac{10^{-6}}{3}\right) \Leftrightarrow t = -\dfrac{1}{0.069}\ln\left(\dfrac{10^{-6}}{3}\right) \approx 216$, so there is 1 μ g of ^{91} Kr remaining after about 216 seconds, or 3.6 minutes.

10. Let the subscripts J and P represent the two earthquakes. Then we have

$M_J = \log\left(\dfrac{I_J}{S}\right) = 6.4 \Leftrightarrow 10^{6.4} = \dfrac{I_J}{S} \Leftrightarrow 10^{6.4}S = I_J$. Similarly,

$M_P = \log\left(\dfrac{I_P}{S}\right) = 3.1 \Leftrightarrow 10^{3.1} = \dfrac{I_P}{S} \Leftrightarrow 10^{3.1}S = I_P$. So

$\dfrac{I_J}{I_P} = \dfrac{10^{6.4}S}{10^{3.1}S} = 10^{3.3} \approx 1995.3$, and so the Japan earthquake was about 1995 times more intense than the Pennsylvania earthquake.

14. (a) The initial amplitude is 16 in and the frequency is 12 Hz, so a function describing the motion is $y = 16e^{-0.1t}\cos 24\pi t$.

(b)

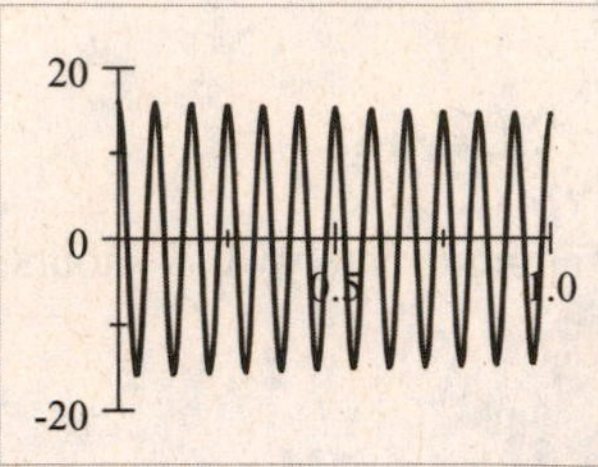

1. (a)

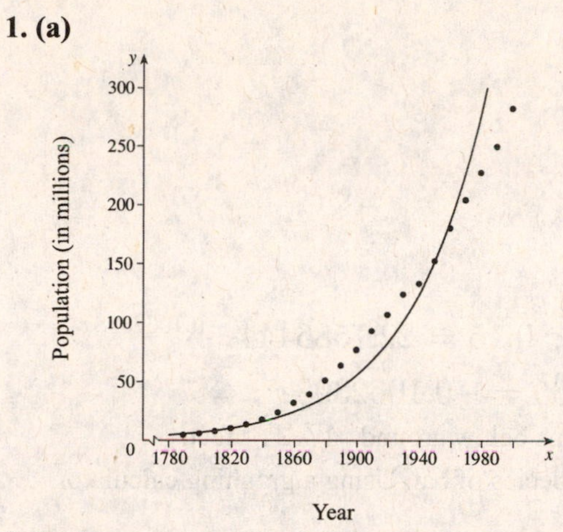

(b) Using a graphing calculator, we obtain the model $y = ab^t$, where $a = 1.1806094 \times 10^{-15}$ and $b = 1.0204139$, and y is the population (in millions) in the year t. Substituting $t = 2010$ into the model of part (b), we get $y = ab^{2010} \approx 515.9$ million. According to the model, the population in 1965 should have been about $y = ab^{1965} \approx 207.8$ million. The values given by the model are clearly much too large. This means that an exponential model is *not* appropriate for these data.

3. (a) Yes.

(b)

Year t	Health Expenditures E ($ bn)	$\ln E$
1970	74.3	4.30811
1980	251.1	5.52585
1985	434.5	6.07420
1987	506.2	6.22693
1990	696.6	6.54621
1992	820.3	6.70967
1994	937.2	6.84290
1996	1039.4	6.94640
1998	1150.0	7.04752
2000	1310.0	7.17778
2001	1424.5	7.26158

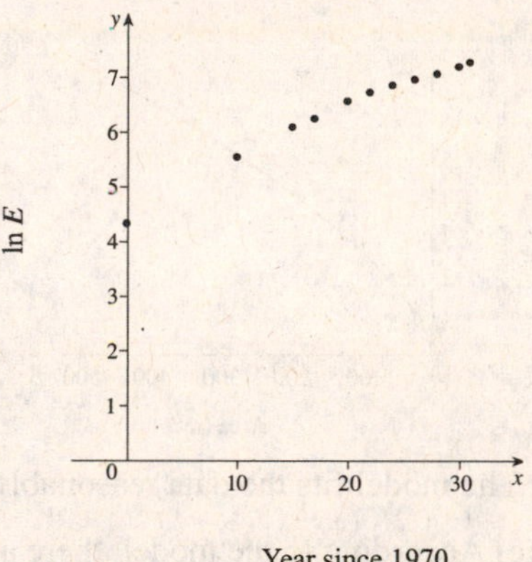

Yes, the scatter plot appears to be roughly linear.

(c) Let t be the number of years elapsed since 1970. Then $\ln E = 4.551437 + 0.09238268t$, where E is expenditure in billions of dollars.

(d) $E = e^{4.551437+0.09238268t} = 94.76849e^{0.09238268t}$

(e) In 2009 we have $t = 2009 - 1970 = 39$, so the estimated 2009 health-care expenditures are $94.76849e^{0.09238268(39)} \approx 3478.5$ billion dollars.

5. (a) Using a graphing calculator, we find that $I_0 = 22.7586444$ and $k = 0.1062398$.

(b)

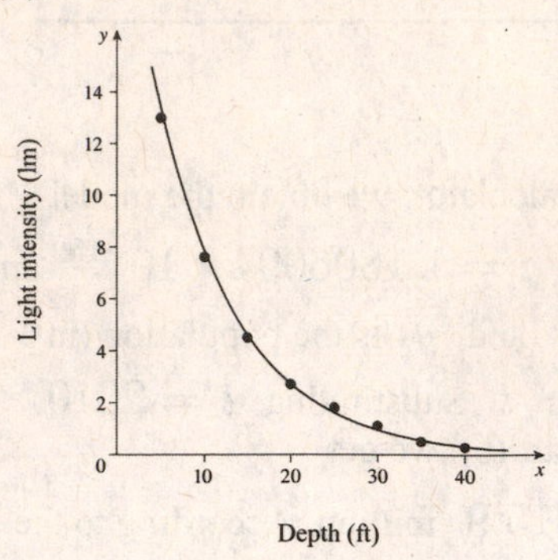

(c) We solve $0.15 = 22.7586444e^{-0.1062398x}$ for x : $0.15 = 22.7586444e^{-0.1062398x}$ $\Leftrightarrow$ $0.006590902 = e^{-0.1062398x}$ $\Leftrightarrow$ $-5.022065 = -0.1062398x$ $\Leftrightarrow$ $x \approx 47.27$. So light intensity drops below 0.15 lumens below around 47.27 feet.

7. (a) Let A be the area of the cave and S the number of species of bat. Using a graphing calculator, we obtain the power function model $S = 0.14A^{0.64}$.

(b)

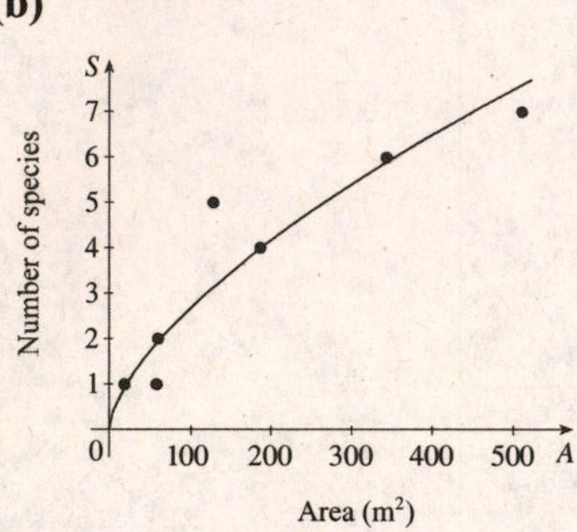

The model fits the data reasonably well.

(c) According to the model, there are $S = 0.14(205)^{0.64} \approx 4$ species of bat living in the El Sapo cave.

9. (a)

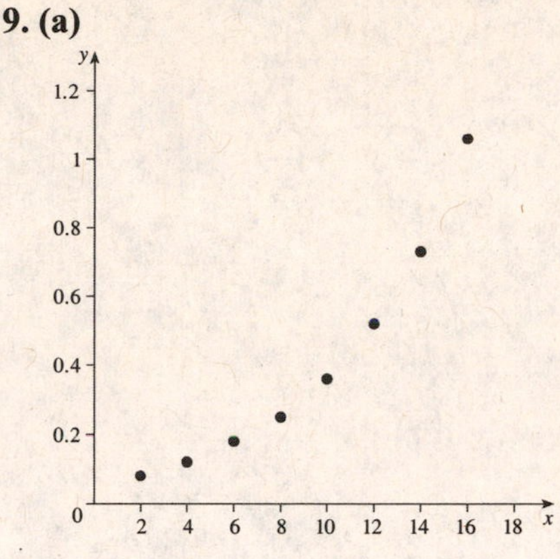

(b)

x	y	$\ln x$	$\ln y$
2	0.08	0.69315	−2.52573
4	0.12	1.38629	−2.12026
6	0.18	1.79176	−1.71480
8	0.25	2.07944	−1.38629
10	0.36	2.30259	−1.02165
12	0.52	2.48491	−0.65393
14	0.73	2.63906	−0.31471
16	1.06	2.77259	0.05827

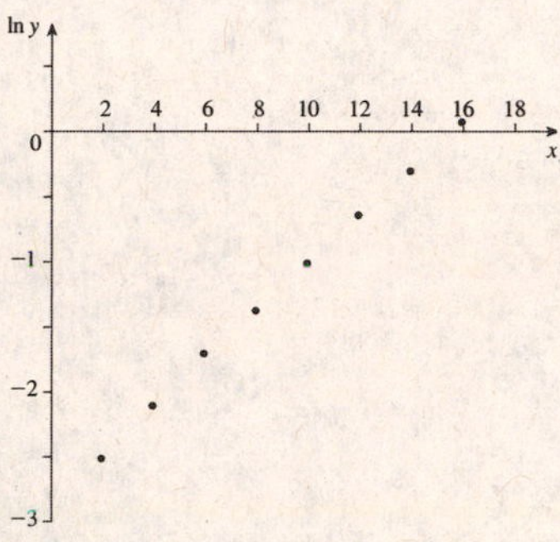

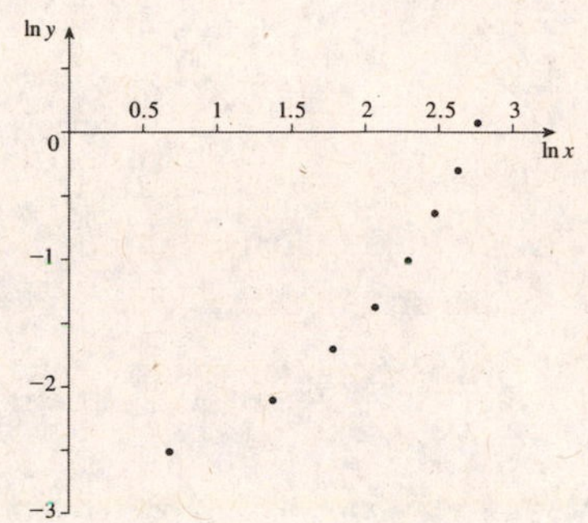

(c) The exponential function.

(d) $y = a \cdot b^x$ where $a = 0.057697$ and $b = 1.200236$.

11. (a) Using the `Logistic` command on a TI-83 we get $y = \dfrac{c}{1 + ae^{-bx}}$ where $a = 49.10976596$, $b = 0.4981144989$, and $c = 500.855793$.

(b) Using the model $N = \dfrac{c}{1 + ae^{-bt}}$ we solve for t . So $N = \dfrac{c}{1 + ae^{-bt}}$ $\Leftrightarrow$ $1 + ae^{-bt} = \dfrac{c}{N}$ $\Leftrightarrow$ $ae^{-bt} = \left(\dfrac{c}{N}\right) - 1 = \dfrac{c - N}{N}$ $\Leftrightarrow$ $e^{-bt} = \dfrac{c - N}{aN}$ $\Leftrightarrow$ $-bt = \ln(c - N) - \ln aN$ $\Leftrightarrow$ $t = \dfrac{1}{b}\left[\ln aN - \ln(c - N)\right]$. Substituting the values for a , b , and c , with $N = 400$ we have $t = \frac{1}{0.4981144989}\left(\ln 19643.90638 - \ln 100.855793\right) \approx 10.58$ days.